USING SPSS FOR WINDOWS AND MACINTOSH

FIFTH EDITION

USING SPSS FOR WINDOWS AND MACINTOSH

Analyzing and Understanding Data

Samuel B. Green
Arizona State University

Neil J. Salkind
University of Kansas

Upper Saddle River, New Jersey 07458

Library of Congress Cataloging-in-Publication Data

Green, Samuel B.
Using SPSS for Windows and Macintosh : analyzing and understanding data / Samuel B. Green, Neil J. Salkind.—5th ed.
p. cm.
Includes bibliographical references and index.
ISBN-13: 978-0-13-189025-1
ISBN-10: 0-13-189025-5
1. Social sciences—Statistical methods—Computer programs. I. Salkind, Neil J. II. Title.
HA32.G737 2007
300.285'53—dc22

2007020611

Executive Editor: Jeff Marshall
Editorial Director: Leah Jewell
Editorial Assistant: Aaron Talwar
Senior Marketing Manager: Jeanette Koskinas
Marketing Assistant: Laura Kennedy
Project Manager (editorial): LeeAnn Doherty
Associate Managing Editor (production): Maureen Richardson
Production Liaison: Fran Russello
Senior Operations Specialist: Sherry Lewis
Cover Design: Kiwi Design
Cover Illustration/Photo: Getty Images, Inc.
Composition/Full-Service Project Management: GGS Book Services/Saraswathi Muralidhar
Printer/Binder: Bind-Rite Graphics

Pearson Education LTD. London
Pearson Education Singapore, Pte. Ltd
Pearson Education, Canada, Ltd
Pearson Education–Japan
Pearson Education Australia PTY, Limited
Pearson Education North Asia Ltd
Pearson Educación de Mexico, S.A. de C.V.
Pearson Education Malaysia, Pte. Ltd
Pearson Education, Upper Saddle River, New Jersey

10 9 8 7 6 5 4 3

ISBN 13: 978-0-13-189025-1
ISBN 10: 0-13-189025-5

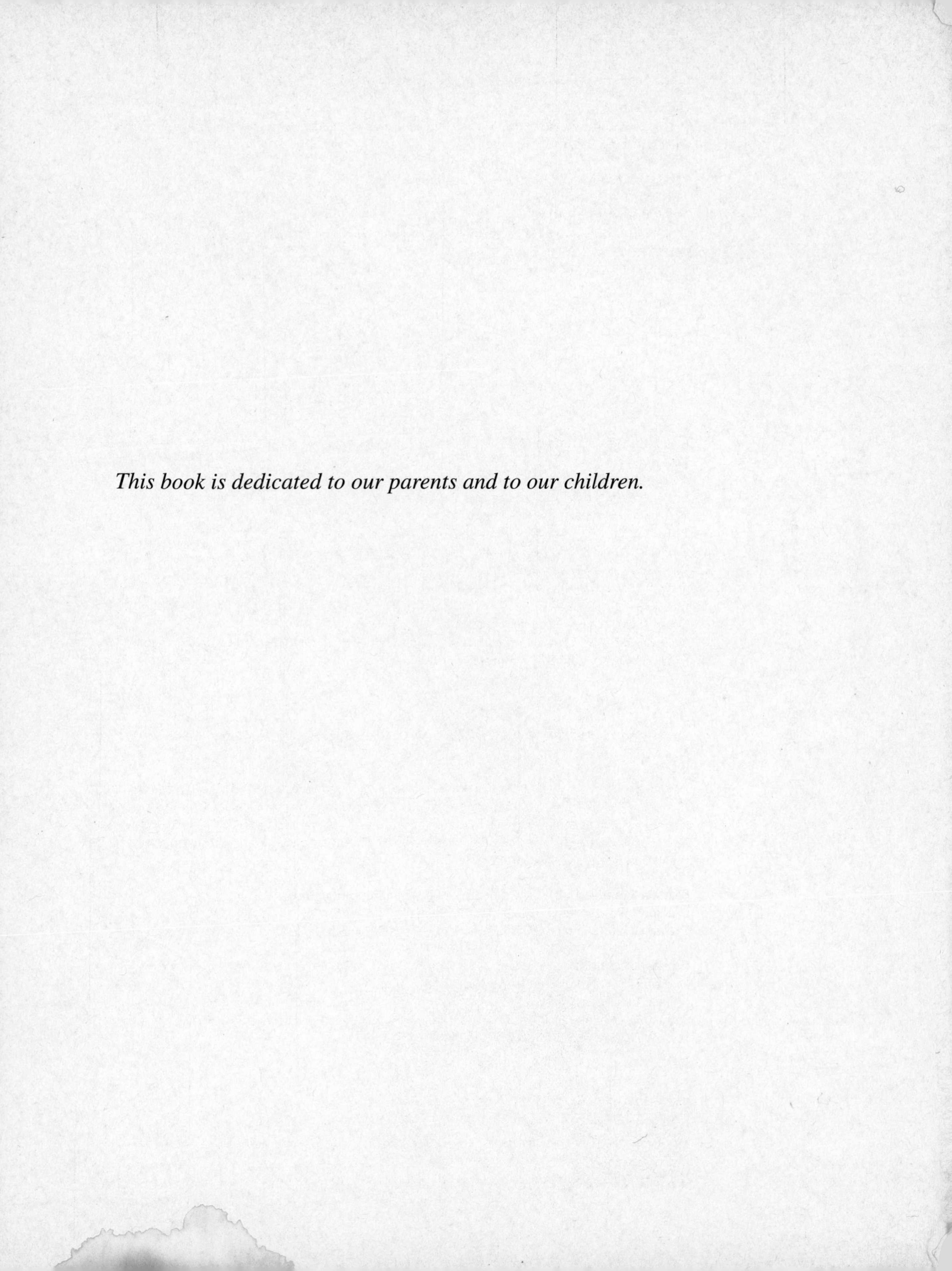

This book is dedicated to our parents and to our children.

PREFACE

The development of easy-to-use statistical software like SPSS has changed the way statistics is being taught and learned. No longer do students have to learn a system of elaborate code to conduct simple or complex analyses. Instead, students simply enter their data into an easy-to-use Data Editor. They can then select items from a drop-down menu to make appropriate transformations of variables, click options from another menu to create graphs of distributions of variables, select among various statistical analyses by clicking on appropriate options, and more. With a minimal amount of time and effort, the output is displayed, showing the results.

Researchers also have benefited from applications like SPSS. They do not have to spend time reacquainting themselves with the ins and outs of a statistical software package or learning new programs for conducting analyses that take hours to master. They also do not have to teach assistants how to write code to produce analyses, or examine and reexamine code that has produced error messages that do not really indicate what is wrong. Everyone can just point and click. More sophisticated users can use the syntax features.

In general, programs like SPSS have made life easier for students who are learning statistics, for teachers who are teaching statistics, and for researchers who are applying statistics. Nevertheless, many users of these programs find "doing statistics" an arduous, unenjoyable task. They still are faced with many potential obstacles, and they feel overwhelmed and stressed rather than challenged and excited about the potential for mastering these important skills.

What are some of the obstacles that students, in particular, face when they are trying to conduct statistical analyses with SPSS?

- Obstacle 1: Although SPSS is easy to use, many students and first-time users find it very complex. They have to learn how to input data into the Data Editor, save and retrieve data, make transformations to data, conduct analyses, manipulate output, create graphs, edit graphs, and so on.
- Obstacle 2: Students can feel helpless. Although they can point and click, they are frequently confronted with new dialogue boxes with many decisions to make. Their instructor does not have the time to talk about each of the options, so students feel as if they are making uninformed decisions.
- Obstacle 3: The amount of output and numbers produced by any statistical procedure is enough to cower most researchers if they are forced to explain their meaning. How can students who are taking statistics for the first time feel confident about interpreting output from an SPSS procedure? In trying to understand output, they are likely to face language problems. For example, "What is a Sig. F? Is it the same as the *p* value that the instructor is talking about? No, it couldn't be, or she or he would have told us."

Researchers, graduate students, and more advanced undergraduate students are going to face additional obstacles.

- Obstacle 4: Users can think of a number of different ways to analyze their data, but they are unsure about which way would yield the most understanding of their results and not violate the assumptions underlying the analyses.
- Obstacle 5: Even if users make all good decisions about statistical approaches and understand the output, they still must write a Results section that conforms to the American Psychological Association (APA) format.

Using SPSS for Windows and Macintosh: Analyzing and Understanding Data was written to try to help readers overcome all of the obstacles discussed above.

Part I, "Introducing SPSS," was written to address Obstacle 1, while Part II, "Working with SPSS Procedures," was designed to address the other four obstacles.

Part I, "Introducing SPSS," consists of 17 lessons divided into four units. It guides students through the most basic of SPSS techniques and uses a step-by-step description.

Unit 1, "Getting Started with SPSS," shows the student how to get started using SPSS, including a survey of the main menus, a description of how to use SPSS Help, and a brief tour of what SPSS can do.

Unit 2, "Creating and Working with Data Files," goes through the steps of defining variables, showing how data are entered and edited, how to use the Data Editor and the data view screens, how to print SPSS data files, and how to import and export information to and from SPSS.

Unit 3, "Working with Data," describes how to find and replace data, recode and compute values, sort data, and merge and split files.

Unit 4, the final unit in Part I, titled "Working with SPSS Charts and Output," teaches the student how to create and enhance SPSS charts as well as how to work with SPSS output including pivot tables. SPSS Windows (version 15) and Macintosh (version 13) differ in the way that graphics are created and edited, and, thus, there is a separate section covering each, Unit 4A for Windows and Unit 4B for the Macintosh. SPSS is becoming increasingly cross-platform and if you know the Windows version, you can easily adapt to the Macintosh version (and vice versa).

Part II, "Working with SPSS Procedures," consists of 27 lessons, divided into six units. Each unit presents a set of statistical techniques and a step-by-step description of how to conduct the statistical analyses. This is not, however, a "cookbook" format. We provide extensive substantive information about each statistical technique, including a brief discussion of the statistical technique under consideration, examples of how the statistic is applied, the assumptions underlying the statistic, a description of the effect size for the statistic, a sample data set that can be analyzed with the statistic, the research question associated with the data set, step-by-step instructions for how to complete the analysis using the sample data set, a discussion of the results of the analysis, a visual display of the results using SPSS graphic options, a Results section describing the results in APA format, alternative analytical techniques (when available), and practice exercises.

Unit 5, "Creating Variables and Computing Descriptive Statistics," shows how to create new variables from existing ones and shows the basic procedures for describing qualitative and quantitative variables.

Unit 6, "*t* Test Procedures," focuses on comparing means and shows how to use a variety of techniques, including independent and dependent *t* tests and the one-sample *t* test.

Unit 7, "Univariate and Multivariate Analysis-of-Variance Techniques," focuses on the family of analysis-of-variance techniques, including one-way and two-way analyses of variance, analysis of covariance, and multivariate analysis of variance.

Unit 8, "Correlation, Regression, and Discriminant Analysis Procedures," includes simple techniques such as bivariate correlational analysis and bivariate regression analysis, as well as more complex analyses such as partial correlational analysis, multiple linear regression, and discriminant analysis.

Unit 9, "Scaling Procedures," focuses on factor analysis, reliability estimation, and item analysis.

Unit 10, "Nonparametric Procedures," discusses a variety of nonparametric techniques, including such tests as the binomial, one-sample chi-square, Kruskal-Wallis, McNemar, Friedman, and Cochran tests.

New to This Edition

Version 15 of SPSS for Windows and version 13 for the Macintosh offer additional features of great value. For more details about the additional features, refer to the SPSS Web site http://www.spss.com/spss/whats_new_base.htm.

This fifth edition of *Using SPSS for Windows and Macintosh* includes the following additional coverage:

- Use of the Chart Builder to enhanced the creation and modification of charts
- Exporting of SPSS output to PDF format
- Increased flexibility in exporting of SPSS files to Excel and other applications
- New techniques for creating and editing charts both for the Windows and Macintosh versions
- Easier methods are presented for conducting tests of simple main effects for two-way analysis of variance.
- Methods are introduced for assessing differences between groups at specific levels of the covariate when the assumption of homogeneity of slopes is violated with analysis of covariance.

The Data Files on the Web

All the data files that you will need to work through the lessons in *Using SPSS for Windows and Macintosh* are available on the Web at http://www.prenhall.com/greensalkind. Part I uses several data files, among them one named Crab Scale Results and another named Teacher Scale Results. These will be introduced as you work through the first 18 lessons. They can also be seen in Appendix A.

Two types of data files are available in the lessons in Part II. The first are data files that may be used when learning particular SPSS procedures, such as paired-samples, *t* test, or factor analysis. These files can be easily identified since they are named, for example, *Lesson 23 Data File 1 or Lesson 36 Data File 1.* Also used in the second half of the book are data files for completing exercises at the end of lessons. These are named, for example, *Lesson 23 Exercise File 1 or Lesson 36 Exercise File 2.*

Please note that the website does not contain any executable SPSS data files. You need to have access to SPSS to use these file, as most users of this book will, at their school, company, or other institution. SPSS (at www.spss.com) offers a wide price range packages, including those for students.

Other Features of the Book

After This Lesson You Will Know

In Part I, at the beginning of each lesson, you will see a list of objectives—skills that you will master when you successfully complete the content of the lesson and work through all of the exercises in the lesson. These advanced objectives indicate what you can expect, and what is expected of you.

Key Words

Also in Part I, at the beginning of each lesson, there is a listing of key words that will be introduced and defined for the first time in the lesson. These words will be in boldface type the first time they are used.

Typing Conventions

There is only one typing convention you must attend to throughout this book. A sequence of actions is represented by what options are selected from what menu, connected by an arrow like this →.

For example, if a certain procedure requires clicking on the File menu and then clicking the New option, it would be represented as follows.

1. Click **File → New**.

Examples

Each lesson includes step-by-step procedures, with copious illustrations of screen shots, for successfully completing a technique with sample data. Exercises at the end of each lesson allow you to practice what you have learned.

Tips

Some of the lessons contain tips (in the margins) that will help you learn SPSS and will teach you shortcuts that make SPSS easier to use.

System Requirements for SPSS 15 for Windows

If you are using SPSS 15 for Windows, then your system must meet the following minimal requirements. Keep in mind that version 15 of SPSS has not been "certified" (the official action) by SPSS to be used with Windows Vista although it does work with Vista.

- Operating system: Microsoft® Windows XP or 2000
- Hardware: Intel® Pentium®-compatible processor
- Memory: 256 MB RAM minimum
- Minimum free drive space: 400 MB
- VGA Monitor
- Web browser: Internet Explorer 6

System Requirements for SPSS 13 for Mac OS X

If you are using SPSS 13 for Macintosh, then your system must meet the following minimal requirements:

- Operating system: Mac OS X version 10.3.9 running Java™ 1.4.2
- Processor: G4 667 MHz
- Memory: 256 MB RAM (512 MB recommended for installations using Mac OS X Tiger™)
- Monitor: Color monitor, 1024 × 768 resolution

Interestingly, SPSS *does support* the new Intel®-based Mac hardware for any version of SPSS on Intel-based Macintosh machines. SPSS 15.1 for Mac OS X (due in the summer of 2007) will be fully compatible with the Intel®-based Mac hardware.

A Note about SPSS for Macintosh

SPSS 15 for Windows and SPSS 13 for Macintosh differ in the tools used to create and edit graphics (as described in Unit 4). Otherwise, they are very similar, and the methods described in this book generally apply, regardless of platforms.

There are a few differences between platforms that should be kept in mind.

1. The Ctrl + Z key combination undoes an operation for Windows, while the Options + Z key combination undoes an operation for Macintosh.
2. When you select more than one item in a dialog box in the Macintosh version, you hold down the Options key rather than the Ctrl key, as done in the Windows version.
3. Options is not available on the Edit menu in the Macintosh version, while it is available in the Windows version.
4. The right-click options are more limited in the Macintosh version.

Acknowledgments

No book is ever the work of only the authors. *Using SPSS for Windows and Macintosh* was first contracted with Chris Cardone, whom we would like to thank for giving us the opportunity to undertake the project. Chris remains a good colleague and a better friend.

We would also like to thank the following people for reviewing the manuscript: Andrew Supple, UNC Greensboro; Brad Chilton, Tarleton State University; Sharon Hill, University of Tennessee at Chattanooga; Notis Pagiavlass, Embry-Riddle Aeronautical University; Seth Hirshorn, University of Michigan at Dearborn; Ami Spears, Mercer University; and Gerene Staratt, Barry University.

Thank you for using this book. We hope it makes your SPSS activities easy to learn, fun to use, and helpful. Should you have any comments about the book (good, bad, or otherwise), feel free to contact us at of the e-mail addresses listed below.

Samuel B. Green
samgreen@asu.edu

Neil J. Salkind
njs@ku.edu

USING SPSS FOR WINDOWS AND MACINTOSH

Introducing SPSS

PART I

Getting Started with SPSS

UNIT 1

You're probably familiar with how other Windows applications work, and you will find that many SPSS features operate exactly the same way. You know about dragging, clicking, double-clicking, and working with Windows. If you don't, you can refer to one of the many basic operating systems books available for Windows or the Macintosh. We assume that the user is familiar with the basic operating systems skills such as clicking with a mouse, dragging objects, naming files, etc.

In this first unit, we introduce you to SPSS, beginning with how to start SPSS, and walk you through a tour so that you know some of the most important features of SPSS.

In Lesson 1, "Starting SPSS," the first of four lessons in this unit, you will find out how the SPSS Windows group is organized and how you start SPSS.

In Lesson 2, "The SPSS Main Menus and Toolbar," we introduce you to the opening SPSS window, point out the various elements in the window, and explain what they do. The main menus in the SPSS window are your opening to all the SPSS features you will learn about in *using SPSS*. We also introduce you to the toolbar, a collection of icons that perform important tasks with a click of the mouse.

Lesson 3, "Using SPSS Help," introduces you to SPSS online help. If you've ever used another Windows application, you know how handy it is to have this type of help immediately available and how it can get you through even the most difficult procedures.

In the last lesson in Unit 1, "A Brief SPSS Tour," we provide a simple example of what SPSS can do including simple analysis, the use of Data View and Variable View, and the creation of a chart. Here we'll whet your appetite for the terrific power and features of SPSS and what is in store for you throughout the book.

LESSON

1 Starting SPSS

After This Lesson, You Will Know

- How to start SPSS
- What the opening SPSS screen looks like

Key Words

- Data Editor
- Data View
- Variable View
- Viewer

With this lesson, you will start your journey on learning how to use SPSS, the most powerful and easiest-to-use data analysis package available.

Keep in mind that throughout these lessons we expect you to work along with us. It's only through hands-on experiences that you will master the basic and advanced features of this program.

Starting SPSS

SPSS is started by clicking the icon (or the name representing the program) that represents the application either on the Start menu in Windows or on the desktop on the Macintosh. You can also access the SPSS icon through Windows Explorer or in the Applications folder on your Macintosh.

TIP

To place SPSS on the Windows Start menu, open Windows Explorer, locate the SPSS executive file (spss.exe), and drag it on to the Start menu. To place it on the Mac desktop, just locate it on the hard drive (in the Applications folder) and drag it onto the desktop.

The SPSS Opening Window

As you can see in Figure 1, the opening window presents a series of options that allow you to select from running the SPSS tutorial, entering data, posing an established query, creating a new query by using the Database Wizard, or opening an existing source of data (an existing file). Should you not want to see this screen each time you open SPSS, then click on the "Don't show this dialog in the future" box in the lower left corner of the window.

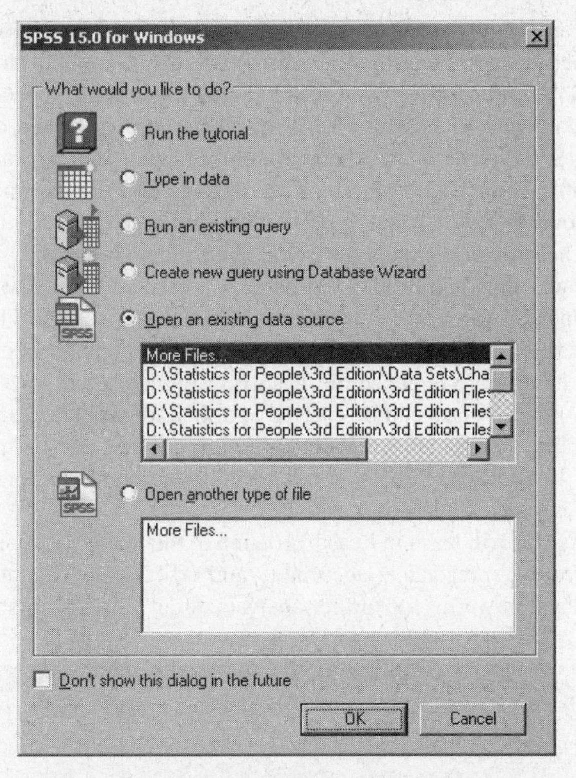

Figure 1. **The SPSS for Windows opening dialog box.**

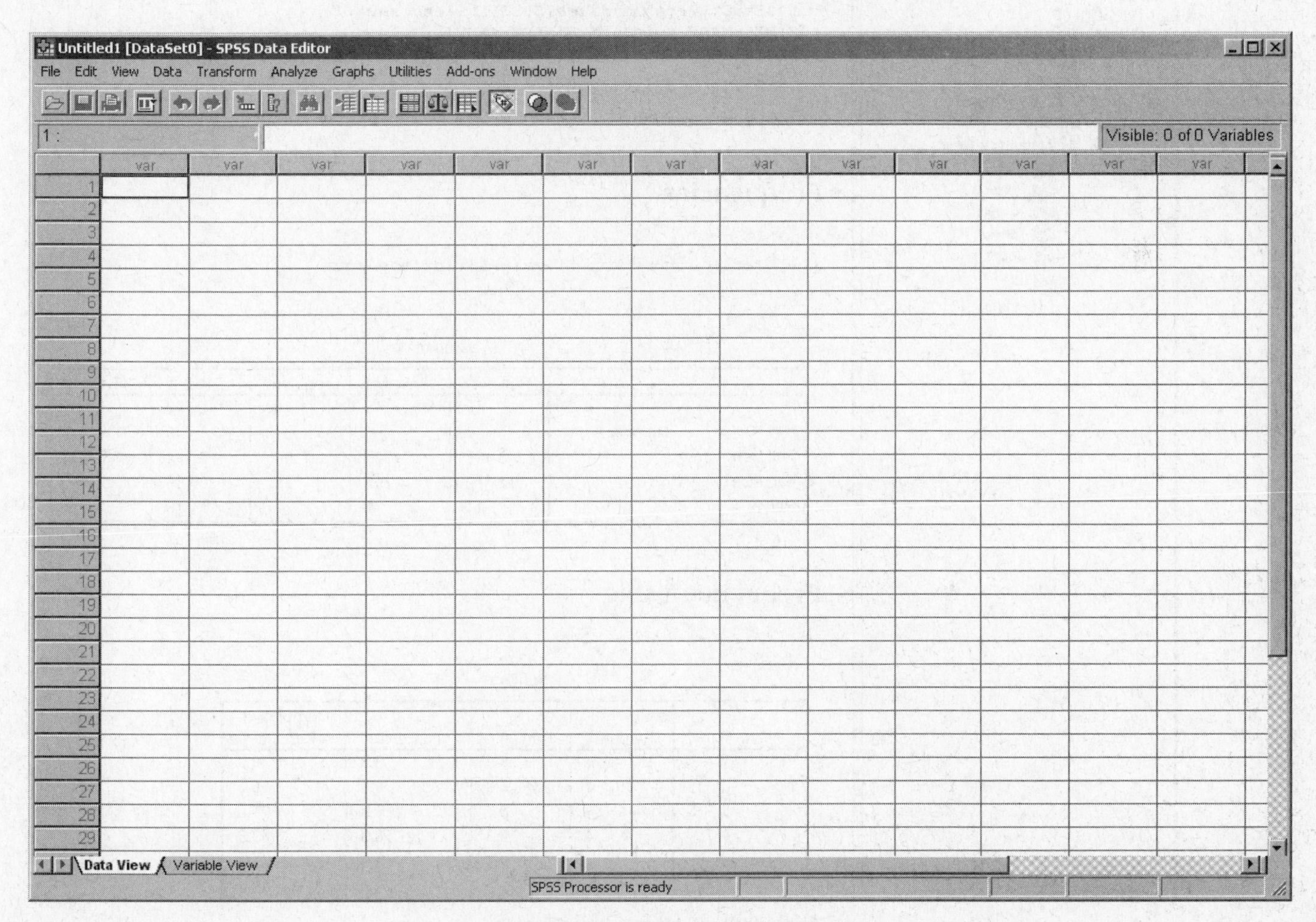

Figure 2. **The SPSS Data View window.**

For our purposes, we will click the **Type in data** option since it is likely to be the one you first select upon opening and learning SPSS. Once you do this, the **Data View** window (also called the **Data Editor**) you see in Figure 2 (on page 3) becomes active. This is where you enter data you want to use with SPSS once that data has been defined. Although you cannot see it when SPSS first opens, there is another open (but not active) window as well. This is the **Variable View** where variables are defined and the parameters for those variables are set. We will cover Data View and Variable View in Lesson 5.

The **Viewer** displays statistical results and charts that you create. An example of the Viewer window is shown in Figure 3. A data set is created in the Data Editor, and once the set is analyzed or graphed, you examine the results of the analysis in the Viewer.

If you think the Data Editor is similar to a spreadsheet in form and function, you are right. In form, it certainly is, since the Data Editor consists of rows and columns just like in Excel and Lotus 1-2-3. Values can be entered and then manipulated. In function as well, the Data Editor is much like a spreadsheet. Values that are entered can be transformed, sorted, rearranged, and more. In addition, SPSS can use formulas to compute new variables and values from existing ones, as you will learn in Lesson 12.

As you will learn in Lesson 10, one of the many conveniences of SPSS is its ability to import data from a spreadsheet accurately and efficiently. This ability makes SPSS particularly well suited and powerful for further analysis of data already in spreadsheet form.

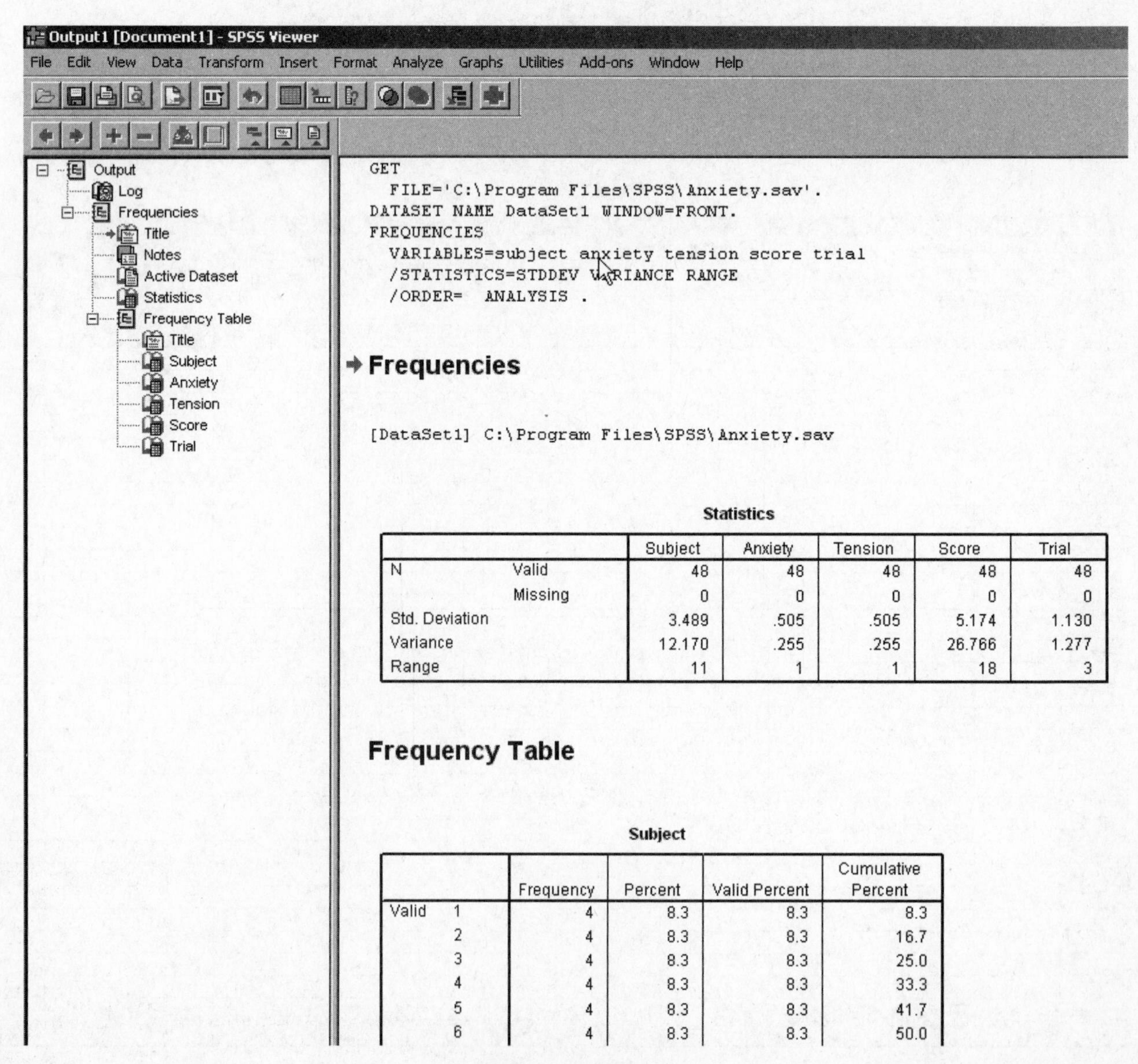

***Figure 3*. The Viewer.**

LESSON 2

The SPSS Main Menus and Toolbar

After This Lesson, You Will Know

- Each of the SPSS main menus and what the commands on the menus do
- The icons on the toolbar

Key Words

- Add-ons menu
- Analyze menu
- Data menu
- Edit menu
- File menu
- Graphs menu
- Open dialog box
- Status bar
- Toolbar
- Transform menu
- Utilities menu
- View menu

Menus are the key to operating any Windows or Mac application, and that is certainly the case with SPSS. Its 11 main menus, including the standard Windows and Help menus for the Windows version, and the 10 main menus for the Mac version, including the standard Windows and Help menus, provide access to every tool and feature that SPSS has to offer. In this lesson, we will review the contents of each of these menus and introduce you to the toolbar, a set of icons that takes the place of menu commands. The icons make it quick and easy to do anything, from saving a file to printing a chart.

The SPSS Main Menus

SPSS comes to you with 11 main menus, as you can see in the opening screen in Figure 4 (on page 6). Although you think you may know all about the File menu and what options are available on it, stick with us through the rest of this lesson to see exactly what the File menu, and the other nine menus, can do for you.

TIP

Want to set the number of files that SPSS lists on the File menu? Click **Edit Options** and change the value in the Recently used file list box.

Figure 4. **The SPSS Main Menus.**

The File Menu

TIP

When items on a menu appear dimmed, it means they are not available.

The purpose of the **File menu** (Figure 5) is to, obviously, work with files. Using the options on this menu, you create new files, open existing ones, save files in a variety of formats, display information about a file, print a file, and exit SPSS. The File menu can also list recently used data files (Recently Used Data) and other recently used files (Recently Used Files), so you can quickly return to a previous document.

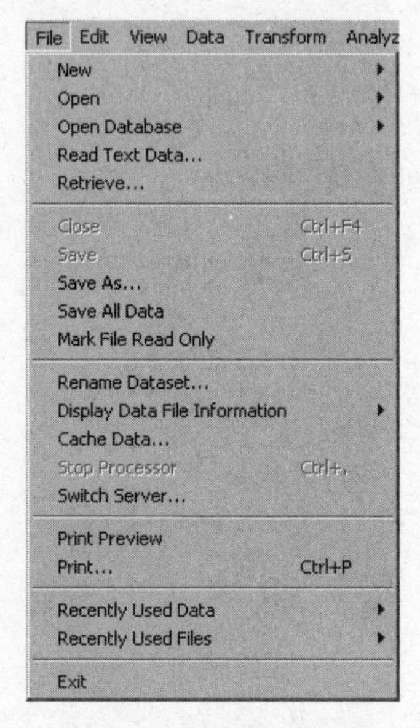

Figure 5. **The File menu.**

TIP

Perhaps the most valuable SPSS command (which is available on many Windows-based applications) is the CTRL+Z key combination which reverses the last data entry you made.

For example, when it comes time to start working with the file named Teacher Scale Results, you would select Open from the File menu and then select the file name from the **Open dialog box**. You will learn more about this process in Lesson 7.

The Edit Menu

When it comes time to cut or copy data and paste it in another location in the current, or another, data file, you will go to the **Edit menu**. You will also seek out options on the Edit menu to search for data or text, replace text, and set SPSS preferences (or default settings). All these activities and more are found on the Edit menu shown in Figure 6.

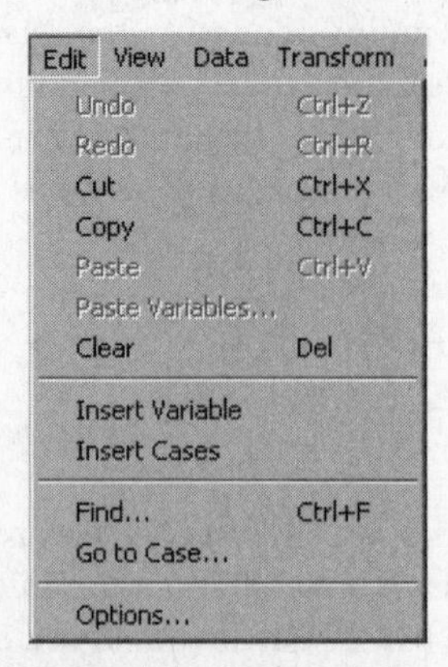

Figure 6. **The Edit menu.**

For example, if you wanted to find what Mary Jones scored on test 1, you could use the Find menu command to search for "Mary Jones" and then read across the file to find her score on the variable named test 1. You would use the Find command on the Edit menu to search for that information.

TIP

If you use labels for your variables, make sure that the Value Labels option is checked in the View menu. Otherwise, they may be in effect, but they will not be visible.

The View Menu

Here's a chance to customize your SPSS desktop. Using various commands on the **View menu**, you can choose to show or hide toolbars, **Status bar**, and grid lines in the Data Editor; change fonts; and use Value Labels. You can see these commands in Figure 7.

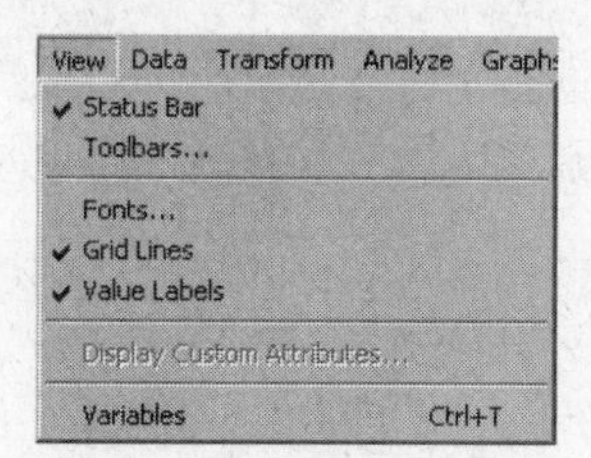

Figure 7. **The View menu.**

For example, if we didn't want to use labels for variables or grid lines, we would be sure that these options (Value Labels amd Grid Lines) were not selected.

The Data Menu

Variables and their values are the central element in any SPSS analysis, and you need powerful tools to work with variables. You have them on SPSS. As you can see in Figure 8, on the **Data menu** there are commands that allow you to define variable properties, insert cases, go to a specific case, merge and aggregate files, and assign weight to cases as you see fit.

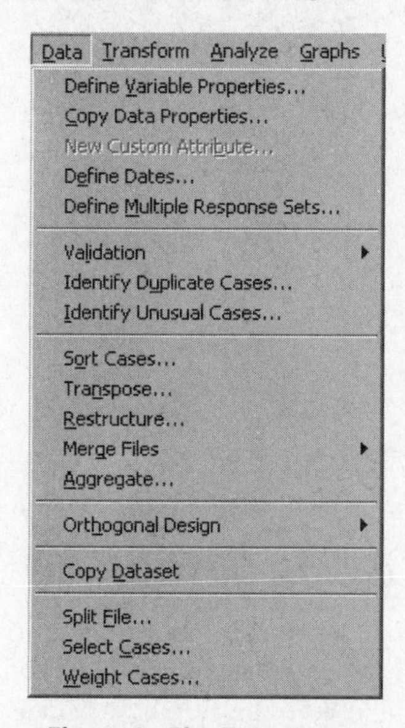

Figure 8. **The Data menu.**

For example, if we want to sort cases, this is the menu you would use and the Sort Cases option is the menu option that would be selected.

The Transform Menu

There will be times when a variable value needs to be transformed or converted to another form or another value. That's where the commands on the **Transform menu** you see in Figure 9 (on page 8) come in handy. On this menu, you will find commands that allow you to compute new values, create a set of random values, recode values, replace missing values, and do more.

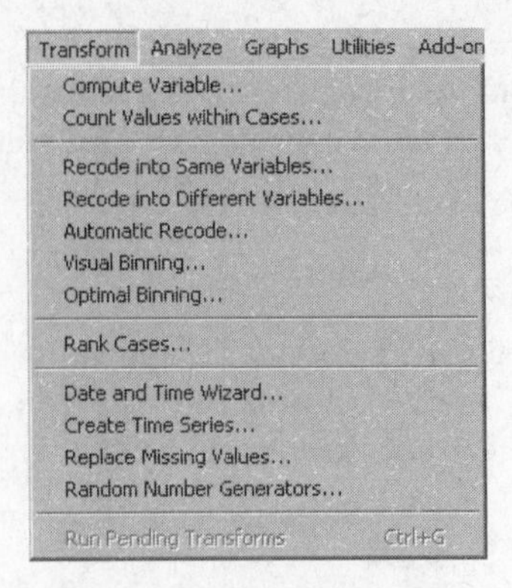

Figure 9. **The Transform menu.**

For example, using the Compute Variable command on the Transform menu, you could easily compute a new variable that represents the mean of a set of items.

The Analyze Menu

Here's the meat-and-potatoes menu! As you can see in Figure 10, there are 21 different options on the **Analyze menu** that lead to almost any statistical analysis technique you might want to use. These range from a simple computation of a mean and standard deviation to time series analysis and multiple regression.

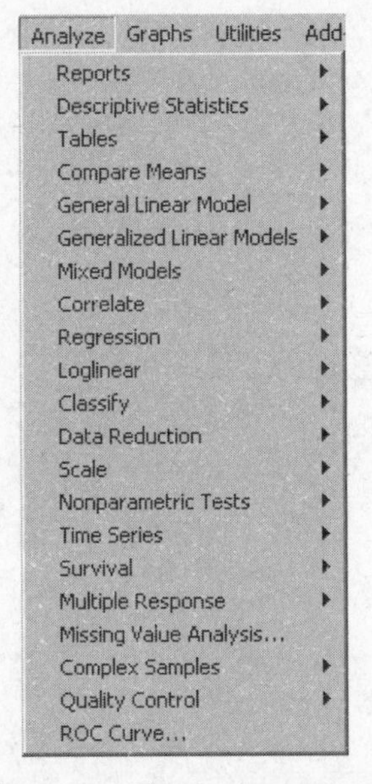

Figure 10. **The Analyze menu.**

For example, if you wanted to determine if there is a significant difference between the average rating that Professor 6 received on a teaching evaluation form versus the average rating received by Professor 4, you could look to the Compare Means option on the Analyze menu.

The Graphs Menu

Want to see what those numbers really look like? Go to the **Graphs menu** where you can create a bar, line, area, and some 16 other types of graphs. Graphs make numbers come alive, and you should pay special attention to Unit 4, where we show you how to create, edit, and print them. Take a look at Figure 11 to see what graph options are available. With version 15, you also have the opportunity to use the Interactive menu command (where SPSS walks you through the creation of a graph), or the Legacy Dialogs menu command where the SPSS interface from version 14 is accessible.

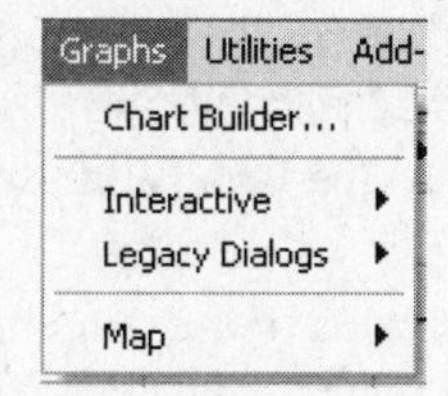

***Figure 11.* The Graphs menu.**

For example, if you want to see test scores as a function of gender, a bar graph (on the Graphs menu) could do it quite nicely.

The Utilities Menu

Here you can find out information about variables and files, and you can define and use sets of variables. You can see these options in Figure 12 on the **Utilities menu**.

For example, the Variables option tells us the specifics about each variable, including the name, label type, and more.

***Figure 12.* The Utilities menu.**

The Add-ons Menu

Add-ons is a kind of catchall menu for commands that do not conveniently fit elsewhere. For example, there is information on SPSS consulting and SPSS tutorials.

The Window and Help Menus

These two menus function much like any other Windows application menus. The **Window menu** helps you switch from one window to another and minimize the SPSS Data Editor or Viewer.

The **Help menu** provides online help. We will focus on the Help menu in the next lesson.

The SPSS Toolbar and Status Bar

What's the easiest way to use SPSS? Clearly, the easiest way is through the use of the **toolbar**, the set of icons that are underneath the menus. You can see the Data View toolbar in Figure 13 (on page 10) and a description of what each icon, which represents a command on an SPSS menu, does. Click on the icon, and the command is performed. So, instead of going to the Utilities menu to select variables, you can just click on the Variables icon on the toolbar. Table 1 presents each toolbar icon, its title, and what it does.

TIP

You can always find out what a toolbar icon represents by placing the mouse cursor on top of the icon. A toolbar tip (such as Save File) will appear.

***Figure 13.* The Data Editor toolbar.**

Different screens have different toolbars. For example, as you will see in Lesson 15, when you create a chart, a new set of icons becomes available on the toolbar.

Table 1
Toolbar Icons

Icon	Title	What it does
	Open File	Opens an already created file
	Save File	Saves a new or already created file
	Print	Prints a file
	Dialog Recall	Recalls the last used dialog box
	Undo	Undoes a change in formatting or data entry
	Redo	Reenters a previous change
	GoTo Case	Goes to a numbered case
	Variables	Provides information about a variable
	Find	Finds a record
	Insert Cases	Inserts a case in the data file
	Insert Variables	Inserts a new variable into the data file
	Split File	Splits a file along a defined variable
	Weight Cases	Weights cases
	Select Cases	Selects a set of cases by using a certain criterion
	Value Labels	Turns labels on and off
	Use Sets	Creates sets of variables
	Show All Variables	Shows all the variables in the data set

TIP

If you are performing an analysis and nothing seems to be happening, look in the Status Bar at the bottom of the SPSS Data Windows before you conclude that SPSS or Windows has locked up. You should be able to see a message in the Status Bar telling you what SPSS is doing.

Another useful tool is the Status Bar located at the bottom of the SPSS window. Here, you can see a one-line report as to what activity SPSS is currently involved in. Messages such as "SPSS Processor is ready" tell you that SPSS is ready for your directions or input of data. Or, "Running Means" tells you that SPSS is in the middle of the procedure named Means.

The Data Files

The sample files available at http://www.prenhall.com/greensalkind are important in learning how to use SPSS. Throughout Part I of the text, titled "Introducing SPSS," we will use two separate sets of data to illustrate various SPSS features, such as entering and working with data. A detailed description of each of these files is shown in Appendix A.

The Crab Scale File

The first data set is a collection of scores on the Crab Scale and some biographical information for 10 college professors who completed a measure of crabbiness. Table 2 gives a summary of the variables, their definition, and their range of values.

Table 2
Crab Scale Summary

Variable	Definition	Range
id_prof	professor's identification number	1 through 10
sex_prof	professor's gender	1 or 2
age	professor's age	33 through 64
rank	professor's rank	Assistant, Associate, or Full Professor
school	professor's school	Liberal Arts, Business School
crab1	score on item 1 on the Crab Scale	
crab2	score on item 2 on the Crab Scale	
crab3	score on item 3 on the Crab Scale	
crab4	score on item 4 on the Crab Scale	
crab5	score on item 5 on the Crab Scale	
crab6	score on item 6 on the Crab Scale	

The Crab Scale includes the following six items:

1. I generally feel crabby if someone tries to help me.
2. I generally feel happy when I watch the news.
3. I generally feel crabby when I watch mothers and fathers talk baby talk to their babies.
4. I generally feel happy when I am able to make sarcastic comments.
5. I generally feel crabby when I am on a family vacation.
6. I generally feel happy when I am beating someone at a game.

A teacher responds to each item on the following 5-point scale:

1. Totally agree
2. Agree
3. In a quandary
4. Disagree
5. Vicious lies

The Crab Scale yields two scores:

1. *The Cross-Situational Crab Index:* This index tries to assess whether individuals refuse to be happy regardless of the situation.
2. *The True Crab Scale:* This index attempts to assess whether an individual is acting in a true crablike fashion: crabby when confronted with a pleasant stimulus and happy when confronted with an unpleasant stimulus.

Items 1 through 6 are summed to yield a total score. For the Cross-Situational Crabbiness Index, all scores are summed together. Items 2, 4, and 6 are happiness items, and the scores on these items must be reversed so that higher scores indicate more crabbiness, as shown at the top of page 12.

Original Scoring	Recoded Scoring
1	5
2	4
3	3
4	2
5	1

You can see the actual set of data in Appendix A. This data file is saved on your SPSS as Crab Scale Results at http://www.prenhall.com/greensalkind.

The Teacher Scale File

No teacher escapes being rated by students. The second set of data we will deal with in Part I of *Using SPSS for Windows and the Macintosh: Analyzing and Understanding Data* is a set of responses by students concerning the performance of these 10 professors.

The second data set is a collection of scores on the Teacher Scale and some biographical information for 50 students who completed the Teacher Scale. Scores on the Teacher Scale that make up this sample file are also shown in Appendix A. They are contained under the file name Teacher Scale Results at http://www.prenhall.com/greensalkind.

Table 3 shows the biographical information we collected on each student and their responses to the 5-point scale. We will be using them in examples throughout this part of the book.

Table 3
Teacher Scale Summary

Variable	Name	Range
id_stud	student's identification number	1 through 50
id_prof	professor's identification number	1 through 5
sex_stud	student's gender	1 or 2
teacher1	score for item 1 on the Teacher Scale	1 through 5
teacher2	score for item 2 on the Teacher Scale	1 through 5
teacher3	score for item 3 on the Teacher Scale	1 through 5
teacher4	score for item 4 on the Teacher Scale	1 through 5
teacher5	score for item 5 on the Teacher Scale	1 through 5

The Teacher Scale contains the following five items:

1. I love that teacher.
2. My teacher says good stuff.
3. The teacher has trouble talking.
4. The teacher is a jerk.
5. My teacher made the boring lively, the unthinkable thinkable, the undoable doable.

Items 3 and 4 must be reversed so that higher scores indicate effectiveness as follows.

Original Scoring	Recoded Scoring
1	5
2	4
3	3
4	2
5	1

LESSON

3 Using SPSS Help

After This Lesson, You Will Know

- About the contents of the SPSS Help menu
- How to use the F1 function key for help
- What Help options are available
- How to use the Find option to search for particular words

Key Words

- Contents
- F1 function key
- Favorites
- Find
- Help menu
- Index
- Search
- SPSS Help
- Topics

If you need help, you've come to the right place. SPSS offers help that is only a few mouse clicks away. It is especially useful when you are in the middle of creating a data file and need information about a specific SPSS feature. Help is so comprehensive that even if you're a novice SPSS user, SPSS Help can show you the way.

How to Get Help

You can get help in SPSS in several ways. The easiest and most direct is by pressing the **F1 function key** or by using the **Help menu**. As you can see, there are 10 options on the Help menu. The easiest and most direct way to get help with the Macintosh version of SPSS is by pressing the **Options** + ? Key combination.

- **Topics** list the topics for which you can get help. You can click on any one of these. As you enter the topic, SPSS searches its internal database to find what you need.
- **Tutorial** takes you through a step-by-step tutorial for major SPSS topics.

- **Case Studies** provides the scenarios for the various topics that SPSS covers. For example, if you access the t-test case study, you'll find a general example for when this statistical procedure is relevant and how it is used.
- **Statistics Coach** walks you through the steps you need to determine what type of analysis you want to conduct.
- **Command Syntax Guide** provides you with information on SPSS programming language. (Note: This option does not appear in the Macintosh version of SPSS.)
- **SPSS Home Page** uses your Internet connection to take you to the home page of SPSS on the Internet. Syntax Guide provides you with help using SPSS's powerful syntax feature.
- **About** tells you the version of SPSS that you are currently using (not for the Mac!).
- **License Authorization Wizard** allow you to seek authorization for your license directly from SPSS via your Internet connection.
- **Register Product** helps you register your copy of SPSS (not for the Mac here either).
- **Check for Updates** tells you what version of SPSS you are using and allows you to check with SPSS to see if there is an update to your current version available.

TIP

Want help on Help? Click **Help ⟶ Topics**, and then click the **Index** tab as you enter Help. You'll see an explanation of various options.

Using the F1 Function Key

The F1 function key is a quick way to get help, and it opens the Base System Help dialog box you see in Figure 14. It is not context sensitive. In other words, it produces the same outcomes as the **Help ⟶ Topics** click combination. Here, you can click **Contents**, **Index**, **Search**, and **Favorites**. (If you are using the Mac version, then you have to click **Help ⟶ Topics**.)

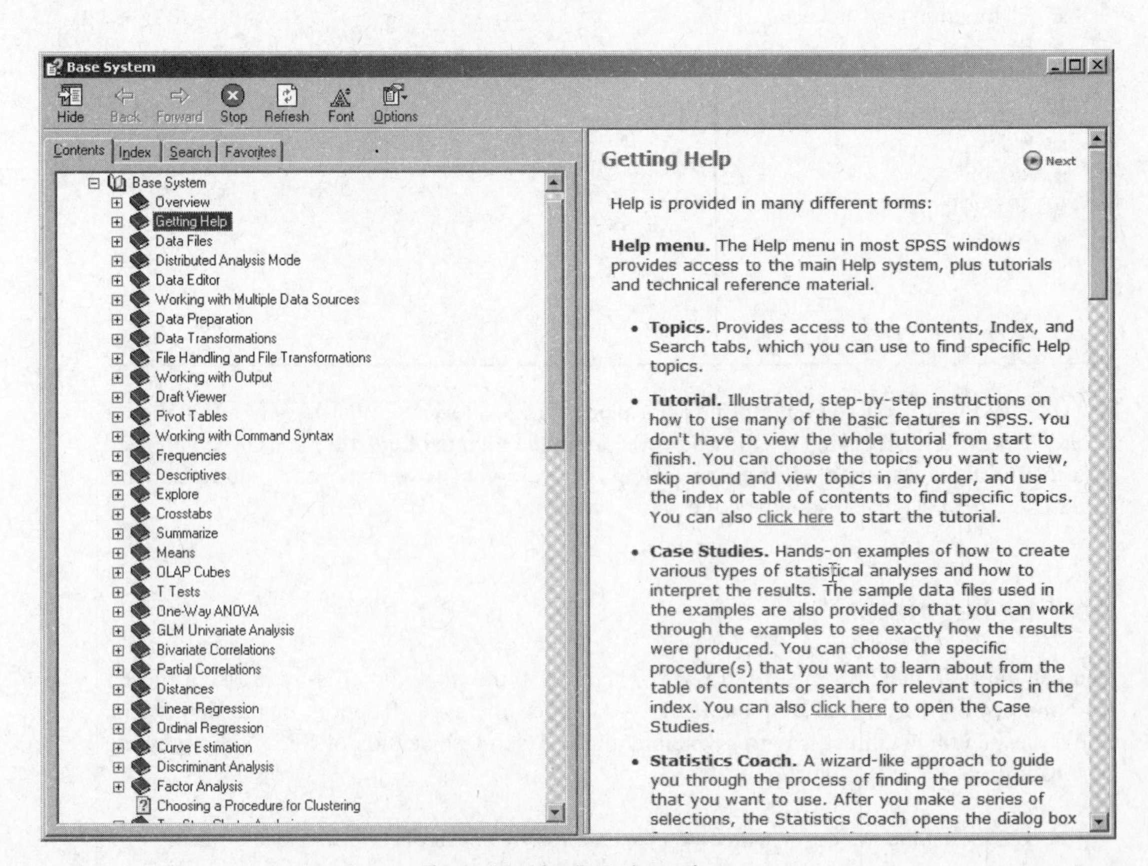

***Figure 14.* The F1 Help options.**

Using the Contents Option

The Contents tab presents the major headings for help. Double-clicking on any heading provides a list of possible topics that you might want to consult for the help you need. For example, if you want help on how to compare the means of two samples, you would follow these steps:

1. Press **F1**.
2. Click the **Contents** tab.
3. Double-click **T Tests**, and you will see a list of the topics within this general heading.
4. Double-click (for example) the topic labeled **Independent-Samples T Test**.
5. Then, on the right-hand side of the Help screen, click **To Obtain an Independent-Samples T Test**. As shown in Figure 15, you will see the help screen that guides you through the procedure.

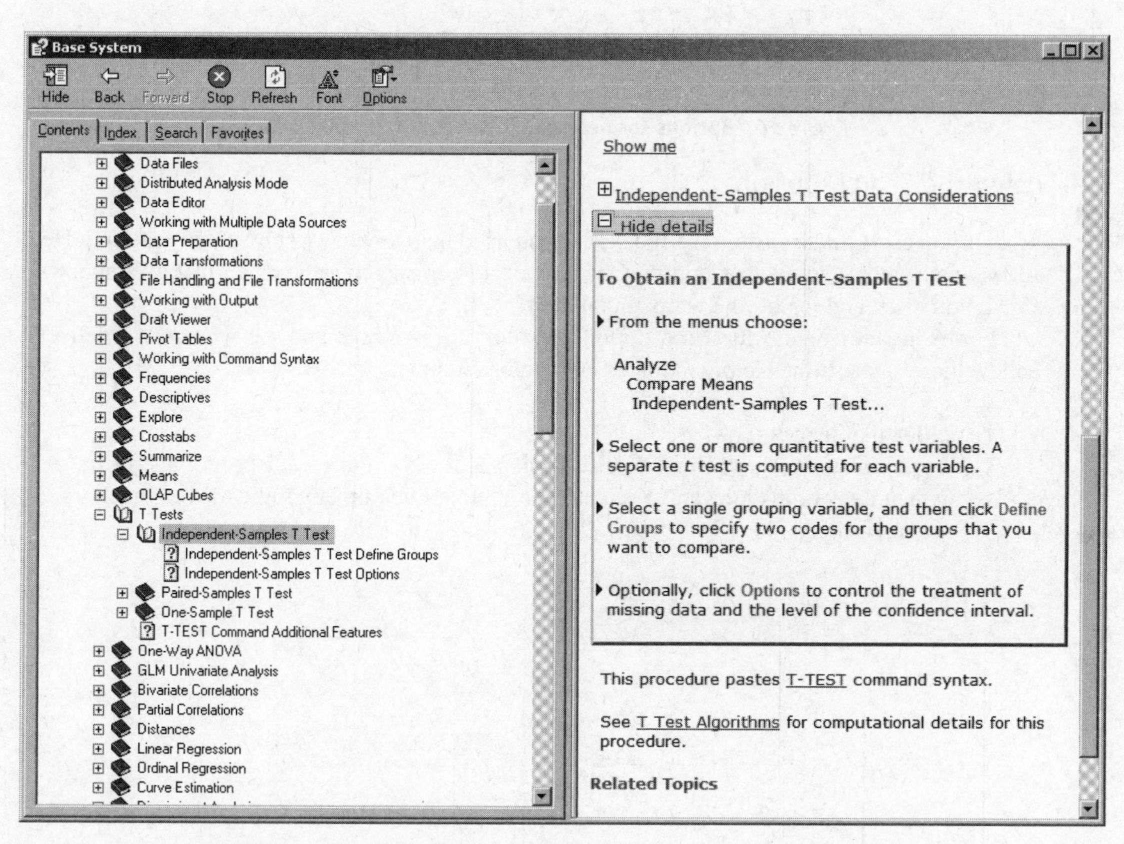

***Figure 15.* Getting help on a particular topic using the Contents tab.**

Using the Index Option

The Index tab provides an alphabetical listing of all the topics in SPSS. To find help on a particular topic, follow these steps. For example, here's how you could use the Index option to find help on means.

1. Click the **Index** tab.
2. Type **mean**. As you enter the letters of the term for which you need help, **SPSS Help** tries to identify the topic listing. You can then get help by double-clicking on any topic within the more general topic of mean. In Figure 16 (on page 16), you can see a listing of topics on the mean.

TIP

At the bottom of each window of help, there is usually a series of related topics. Just click on any one to take you to that related topic.

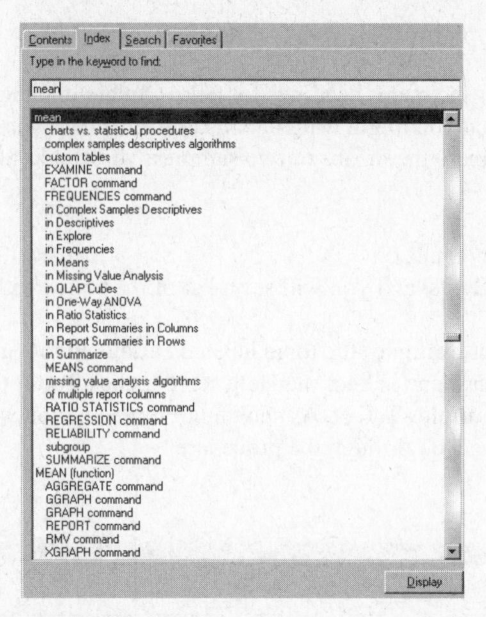

Figure 16. **Options for help under the general topic of mean.**

Using the Search Option

> **TIP**
>
> Right-click on any help topic and print it out to create your own personal notebook of topics for which you can return often for assistance (a great way to create a help booklet on particular topics as well).

What if you can't find a term in the Index, but you need help anyway? The Search option in Help allows you to enter any words that may be part of a Help screen rather than just a category. In effect, you are searching all the words in all the topics.

For example, let's use the **Find** option and the term *variance* and see what SPSS delivers. Follow these steps. If the Help window is open, close it now.

1. Click the **Search** tab.
2. Type **variance**, and then click **List Topics** and SPSS returns a list of topics for which the word is relevant. You can then double-click on any topic area to display extensive help.

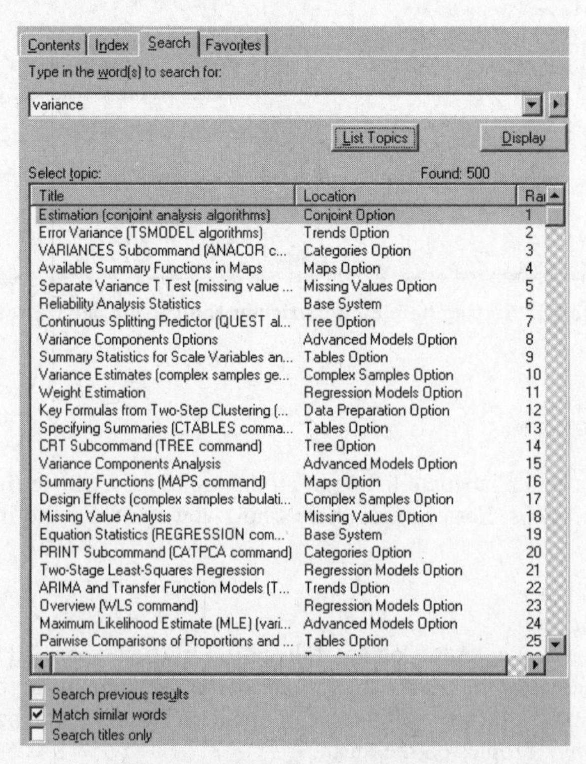

Figure 17. **Using the Search feature.**

In Figure 17, you can see the topics related to the word *variance* for which you can get additional help by double-clicking on any of the items. Note that you can also produce the results from a previous search, match similar words, and search for titles only.

If you know exactly what you're looking for, you can be specific in your search, but be careful because specificity can cause you to miss what you are looking for if you are not sure of your topic's exact terminology.

TIP

Unlike other Windows applications you may be used to, SPSS Help is not context sensitive. No matter what part of a window or a procedure you are in, pressing F1 does not take you to help about that procedure. Rather, you will still have to identify what you want help on to receive help.

Using the Favorites Option

The **Favorites** option allows you to automatically save those topics for which you have sought help and may again in the future. When you see the Help that you want in the right-hand side of the Help screen, click the **Favorites** tab and then click **Add** and that topic will be added to your favorites.

LESSON

4 A Brief SPSS Tour

After This Lesson, You Will Know

- How to open a file
- Some of the basic functions that SPSS can perform

Key Words

- Fonts
- Open

In the first three lessons of *Using SPSS*, you learned how to start SPSS, took a look at the SPSS main menus, and found out how to use SPSS Help.

Now sit back and enjoy a brief tour of SPSS activities. Nothing fancy here. Just creating some new variables, simple descriptions of data, a test of significance, and a graph or two. What we're trying to show you is how easy it is to use SPSS. We will spend all of Part II concentrating on exactly how to use SPSS analytic procedures.

Opening a File

TIP

If you want to open saved output, then you have to select the Output option on the File → Open menu, or the Syntax or Script option or whatever you want to open.

You can enter your own data to create a new SPSS data file, use an existing file, or even import data from such applications as Microsoft Excel or Lotus into SPSS. Any way you do it, you need to have data to work with. In Figure 18 (on page 19), you can see we have a screen full of data, part of the Teacher Scale Results file mentioned earlier and available at http://www.prenhall.com/greensalkind. We opened it using the **Open** option on the File menu.

If you use a spreadsheet application, then the structure of the data file appears familiar; it has rows and columns, with cases for rows and variables for columns.

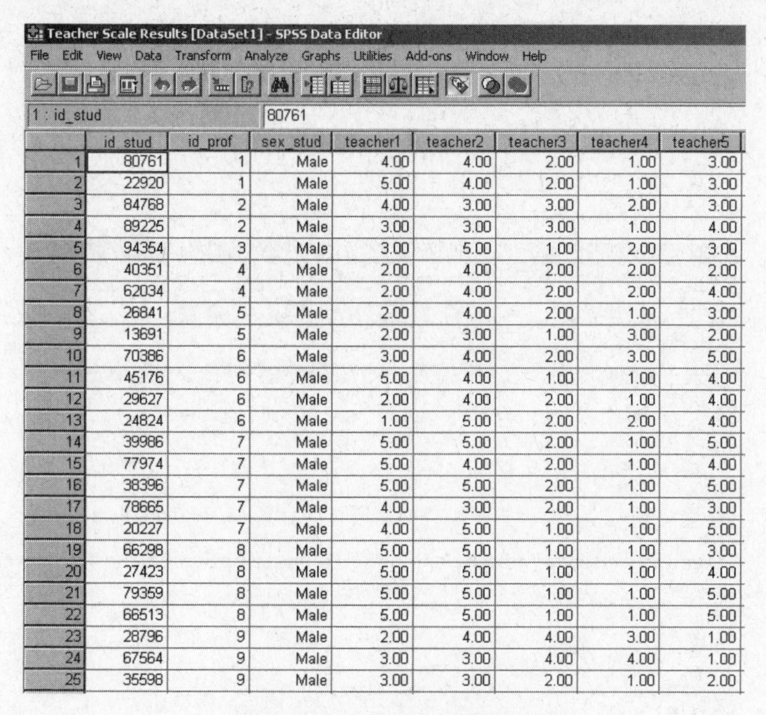

	id_stud	id_prof	sex_stud	teacher1	teacher2	teacher3	teacher4	teacher5
1	80761	1	Male	4.00	4.00	2.00	1.00	3.00
2	22920	1	Male	5.00	4.00	2.00	1.00	3.00
3	84768	2	Male	4.00	3.00	3.00	2.00	3.00
4	89225	2	Male	3.00	3.00	3.00	1.00	4.00
5	94354	3	Male	3.00	5.00	1.00	2.00	3.00
6	40351	4	Male	2.00	4.00	2.00	2.00	2.00
7	62034	4	Male	2.00	4.00	2.00	2.00	4.00
8	26841	5	Male	2.00	4.00	2.00	1.00	3.00
9	13691	5	Male	2.00	3.00	1.00	3.00	2.00
10	70386	6	Male	3.00	4.00	2.00	3.00	5.00
11	45176	6	Male	5.00	4.00	1.00	1.00	4.00
12	29627	6	Male	2.00	4.00	2.00	1.00	4.00
13	24824	6	Male	1.00	5.00	2.00	2.00	4.00
14	39986	7	Male	5.00	5.00	2.00	1.00	5.00
15	77974	7	Male	5.00	4.00	2.00	1.00	4.00
16	38396	7	Male	5.00	5.00	2.00	1.00	5.00
17	78665	7	Male	4.00	3.00	2.00	1.00	3.00
18	20227	7	Male	4.00	5.00	1.00	1.00	5.00
19	66298	8	Male	5.00	5.00	1.00	1.00	3.00
20	27423	8	Male	5.00	5.00	1.00	1.00	4.00
21	79359	8	Male	5.00	5.00	1.00	1.00	5.00
22	66513	8	Male	5.00	5.00	1.00	1.00	5.00
23	28796	9	Male	2.00	4.00	4.00	3.00	1.00
24	67564	9	Male	3.00	3.00	4.00	4.00	1.00
25	35598	9	Male	3.00	3.00	2.00	1.00	2.00

Figure 18. **An opened SPSS file.**

Working with Appearance

TIP

A quick way to change the appearance of data in the Data Editor is to right-click in any cell and click **Grid Font.** Then select the font, size, and style.

Everyone has his or her preference on how things look. In Figure 18, the data appear in a sans serif font named Arial 10 point (with one point being equal to 1/72 of an inch) in size. For whatever reason, you may want to select another font and change the style in which the values or headings appear.

Simply by using the **Fonts** option on the View menu, we changed the font to Times New Roman 10 point. We also used the Grid Lines option on the View menu to turn off the grid lines. The partial screen showing the data file appears in Figure 19.

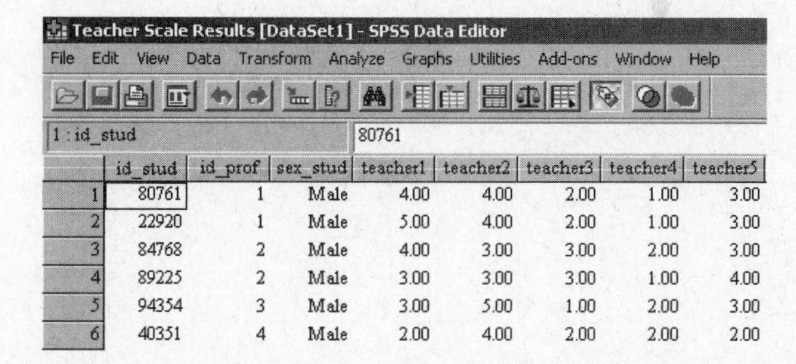

	id_stud	id_prof	sex_stud	teacher1	teacher2	teacher3	teacher4	teacher5
1	80761	1	Male	4.00	4.00	2.00	1.00	3.00
2	22920	1	Male	5.00	4.00	2.00	1.00	3.00
3	84768	2	Male	4.00	3.00	3.00	2.00	3.00
4	89225	2	Male	3.00	3.00	3.00	1.00	4.00
5	94354	3	Male	3.00	5.00	1.00	2.00	3.00
6	40351	4	Male	2.00	4.00	2.00	2.00	2.00

Figure 19. **Working with font options to change the appearance of a data set.**

Creating a New Variable

As you know, the Teacher Scale is a five-item measure of teacher effectiveness. Let's assume that we want to compute a new variable that is the average of the five items, so that each student's ratings have an average score.

To create a new variable, we use the Compute Variable option on the Transform menu. The finished result (the new variable named average) is shown in Figure 20.

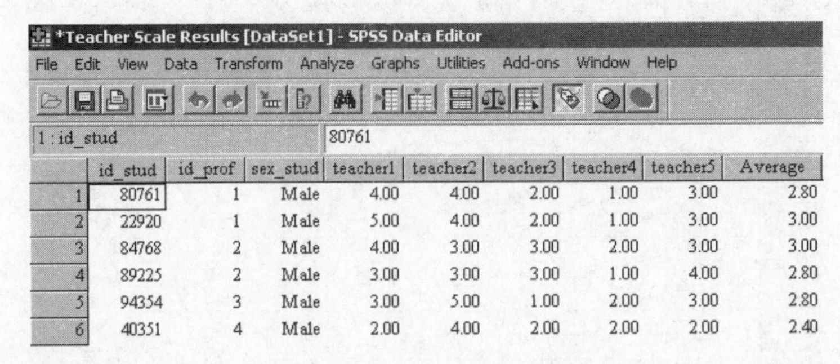

	id_stud	id_prof	sex_stud	teacher1	teacher2	teacher3	teacher4	teacher5	Average
1	80761	1	Male	4.00	4.00	2.00	1.00	3.00	2.80
2	22920	1	Male	5.00	4.00	2.00	1.00	3.00	3.00
3	84768	2	Male	4.00	3.00	3.00	2.00	3.00	3.00
4	89225	2	Male	3.00	3.00	3.00	1.00	4.00	2.80
5	94354	3	Male	3.00	5.00	1.00	2.00	3.00	2.80
6	40351	4	Male	2.00	4.00	2.00	2.00	2.00	2.40

Figure 20. **Creating a new variable by using the Compute option on the Transform menu.**

A Simple Table

Now it's time to get to the reason why we're using SPSS in the first place: the various analytical tools that are available. First, let's say we want to know the general distribution of males and females. That's all, just a count of how many males and how many females are in the total sample.

In Figure 21, you can see an output window that provides exactly the information we asked for, which was the frequency of the number of males and females. We clicked **Analyze → Descriptive Statistics → Frequencies** to compute these values. We could have computed several other descriptive statistics, but just the count is fine for now. Here's

Frequencies

[DataSet1] D:\Using SPSS\SPSS 5th Edition\Data Files 5e\Teacher Scale Results.sav

Statistics

Sex of Student

N	Valid	50
	Missing	0

Sex of Student

		Frequency	Percent	Valid Percent	Cumulative Percent
Valid	Male	28	56.0	56.0	56.0
	Female	22	44.0	44.0	100.0
	Total	50	100.0	100.0	

Figure 21. **A simple SPSS table.**

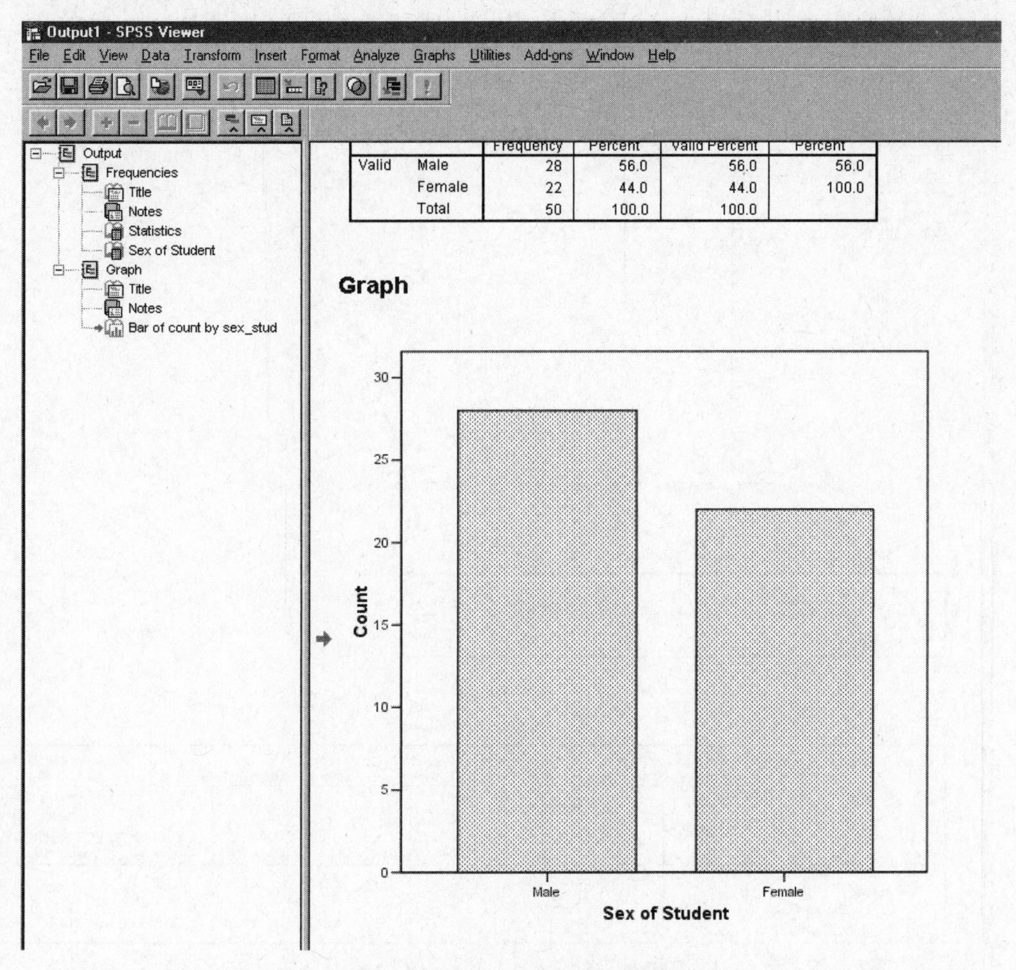

Figure 22. **A simple SPSS chart.**

your first look at real SPSS output. The Viewer window and the outline pane on the left list all the output available.

Guess what? With just another few clicks, we can create the bar chart you see in Figure 22. This graph also shows the special graphing tools on the toolbar in the SPSS Chart Editor. Also, you should notice that as additional items are added to the Viewer, the outline pane to the left of the viewer lists those as well. There are lots of ways to jazz up this graph, many of which we will discuss in Lesson 16.

A Simple Analysis

Let's see if males and females differ in their average Teach scores. This is a simple analysis requiring a t test for independent samples. You may already know about this procedure from another statistics class you have had, or it may be entirely new. The procedure is a comparison between the mean for the group of males and the mean for the group of females.

In Figure 23 (on page 22), you can see a partial summary of the results of the t test. Notice that the listing in the left pane of the Viewer now shows the Frequencies and T-Test procedures listed. To see any part of the output, all we need to do is click on that element, in the left pane of the Viewer, as we did with Frequencies. The large arrow (which appears in red on the screen) indicates which of the various outputs you have highlighted in the output pane.

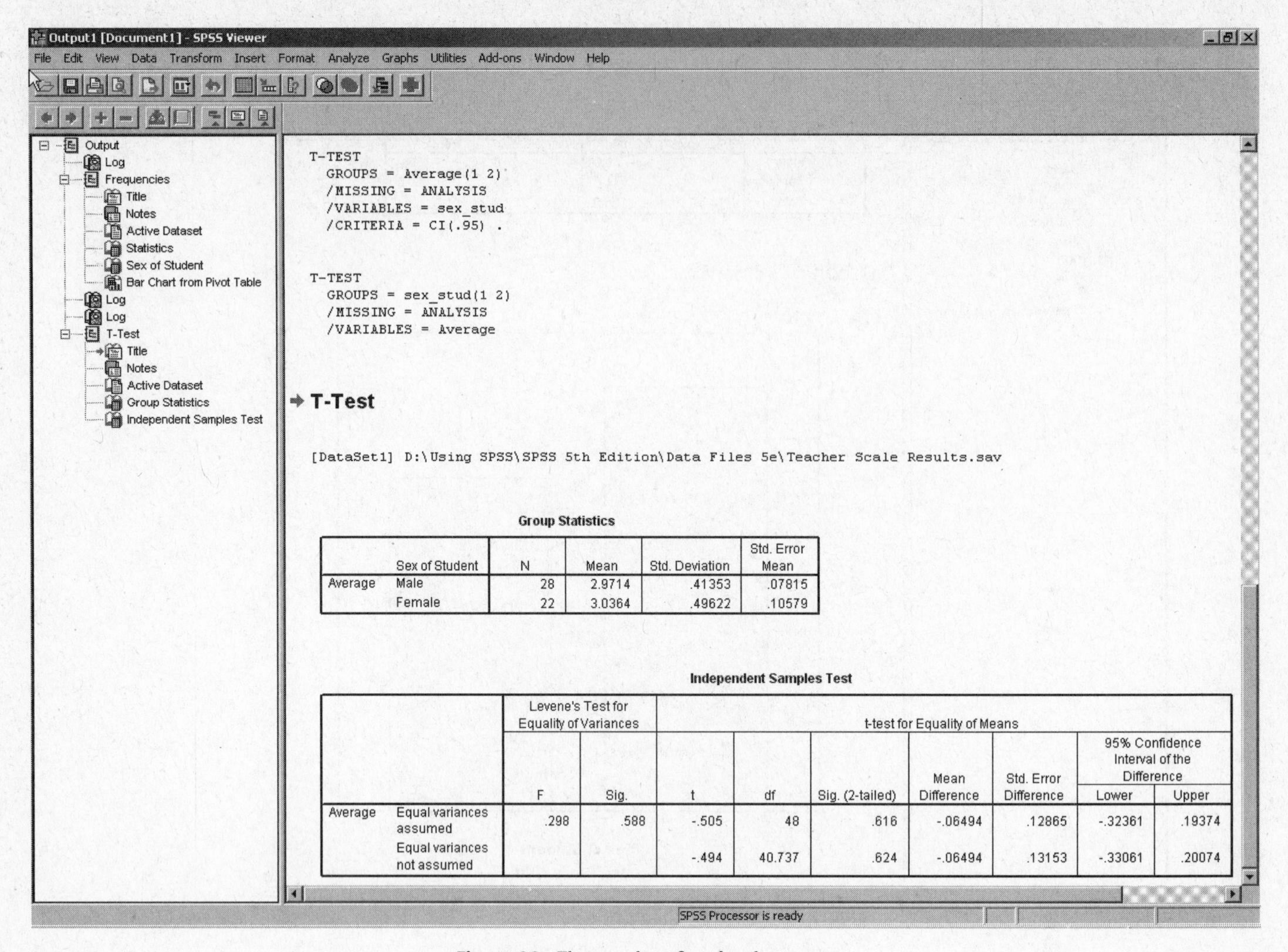

Group Statistics

	Sex of Student	N	Mean	Std. Deviation	Std. Error Mean
Average	Male	28	2.9714	.41353	.07815
	Female	22	3.0364	.49622	.10579

Independent Samples Test

		Levene's Test for Equality of Variances		t-test for Equality of Means					95% Confidence Interval of the Difference	
		F	Sig.	t	df	Sig. (2-tailed)	Mean Difference	Std. Error Difference	Lower	Upper
Average	Equal variances assumed	.298	.588	-.505	48	.616	-.06494	.12865	-.32361	.19374
	Equal variances not assumed			-.494	40.737	.624	-.06494	.13153	-.33061	.20074

***Figure 23*. The results of a simple *t* test.**

PART I

Creating and Working with Data Files

UNIT 2

You've been introduced to SPSS and what SPSS offers. Now it's time to learn how to work with data that you use to create SPSS data files. The seven lessons in this unit will cover important issues such as defining variables; entering and editing data; inserting and deleting cases and variables; selecting, cutting, and pasting data; printing SPSS documents; exiting an SPSS data file; and importing and exporting SPSS data.

These crucial skills lay a foundation for Part II of *Using SPSS for Windows and Macintosh*, where you will learn how to use SPSS procedures. Read through these seven lessons and be sure to work through each step.

LESSON

5 Defining Variables

After This Lesson, You Will Know

- How to open a new data file
- How to define an SPSS variable's characteristics

Key Words

- Column
- Data View window
- Row
- Variable View window

Here's where all the work begins, with the entry of data into the SPSS spreadsheet-like structure, the Data View (as you saw in Figure 18). When you first open SPSS, the Data Editor window is labeled Untitled-SPSS Data Editor, and it's ready for you to enter variables as well as cases. You can just begin entering data in row 1, column 1, and, as you can see in Figure 24, SPSS will record that as the first data point for var00001 (which SPSS will automatically name the column or variable). In SPSS, rows always represent cases and columns always represent variables.

Creating an SPSS New Window

If you are already working within SPSS and want to create a new **Data View window**, follow this step:

Figure 24. **Entering data in the Data View window.**

1. Click **File** ⟶ **New** ⟶ **Data**. You will see a blank Data View window and you are ready to either define or enter data. Note that in past versions of SPSS, when you had one data file open, you would have to save that one before opening a new one. With this version, you can have more than one file open at the same time and easily switch between them.

Having SPSS Define Variables

Now that the new data window is open, you can do one of two things. The first is to enter data into any cell in the Data View window and press the Enter key (which is the standard way of entering data). For example, you could enter the value 87 in row 1, column 1 and press the Enter key. Since SPSS must have a variable name to work with, SPSS will automatically name the variable var00001. If you did this in row 1, column 5, then SPSS would name the variable var00005 and also number the other columns sequentially, as you see in Figure 25.

Note also that right below the main menu in the upper left-hand corner of the Data View window is an indicator of what cell is being highlighted. For example, in Figure 25, you can see that 1:var00005 represents row 1, variable 5.

Figure 25. **SPSS automatically defines variables.**

Custom Defining Variables: Using the Variable View Window

However, a critical part of dealing with data is defining those variables that you intend to enter—and by defining, we mean everything from providing a name for the variable (or the column in the Data View window) to defining the type of variable it is and how many decimal places it will use.

In order to define a variable, one must first go to the **Variable View window** by clicking the Variable View tab at the bottom of the SPSS screen. Once that is done, you will see the Variable View window as shown in Figure 26 and be able to define any one variable as you see fit.

Once in the Variable View window, you can define variables along the following parameters:

- **Name** provides a name for a variable up to eight characters.
- **Type** defines the type of variable such as text, numeric, string, scientific notation, etc.
- **Width** defines the number of characters the column housing the variable will allow.
- **Decimals** defines the number of decimals that will appear in the Data View window.
- **Label** defines a label up to 256 characters for the variable.
- **Values** defines the labels that correspond to certain numerical values (such as 1 for male and 2 for female).
- **Missing** indicates how missing data will be dealt with.

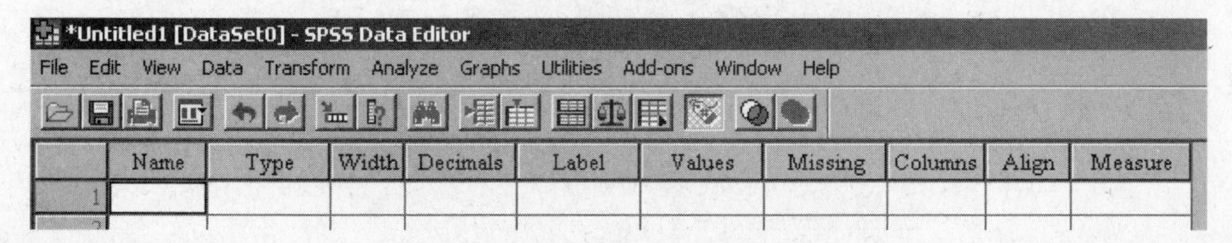

Figure 26. **The Variable View window.**

- **Columns** defines the number of spaces allocated for the variable in the Data View window.
- **Align** defines how the data is to appear in the cell (right, left, or center aligned).
- **Measure** defines the scale of measurement that best characterizes the variable (nominal, ordinal, or interval).

The general way in which these characteristics of a variable are defined is by clicking on the cell for a particular variable and then specifying the particulars of those characteristics for the variable under question.

Each of these variable characteristics is described as follows. We use the file named Crab Scale Results and discuss how we defined these variables. The Variable View of the file is shown in Figure 27.

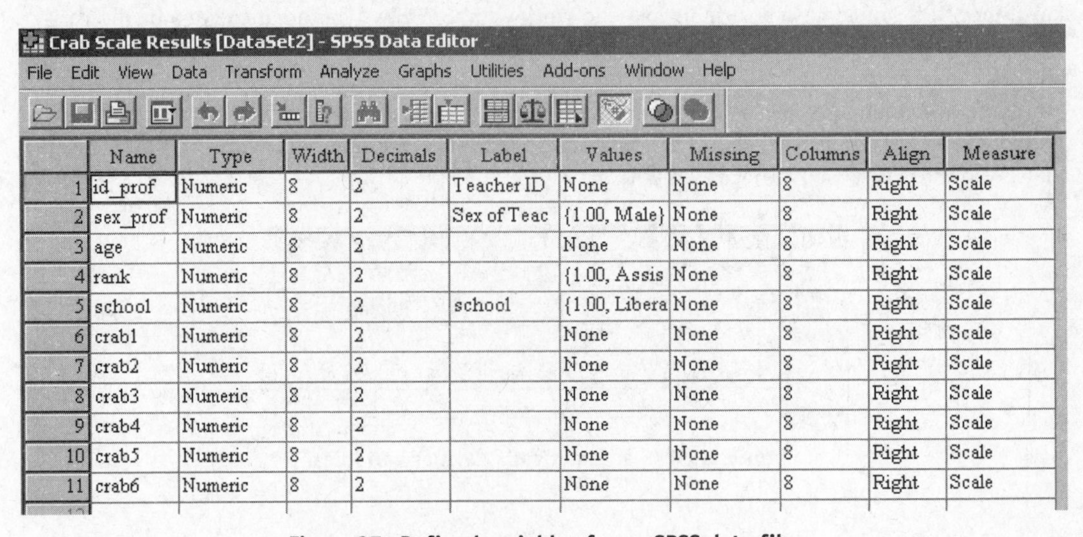

	Name	Type	Width	Decimals	Label	Values	Missing	Columns	Align	Measure
1	id_prof	Numeric	8	2	Teacher ID	None	None	8	Right	Scale
2	sex_prof	Numeric	8	2	Sex of Teac	{1.00, Male}	None	8	Right	Scale
3	age	Numeric	8	2		None	None	8	Right	Scale
4	rank	Numeric	8	2		{1.00, Assis	None	8	Right	Scale
5	school	Numeric	8	2	school	{1.00, Libera	None	8	Right	Scale
6	crab1	Numeric	8	2		None	None	8	Right	Scale
7	crab2	Numeric	8	2		None	None	8	Right	Scale
8	crab3	Numeric	8	2		None	None	8	Right	Scale
9	crab4	Numeric	8	2		None	None	8	Right	Scale
10	crab5	Numeric	8	2		None	None	8	Right	Scale
11	crab6	Numeric	8	2		None	None	8	Right	Scale

Figure 27. **Defined variables for an SPSS data file.**

Defining Variable Names

To define the name of a variable, follow these steps:

1. Click on a cell in the Name column in the Variable View window.
2. Enter the name for the first variable as id_prof and press the **Enter key**.

In general, it's best to stick with letters and numerals for variable names with words separated by an underscore such as id_prof, not id prof. It's not only better, but SPSS does not allow spaces and characters other than letters and numerals.

Defining Variable Types

If you click on a cell in the Type column and the ellipsis in it, you see the Variable Type dialog box in Figure 28. Its variable definitions include numeric (e.g., 3435.45), comma (e.g., 3,435.45), dot

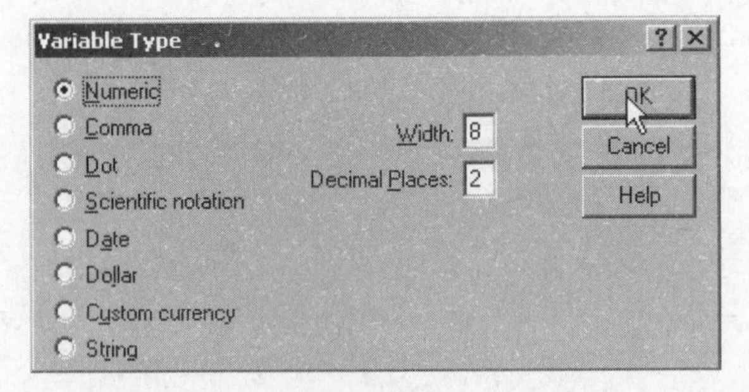

Figure 28. **The Variable Type dialog box.**

(e.g., align, 3.43545), scientific notation (e.g., 3.4E+03), date (e.g., 12-FEB-1996), dollar (e.g., $3,435.45), custom currency (as you design it), or a string (such as William). In our example, all variables are defined as numeric.

Defining Variable Widths

To define the width of a variable, follow these steps:

1. Click on a cell in the Width column for the corresponding variable (in the Variable View window).
2. Enter the number of characters you want to use to define the variable or click on the up and down triangles (see Figure 29) to change the value.

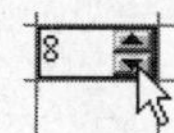

Figure 29. **Setting the width of a variable.**

Defining Variable Decimals

To define the number of decimals used for a variable, follow these steps:

1. Click on the **Decimal** column for the corresponding variable.
2. Enter the number of characters you want to use to define the number of decimals or click on the up and down triangles to change the number in the cell.

Defining Variable Labels

To define the Label used for a variable, follow these steps:

1. Click on the **Label** column for the corresponding variable.
2. Enter up to 256 characters for the label, including spaces. Labels appear in SPSS output and are not visible in the Data View window. For example, in Figure 27, you can see how the variable named sex_prof is accompanied by the label Sex of Teacher.

Defining Variable Values

Values are used to represent the contents of a variable in the Data View window. For example, we will show you how we defined the sex of the teacher by using 1s and 2s, while 1s, 2s, and 3s indicate rank as assistant, associate, and full professor in the Data View window.

In general, it makes a great deal more sense to work with numbers than with string or alphanumeric variables in an analysis. In other words, values such as 1 for assistant professor, 2 for associate professor, and 3 for full professor provide far more information than the actual text that describes the levels of the variable named rank.

Just the same, it sure is a lot easier to look at a data file and see words rather than numbers. Just think about the difference between a data file with numbers representing various levels (such as 1, 2, and 3) of a variable and the actual values (such as assistant professor, associate professor, or full professor).

To define the Values used for a variable, follow these steps:

1. Click on the cell in the Values column for the corresponding variable. In this case, we are using rank as the example.
2. Click on the **ellipsis** to open the Value Labels dialog box as shown in Figure 30 (on page 28).

TIP

Once a variable is defined, you can easily change the value by just clicking on the arrow of a highlighted cell in the Data View window.

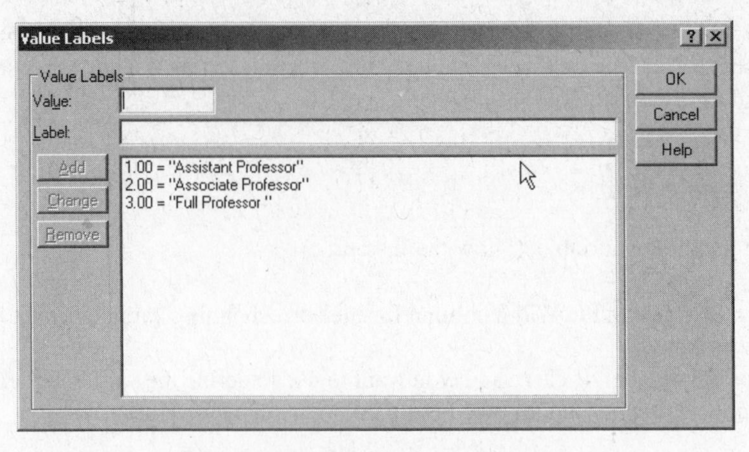

Figure 30. **The Values Labels dialog box.**

3. Enter a **Value** for the variable, such as 1, 2, or 3. This is what will appear in the Data File window.
4. Enter the **Value Label** for the value such as assistant, associate, or full.
5. Click **Add**.
6. Click **Continue**. When you finish your business in the Define Labels dialog box, click **OK** and the new labels will take effect.

There are a few differences in the way the Windows and Mac versions of SPSS treat variable names and variable labels. The rules are summarized as follows.

	Windows version	**Mac version**
Variable Names	Up to 64 characters with upper- and lowercase supported with no spaces allowed	Limited to 8 characters with no spaces allowed
Variable Labels	Uppercase and spaces allowed	Uppercase and spaces allowed

Defining Missing Values for a Variable

To define the values used for a variable, follow these steps:

1. Click on the Missing column for the corresponding variable.
2. Click the **ellipsis** and you will see the Missing Values dialog box. The various options you can select are as follows:
   - The No Missing Values option treats all values as being valid, such that no missing values are present in the data file.
   - The Discrete Missing Values option allows you to enter up to three values for a missing variable.
   - The Range Plus One Discrete Missing Value option will identify as missing all the values between the low and high values you identify, plus one additional value outside the range.

TIP

In defining characteristics, don't confuse variable width with variable columns. Width refers to the number of characters used to define the display in the Data View window. Columns refers to the number of columns used to define the variable.

To change the width of a column (in the Data View or Variable View windows), you can drag on the vertical lines that separate the columns from one another.

Defining Variable Columns

To define the variable Columns Values, follow these steps:

1. Click on the **Columns** column for the corresponding variable.

2. Enter the value you want for the width of the column or click on one of the up or down triangles to set the width.

Defining Variable Alignment

To define the alignment used for a variable, follow these steps:

1. Click on the **Align** column for the corresponding variable.
2. From the drop-down menu, select Left, Right, or Center.

Defining Variable Measure

To define the Measure used for a variable, follow these steps:

1. Click on the **Measure** column for the corresponding variable.
2. From the drop-down menu, select Scale, Ordinal, or Nominal as the level of measurement you want to use for this variable.

In sum, we have defined 11 different variables as shown in Figure 27 (on page 26). Each has particular set of characteristics and any of these can be changed quickly and easily by clicking on the Variable View window tab and then changing the variable attribute you want.

LESSON

6 Entering and Editing Data

After This Lesson, You Will Know

- How to enter data into an SPSS Data Editor window
- About editing data in SPSS

Key Words

- Active cell
- Save Data As dialog box
- Save File As dialog box

You just learned how to define a set of variables. Now it's time to enter data that correspond to the set of variables. We'll also learn how to edit data. The file that you will create is available at http://www.prenhall.com/greensalkind and is named Lesson 6 Data File 1. It consists of 10 cases and three variables (id, sex, and test1), and you can see what it looks like in Figure 31.

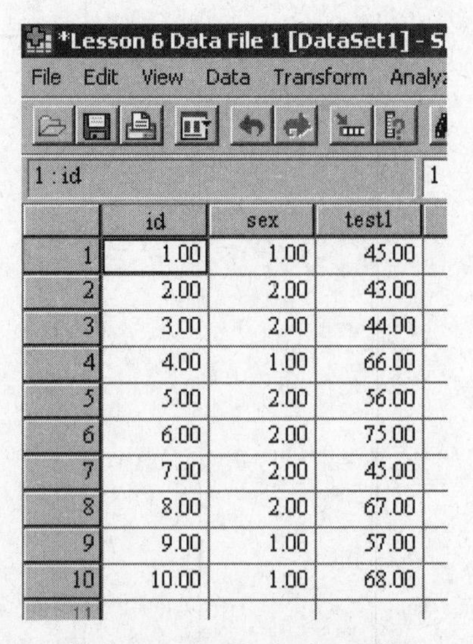

	id	sex	test1
1	1.00	1.00	45.00
2	2.00	2.00	43.00
3	3.00	2.00	44.00
4	4.00	1.00	66.00
5	5.00	2.00	56.00
6	6.00	2.00	75.00
7	7.00	2.00	45.00
8	8.00	2.00	67.00
9	9.00	1.00	57.00
10	10.00	1.00	68.00
11			

Figure 31. **A simple data file.**

Getting Ready for Data

Let's start at the beginning. First, we'll define each of the three variables in the data file you see in Figure 31. Start with an empty Data Editor window.

1. Click the **Variable View** tab in the opening window.
2. In cell 1, row 1, type the variable name id.
3. Press the **Enter** key and you will see SPSS complete the default definition for each of the variable's characteristics.
4. In cell 1, row 2, type the variable name sex.
5. Press the **Enter** key and you will see SPSS complete the default definition for each of the variable's characteristics.
6. In cell 1, row 3, type the variable name test1.
7. Press the **Enter** key and you will see SPSS complete the default definition for each of the variable's characteristics. Each row (representing a variable) should contain the information necessary to define three variables; id, sex, and test1 as you see in Figure 32.

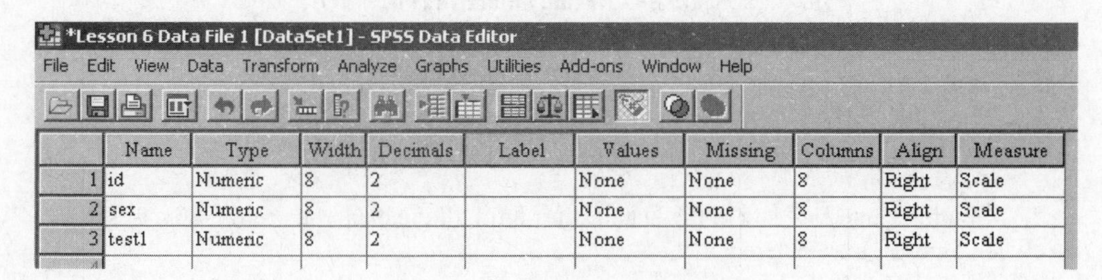

	Name	Type	Width	Decimals	Label	Values	Missing	Columns	Align	Measure
1	id	Numeric	8	2		None	None	8	Right	Scale
2	sex	Numeric	8	2		None	None	8	Right	Scale
3	test1	Numeric	8	2		None	None	8	Right	Scale

Figure 32. **Defining variables.**

8. Click on the **Data View** tab to return to the Data Editor window and you will see the three variables defined as shown in Figure 33.

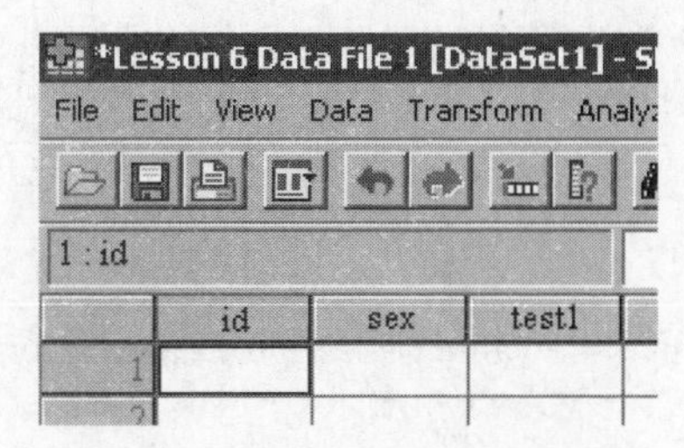

Figure 33. **The defined variables.**

Entering Data

Entering data into the Data Editor window and creating a data file is as simple as entering data in a spreadsheet or creating a table by using a word processor. You just click in the cell where you want to enter data (which then appears as outlined), click, and type away. Let's enter data for the first case.

1. Place the cursor on row 1, column 1.
2. Click once. The cell borders of the individual cell in the data file will be outlined.

TIP

As with any Windows operation, double-clicking on the contents of a cell selects all the material in that cell. You can just double-click and type the replacement.

3. Type **1** for id.
4. Press **Tab** to move to the next variable in the current case. You can also press the down arrow key to move to the next case within the same variable or use the Return (or Enter) key to accomplish the same. SPSS will enter the value 1.00 since the default format is two decimal places. The next cell in the row will then be highlighted. You'll notice that as you enter values into cells in the Data Editor the value appears in the data entry information bar right above the column headings.

The cell in the Data Editor that is highlighted and that shows the value in the information bar is the **active cell**. In Figure 34, the active cell has a value of 1.

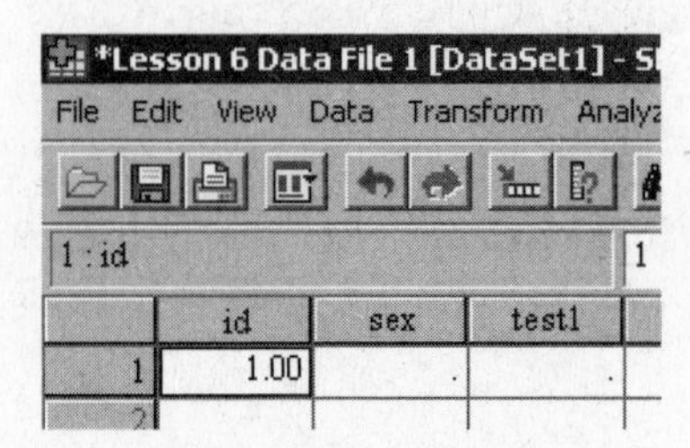

Figure 34. **Noting an active cell.**

5. Type **1** for sex.
6. Press the **TAB** key to move to the column labeled "test1." The Shift + Tab key combination moves one cell to the left.
7. Type **45**.
8. Continue entering data in the Data Editor until all the data you see (all 10 cases) in Figure 31 are entered.

If you make an error, either backspace OR just don't worry about it! You will learn how to correct such errors in the next section in this lesson.

Editing Data

Editing data in a data file is simply a matter of locating the data you want to edit and then changing them. There are several techniques you can use to do this, including a simple click and edit or cutting and pasting. You'll learn about them all in this part of the lesson.

Changing a Cell Value

TIP

How often should you save? You should save after a set amount of work (either time or pages worth). One general rule is to save as often as necessary so that you can easily re-create any work you might lose between saves—every time you finish entering 10 cases or after 30 minutes of work might be a good starting guideline.

Let's say that you want to change the value of test1 for case 3 from 44 to 54. You made an error in data entry and want to correct it. Here's how.

1. Click on the cell you want to change, which in this case is the cell in test1, case 3 with a value of 44. When you click on the cell, its value will appear in the data information bar or the cell editor area.
2. Type **54**. As you begin typing the new value, it replaces the old one.
3. Press the **Enter** key. The new value is now part of the data file.

Editing a Cell Value

You can also edit an existing value without replacing it. For example, you may want to change the test1 value for case 1 from 45 to 40, and you just want to change the last digit rather than replace the entire value. Here's how.

1. Click the cell containing the value you want to edit. In this case, it is the cell for test1, case 1, which contains the value 45. For the Macintosh version, you need to double-click.
2. Click to the right of the last digit in the data entry information bar. A blinking vertical line will appear in the cell editor following the last digit of the cell entry.
3. Backspace once to delete the 5.
4. Type **0**.
5. Press the **Enter** key or move to another cell. The new value is now part of the data file.

If you wanted to change a value such as 4565 to 4665, you would use the mouse pointer (or the backspace key on the blinking cursor in the information bar) to insert the blinking cursor between the first 5 and the 6, press the backspace key once, and then enter the new value. If you wanted to insert a value in an existing one (such as changing 198 to 1298), you would place the cursor where you want the new values to be inserted and just enter them.

Saving a Data File

This is the easiest operation of all, but it may be the most important. As you know from your experience with other applications, saving the files that you create is essential. First, saving allows you to recall the file to work on at a later time. Second, saving allows you to back up files. Finally, you can save a file under a new name and use the copy for a purpose different from the original purpose.

Don't wait to save until your data file is entered and you are sure there are no errors. Why should you save a file before it is entered exactly as you want it? It's a matter of being "safe rather than sorry."

When you are creating a data file document (and before you save it for the first time), the only "copy" of it is stored in the computer's memory. If, for some reason, the computer malfunctions or the power (and the computer) goes off, whatever is stored in this temporary memory will be lost. It literally disappears, whether it is a data file with 3 variables and 10 cases or 300 variables and 10,000 cases. Save data files as you work!

Before saving a file for the first time, however, the first order of business is deciding on a name for the file.

To save a data file, you must assign a unique file name to it. For example, if you were to name a file "data," you might find it confusing, because you will surely be creating more than one file, and "data" describes virtually all of them. Also, should you try to save a file to a directory containing an identically named file with different contents, the new file will overwrite the original file. You lose the original file and all your work.

For these reasons, use a name that describes the contents of the data file. SPSS also automatically attaches the .sav extension to all saved files.

With Windows, file names can be up to 225 characters long, so you shouldn't have any difficulty being sufficiently descriptive, although being wordy has problems too. Try to find a middle ground between describing the file and not having a name so long that it can't easily fit on the screen.

TIP

You can have Windows hide file extensions by going to Windows Explorer, right-clicking, selecting properties, and asking Windows to hide extensions.

To save the data document that is active (which has the data you entered, as you saw in Figure 31 on page 30), follow these steps:

1. Click **File** ⟶ **Save**. When you do this, you will see the **Save Data As dialog box**, which you should be familiar with by now.
2. Select the directory in which you want to save the data. If you are saving to a hard drive and you are working in a computer lab, be sure you have permission. If you are saving to a floppy disk, select drive a or drive b.
3. Enter the file name you want to use in the File Name text box to save the data that has been entered. We used Lesson 6 Data File 1 available at http://prenhall.com/greensalkind.
4. Click **OK**.

TIP

When you do have more than one data file open, the red matrix in the upper left-hand corner of the SPSS window has a green cross superimposed on it as you see here.

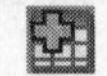

When the file is not selected (and not the active one of several that are opened), just the red matrix appears as shown here.

The data you entered will then be saved as a data file, and the name of the file will appear in the title bar of the Data Editor window. The next time you select Save Data from the File menu, you will not see the Save Data As dialog box. SPSS will just save the changes under the name you originally assigned to the data file. And remember, SPSS will save the data in the active directory. In most cases, that will be the same directory that contains SPSS, a situation you may or may not want.

The Save As option on the file menu allows you to save a file under another name or under another format. For example, if you select the Save As options, you can select from such formats as Excel and Lotus 1-2-3—very handy for exporting your SPSS file to another application.

A significant improvement over earlier versions of SPSS is that you are now able to save, and keep, more than one data file open at the same time. The number of files that you can keep open is closely related to the amount of active memory your computer has available and to switch from one data set to another (or from one window to another), with click on the open data set in the Task bar at the bottom of your Windows display, or use the Alt+Tab key combination to toggle from one data set to another.

Saving a File in a New Location

You can easily select any other location to save a file. Perhaps you want to keep the application and data files separate. For example, many SPSS users have a separate directory for each project. That way, files from different projects don't get mixed together, or you may want to save a file to a floppy disk and then take the disk home for safekeeping.

Here's how to save a file to another location:

1. Click **File → Save As**. You'll see the Save Data As dialog box.
2. Enter the name of the file in the File Name text box.
3. In the Save In section of the Save As dialog box, click the directory or hard drive where you want the file saved.
4. Click **OK**, and the file is saved to the new location.

You can save to any location by clicking a series of drives and directories.

Opening a File

Once a file is saved, you have to open or retrieve it when you want to use it again. The steps are simple.

1. Click **File → Open → Data**, and you will see the Open File dialog box as seen in Figure 35.
2. Find the file you want to open, and double-click on the file name to open it or highlight the file name and click **OK**.

More About Saving

Now that this latest version of SPSS allows you to open more than one data file at a time, there are certain conventions you have to remember. And, some of this is counterintuitive to what you know about using other Windows aplications such as Word and Excel.

1. You can create new data windows, but unless you enter data in them, they will not appear on the screen. So, if you open three new data windows (in addition to what is already opened), the last one will be numbered (and named [DataSet4]) and only this window and the original is open (DataSet windows 2 and 3 just do not appear).

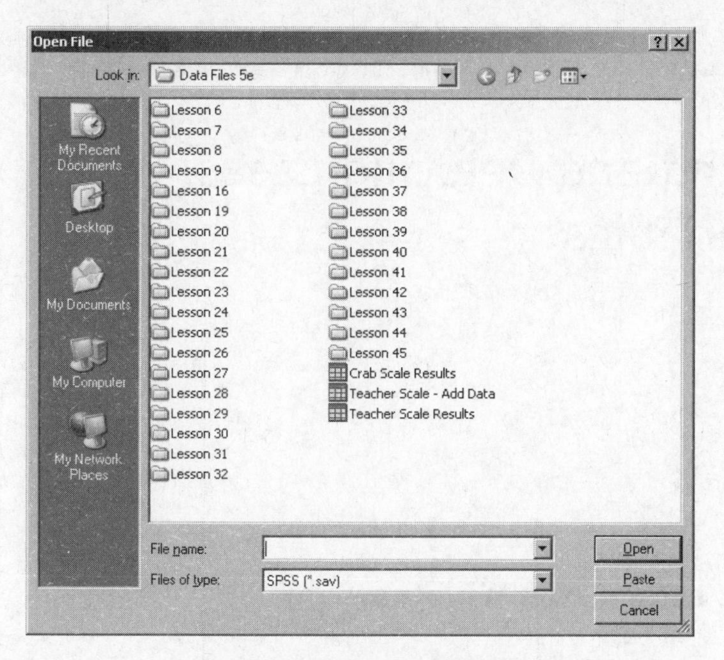

Figure 35. **Opening an SPSS file.**

2. You can use the **File → Close** (or the Ctrl+F4 key combiantion) to close any window but the one that was first opened (and assuming it is populated with data). To close the last Data Editor window, you have to exit SPSS. This means that, unlike most other Windows aplications, you cannot close all the existing windows and still have the SPSS application active.

LESSON

7

Inserting and Deleting Cases and Variables

After This Lesson, You Will Know

- How to insert and delete a case
- How to insert and delete a variable
- How to delete multiple cases and variables

Key Words

- Delete
- Insert

More often than not, the data you first enter to create a data file need to be changed. For example, you may need to add a variable or delete a case or make changes as the result of the validation process (see Lesson 11). You learned in the last lesson how to enter and edit data and then save all the data as a data file. But the simple editing may not be enough.

There are times when you want to add one or more cases, or even an entirely new variable, to a data set. A subject might drop out of an experiment and you may choose to delete a case, or, for measurement or design reasons, an entire variable may no longer be useful and may need to be deleted from a data file. In this lesson, we will show you how to insert and delete both cases and variables.

The SPSS file we'll work with in this lesson is shown in Figure 36. It contains three variables (age in years, social class, and preference for soft drink) and five cases, and it is named Lesson 7 Data File 1 at http://www.prenhall.com/greensalkind.

Inserting a Case

Notice in Figure 36 that the cases are sorted by age in ascending order from the youngest subject (aged 44) to the oldest subject (age 67). Let's say we want to add a new case. We'll place it between the second and third cases.

TIP

When inserting a case, you can select any cell in the row above which you want to insert the case.

1. Select a cell in the row above which you want to **insert** the new case. In this example, it is row 3 since we want the new case to appear right above row 3. You can select any cell in the row.

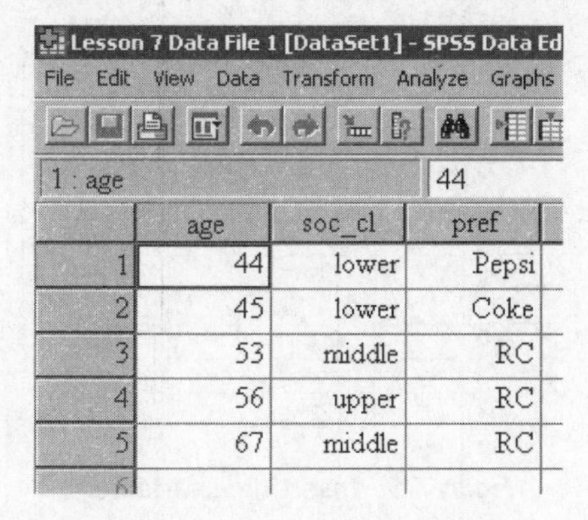

	age	soc_cl	pref
1	44	lower	Pepsi
2	45	lower	Coke
3	53	middle	RC
4	56	upper	RC
5	67	middle	RC

Figure 36. **Lesson 7 Data File 1.**

2. Click **Edit → Insert Cases**, or click the Insert Cases button on the toolbar. As you can see in Figure 37, an entire new row opens up, providing room to enter the data for the new case.

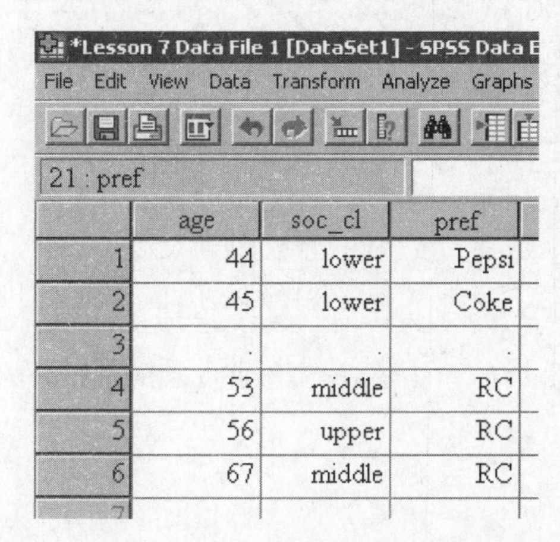

	age	soc_cl	pref
1	44	lower	Pepsi
2	45	lower	Coke
3	.	.	.
4	53	middle	RC
5	56	upper	RC
6	67	middle	RC

Figure 37. **Inserting a new case (not a variable) in a data file.**

You'll notice that when you add a case, dots appear in the blank cells, indicating that there are missing data. Now you can enter the data that make up the new case. Enter any data you want.

Inserting a Variable

Now let's say that you need to add an additional variable to the data file. For example, you want to add the variable weight in pounds so that it appears before the variable labeled "pref," for preference.

1. Select a cell in the column to the right of where you want to insert the new variable. We selected a cell in column 3 (labeled pref) since we want to insert a new variable just to the left of pref.
2. Click **Edit → Insert Variables**, or click the Insert Variable button on the toolbar. As you can see in Figure 38 (on page 38), an entire new column opens up, providing room to enter the data for the new variable. As you can also see, the new variable is automatically named by SPSS as Var00001.

TIP

You can also right-click once a cell has been selected and click **Insert Cases** rather than using the menus.

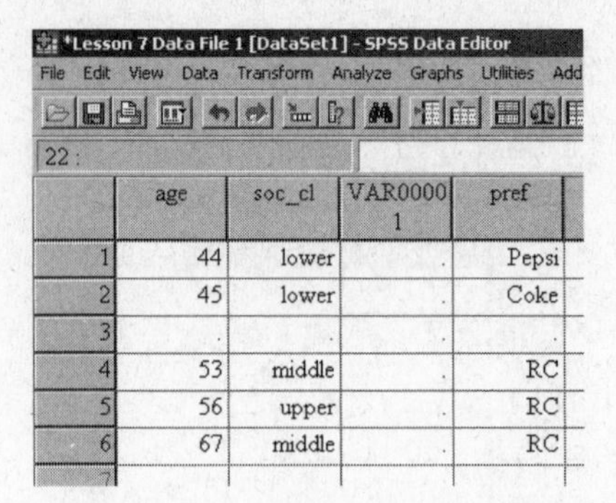

Figure 38. **Inserting a variable.**

You'll notice that when you add a new case or a variable, dots appear in the blank cells, indicating that there are missing cases. You can define the variable (we named it weight) and enter the data for each case. The data file with a new case and variable added, and with data added as well, appears in Figure 39.

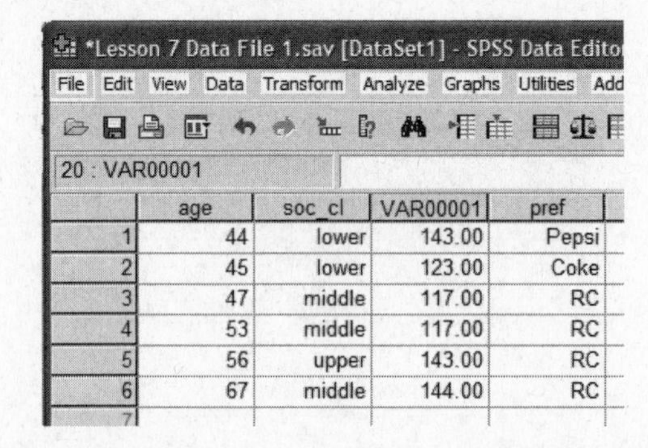

Figure 39. **After a new variable has been inserted.**

Deleting a Case

Now it comes time to learn how to **delete** a case that you no longer want to be part of the data file. For example, let's delete the third case in the data file you now see in Figure 39.

1. Click on the case number (not any cell in the row) to highlight the entire row as you see in Figure 40.

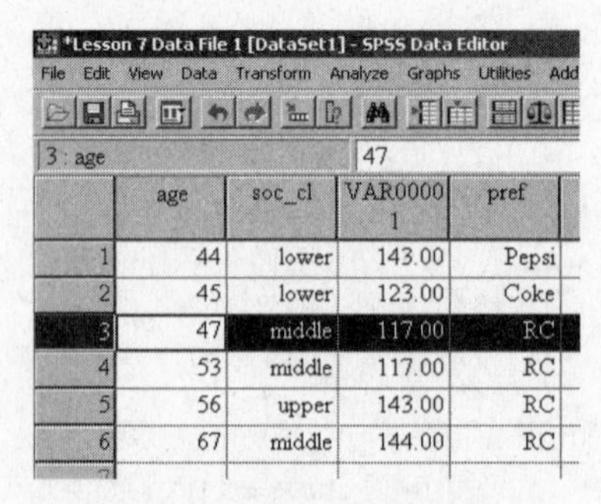

Figure 40. **Deleting a case.**

2. Click **Edit → Clear**. You can also press the **Del** key, or right-click and click **Clear**, and the case will be deleted.

Deleting a Variable

We're sure you know what's next: deleting a variable. Needing to delete a variable is probably a rare occurrence; you would have thought through your data collection procedures well enough to avoid collecting data you don't need. Nonetheless, the situation does arise.

For example, perhaps you find out after the fact that data you collected (such as social class) are confidential and cannot be used. You therefore need to delete the variable across all cases. Here's how.

1. Click on the variable or column to highlight the entire column.
2. Click **Edit → Clear** or press the **Del** key when the row is highlighted. You can also press the **Del** key, and the variable will be deleted, or right-click and click **Clear**.

TIP

Oops! Did you inadvertently delete something? Then select the Undo command on the Edit menu or use the Ctrl → Z key combination to undo the last action.

To delete more than one case or more than one variable, drag over the cases or variables and select them all and then press the **Del** key.

LESSON

8 Selecting, Copying, Cutting, and Pasting Data

After This Lesson, You Will Know

- How to select data
- How Cut, Copy, and Paste can be used in a data file
- How to cut and paste data
- How to copy and paste data
- What the clipboard does, and what its limitations are

Key Words

- Buffer
- Clipboard
- Copy
- Cut
- Paste

This last lesson on editing a data file has to do with some Windows techniques that you may already be familiar with: selecting, cutting, copying, and pasting. In fact, these simple operations can save you time and trouble, especially when you want to move the value of one or more cells from one location to another, or to repeat columns or rows.

Copying, Cutting, and Pasting

With the **Cut**, **Copy**, and **Paste** tools located on the Edit menu, you can easily move cell entries (be it one or more cells) from one part of a data file to another part of the same data file. You can also move cell entries from one data file to another entirely different data file or even to another Windows application!

In general, the steps you take to cut or copy and paste are as follows:

1. Select the data you want to cut or copy by dragging over the data with the mouse pointer. It can be an individual cell, row, column, or more than one row or column. Rows and columns can also be selected by clicking on them at the top (for a column) or at the left (for a row). A range of cells defines a rectangular- or square-shaped set of cells.

TIP

The Shift + arrow key combination can be used to select a range of cells as well. Click in an "anchor" cell, then use the Shift up, down, right, or left arrows to select what you want.

2. Click **Edit ⟶ Cut** or **Copy**. Unfortunately, unlike in other Windows applications, there are no cut, copy, and paste buttons on the SPSS toolbar.
3. Select the destination, or target cell, where you want the data pasted. The destination or target cell you select will act as the upper left-hand anchor if you are cutting or copying more than one cell.
4. Click **Edit ⟶ Paste**.

TIP

There are no cut, copy, or paste buttons on the toolbar, but you can still use the famous Ctrl+C key combination (for copy), Ctrl+V key combination (for paste), and Ctrl+X key combination for cut.

The selected data will appear in the new location. Here is an example of cutting and pasting and copying and pasting. The data that we're using, which you see in Figure 41, is contained in the file named Lesson 8 Data File 1 at http://www.prenhall.com/greensalkind. It consists of five cases of three variables (weight in pounds, height in inches, and strength ranking).

Lesson 8 Data File 1 [DataSet2] - SPSS Data Ed

File Edit View Data Transform Analyze Graphs

19 : height

	weight	height	strength
1	132	68	5
2	155	71	4
3	215	76	1
4	165	72	2
5	178	71	3

Figure 41. **Lesson 8 Data File 1.**

Selecting Data

Before you can cut, copy, or paste anything, you first have to highlight what you want to cut or copy. You are probably used to the most direct way of highlighting data, which is through the drag technique. For example, if you want to highlight a particular section of a data file, just drag the mouse pointer over that section.

Thus, any set of data, cases, or variables can be highlighted by just dragging over the information by using the mouse. An individual cell is automatically highlighted when you click on that cell. A column (or row) is highlighted when you click at the top or left-hand side of that column or row respectively.

What if you're not a mouser? Not to worry. Here's an easy-to-use summary (Table 4) that will tell you how to use the keyboard to select any amount of text you want.

Table 4
Keyboard Strokes for Selecting

To select	Use these keys
one cell down	↓ or the Enter key
one cell up	↑
one cell to the right	→
one cell to the left	←
the first cell in a case	Home
the last cell in a case	End
the first case in a variable	Ctrl + ↑
the last case in a variable	Ctrl + ↓
an entire case	Shift + Space
an entire variable	Ctrl + Space
a block of cells	Shift + ↑ or ↓ or ← or →

Cutting and Pasting

Data are cut from a data file when you want to remove them from their original locations and paste them into another location.

Here's an example of cutting data to rearrange the order in which variables appear. We'll be showing you this activity in Figures 42 and 43 where part of the data file (the strength variable will be moved to the column following the height variable) is to be copied to a new location. Specifically, we'll move the height variable so that it is after strength.

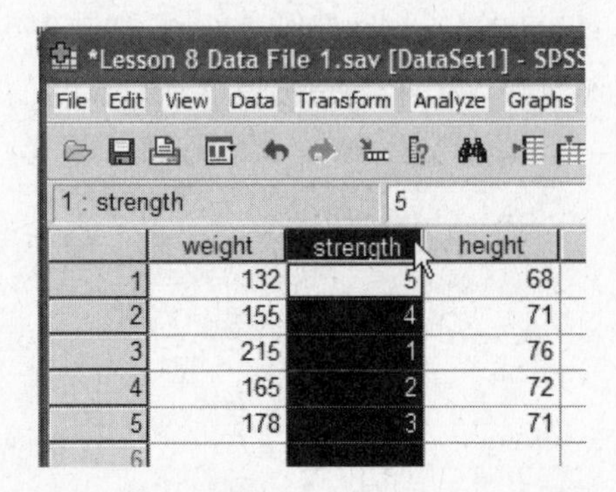

Figure 42. **Selecting a variable to be moved.**

TIP

When cutting, copying, and pasting, SPSS has right-click features where you can highlight and then right-click to select Cut, Copy, or Paste. You can also use keyboard commands to copy (Ctrl+C), cut (Ctrl+X), and paste (Ctrl+V).

1. Highlight the data you want to cut as shown in Figure 42. In this example, we are highlighting the variable (and not just the data in that column) named strength. Remember, when data are cut, they disappear from the original location. When you select, be sure to click on the variable name.
2. Click **Edit ⟶ Cut**. When you do this, the variable height disappears in the Data View window.
3. Move the insertion point to the location where you want to insert, or paste, the data. Click at the top of the column where the new variable name will appear so you highlight the entire column.
4. Click **Edit ⟶ Paste**. The data that were cut will then appear as you see in Figure 43. This is a pretty efficient way of getting data from one place to another. If you want to insert a variable between others, you first have to create a blank variable and then paste in the copied or cut data.

	weight	height	strength
1	132	68	5
2	155	71	4
3	215	76	1
4	165	72	2
5	178	71	3

Figure 43. **Pasting a variable in a new location.**

This is an example of cutting and pasting text within the same data file. Should you want to cut and paste between data files, the process is much the same. Here are the general steps:

1. Highlight the data you want to cut and paste.
2. Click **Edit → Cut**.
3. Open a new data file or the one into which you want the data pasted.
4. Move the insertion point to the location where you want to insert, or paste, the data.
5. Click **Edit → Paste**. Data will appear in the new location.

In Figure 44, you can see the results of pasting data into an Excel document. Notice that only the data and not the column headings are transferred.

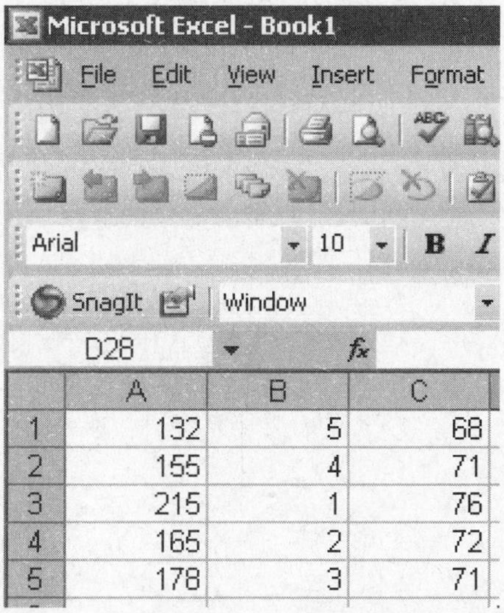

TIP

Want to paste the same data in several locations? Just continue to select the Paste option from the File menu or the right-click menu.

	A	B	C
1	132	5	68
2	155	4	71
3	215	1	76
4	165	2	72
5	178	3	71

Figure 44. **Pasting into an Excel document.**

Copying and Pasting

Copying data is not much different from cutting data, except that the data remain in their original locations. Copying is the ideal tool when you want to duplicate data within a document or between documents.

For example, the weight, height, and strength variables may be part of one data file, and you want to copy only the first two variables (weight and height) to a new data file. You will add additional variables and cases to the new file, but you want to preserve the original.

The steps for copying and pasting are basically the same as cutting and pasting except for one different command. Here are the steps:

1. Highlight the data you want to copy. Remember, when data is copied, the data remains where it was originally located.
2. Click **Edit → Copy**.
3. Move the insertion point to the target cell where you want to insert or paste the copied data.
4. Click **Edit → Paste**. The data that were copied will then appear.

A good use of copy and paste would be to copy data from a master data file that you want to use over and over, don't want to change, and need to borrow from in the future. Using copy and paste will leave the master document intact while you copy what you need.

Where Copied or Cut Data Go

What happens to data that you cut or copy? When you cut or copy data, they are placed in a **buffer**, an area reserved for temporary memory. This buffer is called the **clipboard**. The data are retrieved from the buffer when you select Paste. In fact, whatever information is in the buffer stays there until it is replaced with something else.

For this reason, you can paste the contents of the buffer into documents over and over again until something new has been cut or copied, replacing what was in the buffer. That's why the Paste button is always darkened (or active) since it's ready to paste whatever is in the buffer.

It's useful to remember some things about the clipboard. First, this buffer can hold only one thing at a time. As soon as you cut or copy something new, the new data replace whatever was first in there.

Second, when you quit SPSS, the contents of the clipboard disappear. The clipboard is a temporary place so anything you want to save permanently, you should save to a file with a unique name.

LESSON 9

Printing and Exiting an SPSS Data File

After This Lesson, You Will Know

- How to print a data file
- How to print a selection from a data file
- How to print output from an analysis or a chart
- How to create output as a PDF (Portable Document Format)

Key Word

- Print dialog box

Once you've created a data file or completed any type of analysis or chart, you probably will want to print a hard copy for safekeeping or for inclusion in a report or paper. When your SPSS document is printed and you want to stop working, it's time to exit SPSS. You probably already know the way to exit an application is not by turning off the computer! Rather, you need to give your application (and your operating system) time to do any necessary housekeeping (closing and ordering files, and so forth). In this lesson, we will show you how to print and safely exit SPSS.

Printing with SPSS

Printing is almost as important a process as editing and saving data files. If you can't print, you have little to take away from your work session. You can export data from an SPSS file to another application and print to a file, but getting a hard copy right from SPSS is often more timely and more important.

In SPSS, you can print either an entire data file or a selection from that file. We'll show you how to do both.

Printing an SPSS Data File

Let's begin with data files. It's simple to print a data file.

1. Be sure that the data file you want to print is the active window.

TIP

You can easily print a file with one click; just click the **Print** icon on the toolbar.

2. Click **File** ⟶ **Print**. When you do this, you will see the **Print dialog box** shown in Figure 45.

   As you can see, you can choose to print the entire document or a specific selection (which you would have already made in the Data Editor window), and increase the number of copies from 1 to 9999. (That's the limit.)
3. Click **OK** and whatever is active will print.

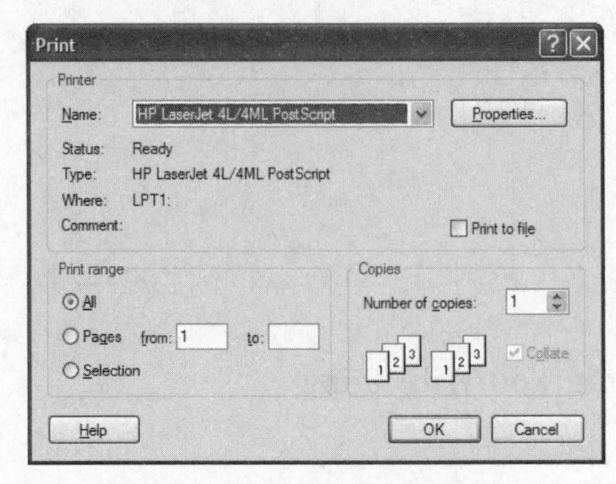

Figure 45. **The Print dialog box.**

TIP

The Properties dialog box (File ⟶ Print ⟶ Properties) allows you to do some pretty cool things. For example, you can print output so that there are two or four pages on a sheet—perfect for reviewing the results of an analysis before it is printed up for the final draft of an article or report. (Note: You will not find this dialog box in the Mac version.)

Look at the printed output in Figure 46, and then read the comments below.

Lesson 9 Data File 1

	weight	height	strength
1	132	68	5
2	155	71	4
3	215	76	1
4	165	72	2
5	178	71	3

12/05/05 09:41:15 AM 1/1

Figure 46. **SPSS output.**

First, you can see the title of the name assigned to the data file at the top of the page. In this case, it's Lesson 9 Data File 1. This is a convenient reminder of what's being printed.

Second, a footer with the date and time and page numbers is automatically generated and printed at the bottom of the page.

Third, the default is to print gridlines. If you do not want the grid lines to print, click **Grid Lines** from the View menu (so it is not checked) and the grid lines will not appear, either on screen or on the hard copy. If you want value labels printed, then select Value Labels from the View menu, and those labels will appear in cells, rather than the actual value you entered.

Finally, the default for the orientation is to print in Portrait mode, which means the short edge of the paper is horizontal. If you want to switch to Landscape mode, you have to select Properties from the Print dialog box and then click Landscape on the Paper tab.

Printing a Selection from an SPSS Data File

Printing a selection from a data file follows exactly the steps that we listed for printing a data file, except that in the Print dialog box, you click the Selection option.

1. Be sure that the data you want to print is selected. (Use the techniques we discussed in the last lesson.)

2. Click **File → Print** or click the Print icon on the toolbar. When you do this, you will see the Print dialog box.
3. Click **Selection** if it is not already selected.
4. Click **OK**, and whatever you selected will be printed.

> **TIP**
> You can select more than one item from the Viewer by holding down the Ctrl key as you select different output elements.

The Page Setup menu on the File menu (when the viewer is active) allows you a great deal of flexibility in printing out the contents of the Viewer window. You can set margins and orientation, and even click the options button and create a highly customized header or footer.

Printing from the Viewer Window

When you complete any type of analysis in SPSS or create a chart, the results appear in the Viewer window. To print the output from such a window, follow exactly the same procedures as you did for printing any data file.

Clicking **OK** in the Print dialog box will print all the contents of the Viewer. To print a selection from the Viewer, first make the selection by either clicking on the actual material in the right pane of the Navigator, or clicking on the name of the material you want to print in the left pane of the Viewer. Then select the Print option from the File menu and click **Selection** in the Print dialog box. If the Selection option is dimmed in the Print dialog box, it means that you didn't make a selection and therefore cannot print it.

We will discuss printing graphs in Unit 4.

Creating PDF Documents

A "PDF" is a document created in the Portable Document Format. The advantage of such a document is that it can be easily shared among users and can only be read using Adobe's Acrobat Reader (you can find more about this application and download it for free at http://www.adobe.com/products/acrobat/readermain.html).

To create output (and you can only use this feature for output and nothing in the Variable window), follow these steps:

1. Create the output you want to print.
2. Click **File → Print**.
3. In the Print dialog box, under the Name drop down-menu, click **SPSS PDF Converter**.
4. Provide a name for the file.
5. Click **OK**.

Some things to remember about PDFs and SPSS:

1. You can read a PDF file using the Adobe Acrobat Reader. Don't confuse this with Adobe Acrobat which is the full application which allows you to create and edit PDFs.
2. Increasingly, other applications will allow you to read PDFs as well. Check the application you are using.
3. Even though you cannot prepare a PDF of information in the Data View window, you can still get to the Print dialog box and select the PDF option from the drop-down menu. You'll get an error message when you click OK to try and actually create the PDF.

Exiting SPSS

How do you end your SPSS session and close up shop for the day? Easy.

1. Click **File → Exit**. SPSS will be sure that you get the chance to save any unsaved or edited windows and will then close. Remember that you cannot close the last window that is open without exiting SPSS. (Note: For the Mac, click **File** and then click **Quit**.)

LESSON

10 Exporting and Importing SPSS Data

After This Lesson, You Will Know

- How to export a data file to another application
- How to export a chart to another application
- How to import data from another application

Key Words

- Export
- Import

You already know how to open an SPSS data file by selecting Open from the File menu, locating the file, and clicking **OK**. But what if you created a file in another application, such as Excel or Lotus 1-2-3 or Word for Windows, and want to use that file as a data file in SPSS? Or what if you want to use an SPSS data file or a chart in a report you created with Word for Windows? Does that mean reentering all the data? Not at all. SPSS lets you export and import data, and in this lesson you will learn how to do both.

Getting Started Exporting and Importing Data

TIP

You export data from SPSS to another application and import data from another application to SPSS.

When you **export** an SPSS data file, you are sending it to another application. Just as one country exports products to other countries, so you are exporting data from SPSS to another application.

If you have trouble exporting a file created with another application to SPSS, try and convert the file to a text or ASCII file within the application where it was originally created. SPSS can more easily read ASCII files than any others.

When you **import** data from another application, you are going to use it with SPSS. It may be a file created in Word or Excel or dBase. Just as one country imports products from other countries, so you are importing data from another application to SPSS.

In general, the more commercially popular an application is, the more different types of data formats the program supports. For example, Word can translate more than 30 different data formats (some different verisons of the same aplication such as Excel 2.1 and Excel 97). SPSS was designed for a narrower audience than such a popular word processor, so fewer applications can directly read SPSS files. With the increasing popularity of SPSS and with the release of version 15, however, SPSS is more compatible with other applications than ever before.

Exporting Data

Here's just the situation where you would be interested in exporting data. You have been using SPSS to analyze data, and you want to use the results of the analysis in another application. To do so, you want to export data from SPSS. Let's go through the steps of both exporting data, exporting an analysis of that data, and exporting a chart.

Why might you want to export data? One good example is when you want to create an appendix of the raw data used in the analysis. You could just print the data file, but SPSS is limited in that you cannot use more than eight characters to name a variable and you want the data file to be very descriptive. So, you want longer variable names and you'll export the data to Microsoft Excel, which has much greater flexibility in formatting. And since you are writing your report in Microsoft Word for Windows, an Excel-generated appendix is completely compatible.

The first and most important thing to remember about exporting data to another application is that the SPSS data must be saved in a format that the other application can understand. Fortunately, SPSS can save data files under a wide variety of formats. Just by examining Table 5, you can get a general idea as to what applications data files can be exported to. The SPSS data file must be saved with the appropriate extension for the application you are exporting to.

Table 5
Application Extensions

Application	Extension
1-2-3 Release	.sk1 (v3), .wk1 (v2), .wks (v1)
dBase	.dbf
Excel	.xls
Fixed ASCII	.dat
SAS	.sd, .ssd
SPSS	.sav
SPSS/PC	.sys
SPSS Portable	.por
Stata	.dta
SYLK	.slk
Tab Delimited	.dat

Let's go through some examples, beginning with exporting the file we named Teacher Scale Results to Excel for Windows.

1. Be sure the data file you want to export is active.
2. Click **File ⟶ Save As**.
3. In the Save as type: area, select the format used by the application you are exporting to. When you make the selection, the file extension changes. In this example, it would be .xls (for Excel). Be sure to provide a name for the new file. In this case, it is "scores," and the file will be saved as scores.xls.
4. Click **OK**. SPSS provides a small report confirming that the save has occurred and a new file is created under a new name, such as Excel File.xls, which can now be read by Excel. When you do export a file, SPSS confirms the export by giving you a message in the Viewer window regarding the number of variables and number of cases exported. It's a nice confirmation.
5. Open Excel.
6. In Excel, click **File ⟶ Open**.
7. Open the new exported file. Excel displays the file as shown in Figure 47 (on page 50).

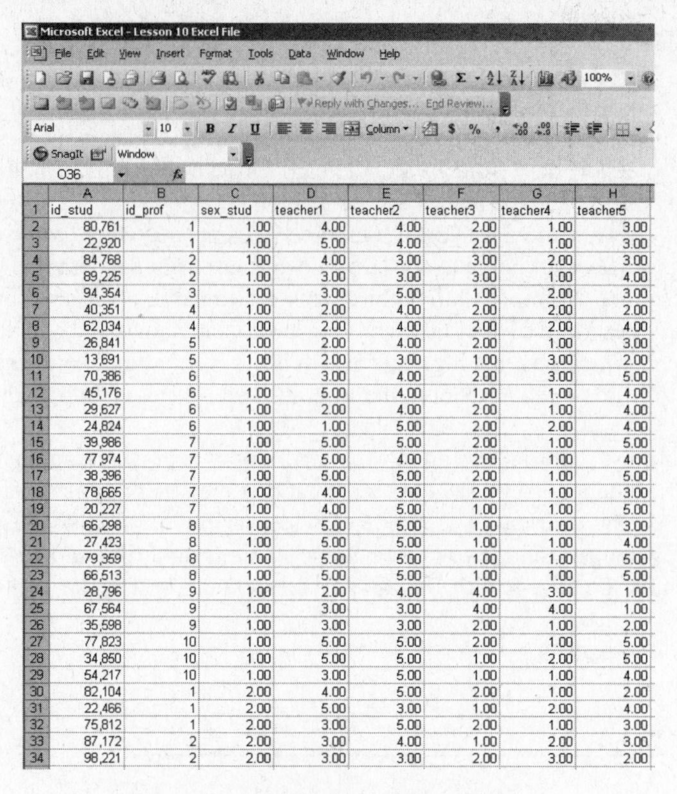

	A	B	C	D	E	F	G	H
1	id_stud	id_prof	sex_stud	teacher1	teacher2	teacher3	teacher4	teacher5
2	80,761	1	1.00	4.00	4.00	2.00	1.00	3.00
3	22,920	1	1.00	5.00	4.00	2.00	1.00	3.00
4	84,768	2	1.00	4.00	3.00	3.00	2.00	3.00
5	89,225	2	1.00	3.00	3.00	3.00	1.00	4.00
6	94,354	3	1.00	3.00	5.00	1.00	2.00	3.00
7	40,351	4	1.00	2.00	4.00	2.00	2.00	2.00
8	62,034	4	1.00	2.00	4.00	2.00	2.00	4.00
9	26,841	5	1.00	2.00	4.00	2.00	1.00	3.00
10	13,691	5	1.00	2.00	3.00	1.00	3.00	2.00
11	70,386	6	1.00	3.00	4.00	2.00	3.00	5.00
12	45,176	6	1.00	5.00	4.00	1.00	1.00	4.00
13	29,627	6	1.00	2.00	4.00	2.00	1.00	4.00
14	24,824	6	1.00	1.00	5.00	2.00	2.00	4.00
15	39,986	7	1.00	5.00	5.00	2.00	1.00	5.00
16	77,974	7	1.00	5.00	4.00	2.00	1.00	4.00
17	38,396	7	1.00	5.00	5.00	2.00	1.00	5.00
18	78,665	7	1.00	4.00	3.00	2.00	1.00	3.00
19	20,227	7	1.00	4.00	5.00	1.00	1.00	5.00
20	66,298	8	1.00	5.00	5.00	1.00	1.00	3.00
21	27,423	8	1.00	5.00	5.00	1.00	1.00	4.00
22	79,359	8	1.00	5.00	5.00	1.00	1.00	5.00
23	66,513	8	1.00	5.00	5.00	1.00	1.00	5.00
24	28,796	9	1.00	2.00	4.00	4.00	3.00	1.00
25	67,564	9	1.00	3.00	3.00	4.00	4.00	1.00
26	35,598	9	1.00	3.00	3.00	2.00	1.00	2.00
27	77,823	10	1.00	5.00	5.00	2.00	1.00	5.00
28	34,850	10	1.00	5.00	5.00	1.00	2.00	5.00
29	54,217	10	1.00	3.00	5.00	1.00	1.00	4.00
30	82,104	1	2.00	4.00	5.00	2.00	1.00	2.00
31	22,466	1	2.00	5.00	3.00	1.00	2.00	4.00
32	75,812	1	2.00	3.00	5.00	2.00	1.00	3.00
33	87,172	2	2.00	3.00	4.00	1.00	2.00	3.00
34	98,221	2	2.00	3.00	3.00	2.00	3.00	2.00

Figure 47. **An SPSS file exported to Excel.**

What if the application you want to export data into is not listed in the previous table? Almost every application can read Tab Delimited or Fixed ASCII data. For example, let's say you wanted to export an SPSS data file into Word for Windows (which saves files with a .doc extension). You could save it as a Tab Delimited file, and then have Word read it, as we did in Figure 48. Now, it

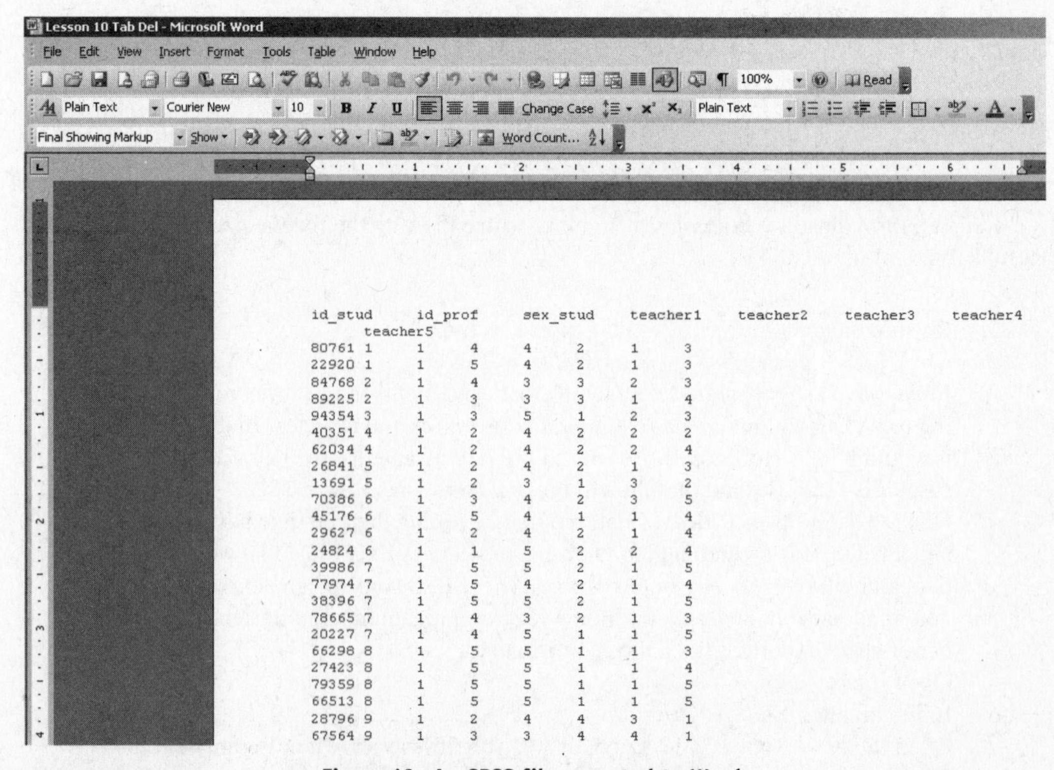

```
id_stud     id_prof     sex_stud     teacher1     teacher2     teacher3     teacher4
     teacher5
80761 1     1     4     4     2     1     3
22920 1     1     5     4     2     1     3
84768 2     1     4     3     3     2     3
89225 2     1     3     3     3     1     4
94354 3     1     3     5     1     2     3
40351 4     1     2     4     2     2     2
62034 4     1     2     4     2     2     4
26841 5     1     2     4     2     1     3
13691 5     1     2     3     1     3     2
70386 6     1     3     4     2     3     5
45176 6     1     5     4     1     1     4
29627 6     1     2     4     2     1     4
24824 6     1     1     5     2     2     4
39986 7     1     5     5     2     1     5
77974 7     1     5     4     2     1     4
38396 7     1     5     5     2     1     5
78665 7     1     4     3     2     1     3
20227 7     1     4     5     1     1     5
66298 8     1     5     5     1     1     3
27423 8     1     5     5     1     1     4
79359 8     1     5     5     1     1     5
66513 8     1     5     5     1     1     5
28796 9     1     2     4     4     3     1
67564 9     1     3     3     4     4     1
```

Figure 48. **An SPSS file exported to Word.**

"ain't purty," but the information is all there. With some changing of font size and spacing, it could appear quite nice and fit well into any document.

Exporting a Chart

SPSS has some pretty terrific charting features. You saw what an SPSS chart looks like in Lesson 4, and you'll learn more about them later in Part I of *Using SPSS*. It won't be uncommon for you to need to create a chart and export it into another application. Let's export a graph into a Word file.

> **TIP**
>
> Copy of Copy Objects? The difference here is that when you use the Copy command, what you paste into a document can be edited. The Copy Objects command copies the material as an object, and it cannot be edited in the new document.

1. Create the graph you want to export. Be sure it is active.
2. Right-click on the actual graph in the right pane of the Viewer and select the **Copy** command.
3. Open the application into which you want to export the chart. In this example, the application is Word.
4. Click **File ⟶ Paste,** or right-click, and then make the selection. The chart should appear in the new document.

Exporting the Results of SPSS Output

Here's the situation. You've just done a simple t test with SPSS, and you want to export the results of that analysis into another document. In effect, output is treated just like text. It can be cut or copied and pasted, and exporting output is identical to what you just did with a chart.

Why retype all that information when you can just paste it right in? Here's how.

1. Run the analysis you want.
2. Right-click on the output you want to export in the right pane of the Viewer and click the **Copy** command.
3. Open the document into which you want to paste the output.
4. Click **Edit ⟶ Paste**, or right-click, and then click **Paste** and the results are as you see in Figure 49.

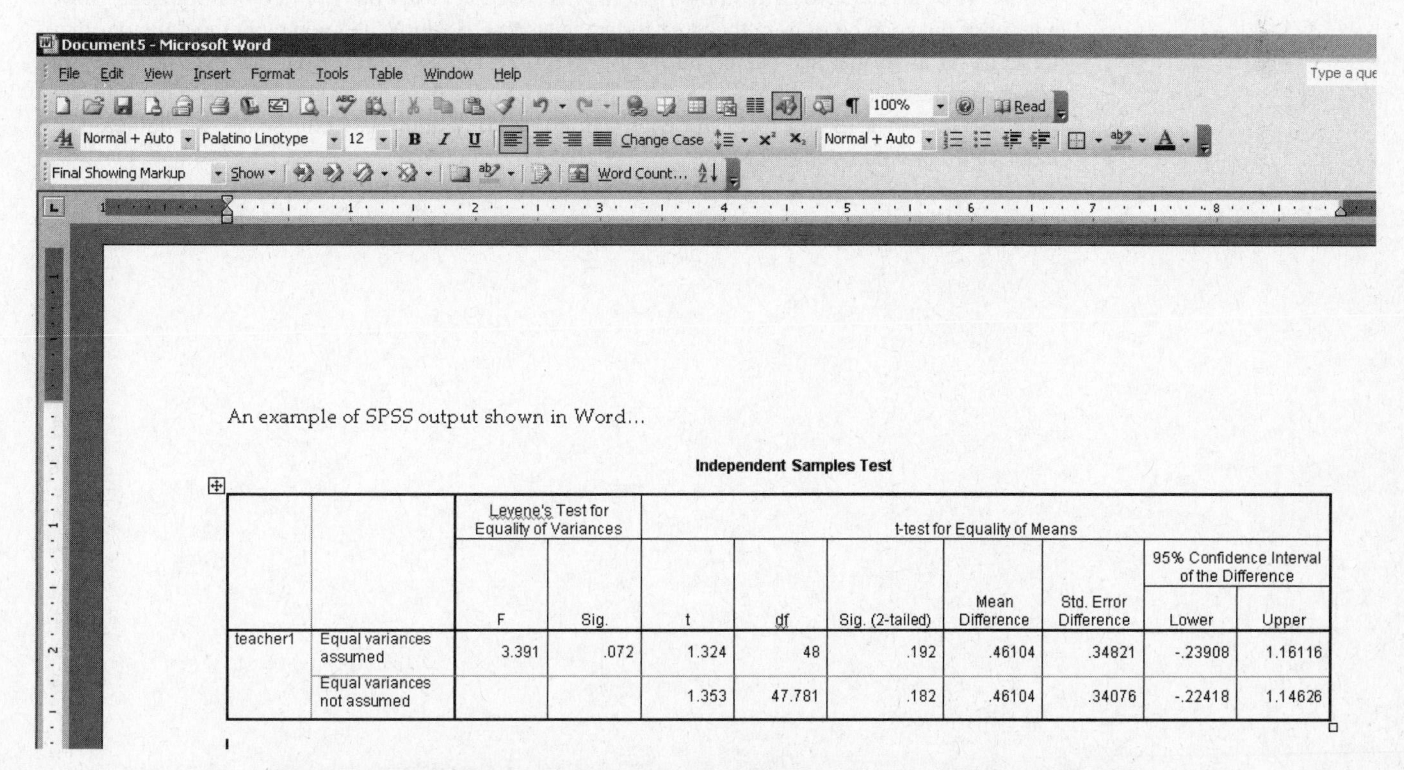

An example of SPSS output shown in Word...

Independent Samples Test

		Levene's Test for Equality of Variances		t-test for Equality of Means					95% Confidence Interval of the Difference	
		F	Sig.	t	df	Sig. (2-tailed)	Mean Difference	Std. Error Difference	Lower	Upper
teacher1	Equal variances assumed	3.391	.072	1.324	48	.192	.46104	.34821	-.23908	1.16116
	Equal variances not assumed			1.353	47.781	.182	.46104	.34076	-.22418	1.14626

Figure 49. **Using the Copy command to paste SPSS output into a different application.**

Once again, here's an important caution. If you copy results as an object (using the Copy Objects menu command), then you will not be able to edit the graph in the new application. An object is a bitmapped image, much like a picture.

However, if you simply **Copy** them (and not use the Copy Objects option), the data will be exported as text and you can edit it, but it will lose it's original format. The best advice? Make your SPSS results exactly as you would like them to appear (by double-clicking on and editing the table within SPSS) and then copy them as an object.

Importing Data

TIP

SPSS creates objects as output that cannot be directly edited when exported. They need to be edited within SPSS or exported with the use of the simple Copy command.

Here's the situation: You are a user of Excel or Lotus 1-2-3, and you've entered your data into one of these spreadsheets. Or perhaps you choose to enter your data into one of the many word processors available, such as Word for Windows or WordPerfect. Now you want to analyze the data with SPSS.

This might be the situation if you did not have access to SPSS when it came time to enter the original data or if you just feel more comfortable using some other application to enter data:

1. Click **File → Open**.
2. Locate the file you want to import, highlight it, select the type of file it is from the drop-down File Type menu, and then click **OK**. When you do this, you will see the data in the SPSS Data Editor window appearing exactly as if you had entered the data directly into SPSS.

Just as you can export SPSS data to other applications, so you can import from a variety of different applications. A word of caution, however. Even though SPSS can read certain file formats (such as .xls for Excel files), it may not mean that SPSS can read the latest version of Excel. Revisions of many products occur so often it is impossible for them all to stay synchronized as far as versions are concerned. So before you start entering loads of data to import, check to see if it's even possible.

The most important thing to remember about exporting and importing is finding the right format. You need to save a file in the right format so it can easily be exported or imported. Almost any set of data can be imported and exported as an ASCII file. When you want the data to remain intact as far as format and such, you'll have to look for a different format and experiment to see which format works best. At the very least, you should be able to simply select the data and import or export it to and from SPSS by using the Copy and Paste commands available on the main menu or when you right-click.

LESSON 11

Validating SPSS Data

After This Lesson, You Will Know

- What the validation process includes
- The differences between different types of validation rules
- How to load validation rules
- How to validate a data set

Key Word

- Single-variable rules

There are many important steps in the data collection and data analysis processes and one of the most important is to make sure that the data that are entered for analysis are entered correctly and the process completed accurately.

For small data sets consisting of less than 100 cases, this validation process can easily be done by visual examination—much like proofreading a manuscript. But, for data sets where there are thousands of cases, the validation of each case and confirmation that it is correctly entered, is a much more burdensome task. That's why the validation feature in version 15.x of SPSS is such a welcome addition.

The basic logic behind the validation process is to apply a set of rules to a data set. When those rules are violated, SPSS indicates such and you can then alter the data as necessary (or not). These rules are available both as a set of predesigned statements or can be designed by the user.

Although the process can be quite complex, we are going to illustrate the most basic procedures using the data set found in the SPSS file that comes with the original installation named *stroke_invalid.sav.* It is located in the Tutorial folder (under sample_files) wherever SPSS was originally installed (on your local computer or on the network you are connected to). We assume that this file is open and active.

TIP

During the validation process you can have SPSS use a predetermined set of validation rules or define your own.

Validating a Data Set

The Validation menu contains three options.

- **Load Predefined Rules** allows the user to use the set of rules that have already been defined by SPSS. These rules are always those that are first applied, but can be edited.

- **Define Rules** allows the user to define a set of rules for this particular data set.
- **Validate Data** initiates the validation process and results in output in an output window.

Loading the Predefined Rules

The first step in the validation process is to either use the predefined rules that SPSS has built into the system or to define your own rules. To validate a data set using predefined rules, follow this step.

1. Click **Data → Validation → Load Predefined Rules.** When you do this, you will see a brief message indicating that the SPSS validation rules will be added from an internal directory. This step is just the beginning of the process since after you load the rules. You then have to validate the data against those rules.

Validating the Data Set Using Predefined Rules

1. Click **Data → Validate Data** and you will see the Validate Data dialog box as shown in Figure 50. It is within this dialog box that you will select those variables (both those to be analyzed and those used to identify cases) for analysis. In this example, we are doing the most simple of validations and only seeing if patient ID is present for all cases entered.

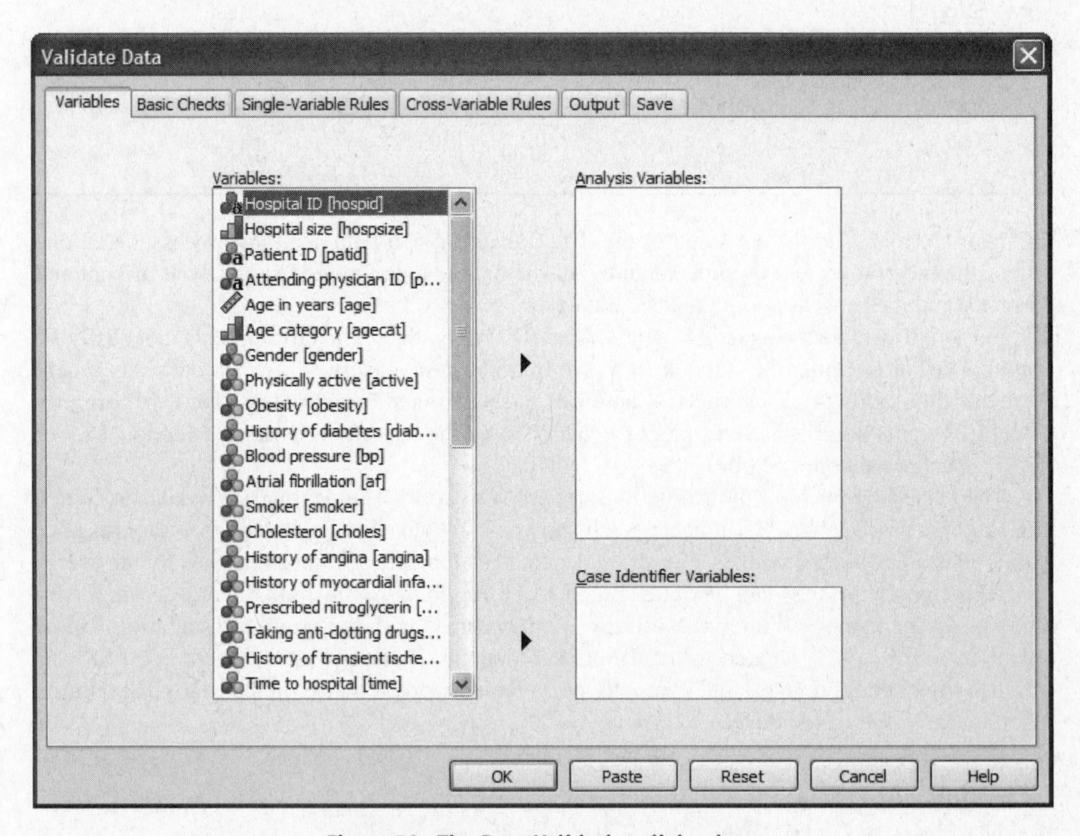

Figure 50. **The Data Validation dialog box.**

2. Click **Patient ID [patid]** and click ▲ to move it to the Case Identifier Variables: box.
3. Click **OK** and the results of the validation are shown in Figure 51. There are several things to more in this output. Some notes about the screen you see in Figure 51.

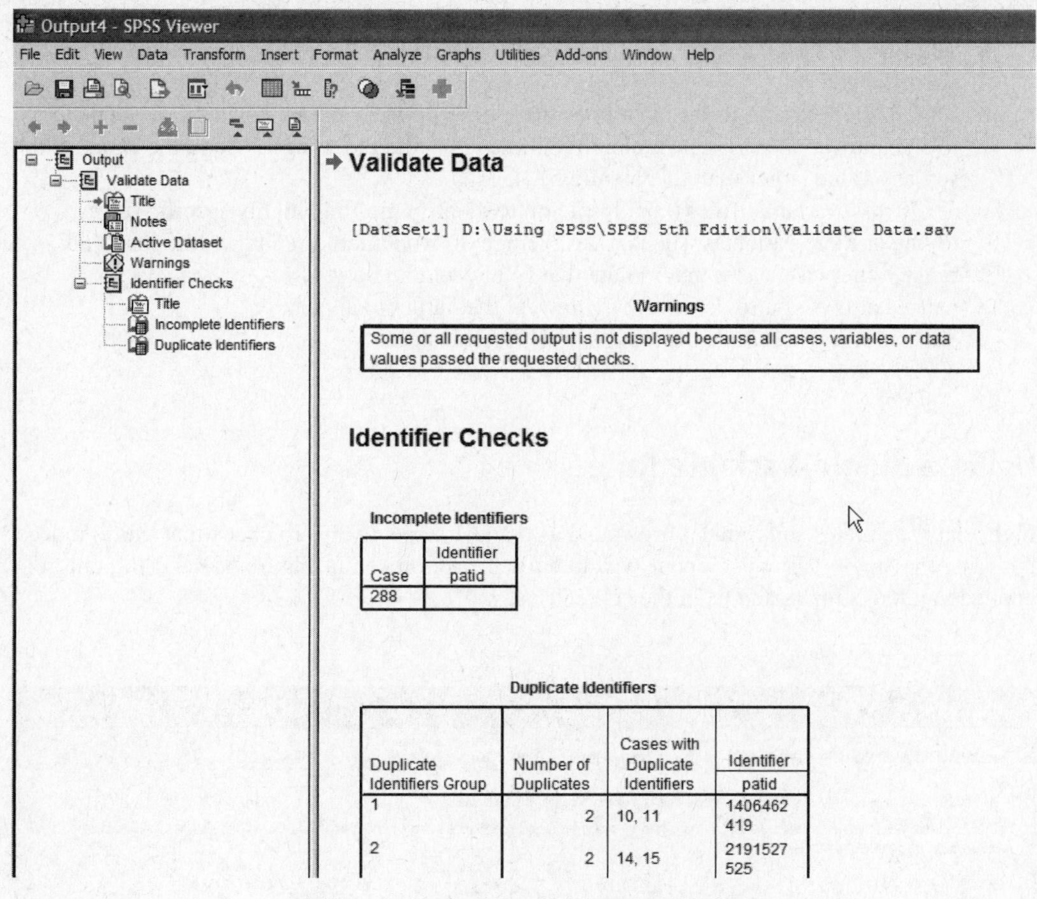

Figure 51. **The results of the data validation.**

- Like other SPSS results, the results of a data validation are placed in an Output window and can be worked with as we described in Lesson 18.
- Since almost all of the predefined rules were not violated, SPSS produces a message to the effect that "Some or all requested output is not displayed. . . ." indicating that many of the predefined checks were passed.
- The Identifier Checks predefined rule however did find some violations. For example, case number 288 does not contain a patient identifier (patid). It would be prudent to then go back and find case 288 and make the needed correction.
- Also, there are several Duplicate Identifiers, or patients who are identified by multiple numbers such as cases 10 and 11 and cases 21 and 22. In other words, these cases were entered more than once. In Figure 52, you can see how cases 21 and 22 contain similar data.

21	PBW	Small	7237535360	616528	53	45-54	Female	No	Yes
22	PBW	Small	7237535360	616528	53	45-54	Female	No	Yes

Figure 52. **Cases with identical data revealed through the data validation process.**

What you see in this output is a simple, but excellent, example of how SPSS can validate a data file given a fixed set of rules.

Other Validate Data Dialog Box Options

In the Validate Data dialog box you see in Figure 50, there are other tabs that are important to know about.

- The Variables tab lists the variables that can be included in the validation analysis.
- The Basic Checks tab allows you to define basic checks that if violated, cause cases to be flagged.
- The Single-Variable Rules tab allows the user to define a rule that is to be applied to a single variable. For example, if you knew that the oldest age is 86 in the data set, any value larger than that should be flagged.
- The Cross-Variable Rules tab allows you to identify multiple highly specific rules to one or more variables, such as the average of two variables being within a certain range, otherwise, cases that violate that rule would be flagged.
- Output allows you to define how you want the output to appear.
- Save allows you to save the results of the set of rules.

Using a Single-Variable Rule

In the data set, males and females are coded as 0 and 1 respectively. To check that this is indeed the case, the single-variable Gender was identified (as you see in Figure 53) and the one rule applied that looks for 1s and 0s in the cell entries.

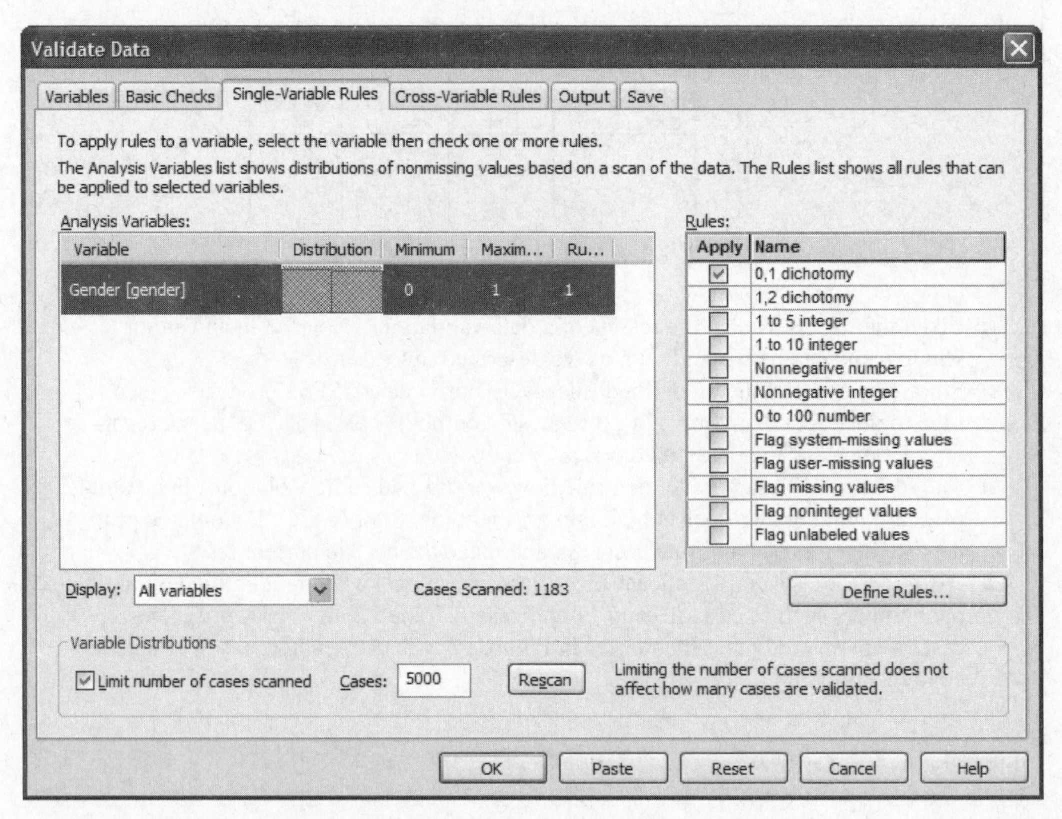

Figure 53. **Defining the rule for a single variable.**

When the rules are defined and OK clicked in the Validate Data dialog box, the results as shown in Figure 54 indicate that there are no cells that violate the rule.

Some observations about the Validate Data dialog box:

- The variable selected in the Variables tab of the Validate Data dialog box (Gender) is shown.
- The distribution for that variable is shown where both the 1s and 0s are represented by bars and appear to be almost equal in frequency.

➔ Validate Data

```
[DataSet1] D:\Using SPSS\SPSS 5th Edition\Validate Data.sav
```

Warnings

Some or all requested output is not displayed because all cases, variables, or data values passed the requested checks.

Figure 54. **The Single-Variable tab of the Validate Data dialog box and the validation of the coding used for gender.**

- The minimum and maximum values are shown.
- The number of rules that will be applied is shown as well. The number of rules that will be applied to the data set in the validation procedure is equal to the number of boxes that are checked in the Rules: text box. In this case only one (0, 1 dichotomy) is checked. If the 1, 2 dichotomy rule were checked, then all cells with a 0 (males) would be identified as not being valid.

Using the validate option takes some practice and a very clear understanding of the data set as to what may, or may not, be acceptable. As with any other SPSS feature, the more that this one is used, the more useful it becomes in that it can be fine-tuned to fit the user's exact needs.

PART I

UNIT 3 Working with Data

In the last two units, you learned the basics of SPSS, including how to start and exit the program and how to create a data file; you also learned to edit that data, print it, work with the import and export features of SPSS, and valid an SPSS data set. Our next step, and the focus of this unit, is to work with SPSS data so their final form best meets your particular data analysis needs. With that in mind, we will show you how to find data, recode it, sort a set of cases, and merge and split files.

LESSON

12

Finding Values, Variables, and Cases

After This Lesson, You Will Know

- How to find variables, cases, and values
- How to find text

Key Words

- Case
- Case sensitive
- Go To Case dialog box
- Search for Data dialog box
- Value
- Variable
- Variables dialog box

Just imagine you have a data file that contains 500 cases and 10 variables. You know that the entry for age (1 of the 10 variables) for case 147 was entered incorrectly. You can either scroll down to find that case or use one of the many find tools that SPSS provides. We'll go through those tools and how to use them in this lesson.

Finding Things

Before we begin showing you how to find things, let's pause for a moment to distinguish among variables, cases, and values. Even if we've covered some of this before, it's a good idea to see all of it in one place.

A **value** (also called a cell entry) is any entry in a cell, or the intersection of a row and a column. Values can be numerical, such as 3, 56.89, or $432.12; alphanumerical; or text, such as Sam, Sara, or Julie.

A **variable** is something that can take on more than one value. Age, weight, name, score on test1, and time measured in seconds are all variables. Variables are defined in SPSS by columns.

A **case** is a collection of values that belong to a unique unit in the data file, such as a person, a teacher, or a school. Cases are defined in SPSS by rows.

TIP

Any search operation requires you to enter the search information in a dialog box, which may block your view of the data editor. You'll have to move the dialog box to view the results of the search. Do this by dragging on the title bar.

With a small data file, you can just eyeball what is on the screen to find what you need. With a large data file, however, you could spend a good deal of time trying to find a case or variable, and even more time trying to find a particular value. If you have 10 variables and 500 cases and no missing data, you have 5,000 cell entries (all of which are values). Trying to find a particular one could take all day! Thus, the SPSS features that can find variables, cases, and values are most welcome.

Finding Variables

For the example in this lesson, we'll use the data file named Crab Scale Results available at http://www.prenhall.com/greensalkind. Open that file now. We'll find the variable named rank.

1. Make sure the data file is active.
2. Click **Utilities → Variables**. When you do this, you will see the **Variables dialog box**, as shown in Figure 55.

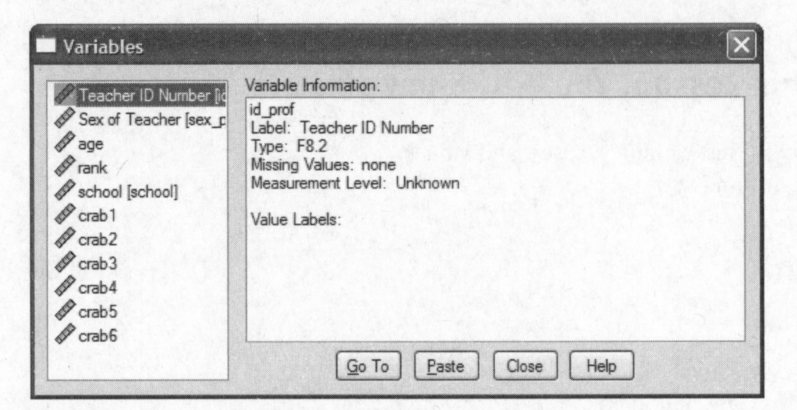

Figure 55. **The Variables dialog box.**

3. Click **Rank**. The Variables dialog box provides a good deal of information about the variable, including its name, label, type, missing values that are included, and value labels (e.g., Assistant Professor, etc.).
4. Click **Go To**, and SPSS will highlight the first cell in the column containing the variable in the Data View window.

You're now ready to perform any type of operation on that variable, such as working with labels, transformation, recoding, and more.

Finding Cases

Now that you know how to find any variable in a data file, it's time to see how you can find any case. In the Crab Scale Results example, let's assume that you need to locate case 8. You can do that by scrolling through the data file, which is simple enough. If the cases number in the hundreds, however, it's not so simple a task. To save you time and trouble, SPSS finds cases like this.

1. Click **Edit → Go To Case**. When you do this, you will see the **Go To Case dialog box**, as shown in Figure 56.

Figure 56. **The Go To Case dialog box.**

2. Type **8**.
3. Click **OK**. SPSS will highlight the row corresponding to the number you entered, for whatever column or variable the cursor is currently located in. For example, if we just found the variable named rank and then (as a separate operation) located case 8, the highlighted cell would be the rank for case 8, which is Assistant Professor.

Finding Values

Finding values is the most useful of all the search tools, but in one important way it also is the most limited. It's the most useful because any data file has more values than it does variables or cases. It's the most limited because it can only search for a value within one variable.

For example, let's say we want to find the case with an age value of 36.

1. Click the column labeled age. You may highlight any cell in the variable (or column) in which you want to search for the value, such as age.
2. Click **Edit → Find**. When you do this, you will see the Find Data in Variable rank dialog box, as you see in Figure 57.
3. Type **36**. You can search either forward or backward through the variable. If you highlighted the first cell in the column, you cannot search backwards since there's no place to go.
4. Click **Find Next** to search through the variable. SPSS will highlight the value when it is found. If SPSS cannot find the value, you will get a Not Found message.

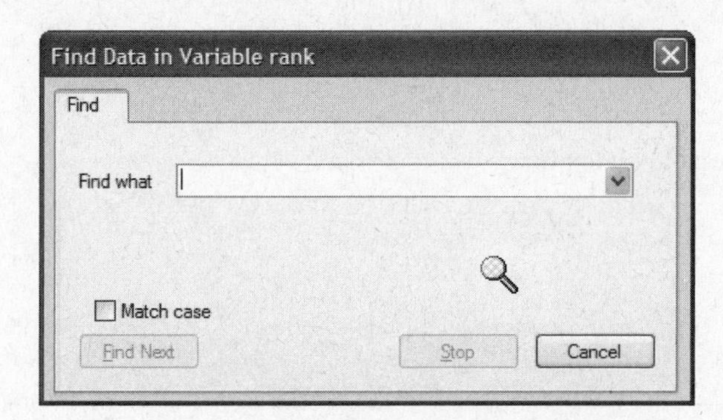

***Figure 57.* The Find Data in Variable dialog box.**

You can also use the Search for Data option to search for values entered as numbers but appearing as text (if they have been assigned labels). For example, if you wanted to find the first occurrence of Female in the variable named sex_prof, you would do as follows:

1. Highlight any cell in the variable (or column) named sex_prof.
2. Click **Edit → Find**.
3. Type **Female** in the **Find Data in Variable: sex_profearch** dialog box.
4. Click **Find Next**. SPSS will highlight the first occurrence of the value Female.

The value Female was not originally entered in the cell. Instead, the value 2 was entered, and through the definition of the variable, the label Female was assigned to that value. If you searched for the value 2 in the same column (sex_prof), SPSS would stop on all occurrences of what you now read as Female. That's because SPSS is looking for data, and not necessarily for numbers or text, and Female and 2 are both data. And, if you entered the value female, SPSS would find Female as well, since it looks for whatever it can find that corresponds to what you enter in the Find Data dialog box.

If you want the search to be **case sensitive** (where upper- and lowercase letters are distinguished from one another, such as Doctor and doctor), then be sure the Ignore Case of Text in Strings box is not checked.

LESSON

13 Recoding Data and Computing Values

After This Lesson, You Will Know

- How to recode data and create new variables
- How to compute new values
- How to recode into Same Variables dialog box
- How to recode into Same Variables: Old and New Values dialog box

Key Words

- Compute
- Compute Variable dialog box
- Formula
- Function
- Recode

There are often situations where you need to take existing data and convert them into a different variable, or to take existing variables and combine them to form an additional variable. For example, you may want to **compute** the mean or standard deviation for a set of variables and enter those values as a separate variable in the data file. Or you may want to **recode** data so that, for example, all values of 10 or greater equal 1, and all values less than 10 equal 0. Your needs for transforming or recoding data will depend on the nature of your data and the purpose of your analysis. In either case, SPSS provides some easy-to-use and powerful tools to accomplish both tasks, which we will show you in this lesson.

Recoding Data

You can recode data in one of two ways. First, you can recode a variable to create a new variable. Or you can recode a variable and modify the variable that has already been entered. In this lesson, you will use the data saved in the file named Crab Scale Results available at http://www.prenhall.com/greensalkind.

As you may remember, the Crab Scale includes the following six items:

1. I generally feel crabby if someone tries to help me.
2. I generally feel happy when I watch the news.

3. I generally feel crabby when I watch mothers and fathers talk baby talk to their babies.
4. I generally feel happy when I am able to make sarcastic comments.
5. I generally feel crabby when I am on a family vacation.
6. I generally feel happy when I am beating someone at a game.

Each item is ranked on the following 5-point scale.

1. Totally agree
2. Agree
3. In a quandary
4. Disagree
5. Vicious lies

Since some of the crabbiness items (2, 4, and 6) are reversed, these items need to be recoded. For example, if someone totally agrees (a value of 1) with item 2, it means they are happy when they watch the news. Since we are measuring crabbiness and since the lower the score the more crabby someone is, the accurate scoring of this item should be reversed, as shown in Table 6.

Table 6
Reversed Crab Scale Scores

Original response	Recoded response
1	5
2	4
3	3
4	2
5	1

Here's how to do just that:

1. Be sure the file named Crab Scale Results is active.
2. Click **Transform → Recode into Same Variables**, since we want the transformed variable to replace the values in the current variable that is being transformed. You should see the Recode into Same Variables dialog box, as shown in Figure 58.

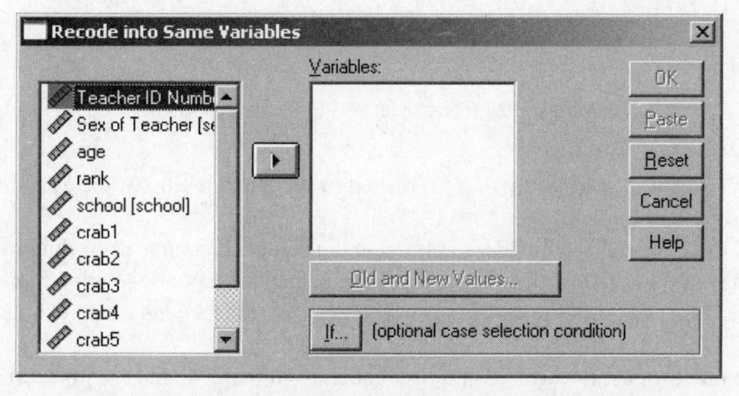

***Figure 58.* The Recode into Same Variables dialog box.**

3. Double-click **crab2** to move the variable into the Variables box (which then becomes the Numeric Variables dialog box).
4. Click the **Old and New Values . . .** button, and you will see the Recode into Same Variables: Old and New Values dialog box, as shown in Figure 59.

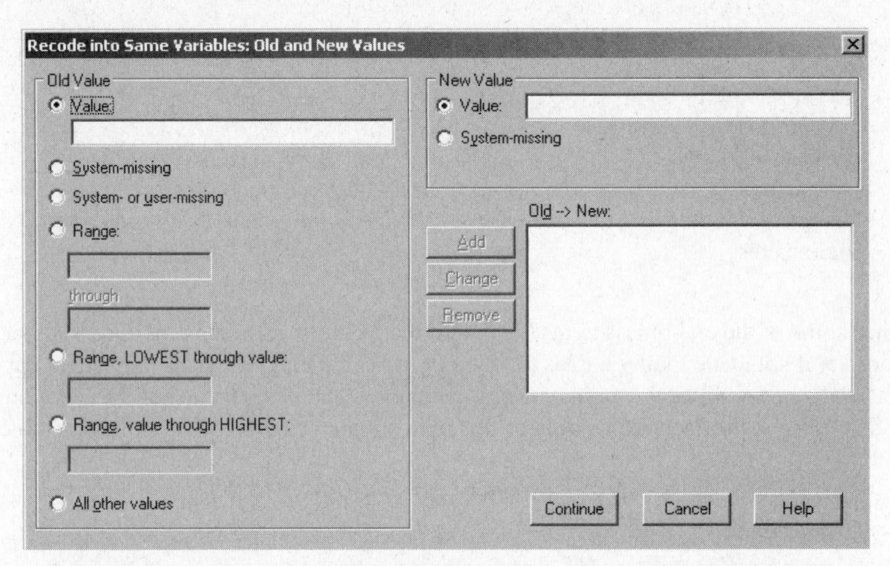

Figure 59. **The Recode into Same Variables: Old and New Values dialog box.**

5. Type **5** in the Value area under Old Value.
6. Type **1** in the Value area under New Value.
7. Click **Add**. The variable is added to the **Old ⟶ New box**, as you see in Figure 60. When SPSS encounters a 5 for variable 2, it will recode it as a 1.

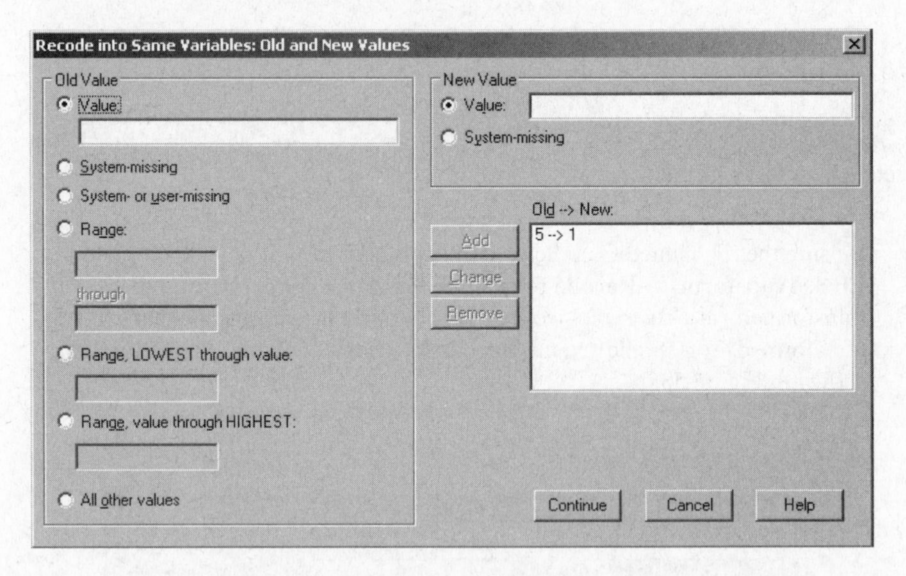

Figure 60. **Transforming Old to New values.**

8. Click **Continue**, and you will be returned to the Recode into Same Variables dialog box.
9. Click **OK**, and the actual values entered in the cells will change according to the way in which they were recoded. For example, if someone responded with a 5 to item 2 on the Crab Scale (as was the case for Professor 3), the recoded value would be 1.

To recode more than one variable at a time (such as items 2, 4, and 6), just add all of them to the Numeric Variables box in the Recode into Same Variables dialog box.

Computing Values

> **TIP**
> A formula is a predesigned function and SPSS comes with 19 different categories, many of which can do exactly what you want.

Computing a value means taking existing data and using them to create a new variable in the form of a numeric expression.

To show you how to do that, we will compute a total score for any one individual on the crabbiness scale based on a custom **formula**. To do this, we will simply add together the scores for variables crab1 through crab5. Keep in mind that SPSS comes with 70 predesigned **functions** (a predesigned formula) that will automatically perform a specific computation.

For example, we could tell SPSS to compute a new variable that is the addition of items crab1, crab2, crab3, crab4, and crab5. Or we could use the SUM function that SPSS provides to accomplish the same end. We'll show you how to do both.

Creating a Formula

Let's create a formula that will add together the values of all five crabbiness items.

1. Click **Transform → Compute Variable.** When you do this, you will see the **Compute Variable dialog box**, as shown in Figure 61.

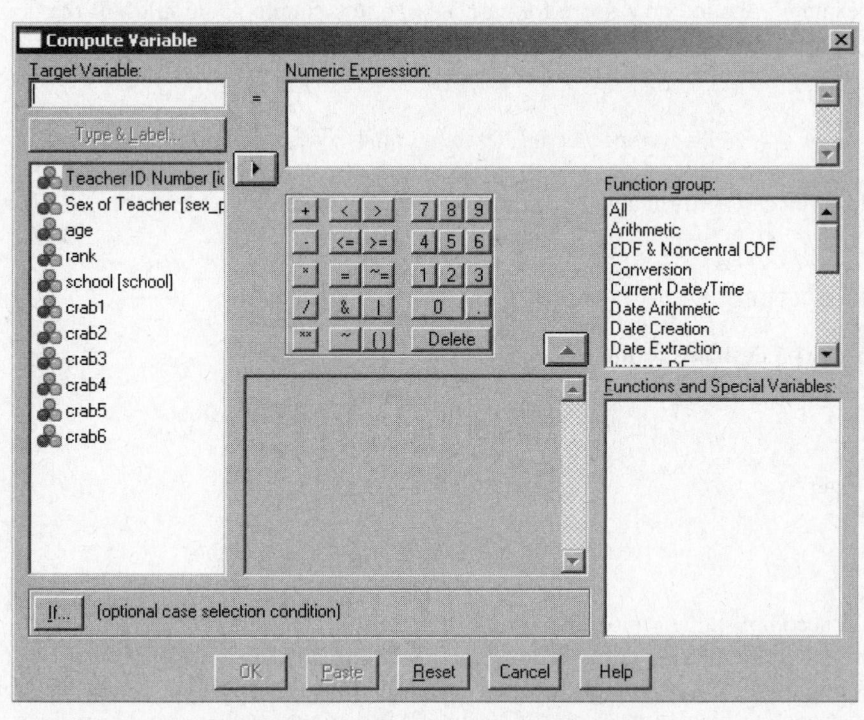

Figure 61. **The Compute Variable dialog box.**

2. Click the **Target Variable text box** and enter the name of the new variable to be created. We named it tot_crab, which is the sum of items 1 through 6 on the Crab Scale.
3. Click the **Numeric Expression text box**.
4. Click **crab1**, the first variable you want to include in the formula.
5. Click the ▶ to add it to the Numeric Expression text box.
6. Type +.
7. Continue adding variables and using the + key to add them together. The completed formula looks like this:

   crab1+crab2+crab3+crab4+crab5+crab6

8. Click **OK**. As you can see in Figure 62 (on page 66), the variable named tot_crab was added to the existing data file.

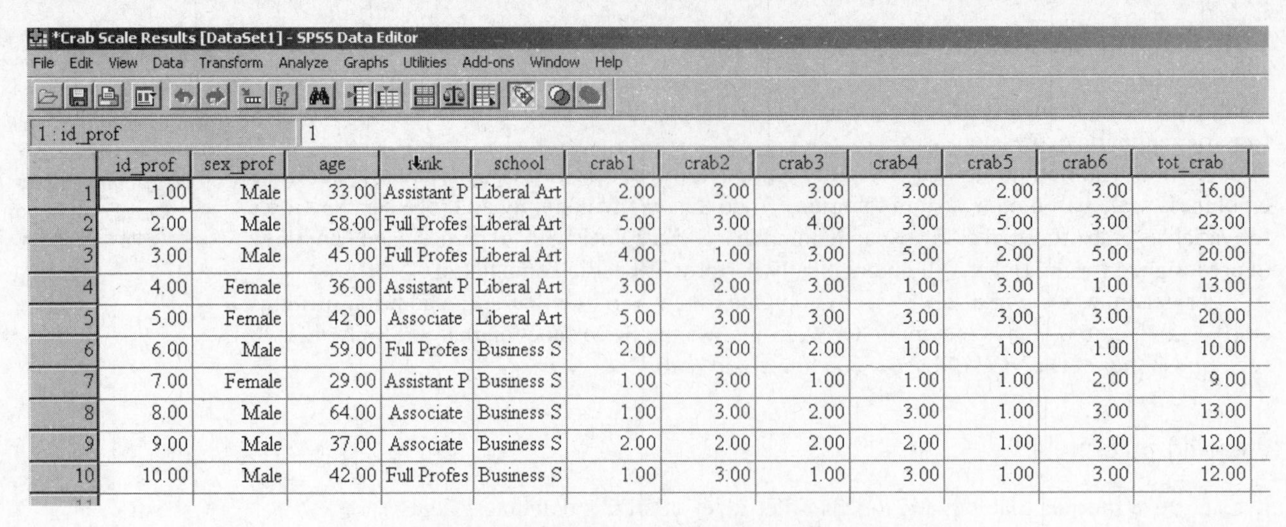

	id_prof	sex_prof	age	rank	school	crab1	crab2	crab3	crab4	crab5	crab6	tot_crab
1	1.00	Male	33.00	Assistant P	Liberal Art	2.00	3.00	3.00	3.00	2.00	3.00	16.00
2	2.00	Male	58.00	Full Profes	Liberal Art	5.00	3.00	4.00	3.00	5.00	3.00	23.00
3	3.00	Male	45.00	Full Profes	Liberal Art	4.00	1.00	3.00	5.00	2.00	5.00	20.00
4	4.00	Female	36.00	Assistant P	Liberal Art	3.00	2.00	3.00	1.00	3.00	1.00	13.00
5	5.00	Female	42.00	Associate	Liberal Art	5.00	3.00	3.00	3.00	3.00	3.00	20.00
6	6.00	Male	59.00	Full Profes	Business S	2.00	3.00	2.00	1.00	1.00	1.00	10.00
7	7.00	Female	29.00	Assistant P	Business S	1.00	3.00	1.00	1.00	1.00	2.00	9.00
8	8.00	Male	64.00	Associate	Business S	1.00	3.00	2.00	3.00	1.00	3.00	13.00
9	9.00	Male	37.00	Associate	Business S	2.00	2.00	2.00	2.00	1.00	3.00	12.00
10	10.00	Male	42.00	Full Profes	Business S	1.00	3.00	1.00	3.00	1.00	3.00	12.00

***Figure 62.* Adding a new variable to a data set.**

For example, the tot_crab score for case 1 is 16, the simple addition of all the crab items (appropriately recoded, we might add). If you wanted the average, you could just edit the formula in the Numeric Expression text box to read

(crab1+crab2+crab3+crab4+crab5+crab6)/6

In fact, you can perform any type of mathematical computation you see fit.

Using a Function

Functions are predesigned formulas. So instead of

crab1+crab2+crab3+crab4+crab5+crab6

you can enter

SUM(crab1,crab2,crab3,crab4,crab5,crab6)

which will accomplish the same thing.

There are more than 70 different functions including arithmetic functions, statistical functions, distribution functions, logical functions, date and time functions, cross-case functions, and string functions. It's beyond the scope of *Using SPSS* to go into each one and provide examples, but regardless of how complex or simple functions are, they are all used the same way.

We'll show you how to use one of the most simple and common functions, SUM, which adds a group of values together.

1. Click **Transform → Compute**. If the dialog box is not clear, click **Reset**.
2. Type **tot_crab** in the Target Variable text box.
3. In the list of functions, scroll down (if necessary) and find the function you want to use.
4. Double-click on that function to place it in the Numerical Expression box.
5. Now add variables by double-clicking them or by typing their names. An alternative to typing in each variable in the list is to use the "to" operator. Instead of crab1, crab2, . . . through crab6, you could type crab1 to crab6 so you have

SUM(crab1 to crab6)

as the operators in the Numeric Expression box. The finished dialog box appears in Figure 63 .

6. Click **OK** and the new variable is created.

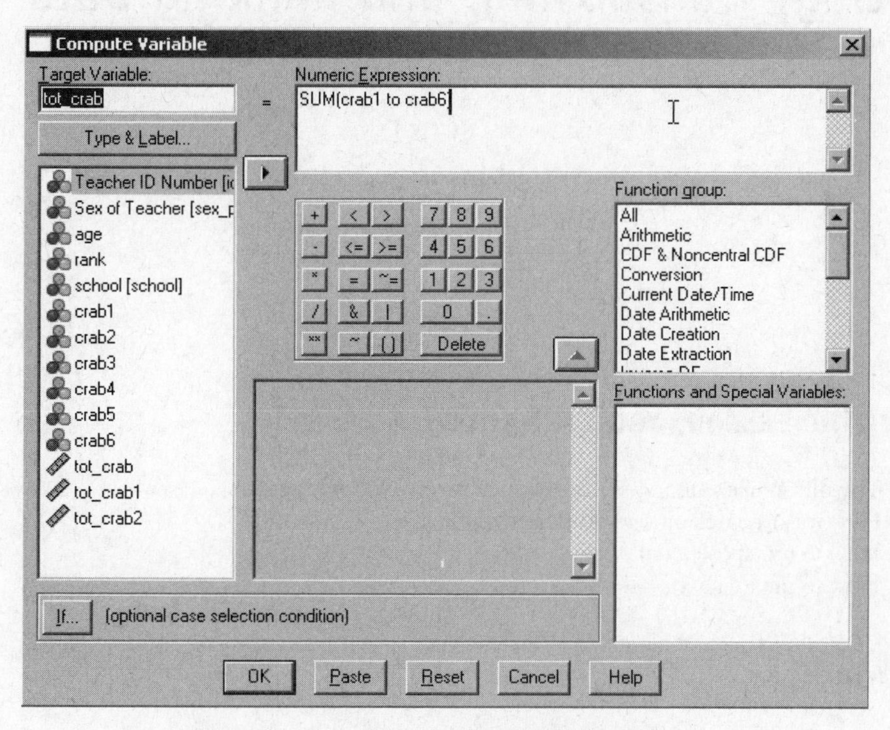

Figure 63. **Entering and using an SPSS function.**

LESSON

14 Sorting, Transposing, and Ranking Data

After This Lesson, You Will Know

- The differences among sorting, transposing, and ranking data
- How to sort cases on one or more variables
- How to transpose a data file
- How to rank data

Key Words

- Data
- Rank
- Sort Cases dialog box
- Transpose dialog box

As you have seen in the last few lessons, working with **data** includes importing, transforming, and recoding data. Working with data also involves sorting, transposing, and ranking data, techniques every SPSS user needs to know to create a data file that exactly fits data analysis needs. We'll look at each of these techniques in this lesson.

Sorting Data

The first of the three data tools is sorting. Sorting data involves reordering of data given the value of one or more variables. Sorting is an invaluable tool when it comes to organizing information.

For example, if you wanted the data for the Crab Scale Results organized by the variable named **rank**, you would sort on that variable. If you wanted to sort on more than one variable, you can do that easily as well, such as a sex within rank sort.

Sorting Data on One Variable

Sorting data is as simple as identifying the variable on which you want to sort and directing SPSS to sort.

1. If it is not already opened, open the file named Crab Scale Results available at http://www.prenhall.com/greensalkind.
2. Click **Sort Cases**. When you do this, you will see the **Sort Cases dialog box**, shown in Figure 64.

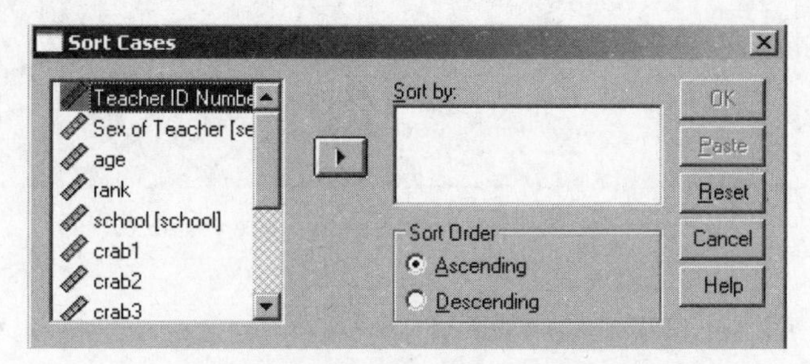

Figure 64. **The Sort Cases dialog box.**

3. Double-click **rank** to move rank from the variable list to the Sort by: text box. The Ascending that appears next to rank in the sort box means that the variable will be sorted in ascending order and ascending is the default condition. If you want, select descending order by clicking the **Descending** button.
4. Click **OK**.

In Figure 65, you can see the data sorted by rank. After the sort, assistant, associate, and full professors are grouped together. As you can see, when alphanumeric information is sorted in ascending order, the sort is from A to Z.

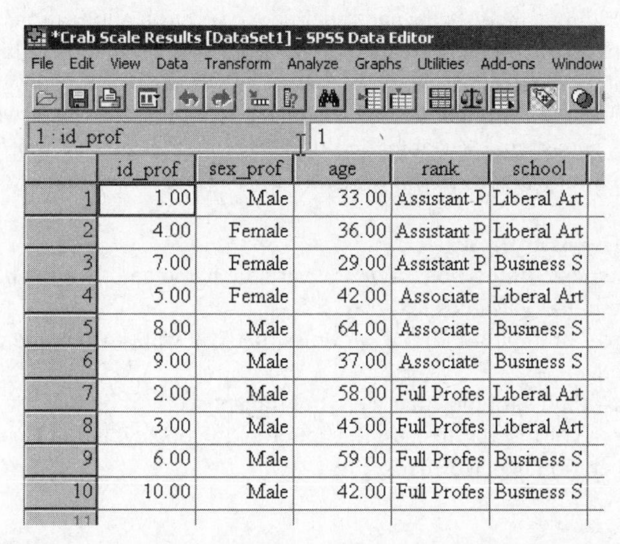

	id_prof	sex_prof	age	rank	school
1	1.00	Male	33.00	Assistant P	Liberal Art
2	4.00	Female	36.00	Assistant P	Liberal Art
3	7.00	Female	29.00	Assistant P	Business S
4	5.00	Female	42.00	Associate	Liberal Art
5	8.00	Male	64.00	Associate	Business S
6	9.00	Male	37.00	Associate	Business S
7	2.00	Male	58.00	Full Profes	Liberal Art
8	3.00	Male	45.00	Full Profes	Liberal Art
9	6.00	Male	59.00	Full Profes	Business S
10	10.00	Male	42.00	Full Profes	Business S

Figure 65. **A completed sort, in ascending order, on the variable rank.**

Sorting Data on More Than One Variable

Sorting data on more than one variable follows exactly the same procedure as the one just outlined, except that you need to select more than one variable to sort on. SPSS will sort in the order they appear in the Sort by list in the Sort Cases dialog box. For example, if you want to sort on gender within rank, then rank should be the first selection in the Sort Cases dialog box and gender (sex_prof) would be the second, as you see in Figure 66 (on page 70).

1. Click **Data → Sort Cases**.
2. Click **rank** to move rank from the variable list to the Sort by: text box.
3. Select whether you want to sort in ascending or descending order.

TIP

When you sort, the results are sorted according to the values in the cells (such as 1 for male and 2 for female) and not the labels. If it were sorted according to the labels, then the sort would be with females first since ascending order for text is A through Z.

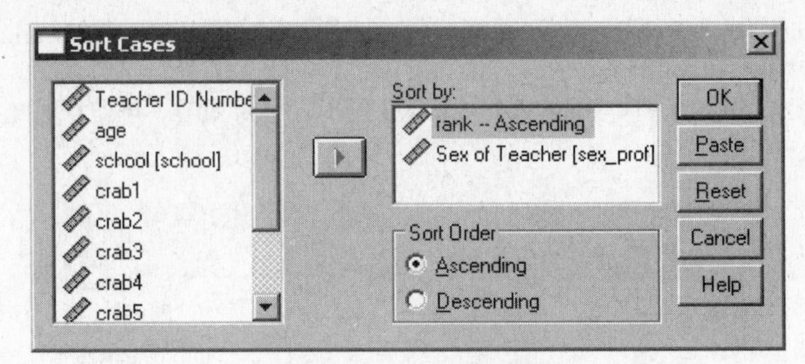

Figure 66. **Sorting on more than one variable.**

4. Double-click **Sex of Teacher** to move sex_prof from the variable list to the Sort by: text box.
5. Select whether you want to sort in ascending or descending order.
6. Click **OK**.

The cases have been sorted on the variable named rank, such that assistant, associate, and full professors are grouped together, and within rank they are sorted on sex_prof.

Transposing Cases and Variables

As you already know, an SPSS data file is constructed with cases represented by rows and variables represented by columns. This is the way that most data are entered in any type of data collection form, including spreadsheets and other data collection applications.

There may be occasions, however, where you want cases listed as columns and variables listed as rows. In other words, you want variables and cases transposed. This is most often the case when you are importing data (see Lesson 10) from a spreadsheet where the data were not entered as cases (rows) and variables (columns). Now that you've imported the data, you want to switch them to fit the conventional SPSS format. Here's how to do it:

TIP

In some dialog boxes, you can double-click to move a variable from the variable list. When you are transposing you cannot do this. Instead, you have to click on the variable name and then move it by clicking ▶.

1. Click **Data → Transpose**.
2. In the **Transpose dialog box**, click the variable name and then the button to insert the variable in the Variables text box.
3. Repeat steps 2 and 3 until all the variables that you want to become cases have been entered.
4. Click **OK**. All the variables that were not transposed will be lost, so if you transpose just some of the variables in the data file, the ones you don't will not appear after the operation is performed.

Assigning Ranks to Data

The final data manipulation technique we will deal with is ranking data. Let's say that we used the Compute feature (on the Transform menu) to create a sum of all Crab scores as we did earlier.

Now let's say that we are not interested in the absolute value of the Crab score, but rather in the relative rank of professors according to their Crab score. Here's how to do it:

1. If you have not done so already, create a new variable (by using the Compute command on the Data menu) named tot_crab, a total of all six Crab item scores.
2. Click **Transform → Rank Cases**. You'll see the Rank Cases dialog box, as shown in Figure 67.

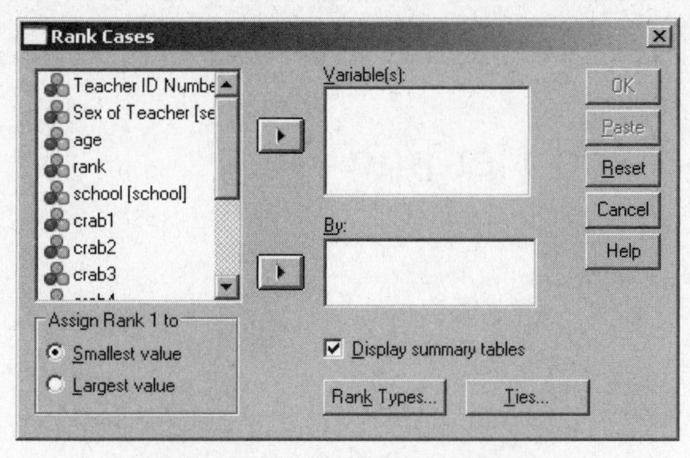

Figure 67. **The Rank Cases dialog box.**

3. Click on **tot_crab** and click on the ▶ button to move the variable into the Variable(s): text box.
4. Click on the **Largest value** button in the Assign Rank 1 to area.
5. Click **OK**, and a new variable, named Rtot_cra, will be created, as you see in Figure 68, reflecting the ranking of all cases (from 1 through 10) on the variable tot_crab. Cases 4 and 8 have the same tot_crab score (13), and, therefore, the same rank (5.5). Cases 9 and 10 also have the same (tied) rank.

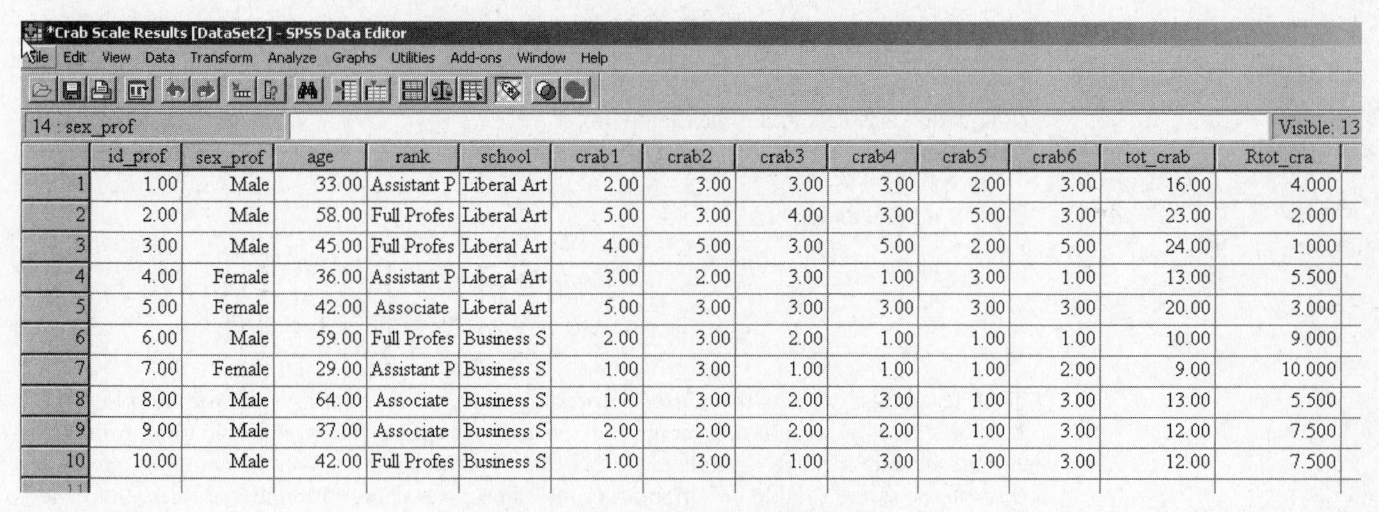

	id_prof	sex_prof	age	rank	school	crab1	crab2	crab3	crab4	crab5	crab6	tot_crab	Rtot_cra
1	1.00	Male	33.00	Assistant P	Liberal Art	2.00	3.00	3.00	3.00	2.00	3.00	16.00	4.000
2	2.00	Male	58.00	Full Profes	Liberal Art	5.00	3.00	4.00	3.00	5.00	3.00	23.00	2.000
3	3.00	Male	45.00	Full Profes	Liberal Art	4.00	5.00	3.00	5.00	2.00	5.00	24.00	1.000
4	4.00	Female	36.00	Assistant P	Liberal Art	3.00	2.00	3.00	1.00	3.00	1.00	13.00	5.500
5	5.00	Female	42.00	Associate	Liberal Art	5.00	3.00	3.00	3.00	3.00	3.00	20.00	3.000
6	6.00	Male	59.00	Full Profes	Business S	2.00	3.00	2.00	1.00	1.00	1.00	10.00	9.000
7	7.00	Female	29.00	Assistant P	Business S	1.00	3.00	1.00	1.00	1.00	2.00	9.00	10.000
8	8.00	Male	64.00	Associate	Business S	1.00	3.00	2.00	3.00	1.00	3.00	13.00	5.500
9	9.00	Male	37.00	Associate	Business S	2.00	2.00	2.00	2.00	1.00	3.00	12.00	7.500
10	10.00	Male	42.00	Full Profes	Business S	1.00	3.00	1.00	3.00	1.00	3.00	12.00	7.500

Figure 68. **The completed ranking of data.**

If you select more than one variable on which cases will be ranked, then a separate and new variable will be created for each one of those rankings, and the relative rankings will be listed as well.

LESSON

15 Splitting and Merging Files

After This Lesson, You Will Know

- How to split a file
- How to merge more than one file and more than one case

Key Words

- Add Cases From . . . dialog box
- Add Cases: Read File dialog box
- Add Variables From . . . dialog box
- Add Variables: Read File dialog box
- Merge
- Split
- Split File dialog box

The last skills you need to master before you move on to creating and working with SPSS charts are how to split and merge files. When you **split** a file, you are dividing it into two new and separate files for separate analysis. When you **merge** files, you are combining two files keyed on the same variable or variables, or the same cases using different variables. In this lesson, we'll look at how to do each of these.

Splitting Files

Throughout this first part of *Using SPSS*, you have used the Crab Scale Results and Teacher Scale Results to perform simple and complex procedures. We'll use the file named Teacher Scale Results to demonstrate how to split a file.

In general, you split a file when you want to create two separate files, both having at least one variable in common. Creating two separate files might be a choice if you want to produce individual listings of data or prepare data in a way that a particular section, split or organized by using a particular variable, can be cut or copied.

For example, in this lesson, we will split the Teacher Scale Results file using the variable named sex_stud into two separate files. One file will be named Sex Males, and the other will be Sex Females. Then we can do analysis of variance or regression on the separate files.

TIP

You can tell that a file has been split when the words Split File On are showing in the right-hand side of the Status Bar.

Splitting a file is not a particularly difficult process. In fact, it's only a matter of selecting the variable on which you want to split.

For example, for the Teacher Scale Results we'll split on sex_stud by following these steps:

1. Open the file named Teacher Scale Results located at http://www.prenhall.com/greensalkind.
2. Click **Data → Split File**. When you do this, you will see the **Split File dialog box**, as shown in Figure 69.

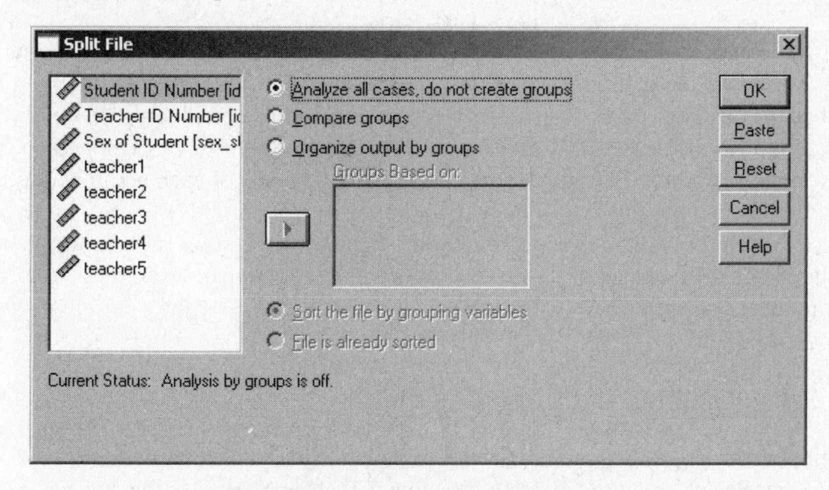

***Figure 69.* The Split File dialog box.**

3. Click **Organize Output By Groups** since you are presumably splitting the files to create data for two separate analyses.
4. Double-click **Sex_of Student.**
5. Click **OK**. The file will then be organized such that all male students are grouped together as are all female students.

TIP

Just as you can sort on more than one variable, you can split on more than one as well. Just enter them into the Split File dialog box in the order you want the entire file split.

Once the split is completed, SPSS will not create two physically separate files. Rather, for the rest of the SPSS session, SPSS will perform every procedure as if the file was physically split into two separate files. A simple descriptive analysis results in descriptive statistics for both males and females.

For example, if you calculate simple descriptive statistics on the variable named teacher1, then you will get separate output for both males and females without having to identify an independent variable named sex_stud. The results of such an analysis are shown in the Viewer in Figure 70. Notice that the analysis was automatically done for both males and females.

Sex of Student = Male

Descriptive Statistics[a]

	N	Minimum	Maximum	Mean	Std. Deviation
teacher1	28	1.00	5.00	3.6429	1.31133
Valid N (listwise)	28				

a. Sex of Student = Male

Sex of Student = Female

Descriptive Statistics[a]

	N	Minimum	Maximum	Mean	Std. Deviation
teacher1	22	1.00	5.00	3.1818	1.09702
Valid N (listwise)	22				

a. Sex of Student = Female

***Figure 70.* Separate analysis on a file split by one variable.**

Any analysis is performed on each split of the file. For example, if you split on sex_stud (2 levels) and id_prof (10 levels), you would have 20 different descriptive analyses.

Merging Files

Merging files is just as useful as splitting files. Files can be merged, or combined, in two different ways.

First, you can combine two files that contain the same variable or variables but different cases. This would be the setting if more than one researcher were collecting data on the same variables but for different cases (or subjects), and you need to combine the two data files. For example, the Teacher Scale Results survey could have been administered to more than one group of students, and we eventually might want to combine the results of each administration.

Second, you can combine two files that contain different variables but the same cases. For example, you might be having one person collect data with a certain sample and then another person collecting a different set of data on the same sample. Then you want these two combined. Let's look at these possibilities.

Merging Same Variables and Different Cases

In this example, we are going to combine, merge, or add (your choice of words) a new set of five cases of Teacher Scale data to the existing file named Teacher Scale Results. You can see the new data in Figure 71. These data are saved as Teacher Scale Results—Add Data on your data disk. Be sure that the file named Teacher Scale Results is still open, or if not, open it now.

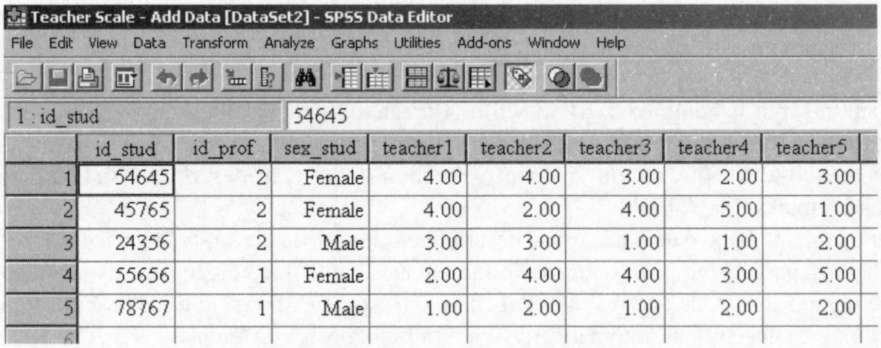

Teacher Scale - Add Data [DataSet2] - SPSS Data Editor

File Edit View Data Transform Analyze Graphs Utilities Add-ons Window Help

1 : id_stud 54645

	id_stud	id_prof	sex_stud	teacher1	teacher2	teacher3	teacher4	teacher5
1	54645	2	Female	4.00	4.00	3.00	2.00	3.00
2	45765	2	Female	4.00	2.00	4.00	5.00	1.00
3	24356	2	Male	3.00	3.00	1.00	1.00	2.00
4	55656	1	Female	2.00	4.00	4.00	3.00	5.00
5	78767	1	Male	1.00	2.00	1.00	2.00	2.00
6								

***Figure 71.* New data to be merged with other data.**

In this example, one researcher collected data on 50 students, and another researcher collected data on five students. The Teacher Scale Results should be open.

1. Click **Data → Merge Files**, then click **Add Cases**. When you do this, you will see the **Add Cases to Teacher Scale Results** (note that SPSS will add the name of the open file to which you are adding to the dialog box title), as shown in Figure 72 (at the top of the page 76). There are two options.

   If there is another file open to be added to an existing file, you can specify that in the dialog box by double-clicking on that filename.

   If another file is not open, you can browse the file you want to add, which is the case in this example.
2. Click the **Browse** button in the Add Cases to Teacher Scale Results dialog box and find the file names Teacher Scale—Add Data.

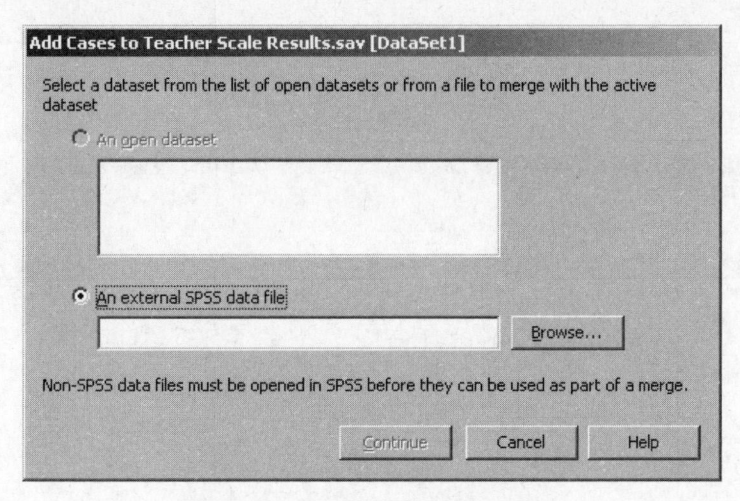

Figure 72. **The Add Cases: Read File.**

3. Click **Continue**. You will see the **Add Cases from Teacher Scale** dialog box as shown in Figure 73.

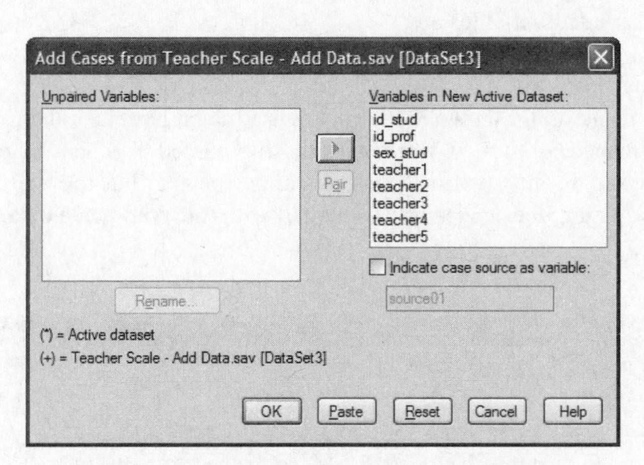

Figure 73. **The Add Cases from . . . dialog box.**

4. Click **OK**, and the file named Teacher Scale Results—Additional Data is added to the file named Teacher Scale Results, and the total file now has 55 cases (50 cases from Teacher Scale Results and 5 from Teacher Scale Results—Additional Data).

Merging Different Variables and Same Cases

Now we have the example of two files that contain a different set of variables for the same cases. For example, we want to combine the Teacher Scale Results—Additional Data and Crab Scale Results files where the common variable is id_prof.

What SPSS will do in this case is combine cases in the Teacher Scale Results file with cases in the Crab Scale Results file, using the variable named id_prof as a common link.

1. Be sure that the original file named Teacher Scale Results is open.
2. Click **Data → Merge Files**, then click **Add Variables**. You will see the **Add Variables: Read File** dialog box.

3. Browse for the file named Crab Scale Results using an external SPSS data file option as shown in Figure 74.

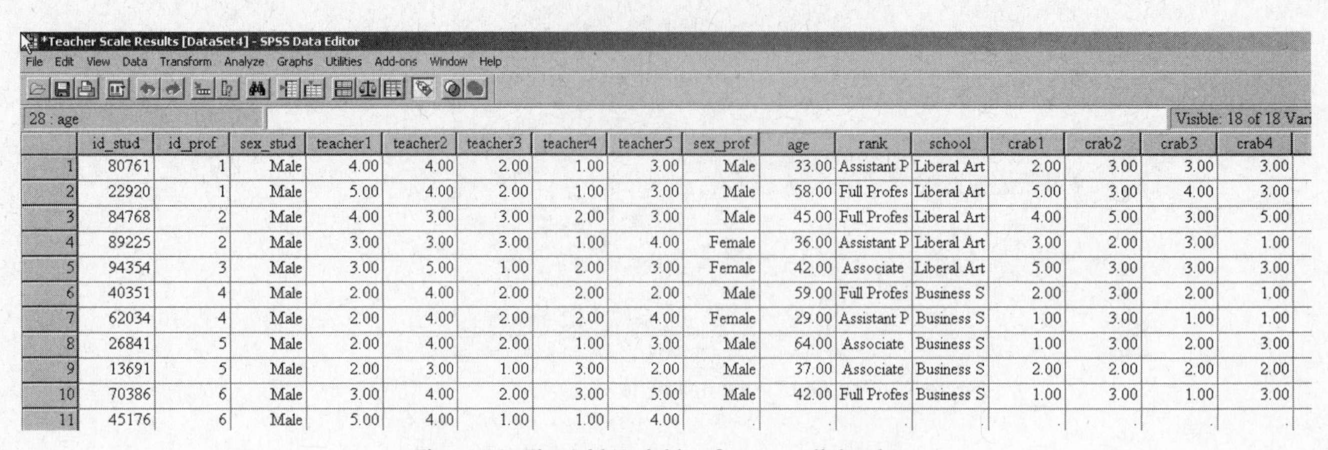

*Teacher Scale Results [DataSet4] - SPSS Data Editor

	id_stud	id_prof	sex_stud	teacher1	teacher2	teacher3	teacher4	teacher5	sex_prof	age	rank	school	crab1	crab2	crab3	crab4
1	80761	1	Male	4.00	4.00	2.00	1.00	3.00	Male	33.00	Assistant P	Liberal Art	2.00	3.00	3.00	3.00
2	22920	1	Male	5.00	4.00	2.00	1.00	3.00	Male	58.00	Full Profes	Liberal Art	5.00	3.00	4.00	3.00
3	84768	2	Male	4.00	3.00	3.00	2.00	3.00	Male	45.00	Full Profes	Liberal Art	4.00	5.00	3.00	5.00
4	89225	2	Male	3.00	3.00	3.00	1.00	4.00	Female	36.00	Assistant P	Liberal Art	3.00	2.00	3.00	1.00
5	94354	3	Male	3.00	5.00	1.00	2.00	3.00	Female	42.00	Associate	Liberal Art	5.00	3.00	3.00	3.00
6	40351	4	Male	2.00	4.00	2.00	2.00	2.00	Male	59.00	Full Profes	Business S	2.00	3.00	2.00	1.00
7	62034	4	Male	2.00	4.00	2.00	2.00	4.00	Female	29.00	Assistant P	Business S	1.00	3.00	1.00	1.00
8	26841	5	Male	2.00	4.00	2.00	1.00	3.00	Male	64.00	Associate	Business S	1.00	3.00	2.00	3.00
9	13691	5	Male	2.00	3.00	1.00	3.00	2.00	Male	37.00	Associate	Business S	2.00	2.00	2.00	2.00
10	70386	6	Male	3.00	4.00	2.00	3.00	5.00	Male	42.00	Full Profes	Business S	1.00	3.00	1.00	3.00
11	45176	6	Male	5.00	4.00	1.00	1.00	4.00		.			.	.	.	.

Figure 74. **The Add Variables from . . . dialog box.**

4. Click **Continue** and you will see the **Add Variables to Teacher Scale Results.**
5. SPSS has already identified that id_prof is common to the two files and has listed it in the Excluded Variables area.
6. Click **OK**.

The result of this merging, shown in Figure 75, is a file larger than either of the two files that share the variable named id_prof. If you recall, the file named Teacher Scale Results does not contain any demographic information, such as gender or age, but the merged file including Teacher Scale information does, since it was merged with the Crab Scale data.

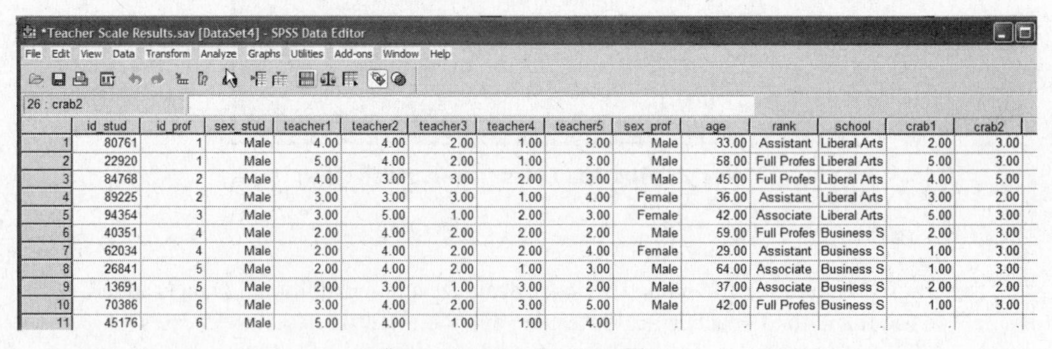

*Teacher Scale Results.sav [DataSet4] - SPSS Data Editor

	id_stud	id_prof	sex_stud	teacher1	teacher2	teacher3	teacher4	teacher5	sex_prof	age	rank	school	crab1	crab2
1	80761	1	Male	4.00	4.00	2.00	1.00	3.00	Male	33.00	Assistant	Liberal Arts	2.00	3.00
2	22920	1	Male	5.00	4.00	2.00	1.00	3.00	Male	58.00	Full Profes	Liberal Arts	5.00	3.00
3	84768	2	Male	4.00	3.00	3.00	2.00	3.00	Male	45.00	Full Profes	Liberal Arts	4.00	5.00
4	89225	2	Male	3.00	3.00	3.00	1.00	4.00	Female	36.00	Assistant	Liberal Arts	3.00	2.00
5	94354	3	Male	3.00	5.00	1.00	2.00	3.00	Female	42.00	Associate	Liberal Arts	5.00	3.00
6	40351	4	Male	2.00	4.00	2.00	2.00	2.00	Male	59.00	Full Profes	Business S	2.00	3.00
7	62034	4	Male	2.00	4.00	2.00	2.00	4.00	Female	29.00	Assistant	Business S	1.00	3.00
8	26841	5	Male	2.00	4.00	2.00	1.00	3.00	Male	64.00	Associate	Business S	1.00	3.00
9	13691	5	Male	2.00	3.00	1.00	3.00	2.00	Male	37.00	Associate	Business S	2.00	2.00
10	70386	6	Male	3.00	4.00	2.00	3.00	5.00	Male	42.00	Full Profes	Business S	1.00	3.00
11	45176	6	Male	5.00	4.00	1.00	1.00	4.00		.			.	.

Figure 75. **Merging two files.**

PART I

UNIT 4A

Working with SPSS Graphs and Output for Windows

A picture surely is worth a thousand words, and nowhere is this more true than with numbers. You could look at a set of data all day long and not really see what the nature of the relationship is between variables until a graph or a table is created. In this last part of introducing you to SPSS basics, we will go through the steps of creating a graph and modifying one, plus some tips on using the Output Navigator window. Note that there are many different tools available to create graphs in SPSS and we will only be concentrating on the tools contained in the Legacy Dialogs option on the Graphs menu.

This section of Part I deals with creating graphs and output for the Windows version (15.x) of SPSS. Unit 4B covers creating graphs and output for the Macintosh version of SPSS (13.x).

LESSON

16A Creating an SPSS Graph

After This Lesson, You Will Know

- How to create a simple line graph
- About some of the different graphs that you can create with SPSS

Key Words

- Area graph
- Bar graph
- Define Simple Line dialog box
- Line Charts dialog box
- Pie graph
- .spo

SPSS offers you just the features to create graphs that bring the results of your analyses to life. In this lesson, we will go through the steps to create several different types of graphs and provide examples of different graphs. In Lesson 17A, we'll show you how to modify a graph by adding a graph title, adding labels to axes, modifying scales, and working with patterns, fonts, and more. Throughout this unit of *Using SPSS*, we'll use the words *graph* and *chart* interchangeably.

TIP

Although the main menu for creating a graph in SPSS is labeled Graphs, you may also notice that the word *chart* is used in dalog boxes. No matter, they both mean the same thing—the visual presentation of data.

Creating a Simple Graph

The one thing that all graphs have in common is that they are based on data. Although you may import data to create a graph, in this example we'll use the data you see here to create a line graph of test scores by grade. These data are available at http://www.prenhall.com/greensalkind in the file named Lesson 16 Data File 1.

Grade	Score
1	45
2	56
3	49
4	57
5	67
6	72

Creating a Line Graph

The steps for creating any graph are basically the same. You first enter the data you want to use in the graph, select the type of graph you want from the Graph menu, define how the graph should appear, and then click **OK**. Here are the steps we followed to create the graph you see in Figure 76.

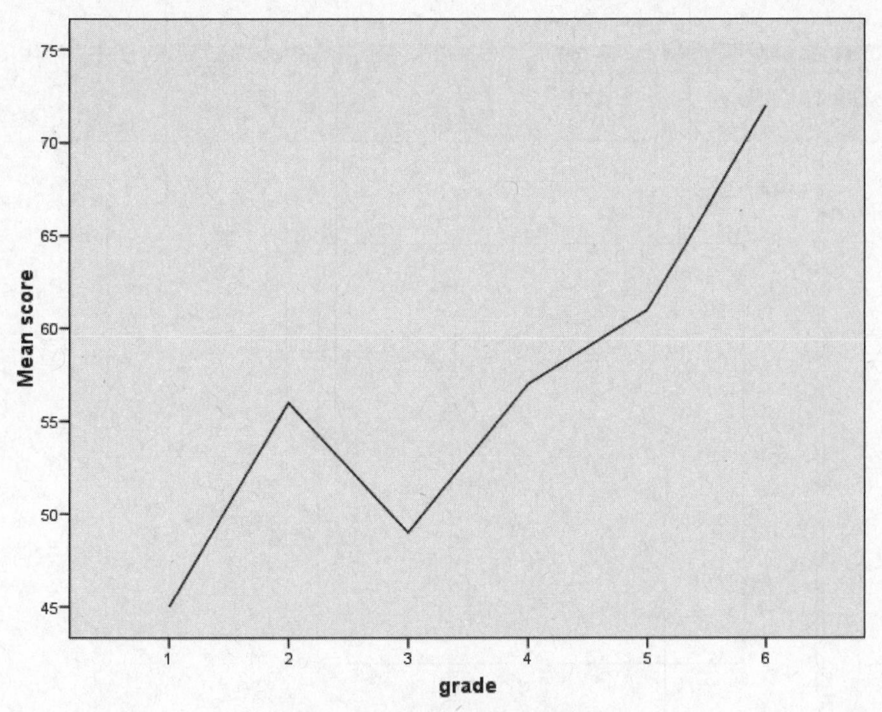

Figure 76. **A simple line graph.**

1. Enter the data you want to use to create the graph.
2. Click **Graphs ⟶ Legacy Dialogs ⟶ Line**. When you do this, you will see the **Line Charts dialog box**, as shown in Figure 77.

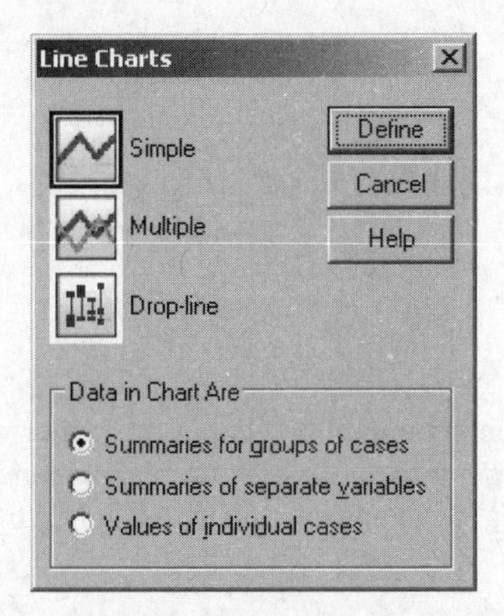

Figure 77. **The Line Charts dialog box.**

3. Click **Simple** and then click **Define**. You will see the **Define Simple Line dialog box**.
4. Click **grade**, click ▶ to move the variable to the Category Axis: area.
5. Click the **Other Statistic** button.

6. Click **score**, and then click ▶ to move the variable to the Variable: area. Now is the time to enter a title or subtitle in any graph by clicking the **Titles** button in the Define dialog box and entering what titles, subtitles, and footnotes you want.
7. Click **OK**, and you see the graph in the Output Navigator, as shown in Figure 78.

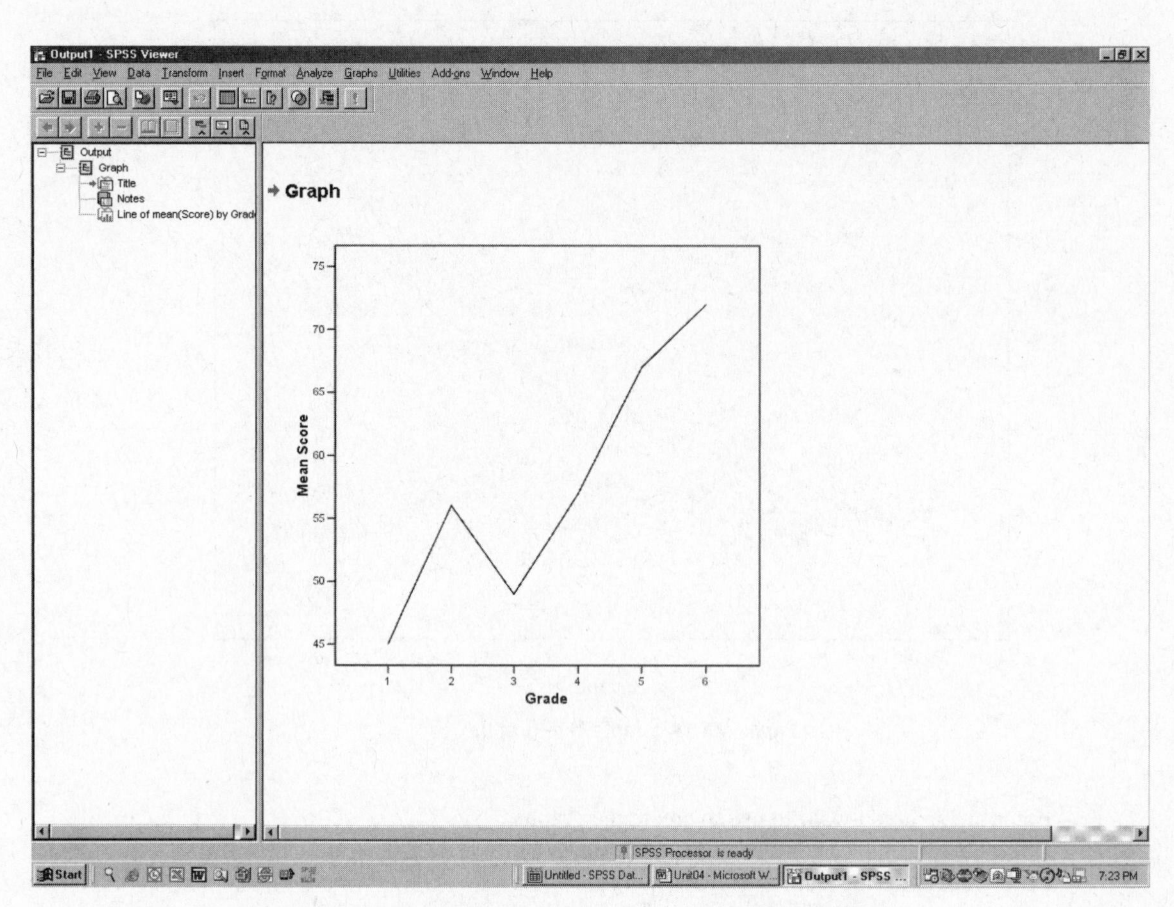

Figure 78. **A simple line graph in an Output Navigator window.**

This is the default graph with no additions, changes, or edits.

Saving a Graph

A graph is only one possible component of the Output Navigator window. The graph is not a separate entity that stands by itself, but part of all the output generated and displayed in the Viewer. To save a graph, you have to save the entire contents of the Output Navigator. To do so, follow these steps:

1. In the Output Navigator window, click **File ⟶ Save**.
2. Provide a name for the Output Navigator window.
3. Click **OK**. The Output Navigator is saved under the name that you provide with an **.spo** extension (attached to all Viewer files). (See Figure 79 on page 81.) If you want to open this file later, you have to select this extension from the Files of type drop-down menu.

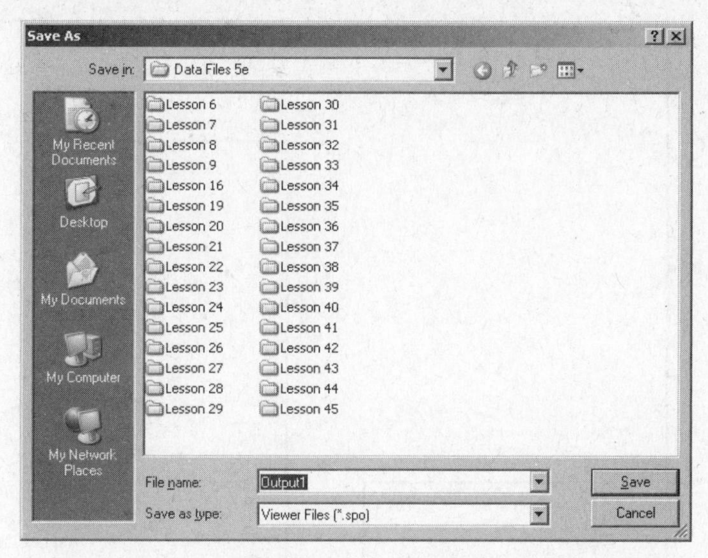

Figure 79. **Output is always saved with an .spo extension.**

Printing a Graph

There are several ways to print a graph. The first way is to print from the Output Navigator by selecting the graph and then selecting Print.

1. Be sure that the graph you want to print is in the active Viewer window.
2. Click on the graph so that it is selected.
3. Click on the Print icon on the Toolbar or click **File ⟶ Print** and then **OK**. The graph will be printed.

TIP

You can print only one or more elements in the Viewer by first selecting the element by clicking on it once and then selecting File ⟶ Print, clicking **Selection**, and then clicking **OK.** To select more than one, hold down the Ctrl key as you click.

Different SPSS Graphs

SPSS offers 11 different types of graphs all of which you can see at **Graphs ⟶ Legacy Dialogs ⟶ Line**.

Although some types of graph may be used in lieu of others, each has its own purpose. The purpose depends on the type of data being graphed as well as the research question being asked. It's beyond the scope of this book to detail each graph and all the possible combinations (for help, just click on any icon in the Graphs Builder help area), but the following are examples of some of the more simple graphs. Keep in mind that there is an almost infinite number of variations of the graphs you can create by using SPSS. You'll learn about modifying graphs in Lesson 17A. All sample graphs that follow appear as they were first created, with no modification.

The Bar Graph

A **bar graph** represents values as separate bars with each bar corresponding to the value in the Data Editor. Bar graphs are often used because they are easy to understand and require little attention to detail. In Figure 80 (on page 82), you can see an example of a bar graph of the number of males and females.

The Area Graph

An **area graph** represents the proportions of the whole that each data point contributes. In Figure 81 (on page 82), you can see an example of average product preference as a function of earning power (high, moderate, and low).

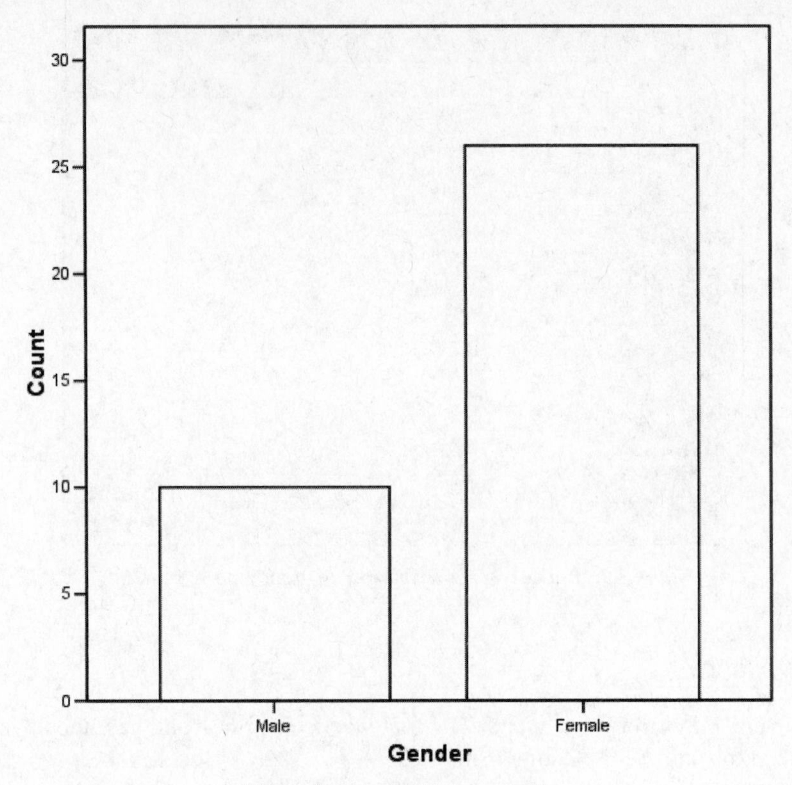

Figure 80. **A simple bar graph.**

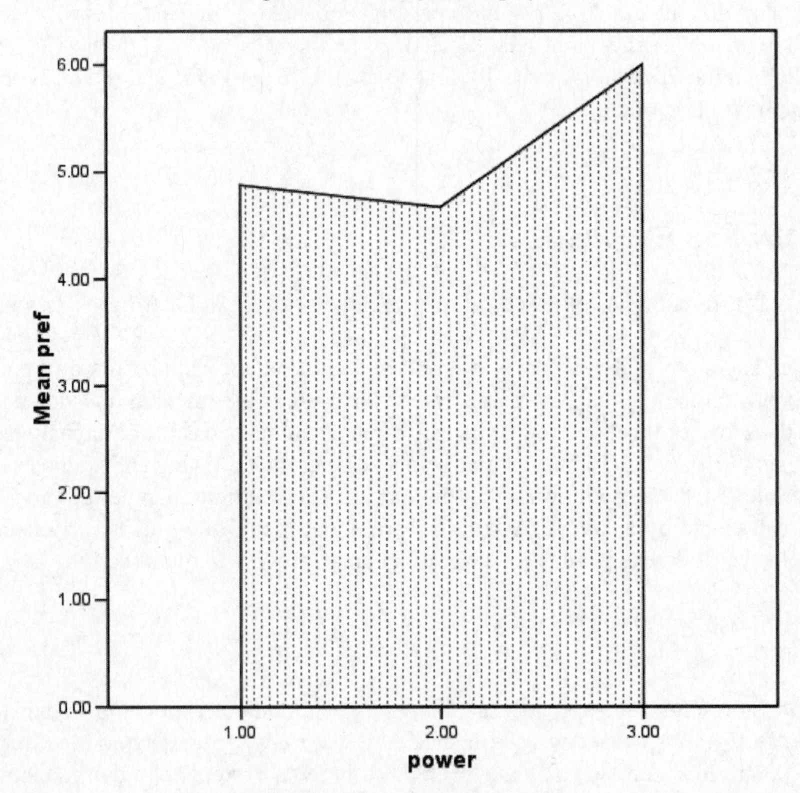

Figure 81. **A simple area graph.**

The Pie Graph

The **pie graph** is a circle divided into wedges, with each wedge representing a proportion of the whole. Pie graphs provide a clear picture of the relative contribution that a data point makes to the overall body of information. In Figure 82, you can see a simple pie graph for expenses (rent, insurance, and miscellaneous).

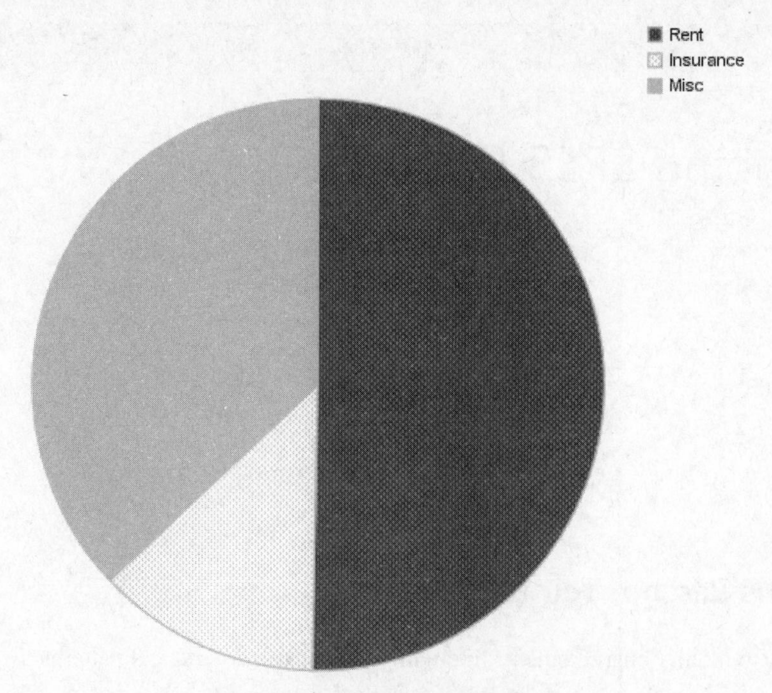

Figure 82. **A simple pie graph.**

Keep in mind that when you first create a graph, all the SPSS defaults are used. You can easily change type size and font and many other characteristics. You will learn how to do that in the next lesson.

LESSON

17A Enhancing SPSS Graphs

After This Lesson, You Will Know

- How to modify chart elements, including titles, axes, colors, and patterns

Key Words

- Category Axis
- Category Axis dialog box
- Colors dialog box
- Pattern dialog box
- Scale Axis
- Scale Axis dialog box
- SPSS Chart Editor
- Text Styles dialog box
- Titles dialog box

A picture might be worth a thousand words, but if your pictures or graphs don't say what you want, what good are they? Once you create a chart, as we showed you in the last lesson, you can finish the job by editing the chart to reflect exactly your meaning. Color, shapes, scales, fonts, text, and more can be altered, and that is what you'll do in this lesson. We'll be working with the line chart that was first shown to you in Figure 76 (on page 79).

Modifying a Chart

The first in the modification of any chart is to double-click anywhere inside of the chart in the Output Navigator to access the **SPSS Chart Editor** and then click the **maximize** button on the Application title bar. As you can see in Figure 83 (on page 85), there is a Chart toolbar containing a set of buttons across the top of the chart. In Figure 84 (on page 85), you can see each button better and what each represents.

One cautionary note before we begin. The most basic reason for modifying a chart should be to increase the chart's ability to communicate your message. If you get caught up in Chart Junk (a close cousin to Font Junk), you'll have a mess on your hands. Modify to make better, not to show your reader that you know how to use SPSS.

TIP

Graphs can be easily copied (use the Edit menu) and pasted into other Windows applications.

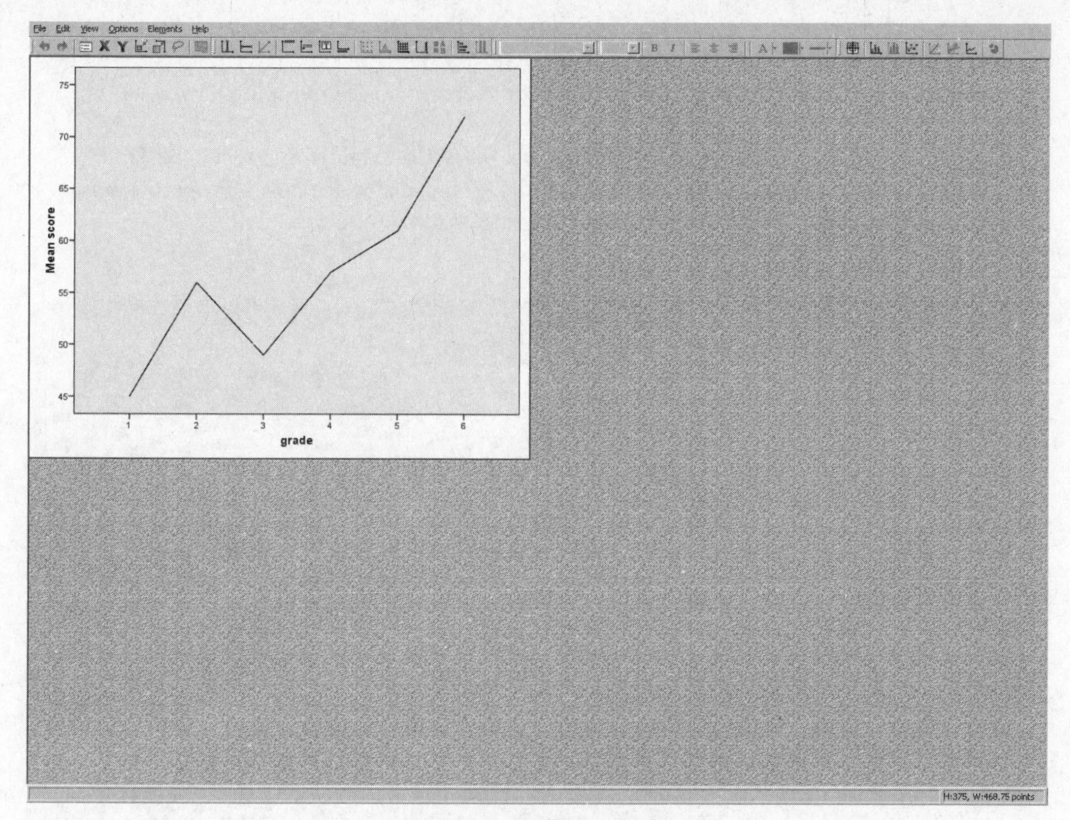

Figure 83. **The Chart Editor.**

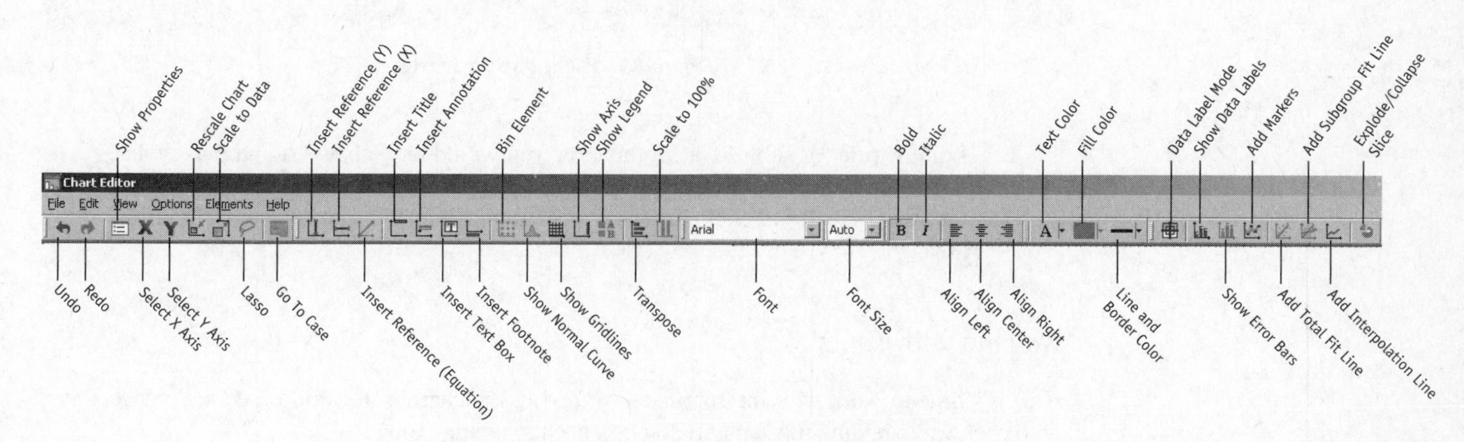

Figure 84. **The Chart Editor tools and what they do.**

Some instructional notes before we begin. The general strategy for changing any element in a chart is to select the chart element and then make the change. You select a chart element much as you select any Windows element, by clicking on it. If you single-click, you then must select the Menu command for the item you want to change. If you double-click, in some cases, you will go immediately to the dialog box in which you can make the change. And, there are usually several different ways to change a chart element. You can usually double-click on the element (such as the Y-axis) and make approriate changes or use a corresponding tool on the toolbar. And, most parts of a graph have a Properties dialog box associated with it where you can fine-tune graph edits.

TIP

The first step in changing any graph element is to click on that element in the Chart Editor. The second step is to make the changes in the Properties box associated with that element. The final step is to click the **Apply** button.

Working with Titles and Subtitles

Our first task is to modify the title and subtitle for the chart you see in Figure 85.

1. With the Chart Editor open, click **Options ⟶ Title.** When you do this, the Title area is highlighted, as you see in Figure 85. You can also see how a Properties box opens up where you can also make adjustments to the title.

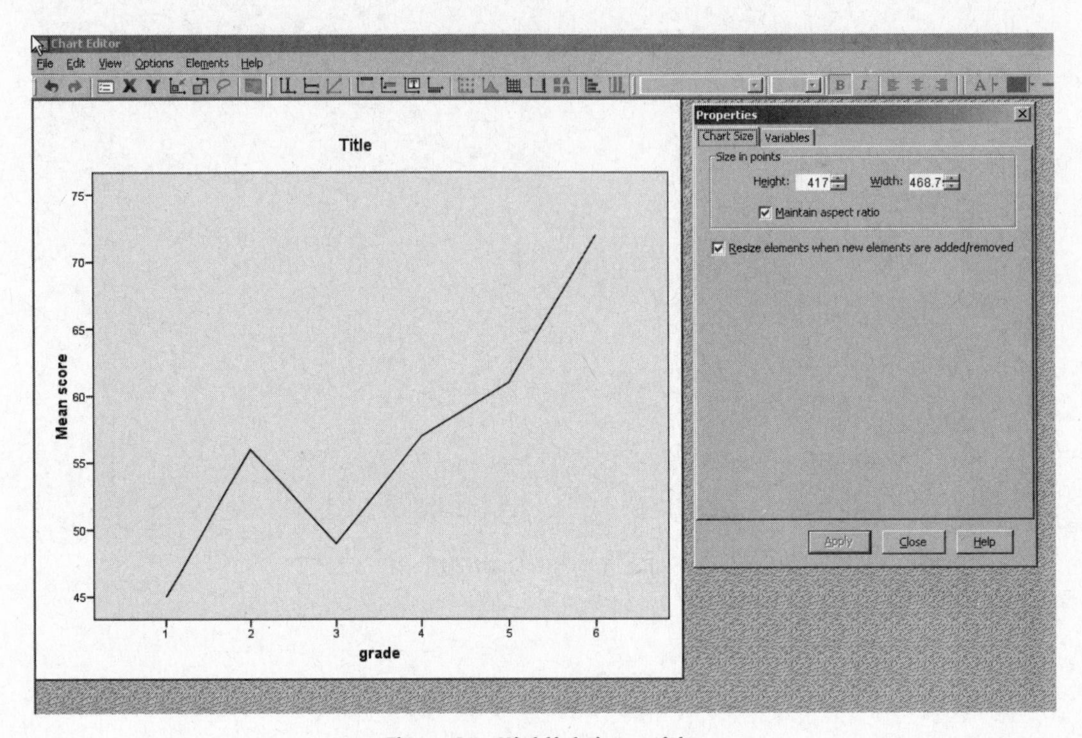

Figure 85. **Highlighting a title.**

2. Edit the title (or subtitle or footnote) as you would any other text and make whatever edits as necessary.
3. Once you are done, click anywhere else in the Chart Editor outside of the title area.

Working with Fonts

Now it's time to work with the font used for text in the chart. You can do this one of two ways, with each way yielding the same dialog box used to change fonts:

1. Select the area of the chart containing the text you want to change. When you select text, it appears with a solid line around it. In this example, we selected the title of the chart, Score by Grade. Once selected, it appears with a box around it.
2. To change a font's properties, double-click on the title and you will see the Properties box, as shown in Figure 86.
3. Make the changes that you want. In this example, we changed it to 14-point Arial Italic.
4. Click the **Apply** button, and the font will change in the chart. The Properties dialog box will remain on the screen until you close it. That way you can change the chart size and other characteristics of the chart without having to reselect the Properties dialog box.

TIP

You can also access the Properties window by using the CTRL+T key combination after you have highlighted the element you want to modify.

You can highlight more than one text area at a time by holding down the CTRL key as you click with the mouse.

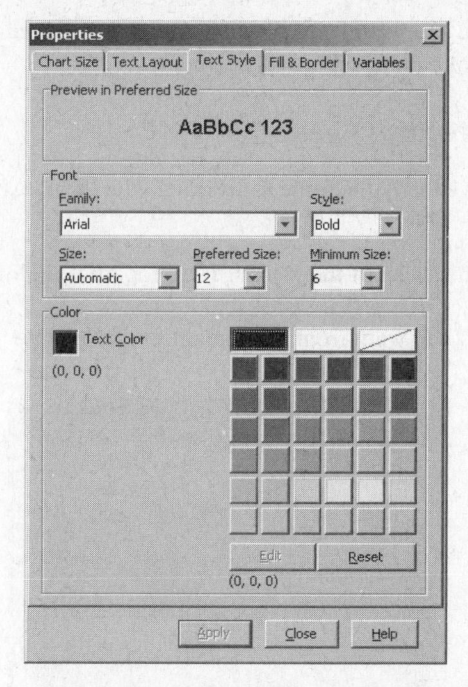

Figure 86. **The Properties dialog box.**

Working with Axes

The *X*- and *Y*-axes provide the calibration for the independent (usually the *X*-axis) variable and the dependent (usually the *Y*-axis) variable. SPSS names the *Y*-axis the **Scale Axis** and the *X*-axis the **Category Axis**. Each of these axes can be modified in a variety of ways. To modify either axis, double-click on the title of the axis and you will see the Properties dialog box for that axis.

How to Modify the Scale (*Y*) Axis

For example, to modify the *Y*-axis, follow these steps:

1. Double-click on the line representing the *Y*-axis (the axis will turn blue) and you will see the Properties dialog box as shown in Figure 87.

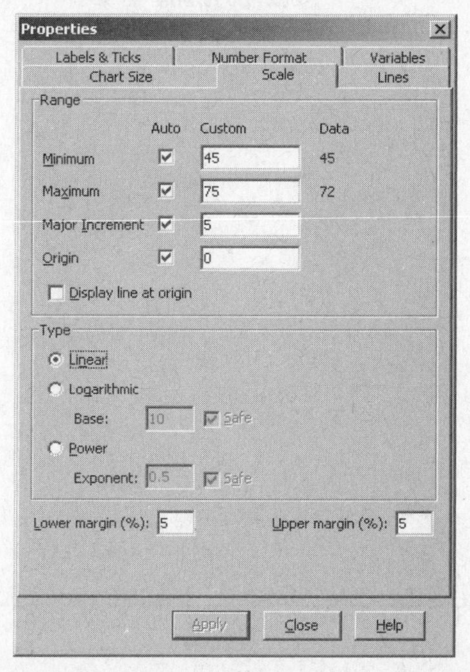

Figure 87. **The Properties dialog box for the *Y*-axis.**

2. Select the options you want from the dialog box. We clicked the **Number Format** tab and made sure that there were no decimals.

How to Modify the Category (*X*) Axis

Working with the *X*-axis is exactly the same as working with the *Y*-axis. Here's how the *X*-axis is modified:

1. Double-click on the label of the *X*-axis. The Properties dialog box opens as shown in Figure 88.
2. Select the options you want from the Properties dialog box.

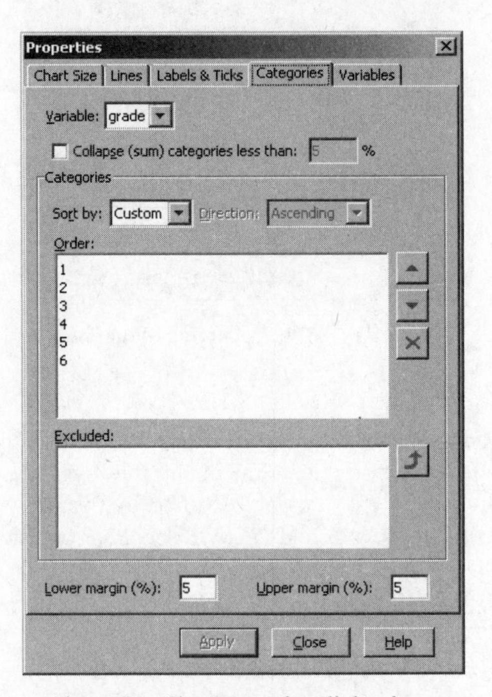

Figure 88. **The Properties dialog box.**

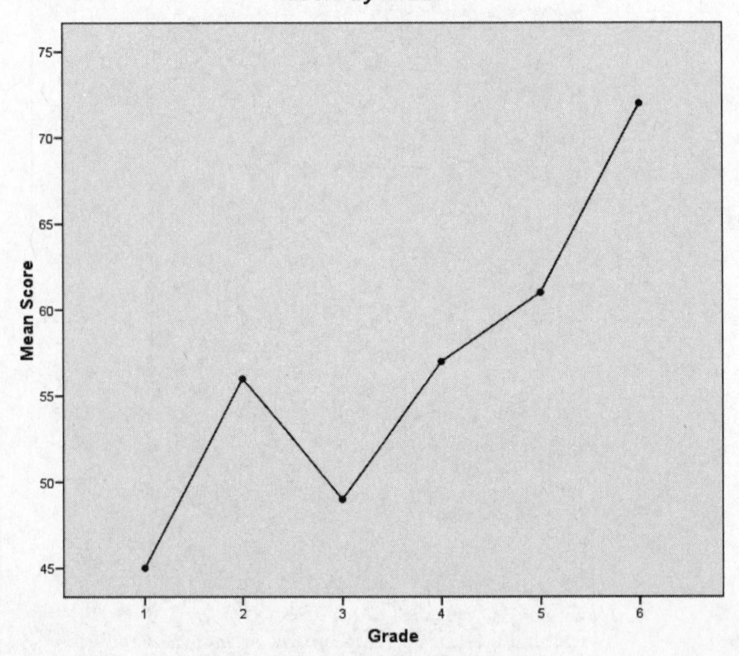

Figure 89. **A modified line chart.**

Our finished and modified line chart (with a few more changes) is shown in Figure 89, and although none of these changes are dramatic, they do show how easy these enchancement tools are to use.

We got back to the output window by closing the Chart Editor when we were done modifying the chart. The only other change we made was left justifying the footnote and italicizing it by using the font options we saw in the Properties dialog box once we double-clicked on the footnote.

Working with Patterns and Colors

Time to have some fun (and be practical besides). When SPSS creates a chart, and if you have a color monitor, it assigns different colors or symbols to different elements in a chart. We'll use two tools, one to work with patterns and one to work with colors. We'll use a simple bar chart, as shown in Figure 90, as an example.

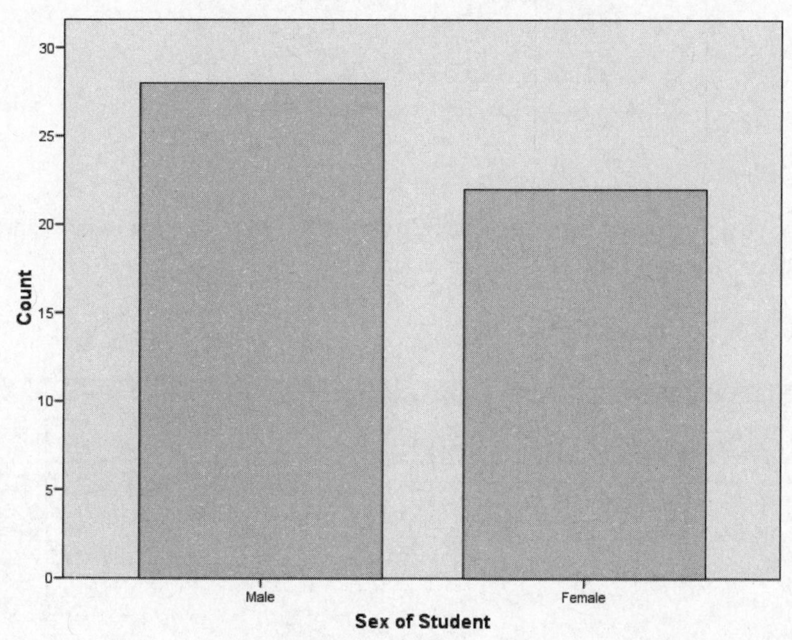

Figure 90. **A simple bar chart.**

Changing Patterns

To change the patterns of any element in a chart, follow these steps:

1. Double-click to open the Chart Editor.
2. Double-click on the chart element you want to select (in this case, either of the two bars) to open the Properties dialog box shown in Figure 91 (on page 90).
3. Click the **Fill & Border** tab.
4. Click the **Pattern** drop-down menu and select the pattern you want.
5. Click **Apply**, and the pattern used in the bars will change. In the bar chart you see in Figure 92 (on page 90), the dotted pattern was used for the Gender variable.

Now we have one more step. We will eliminate the color from the bars, since printing in color (your printer will see the colors as being black or a shade of gray) uses a lot of toner.

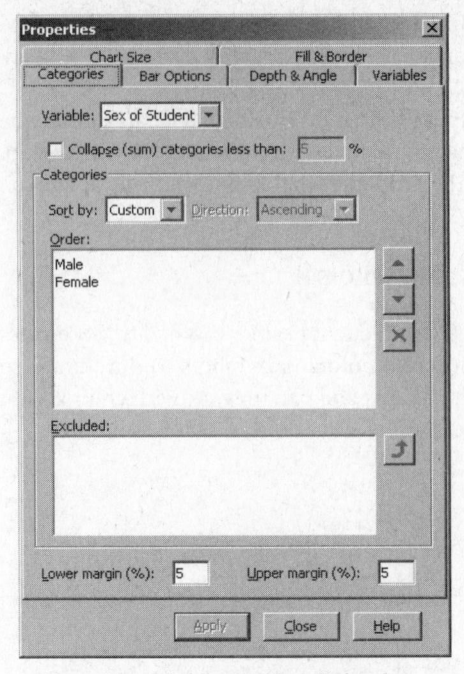

Figure 91. **The Properties dialog box.**

Changing Colors

To change the color of any element in a chart, follow these steps (we are in the same Properties dialog box):

1. Click the **Fill** button under Color.

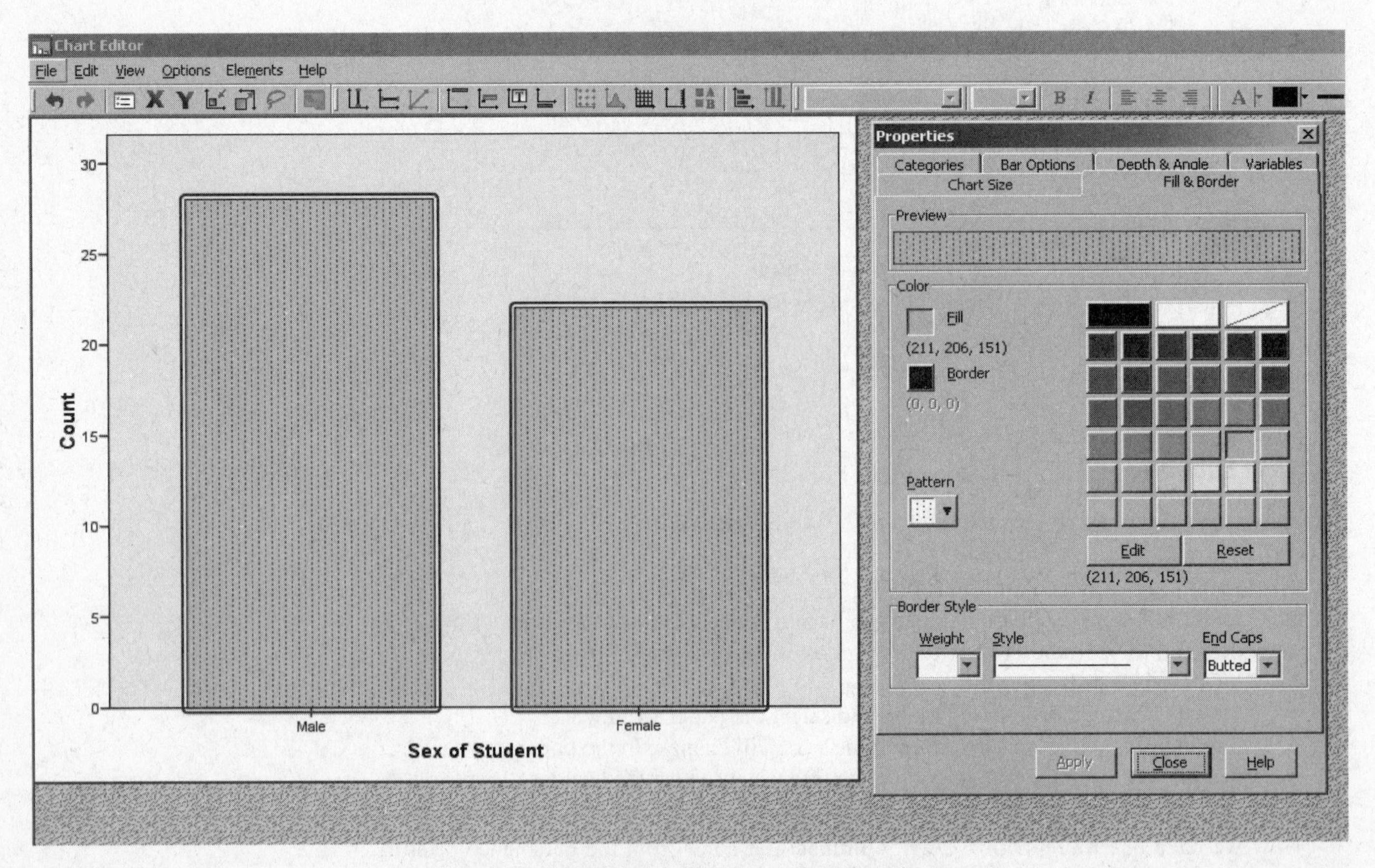

Figure 92. **Changing the fill pattern for a bar chart.**

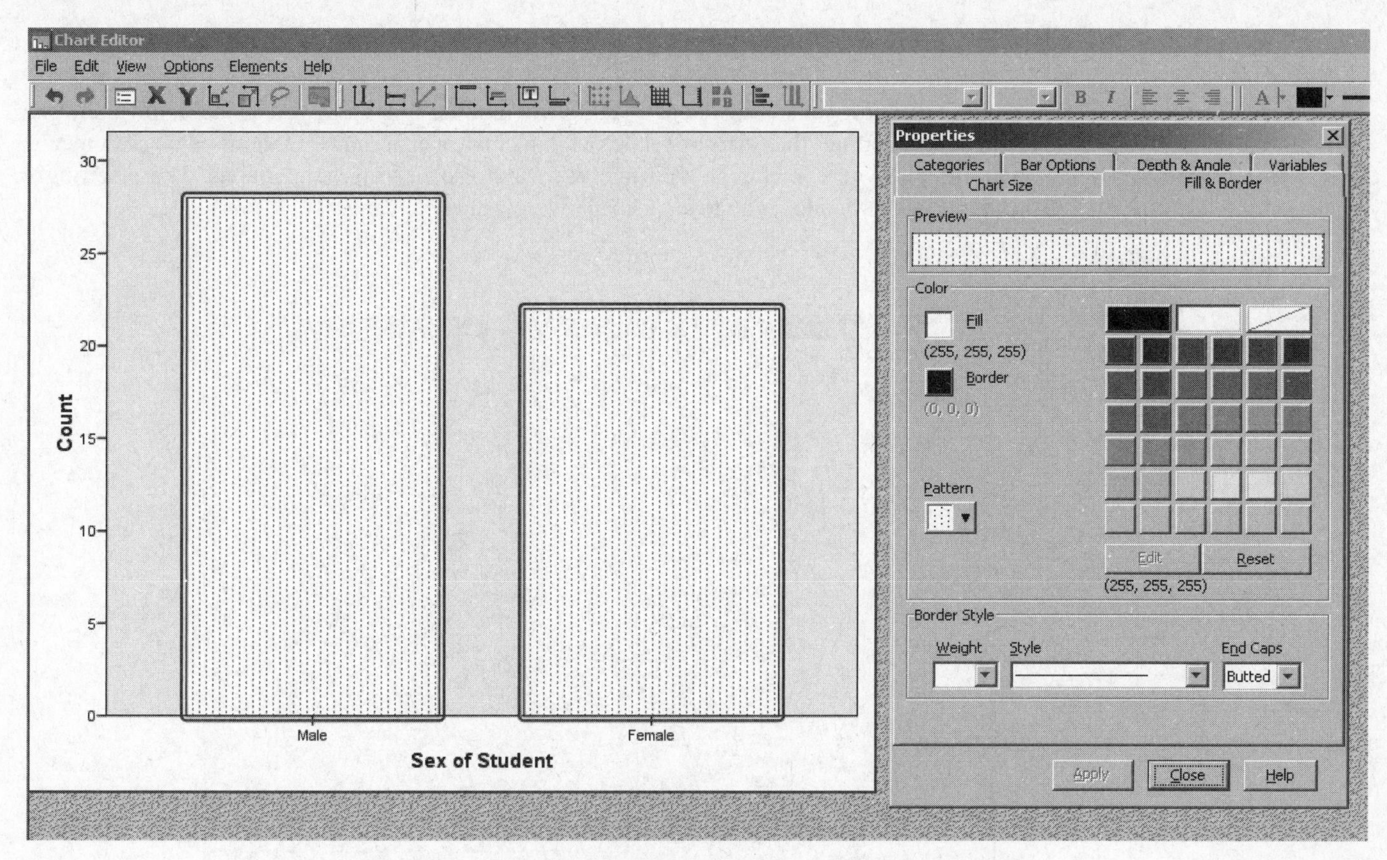

Figure 93. **Changing the pattern and color in a simple bar chart.**

2. Click the color you want to use. In this example, we clicked **white** and then clicked the **Apply** button and the newly formatted chart (before we closed the Chart Editor) appears in Figure 93.

The final chart with modifications is shown in Figure 94.

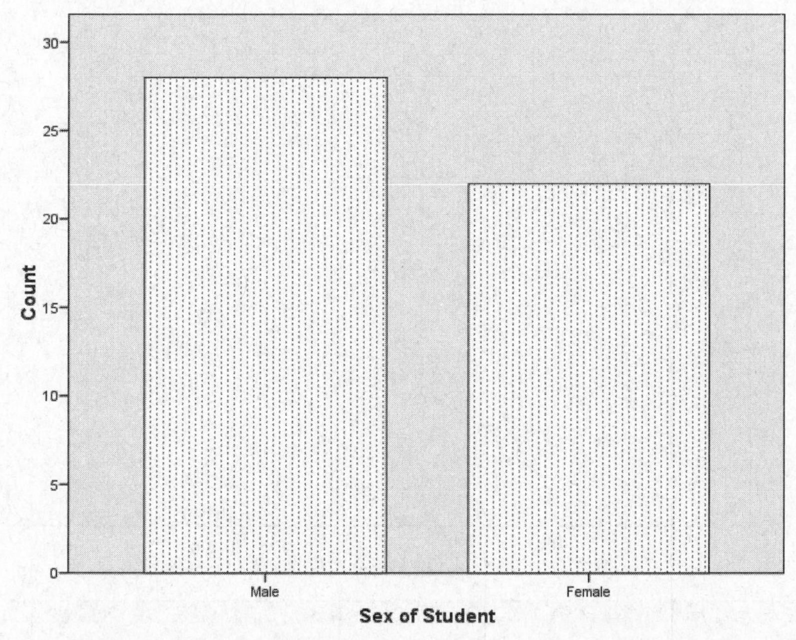

Figure 94. **Changing the color and pattern in a simple bar chart.**

Setting Chart Preferences

Now that you know something about creating and modifying graphs, you should know there are certain default settings that you can set before you create a chart under Options on the Edit menu. Using this tool you can choose a particular format for viewing your graphs. These settings, shown in Figure 95, allow you to set the following visual elements:

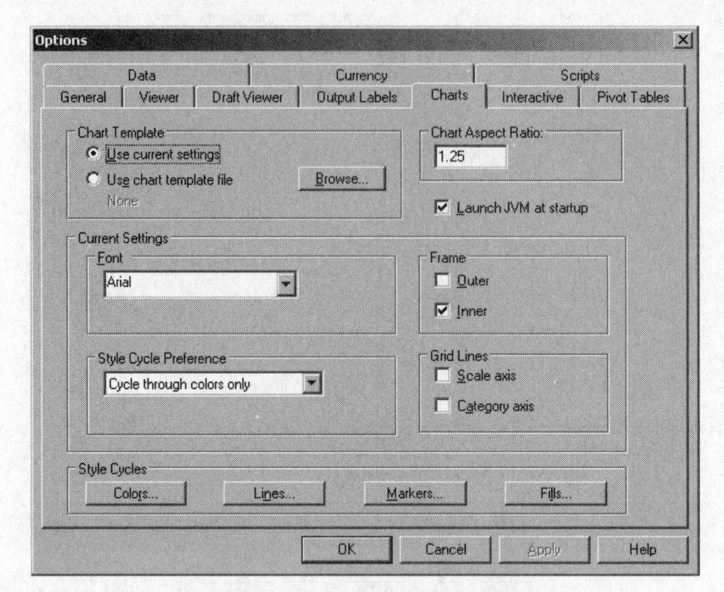

Figure 95. **The Options dialog box.**

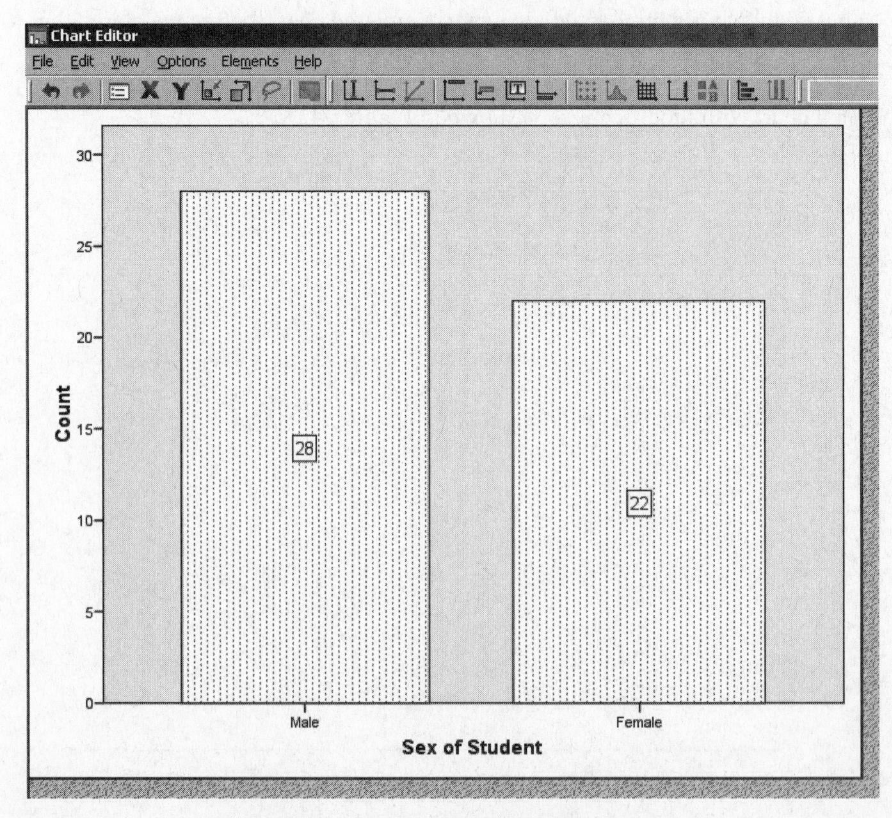

Figure 96. **A more informative bar chart.**

- the font used in the chart
- the nature of how items are filled with patterns or colors
- the height-to-width ratio of the chart
- the presence of a frame (inside or outside)
- the presence of grid lines

For example, if you want the initial chart to use Times New Roman as a font, specify it in the SPSS Options dialog box you access through the Edit menu in the Chart Editor.

Getting Fancy

Before we finish with this bar chart and with your introduction to SPSS graphs, let's add just a few more things to jazz up the bar chart. If the Chart Editor is not open, open it now.

First, let's click on the **Data Label Mode** button on the Chart Editor toolbar. Now place the data label where you want it inside the bar and click only once. Once it appears inside of the bar, you can double-click it and work with the data labels properties. You see the final chart in Figure 96—a bit more informative.

Using a Chart Template and Creating an APA-Style Graph

If you are preparing a chart or graph for publication, you will want it to conform to the style detailed in the fifth edition of the *Publication Manual of the American Psychological Association.*

As you can see, the chart has no title. The title belongs as a caption in the manuscript on the previous manuscript page. You can also see we got rid of the pattern and the bar labels.

If you intend to create graphs according to APA guidelines, you should create a chart template. That way, when you create a new chart, all you need to do is apply the chart template, and the formatting (such as centered axes titles) will be automatic. To create a chart template, follow these steps:

1. Once you complete a chart that fits your style needs and you are still in the Chart Editor window, click **File → Save Chart Template** in the Chart Editor window. You will see the **Save Chart Template** dialog box as shown in Figure 97.

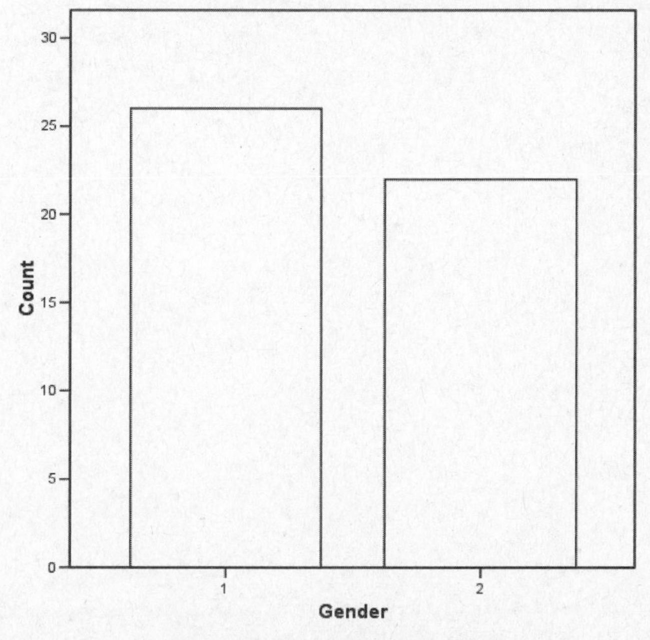

Figure 97. **Creating a template.**

2. Select the options you want to save based on this graph and click **Continue.**
3. Enter a name for the template. The template will automatically have an .sct extension.
4. Click **Save**. The chart template will be saved.

To use a chart template once it has been created, follow these steps:

1. After clicking the **Define** button when creating a chart, click the **Use chart specifications from**: checkbox.
2. Locate the template that you want to use.
3. Click **Open** and the style will be applied to the graph you are developing.

LESSON

18A Using the Viewer and Pivot Tables

After This Lesson, You Will Know

- The use of the two panes in the Viewer
- How to save output
- How to show and hide results in the Viewer window
- How to print output
- How to delete output
- How to move output

Key Words

- Contents pane
- Outline pane

When you complete any type of procedure in SPSS, be it the creation of a simple chart or the results of a complex analysis, the output appears in the Viewer. The Viewer is a separate SPSS screen that lists all operations performed on a set of data in the order in which they were completed.

For example, in Figure 98 (on page 96), you can see that two different types of analysis were done; the first is a bar chart (graph) and the second is a simple description of the data (Descriptives).

As you can see, the screen is titled Output 1 [Document 1]–SPSS Viewer, since it is the first Viewer window created during this work session. Until you save the contents of the Viewer with a new name, the Viewer window will keep the name that SPSS assigned when it was first created. Unless named by you, the names automatically assigned are Output1, Output2, etc.

The Viewer consists of two panes. The **Outline pane** on the left side of the Viewer and the **Results pane** on the right side of the Viewer.

The Outline pane lists in outline form all the analyses that have been completed during the current session. The Results pane contains the results of the analyses. For example, in Figure 98 (on page 96), you can see an outline in the left pane of what is contained in the right pane. In the right Results pane, you find labels for the output along with the corresponding tables, graphs, etc. In any Outline pane, you will find labels (such as Graph, Title, and Notes) and then a description of the specific graph that was created. For now, we'll show you how to save Viewer output, print it, delete elements in the Viewer, and move the elements around so the output best fits your needs.

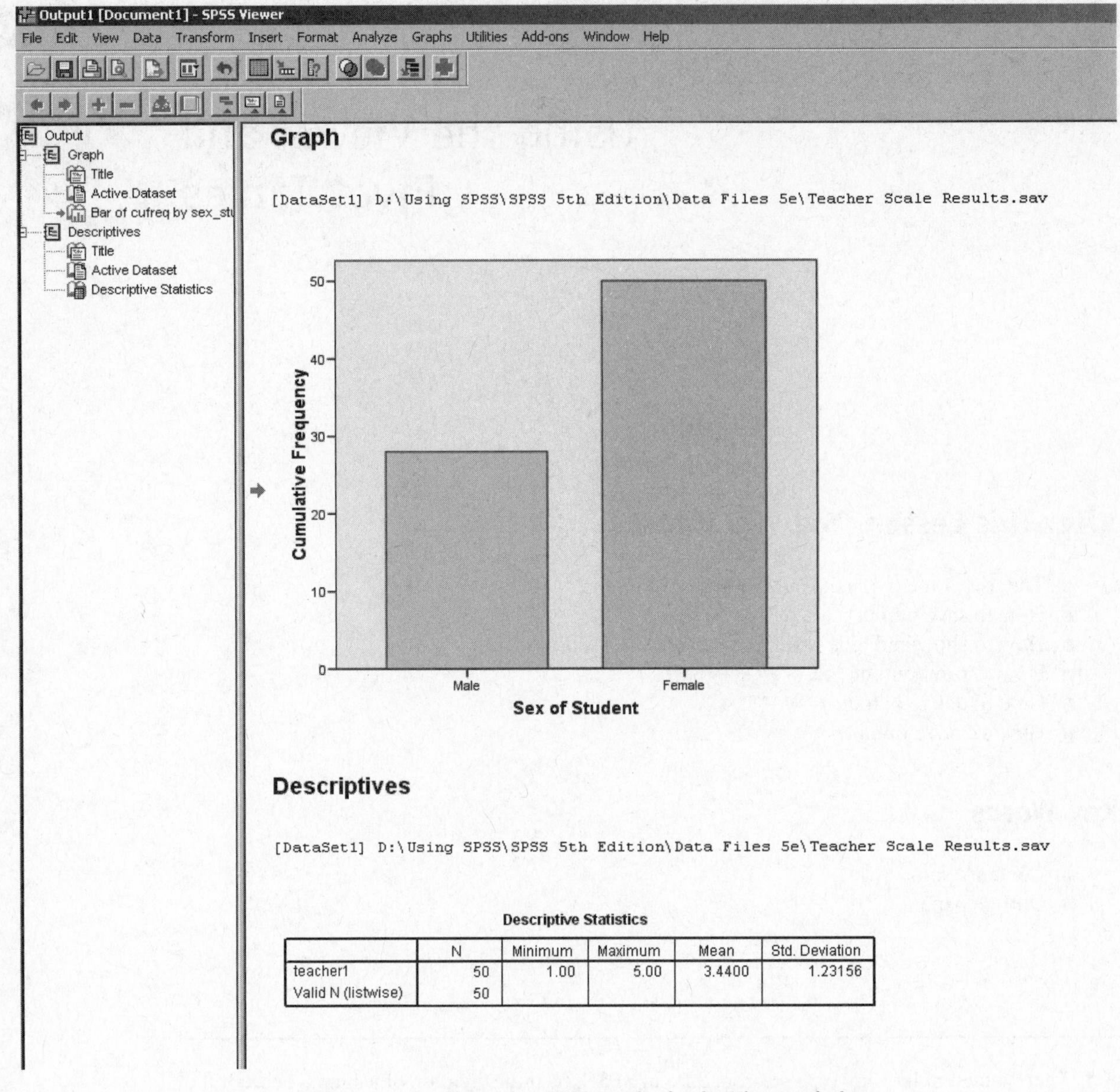

	N	Minimum	Maximum	Mean	Std. Deviation
teacher1	50	1.00	5.00	3.4400	1.23156
Valid N (listwise)	50				

Figure 98. **SPSS results in the Viewer window.**

Saving Viewer Output

Once output is generated, you will probably want to save it for later use or even add it to another work session. To save output from the Viewer, follow these steps:

1. Click **File** ⟶ **Save**.
2. Provide a name for the output.
3. Click **OK**. The output will be saved with the file name that you provided, along with the .spo extension. When you want to open the Viewer you saved, you will have to specify the .spo extension in the File Type drop-down menu in the SPSS Open dialog box.

TIP

The red arrow in the Outline or Results pane indicates which element in the Viewer is active.

To Selectively Show and Hide Results

You can choose to show or hide any results that appear in the Viewer window. You may generate results that you want to hide while you print out others, a convenient way to be selective as you focus your discussion of the results.

To hide a table or chart without deleting it, follow these steps:

1. Click on the item in the Outline pane.
2. Click **View** ⟶ **Hide** and the element that was selected will not longer appear in the Viewer window.

For example, in Figure 99, you will see that *just* the title of the procedure Graph (which appears in Figure 98) in the Outline pane was hidden in the Results pane. Although it is not deleted, it no longer appears in the Results pane. But, as you can see in Figure 99, it still appears in the Outline pane of the Viewer window.

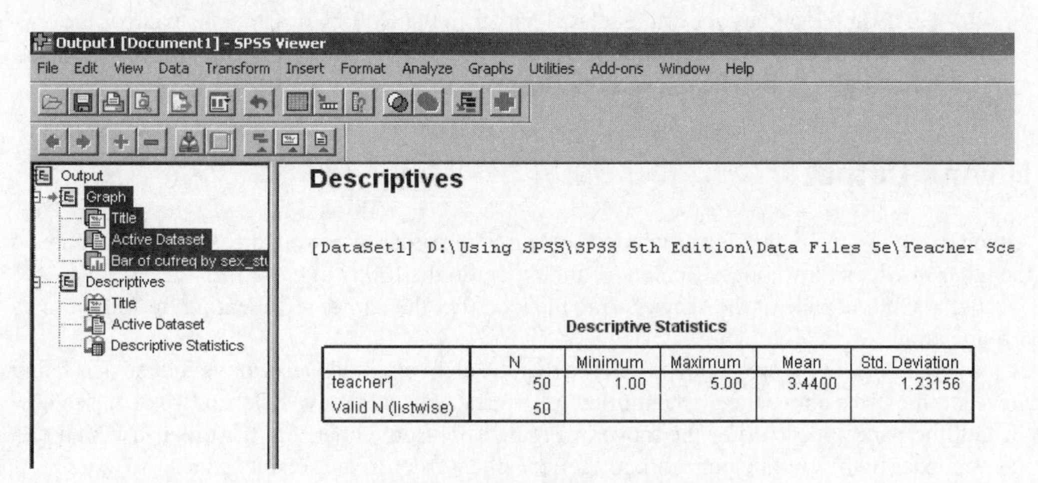

Figure 99. **Hiding SPSS results.**

To show a table or chart once it has been hidden, follow these steps:

1. Click on the item icon in the Outline pane. If you click on the item name, then you will simply highlight the name and the Show option on the View menu will not be active.
2. Click **View** ⟶ **Show**. The item reappears in the **Contents pane**.

You'll want to hide and show output as you decide what you want to appear in a hard copy of the Viewer.

Printing the Contents of the Viewer Window

To print the entire contents of the Viewer, follow these steps:

1. Click **File** ⟶ **Print**.
2. Click **OK**. All the tables and charts in the Viewer will be printed.

Printing a Selection from the Viewer Window

To print a specific selection from the Viewer window such as a table or a chart, follow these steps:

1. Click on the table, chart, or both that you want to print in the Viewer window. You can select more than one element by holding down the Shift or Ctrl key when you click each element.

TIP

If you select an item in the Results pane and press the **Del** key, the item will be deleted and can be recovered using the Ctrl+Z key combination.

2. Click **File** ⟶ **Print**.
3. Click **Selection** ⟶ **OK**. The element or elements you selected will be printed.

Any element surrounded by a black line is selected and will print. You can select as many elements as appear in the Contents pane to print at once.

Deleting Output

To delete output from the Viewer, follow these steps:

1. Click on the output you want deleted, either in the Outline or Contents pane of the Viewer window.
2. Click **Edit** ⟶ **Delete**, or press the **Del** key.

Moving Output

Although the results in the Viewer window appear in the order they were created, you can change the order in which they appear. To change the order, do the following:

In the Outline pane of the Viewer, drag the icon (not the name) representing the output to its new location.

For example, in Figure 100, you can see that the element titled Descriptive Statistics is being moved to the place above the element titled Active Dataset and below Title, and you can see how the Outline pane appears after the move in Figure 101. Remember that the order in which elements appear in the Outline pane directly reflects their order in the Results pane.

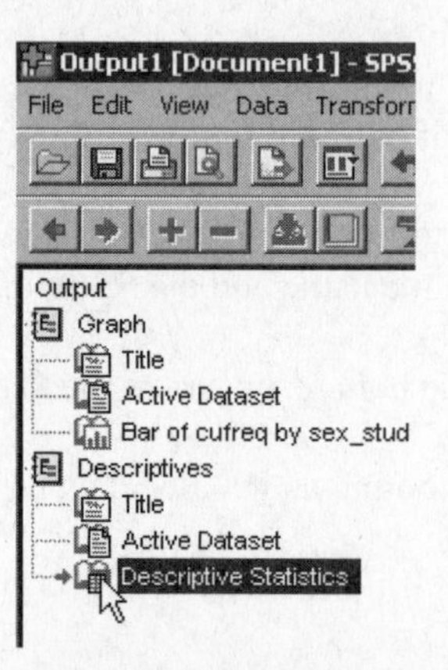

Figure 100. **Moving results in the Outline pane.**

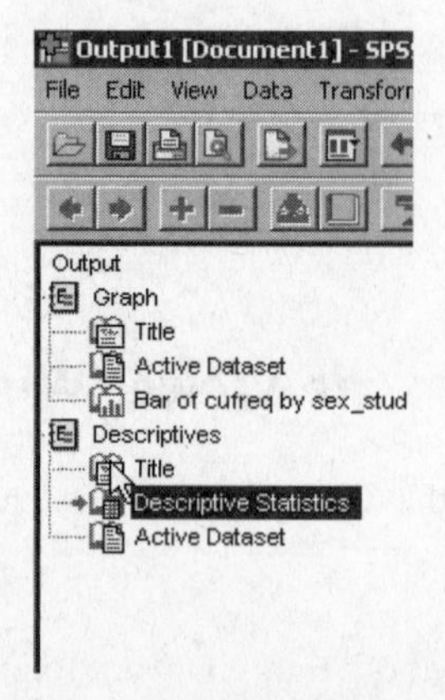

Figure 101. **The result of moving results in the Outline pane.**

An Introduction to Pivot Tables

It's uncommon that an SPSS user would produce output and not have to use at least part of that output in a report, journal article, or an even more ambitious undertaking.

And, much of that activity involves tables. You already know that SPSS can generate tables, but it also contains a facility for manipulating the orignial output to further develop exactly the type of table you need. That's exactly what a **pivot table** is—a table that can be pivoted around a row or column headings to change its appearance. And, you can also get the look that you want with other SPSS tools.

Using the Pivot Tray

To modify a table, follow these steps:

1. Activate the pivot table by right-clicking on the table.
2. Click **SPSS Pivot Table Object → Open**. Then click **Pivot → Pivoting Trays** and you will have both the pivot table and the trays open. Pivot trays are the tools that you use to manipulate the position of rows, columns, and layers in a pivot table (each of which is represented by an icon that looks like this). You can see the pivot table and trays shown in Figure 102 (plus a ToolTip to show you what the particular icon represents in this example).

TIP

You can also edit or modify almost any pivot table element by clicking on it and making the changes and then clicking anywhere in the table to confirm the change.

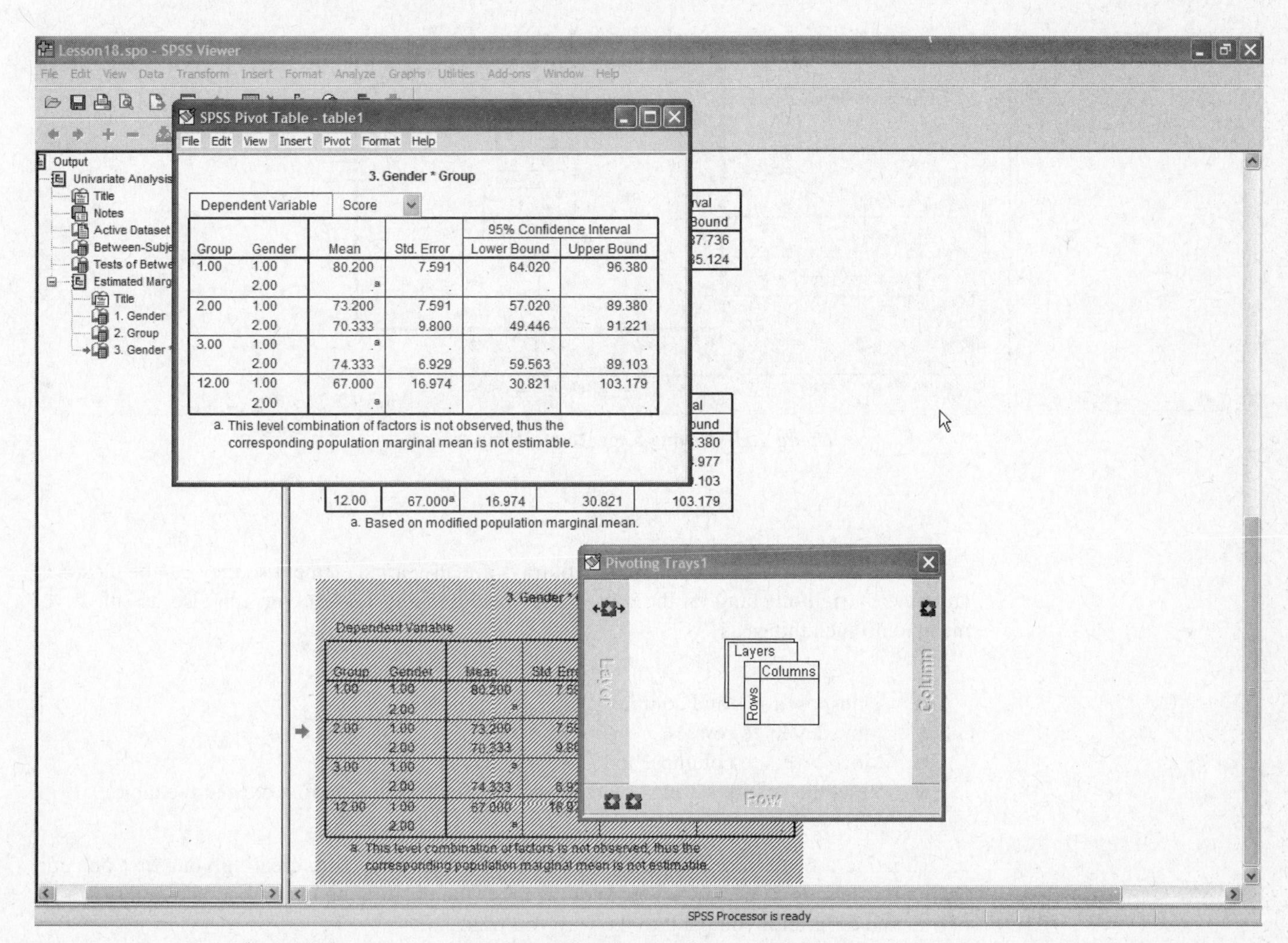

Figure 102. **The pivot table and pivot trays.**

3. The basic idea behind pivot tables is that these icons can be dragged from one layer (row or column) of a table to another to create new look. For example, to move the gender variable from a row to a column, just drag that icon to the column layer as shown in Figure 103.

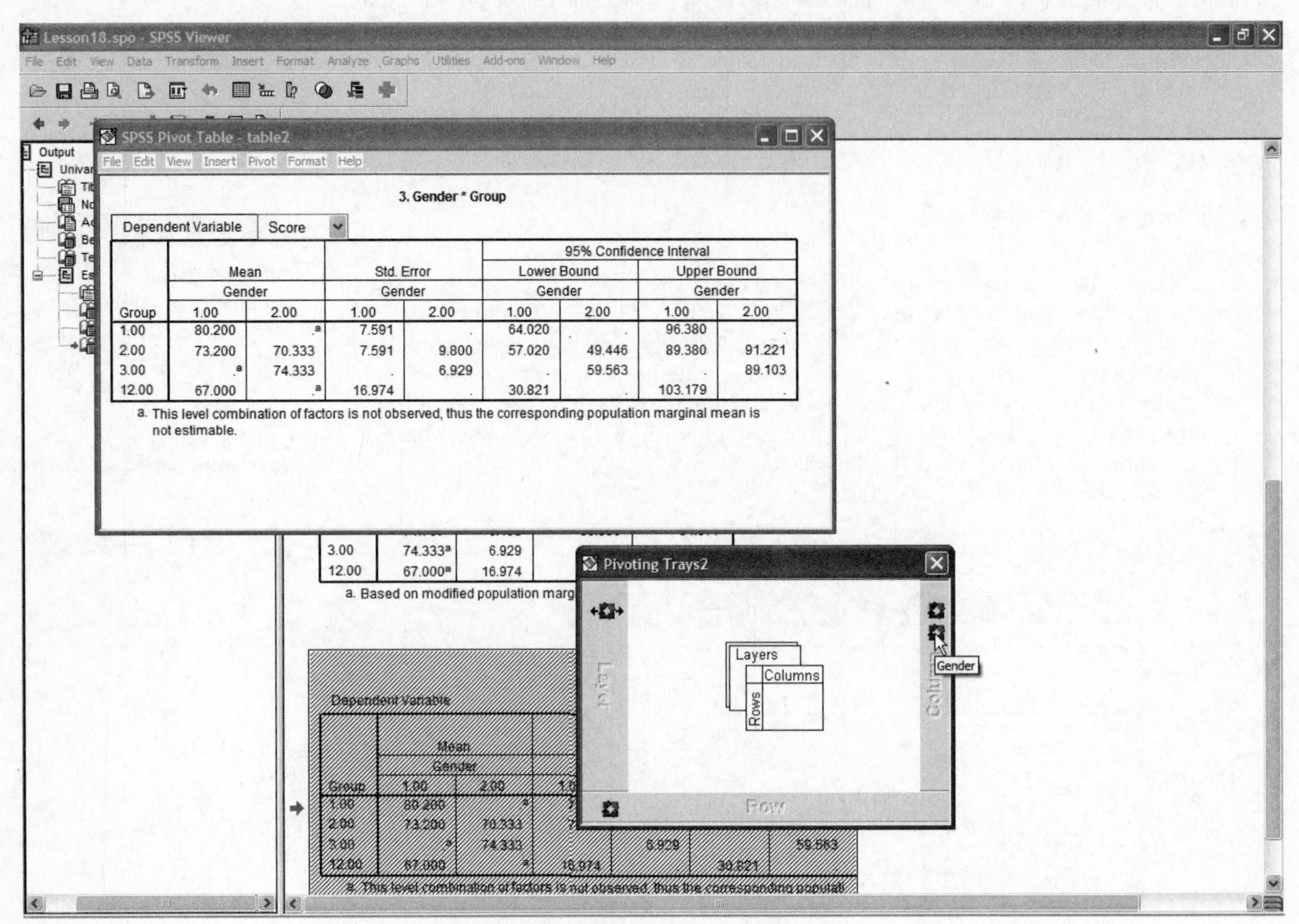

Group	Mean Gender 1.00	Mean Gender 2.00	Std. Error Gender 1.00	Std. Error Gender 2.00	95% CI Lower Bound Gender 1.00	95% CI Lower Bound Gender 2.00	95% CI Upper Bound Gender 1.00	95% CI Upper Bound Gender 2.00
1.00	80.200	.a	7.591	.	64.020	.	96.380	.
2.00	73.200	70.333	7.591	9.800	57.020	49.446	89.380	91.221
3.00	.a	74.333	.	6.929	.	59.563	.	89.103
12.00	67.000	.a	16.974	.	30.821	.	103.179	.

a. This level combination of factors is not observed, thus the corresponding population marginal mean is not estimable.

Figure 103. **Moving a row to a column.**

Once the pivot table is open (the pivot trays are a separate element so they can be closed at any time by right-clicking on the tray window and clicking **Close**), you can also use the Pivot menu to do such things as

- Transpose rows and columns,
- Move layers to rows,
- Move layers to columns, and
- Reset the default settings (which were in effect when you first created the table).

Creating a pivot table is a fairly straightforward process. But, creating your first one and having it match the appearance you want (as far as the positioning of rows and columns) is a bit of trial and error until you get it to the way you want.

Changing Table Appearance

There are two ways to change the appearance of a table. The first is through the use of TableLooks, a set of preformatted table appearances. The second is through the use of Table properties. In both cases, the table needs to be in the edit mode.

Using TableLooks

TableLooks allows you to apply a selection of many different types of appearances to any table. Click **Format → TableLooks** and then select the table appearance that you want to be applied.

For example, in Figure 104, you can see how the Academic (narrow) option was selected from the menu and applied to the table you see in Figure 103.

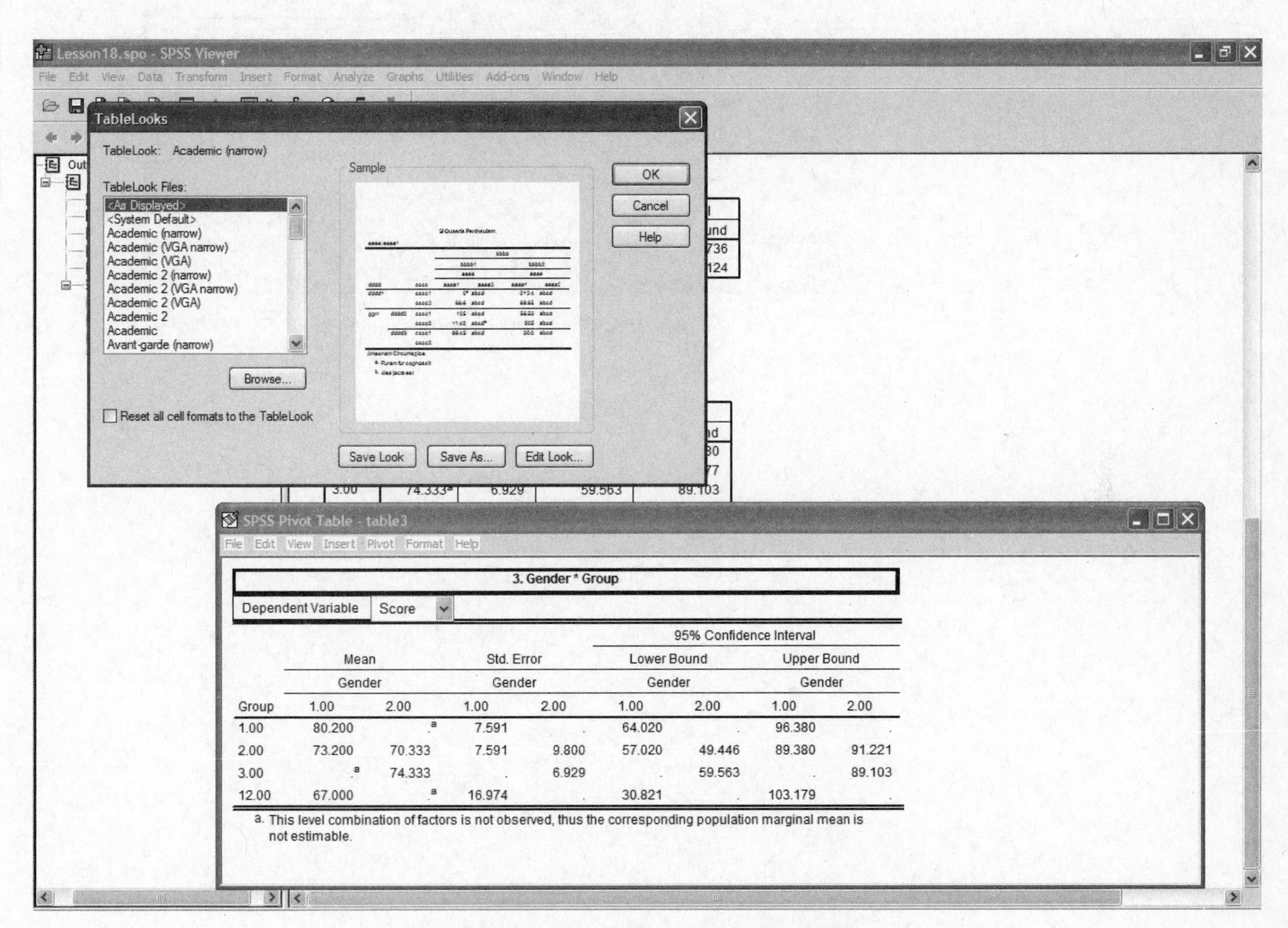

3. Gender * Group

Dependent Variable: Score

	Mean		Std. Error		95% Confidence Interval Lower Bound		95% Confidence Interval Upper Bound	
	Gender		Gender		Gender		Gender	
Group	1.00	2.00	1.00	2.00	1.00	2.00	1.00	2.00
1.00	80.200	.a	7.591	.	64.020	.	96.380	.
2.00	73.200	70.333	7.591	9.800	57.020	49.446	89.380	91.221
3.00	.a	74.333	.	6.929	.	59.563	.	89.103
12.00	67.000	.a	16.974	.	30.821	.	103.179	.

a. This level combination of factors is not observed, thus the corresponding population marginal mean is not estimable.

Figure 104. **Applying One TableLooks options.**

TableLooks has more than 50 different options in the TableLooks dialog box. But do beware, many of them are fancy and communicate more about the colors and the lines than the content.

Using Table Properties

The other way to change the appearance of a table is through the use of the Table Properties option on the Format menu. If you click **Format → Table Properties**, you will see the Table Properties dialog box, as shown in Figure 105, and how you can select any one of several tabs and make changes as you prefer.

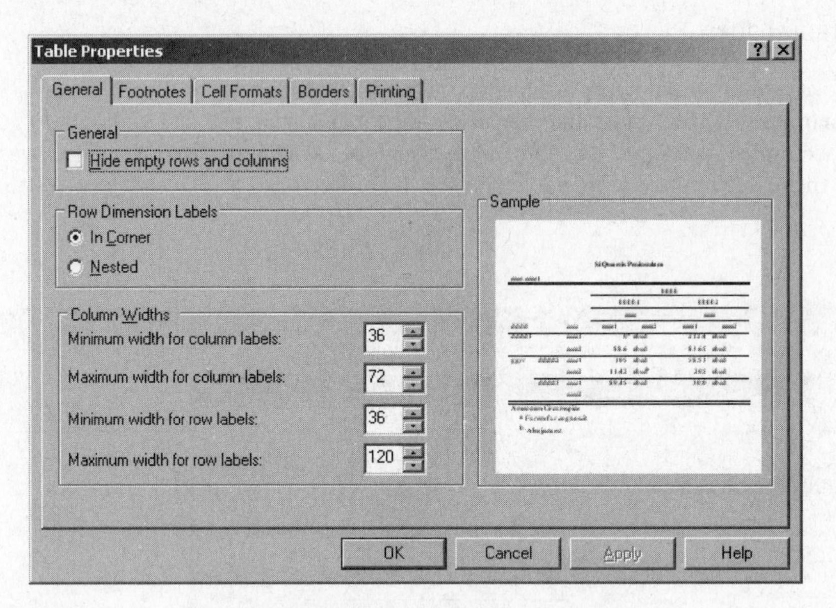

Figure 105. **The Table Properties dialog box.**

PART I

UNIT 4B

Working with SPSS Charts and Output for the Macintosh

A picture surely is worth a thousand words, and nowhere is this more true than with numbers. You could look at a set of data all day long and not really see what the nature of the relationship is between variables until a graph or a table is created. In this last part of introducing you to SPSS basics, we will go through the steps of creating a chart and modifying one, plus some tips on using the Output Navigator window.

This section of Part I deals with creating charts and output for the Macintosh version (13.x) of SPSS. It is very similar in both look and feel to the Windows version number 15.x. Note that the figures in this section are marked with the figure number followed by the letter *m* to indicate that they are for the Mac verison of SPSS. Also note as we pointed out in the introduction, SPSS for the Mac version 13 does not run on the Intel-based Macs.

LESSON

16B Creating an SPSS Chart

After This Lesson, You Will Know

- How to create a simple line chart
- About some of the different charts that you can create with SPSS

Key Words

- Area chart
- Bar chart
- Define Simple Line dialog box
- Line Charts dialog box
- Pie chart
- .spo

SPSS offers you just the features to create charts that bring the results of your analyses to life. In this lesson, we will go through the steps to create several different types of charts and provide examples of different charts. In Lesson 17B, we'll show you how to modify a chart by adding a chart title, adding labels to axes, modifying scales, and working with patterns, fonts, and more. One note of caution: Throughout this unit of Using SPSS, we'll use the words *graph* and *chart* interchangeably.

Creating a Simple Chart

The one thing that all charts have in common is that they are based on data. Although you may import data to create a chart, in this example we'll use the data you see here to create a line chart of test scores by grade. These data are available at http://www.prenhall.com/greensalkind in the file named Lesson 15 Data File 1.

Grade	Score
1	45
2	56
3	49
4	57
5	67
6	72

Creating a Line Chart

The steps for creating any chart are basically the same. You first enter the data you want to use in the chart, select the type of chart you want from the Graph menu, define how the chart should appear, and then click **OK**. Here are the steps we followed to create the chart you see in Figure 76m:

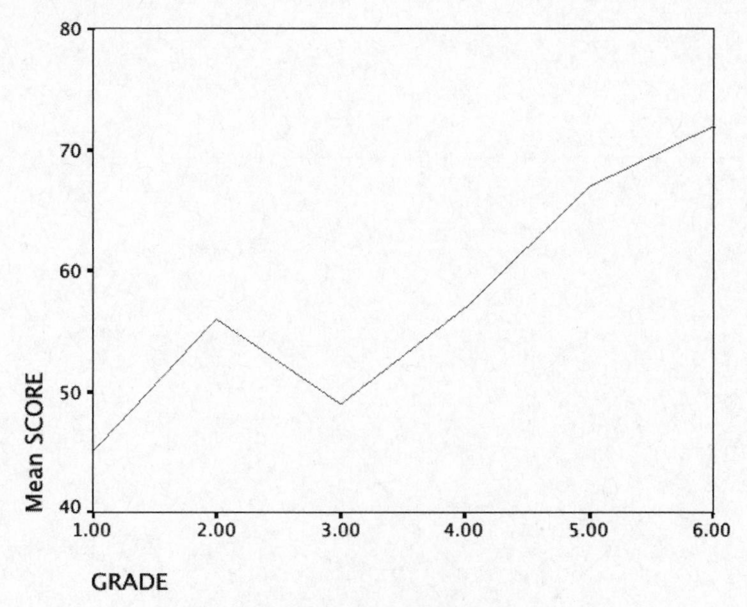

Figure 76m. **A simple line chart.**

1. Enter the data you want to use to create the chart.
2. Click **Graphs → Line**. When you do this, you will see the **Line Charts dialog box**, as shown in Figure 77m.

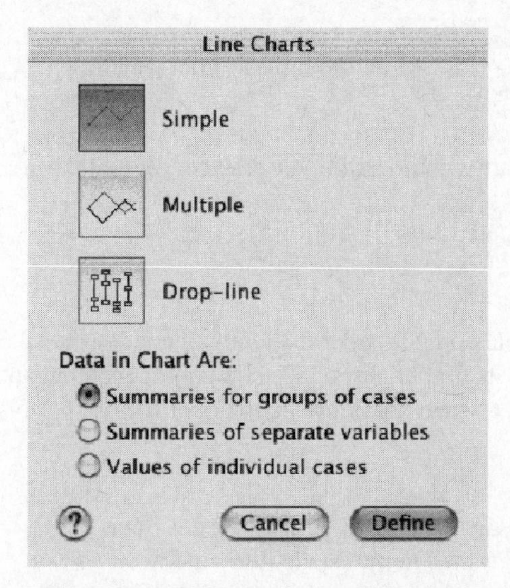

Figure 77m. **The Line Charts dialog box.**

3. Click **Simple**, click **Define**, and you will see the **Define, Simple Line dialog box**.
4. Click **grade**, click ▶ to move the variable to the Category Axis: area.
5. Click the **Other Summary Function** button.

6. Click **score**, and then click ▶ move the variable to the Variable: area.
7. Click **OK**, and you see the chart in the Output Navigator, as shown in Figure 78m.

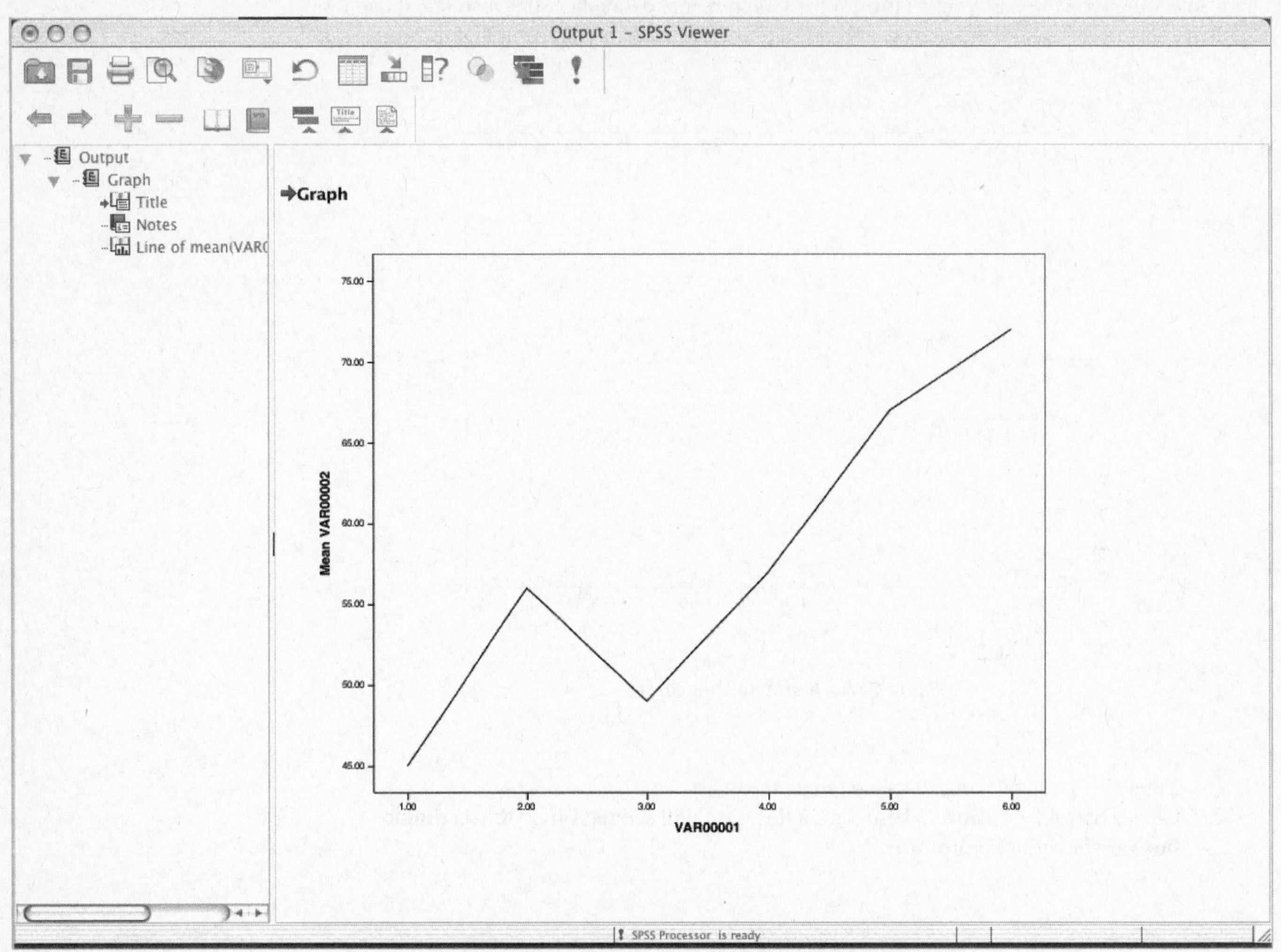

Figure 78m. **A simple line graph in an Output Navigator window.**

This is the default chart with no additions, changes, or edits.

Saving a Chart

A chart is only one possible component of the Output Navigator window. The chart is not a separate entity that stands by itself, but part of all the output generated and displayed in the Viewer. To save a chart, you have to save the entire contents of the Output Navigator. To do so, follow these steps:

1. In the Output Navigator window, click **File** ⟶ **Save**.
2. Provide a name for the Output Navigator window.
3. Click **OK**. The Output Navigator is saved under the name that you provide with an **.spo** extension (attached to all Viewer files). (See Figure 79m.) If you want to open this file later, you have to select this extension from the Files of type drop-down menu.

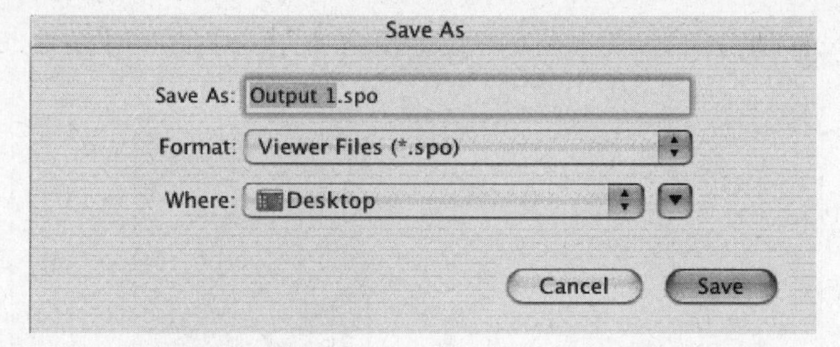

***Figure 79m.* Output is saved with the .spo extension as you can see in this dialog box.**

Printing a Chart

There are several ways to print a chart. The first way is to print from the Output Navigator by selecting the chart and then selecting Print.

1. Be sure that the chart you want to print is in the active Viewer window.
2. Click on the chart so that it is selected.
3. Click on the Print icon on the Toolbar or click **File → Print** and then **OK**. The chart will be printed.

Different SPSS Charts

SPSS for the Mac offers many different types of charts (all of which you can see samples of at GraphsGallery). Although some types of chart may be used in lieu of others, each has its own purpose. The purpose depends on the type of data being charted as well as the research question being asked. It's beyond the scope of this book to detail each chart and all the possible combinations (for help, just click on any icon in the GraphsGallery help area), but the following are examples of some of the simple charts. Keep in mind that there is an almost infinite number of variations of the charts you can create with SPSS. You'll learn about modifying charts in Lesson 17B. All sample charts that follow appear as they were first created, with no modification.

The Bar Chart

A **bar chart** represents values as separate bars with each bar corresponding to the value in the Data Editor. Bar charts are often used because they are easy to understand and require little attention to detail. In Figure 80m (on page 108), you can see an example of a bar chart of the number of males and females.

The Area Chart

An **area chart** represents the proportions of the whole that each data point contributes. In Figure 81m (on page 108), you can see an example of average product preference as a function of earning power (high, moderate, and low).

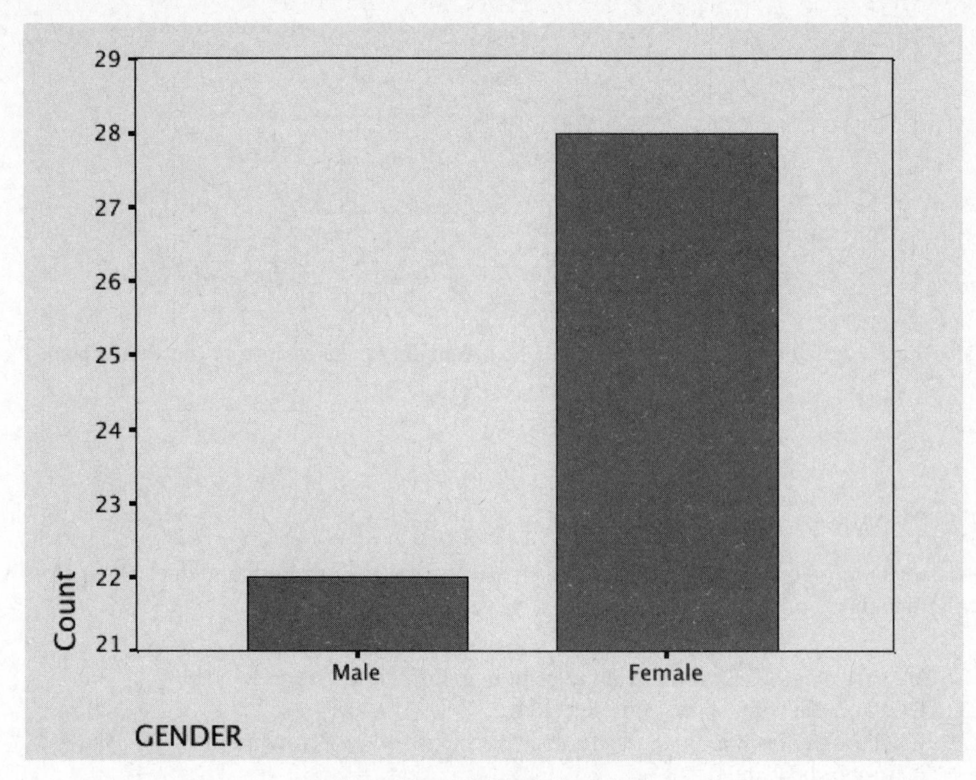

Figure 80m. **A simple bar chart.**

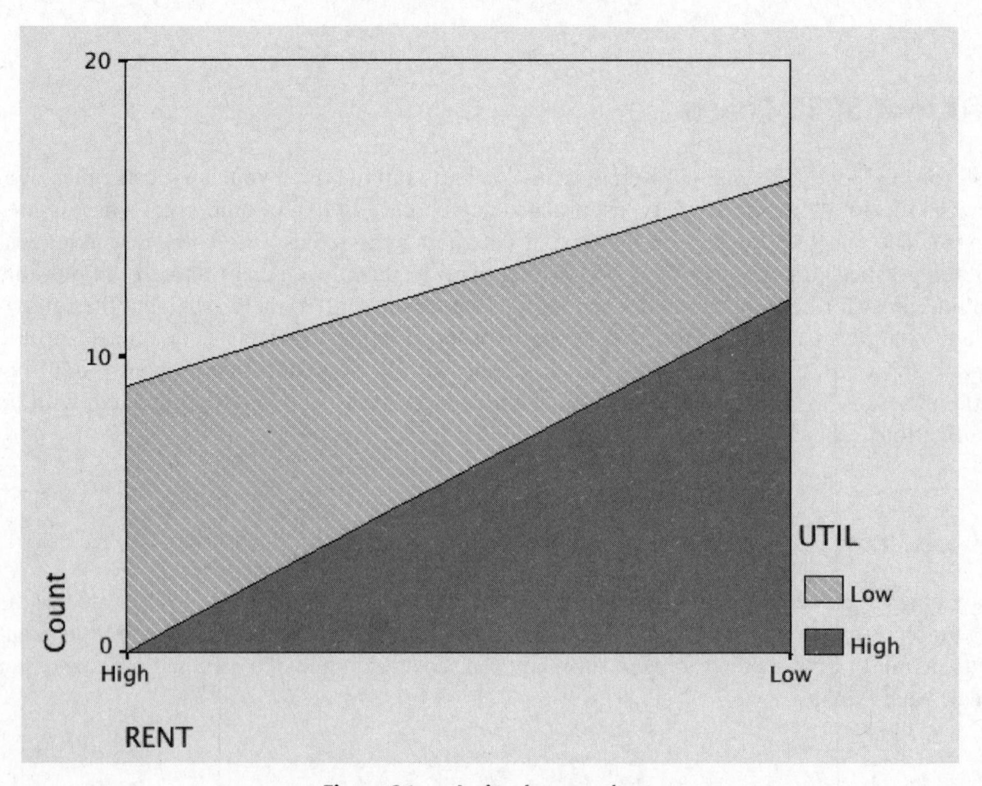

Figure 81m. **A simple area chart.**

The Pie Chart

The **pie chart** is a circle divided into wedges, with each wedge representing a proportion of the whole. Pie charts provide a clear picture of the relative contribution that a data point makes to the overall body of information. In Figure 82m, you can see a simple pie chart for expenses (rent, insurance, and miscellaneous).

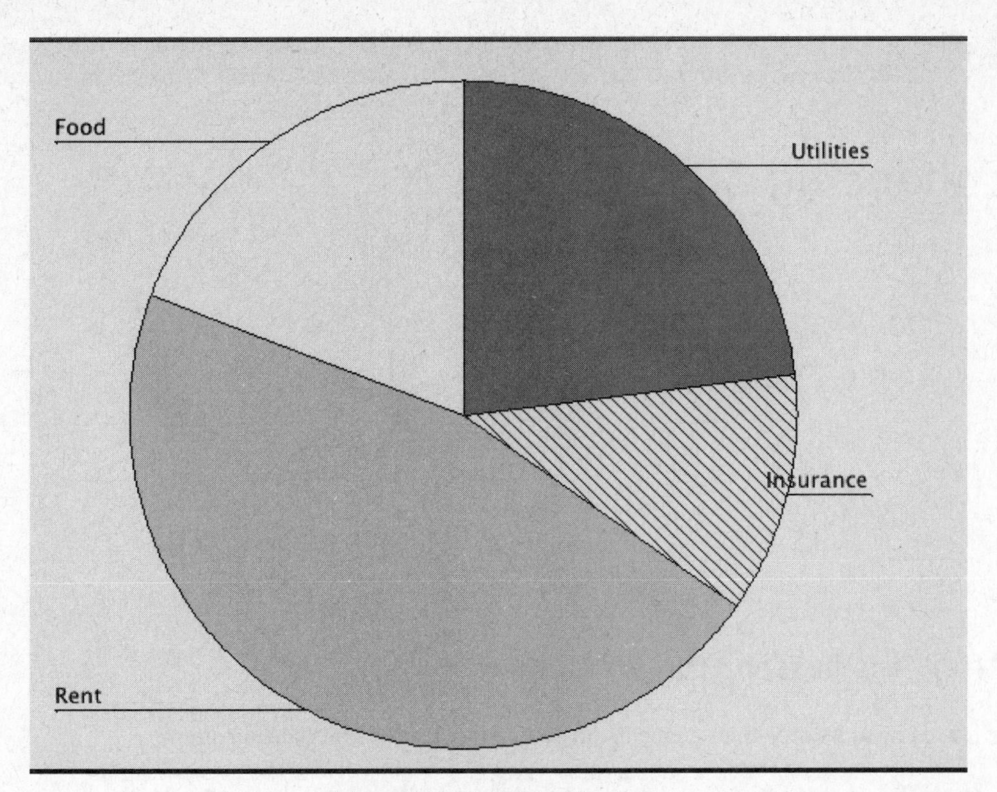

***Figure 82m.* A simple pie chart.**

Keep in mind that when you first create a chart, all the SPSS defaults are used. You can easily change type size and font and many other characteristics. You will learn how to do that in the next lesson.

LESSON

17B Enhancing SPSS Charts

After This Lesson, You Will Know

- How to modify chart elements, including titles, axes, colors, and patterns

Key Words

- Category Axis
- Scale Axis
- SPSS Chart Editor

A picture might be worth a thousand words, but if your pictures or charts don't say what you want, what good are they? Once you create a chart, as we showed you in the last lesson, you can finish the job by editing the chart to reflect exactly your meaning. Color, shapes, scales, fonts, text, and more can be altered, and that is what you'll do in this lesson. We'll be working with the line chart that was first shown to you in Figure 76m (on page 105). Only in this case, we've made all the modifications.

Modifying a Chart

The first step in the modification of any chart is to double-click on the chart in the right pane of the Output Navigator to access the **SPSS Chart Editor** and then click the **maximize** button on the Application title bar. As you can see in Figure 83m (on page 111), there is a Chart toolbar containing a set of buttons across the top of the chart. Place the mouse cursor over a button to learn what it does.

One cautionary note before we begin. The most basic reason for modifying a chart should be to increase the chart's ability to communicate your message. If you get caught up in Chart Junk (a close cousin to Font Junk), you'll have a mess on your hands. Modify to make better, not to show your reader that you know how to use SPSS.

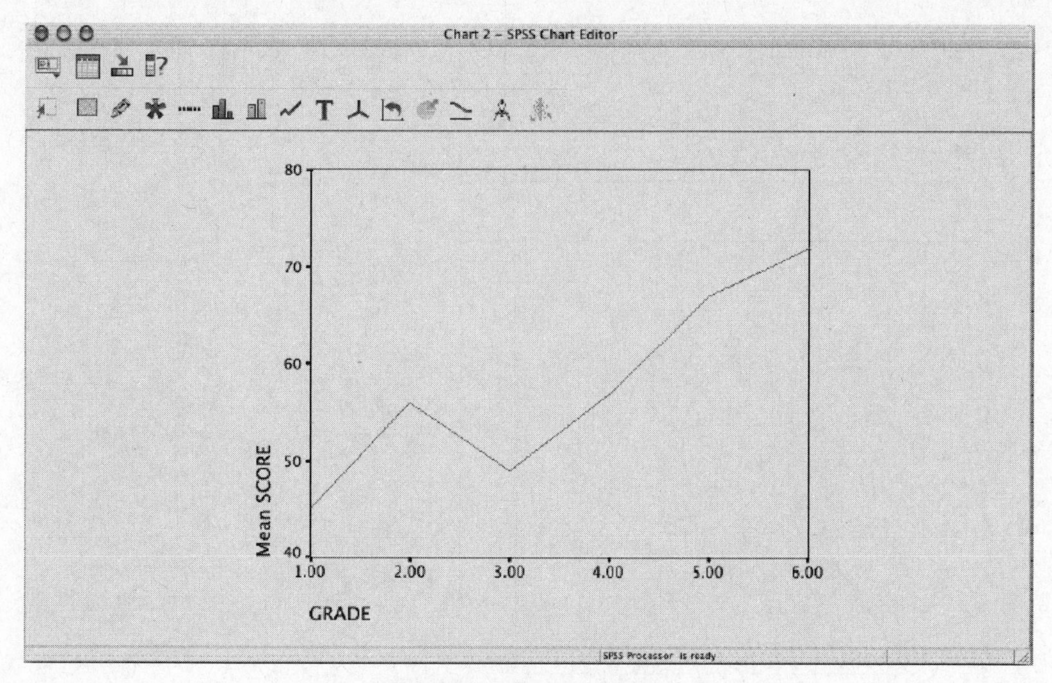

Figure 83m. **The Chart Editor.**

One instructional note before we begin. The general strategy for changing any element in a chart is to select the chart element and in the Properties dialog box (don't worry, SPSS knows what element you are clicking on) make the change. You select a chart element much as you select any Macintosh element, by clicking on it. If you single-click, you then must select the menu command for the item you want to change. If you double-click, in some cases you will go immediately to the dialog box in which you can make the change.

TIP

A definitely forward-thinking feature of SPSS is that dialog boxes often can stay open as you apply changes so that you can immediately see the effect of the change. You can also make several changes without having to open and close the dialog box each time.

Working with Titles and Creating Subtitles

Our first task is to enter the title and subtitle for the chart you see in Figure 76m (on page 105).

1. With the Chart Editor open, click the **Add Title** button on the toolbar and you will see the Properties box as shown along with the Chart Editor in Figure 84m (on page 112), and alone, in Figure 85m (on page 113), through which you can add a title and make the following changes:

   - The size of the chart
   - The font, style, and other text alterations
   - Select from various border and fill options, and
   - Work with the variables in the chart.

Once you have clicked on the Add a Title tool in the Chart Editor, then edit the Title text as you would any other text—click on it and make the necessary changes. If you want to add a subtitle, just add another title using the same button. And using the Propoerties dialog box you saw in Figure 84m, you can change any element you see fit.

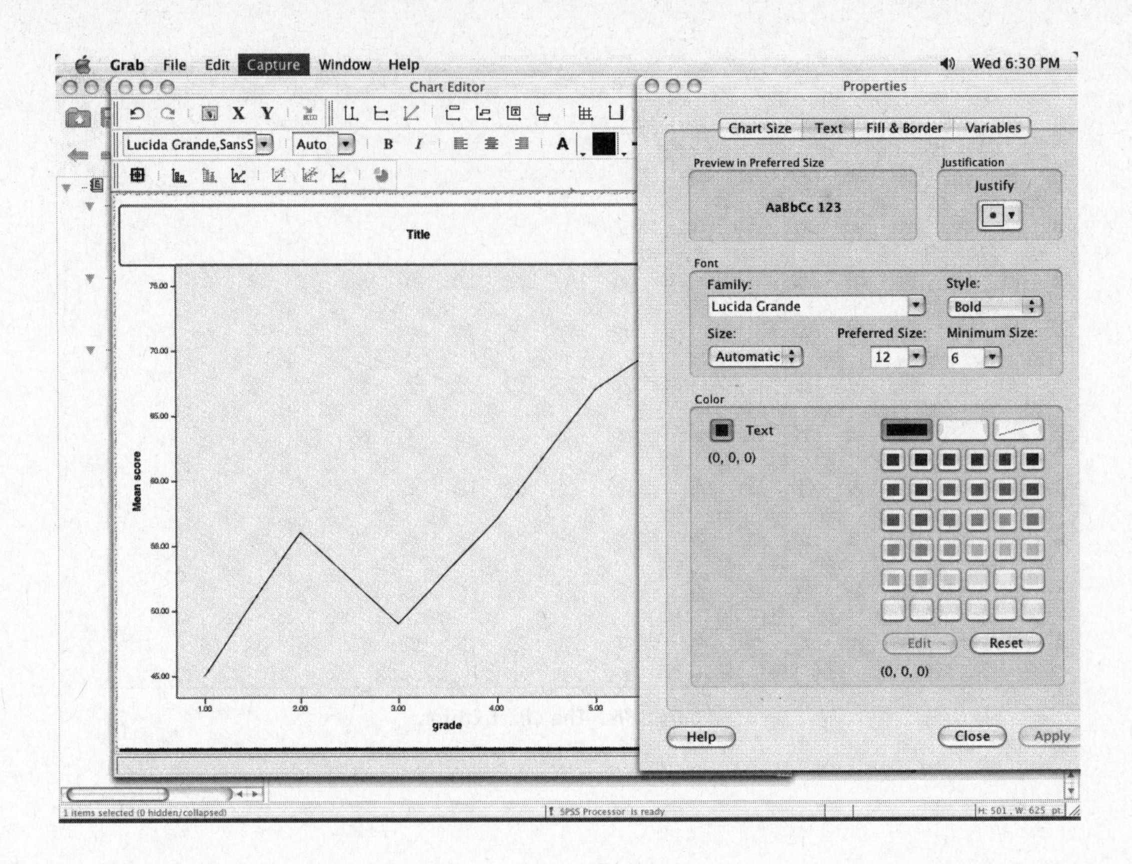

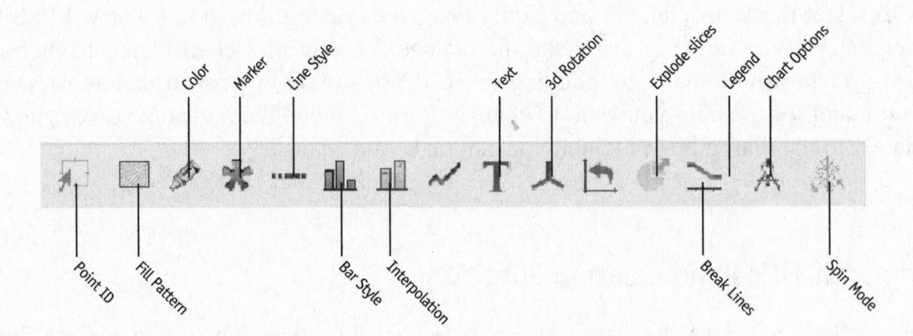

***Figure 84m.* The Chart Editor tools and what they do.**

Working with Frames

SPSS, by default, places a frame around the top and right side of the chart, an inside frame. And if you want, it can also place a frame around the entire chart, called an outside frame. To include or exclude an inner and outer frame, select the option (Inner Frame or Outer Frame or both) from the Chart menu. In our graph, the graph's inner frame was removed, making for a less cluttered appearance.

Working with Axes

The *X*- and *Y*-axes provide the calibration for the independent (usually the *X*-axis) variable and the dependent (usually the *Y*-axis) variable. SPSS names the *Y*-axis the **Scale Axis** and the *X*-axis the **Category Axis**. Each of these axes can be modified in a variety of ways. To modify either axis, double-click on the title of the axis and you will see one of the Properties dialog

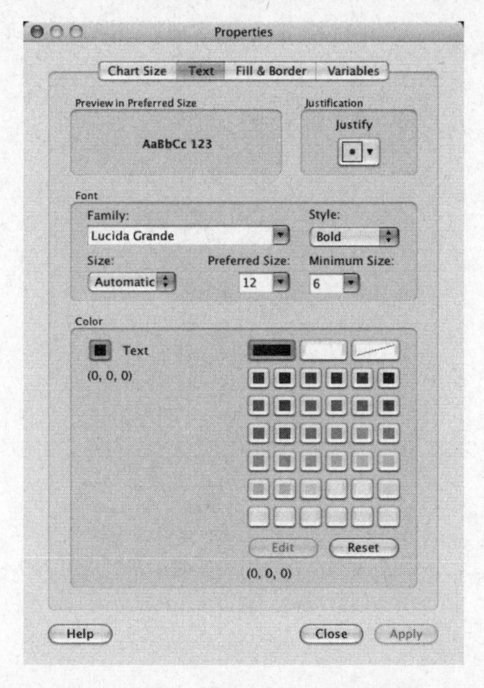

Figure 85m. **The Properties dialog box with the Text tab selected.**

boxes, as shown in Figure 86m. In this example, you can see how the Labels & Ticks tab is visible and changes in whether labels and tick are displayed and where ticks appear can be easily made and applied.

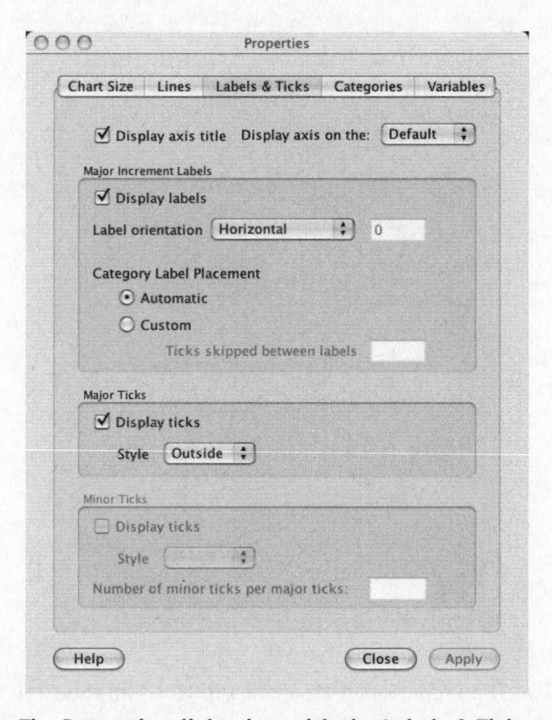

Figure 86m. **The Properties dialog box with the Labels & Ticks tab selected.**

In Figure 87m (on page 114), you can see how the lines associated with a chart can be modified.

Once again, the element that you want modified is double-clicked, producing the Properties dialog box you see in Figure 87m. Here, the weight, style, and end caps of the line can be easily modified. The color of the line can be seen as well.

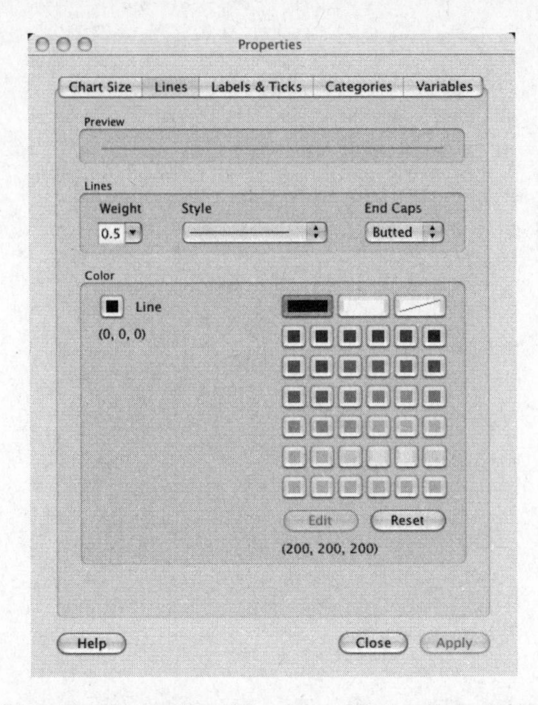

Figure 87m. **The Properties dialog box with the Lines tab selected.**

Another tab, shown in Figure 88m, can be used to modify the titles of the *X*- and *Y*-axes. Click on the axis title of choice and enter a new title, then click **Apply**.

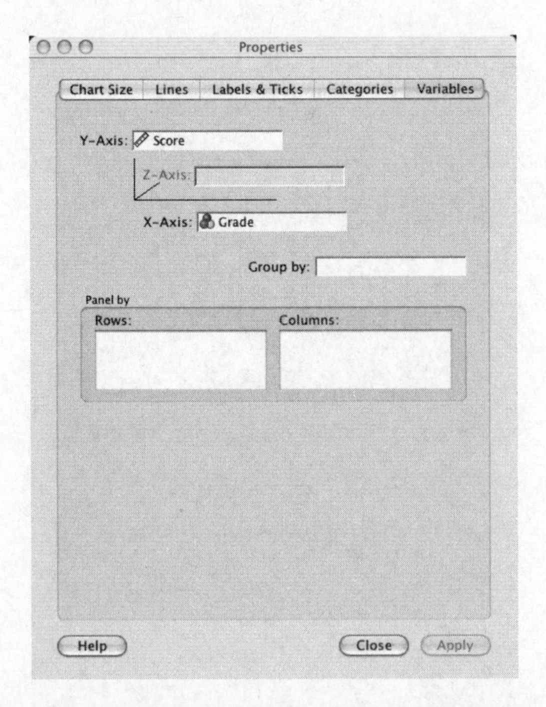

Figure 88m. **The Properties dialog box with the Variables tab selected.**

Working with Patterns and Colors

Time to have some fun (and be practical besides). When SPSS creates a chart, and if you have a color monitor, it assigns different colors or symbols to different elements in a chart. We'll use two tools, one to work with patterns and one to work with colors. We'll use a simple bar chart, as shown in Figure 89m (on page 115), as an example.

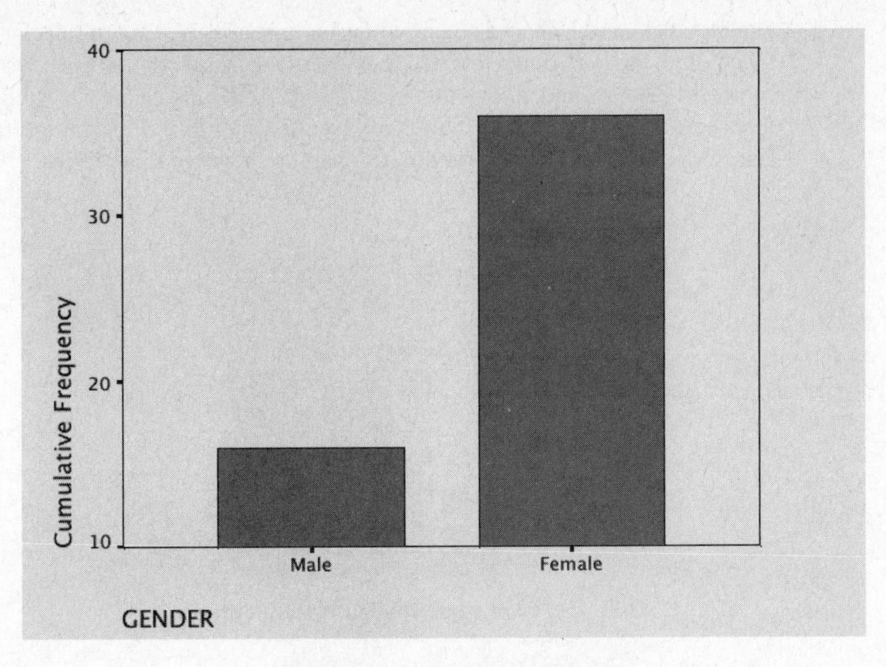

Figure 89m. **A simple bar chart.**

Changing Patterns

To change the patterns of any element in a chart, follow these steps:

1. We'll work with the chart shown in Figure 90m. Double-click to open the Chart Editor.

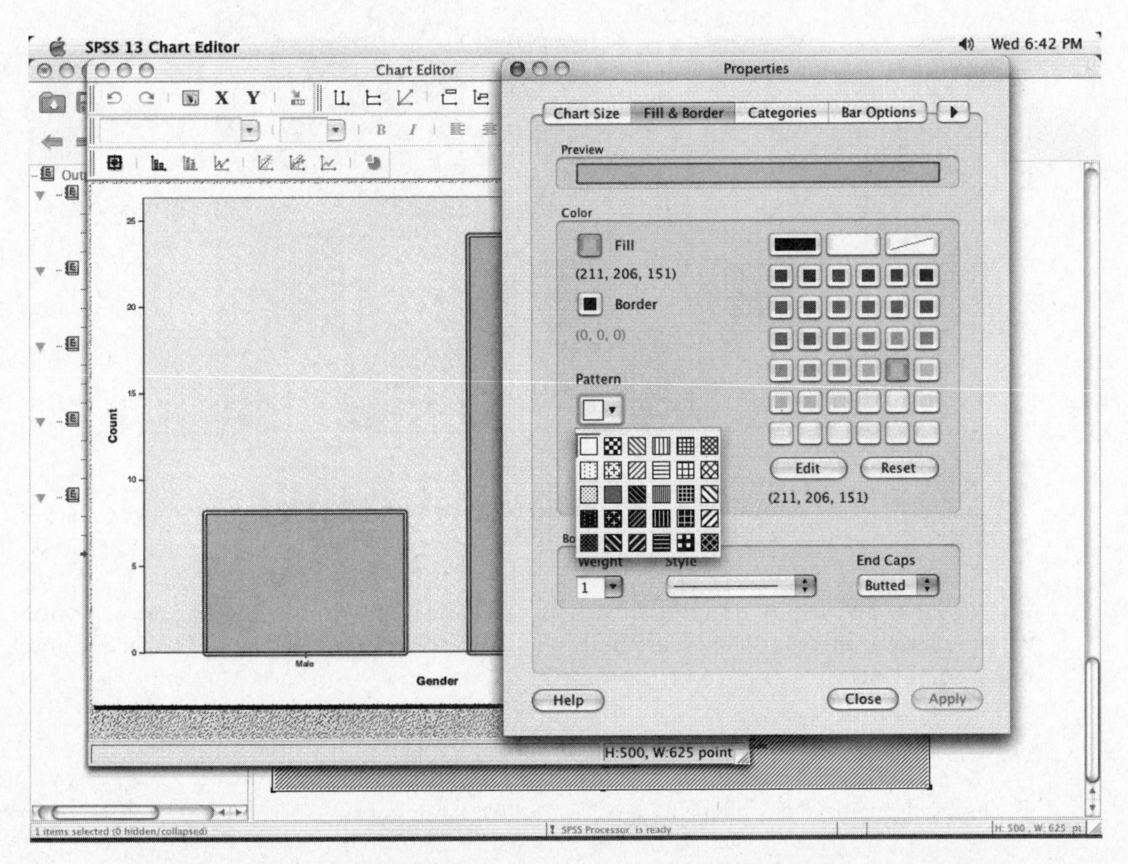

Figure 90m. **The Properties dialog box for modifying the Fill & Border of a bar chart.**

2. Double-click on the chart element you want to select. When you select a chart element, of all the occurrences of that element are selected. In this case, we will click on one of the bars and all the bars are selected. You will see the Chart Editor for that chart and the Properties dialog box for the Chart Size, Fill & Border, Categories and Bar Options for the bar chart as shown in Figure 91m.

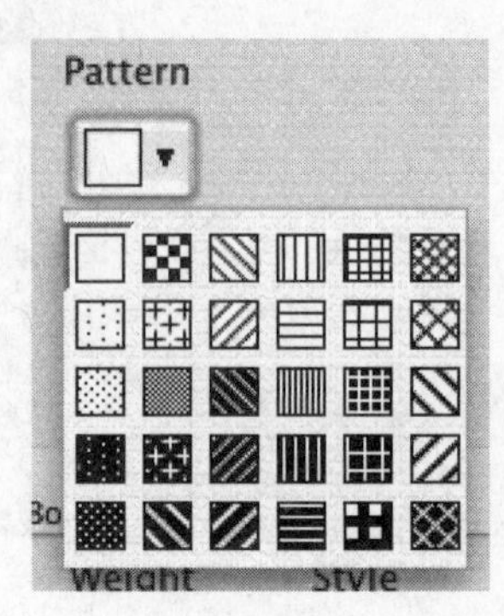

Figure 91m. **The Fill Pattern options.**

The pattern changes, as you see in Figure 92m.

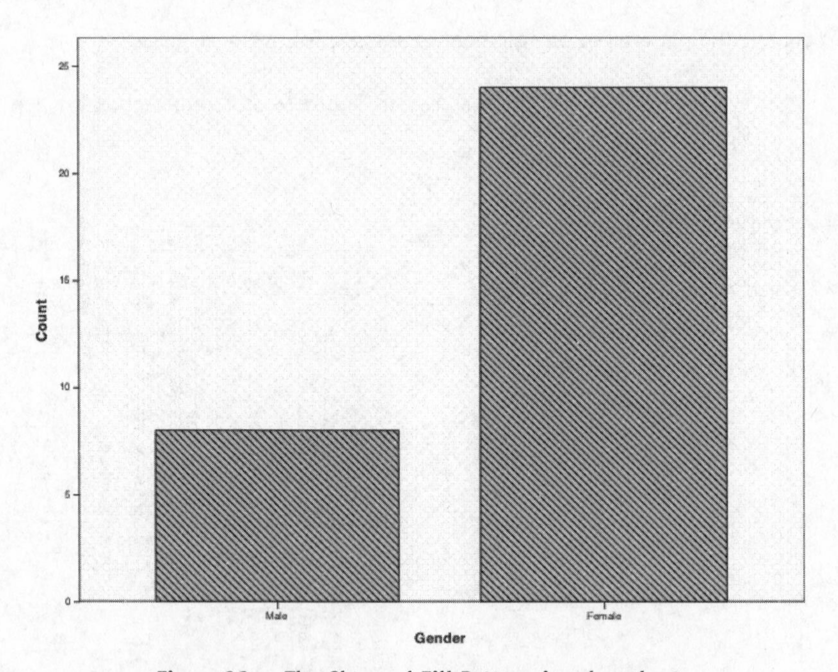

Figure 92m. **The Changed Fill Pattern in a bar chart.**

Now that the patterns are straight, we have one more step. We will eliminate the color from the bars, since printing in color (your printer will see the colors as being black or a shade of gray) uses a lot of toner.

Changing Colors

To change the color of any element in a chart, follow these steps:

1. Click on the chart element you want to select if it is not already selected.

2. In the Properties dialog box as you saw in Figure 90m, focus on the area that allows you to modify the color (on the Fill & Border tab) as you see in Figure 93m.

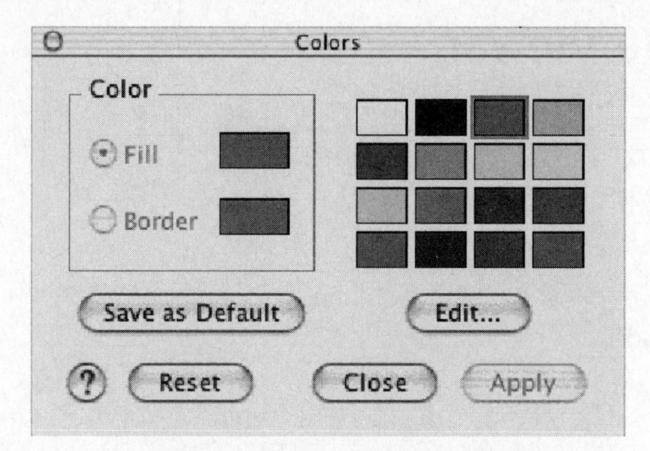

Figure 93m. **The Fill & Border color options.**

As you can see in Figure 93m, you can select a color and then click either Fill or Border to make the changes you want.

LESSON

18B Using the Viewer

After This Lesson, You Will Know

- The use of the two panes in the Viewer
- How to save output
- How to show and hide results in the Viewer window
- How to print output
- How to delete output
- How to move output

Key Words

- Contents pane
- Outline pane

When you complete any type of procedure in SPSS, be it the creation of a simple chart or the results of a complex analysis, the output appears in the Viewer. The Viewer is a separate SPSS screen that lists all operations performed on a set of data in the order in which they were completed.

For example, in Figure 98m, you can see that two different types of analysis were done; the first is a bar chart (graph) and the second is a simple description of the data (descriptives).

As you can see, the screen has an .spo extension—the extension given for all files that contain output. Unless named by you, the names automatically assigned are Output1, Output2, etc.

The Viewer consists of two panes. The **Outline pane** on the left side of the Viewer and the **Results pane** on the right side of the Viewer.

The Outline pane lists in outline form all the analyses that have been completed during the current session. The Results pane contains the results of the analyses. For example, in Figure 98m, you can see an outline in the left pane of what is contained in the right pane. In the right Results pane, you find labels for the output along with the corresponding graphs, descriptives, etc. In any Outline pane, you will find labels (such as Graph, Title, Notes, and then a description of the specific graph that was created). For now, we'll show you how to save Viewer output, print it, delete elements in the Viewer, and move the elements around so the output best fits your needs.

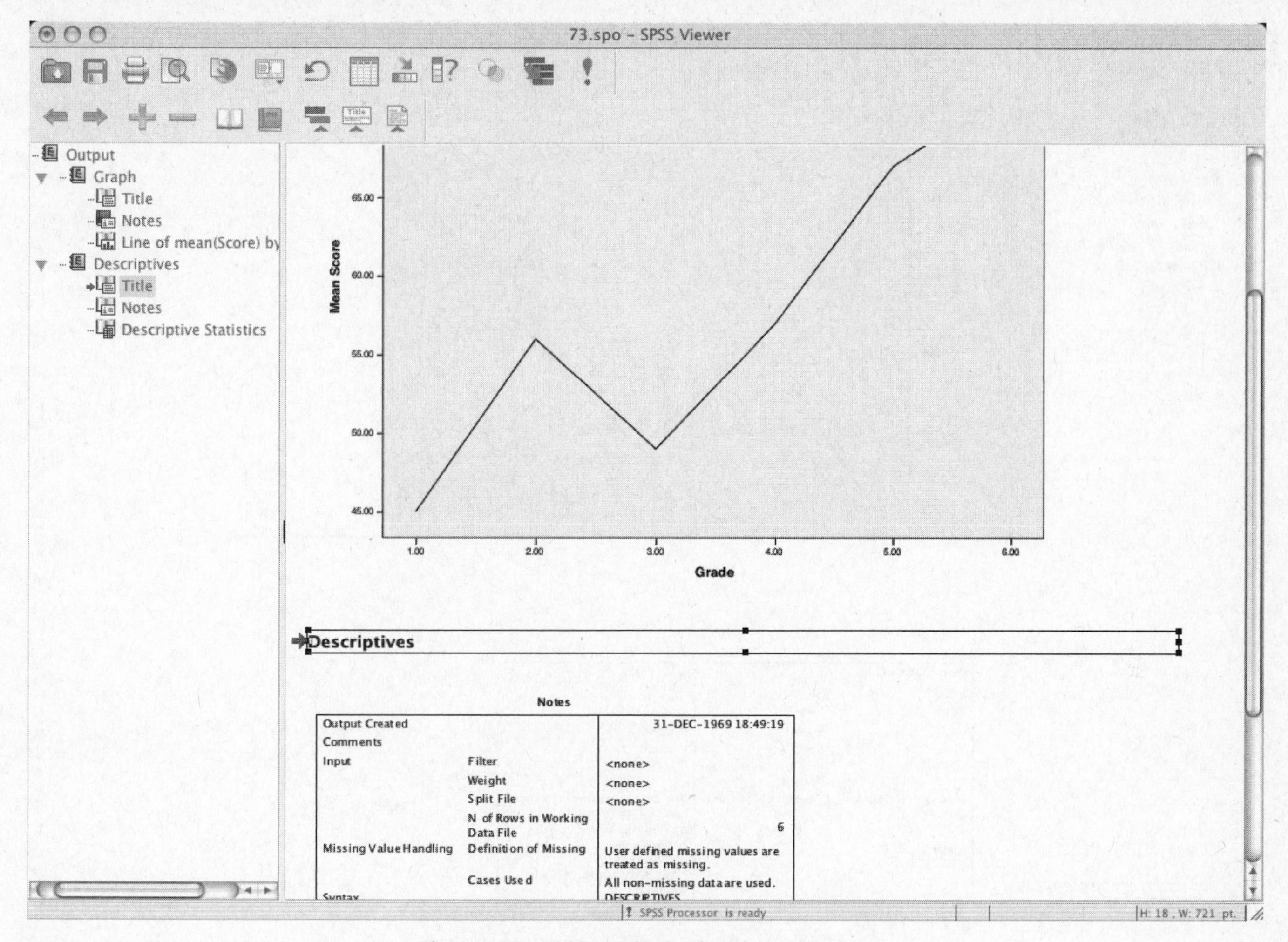

Figure 98m. **SPSS results in the Viewer window.**

Saving Viewer Output

Once output is generated, you will probably want to save it for later use or even add it to another work session. To save output from the Viewer, follow these steps:

1. Click **File ⟶ Save**.
2. Provide a name for the output.
3. Click **OK**. The output will be saved using the file name that you provided, along with the .spo extension. When you want to open the Viewer you saved, you will have to specify the .spo extension in the File Type drop-down menu in the SPSS Open dialog box.

To Selectively Show and Hide Results

You can choose to show or hide any results that appear in the Viewer window. You may generate results that you want to hide while you print out others, a convenient way to be selective as you focus your discussion of the results.

To hide a table or chart without deleting it, follow these steps:

1. Click on the item in the Outline pane.
2. Click **View ⟶ Hide**.

TIP

The red arrow in the Outline or Results pane indicates which element in the Viewer is active.

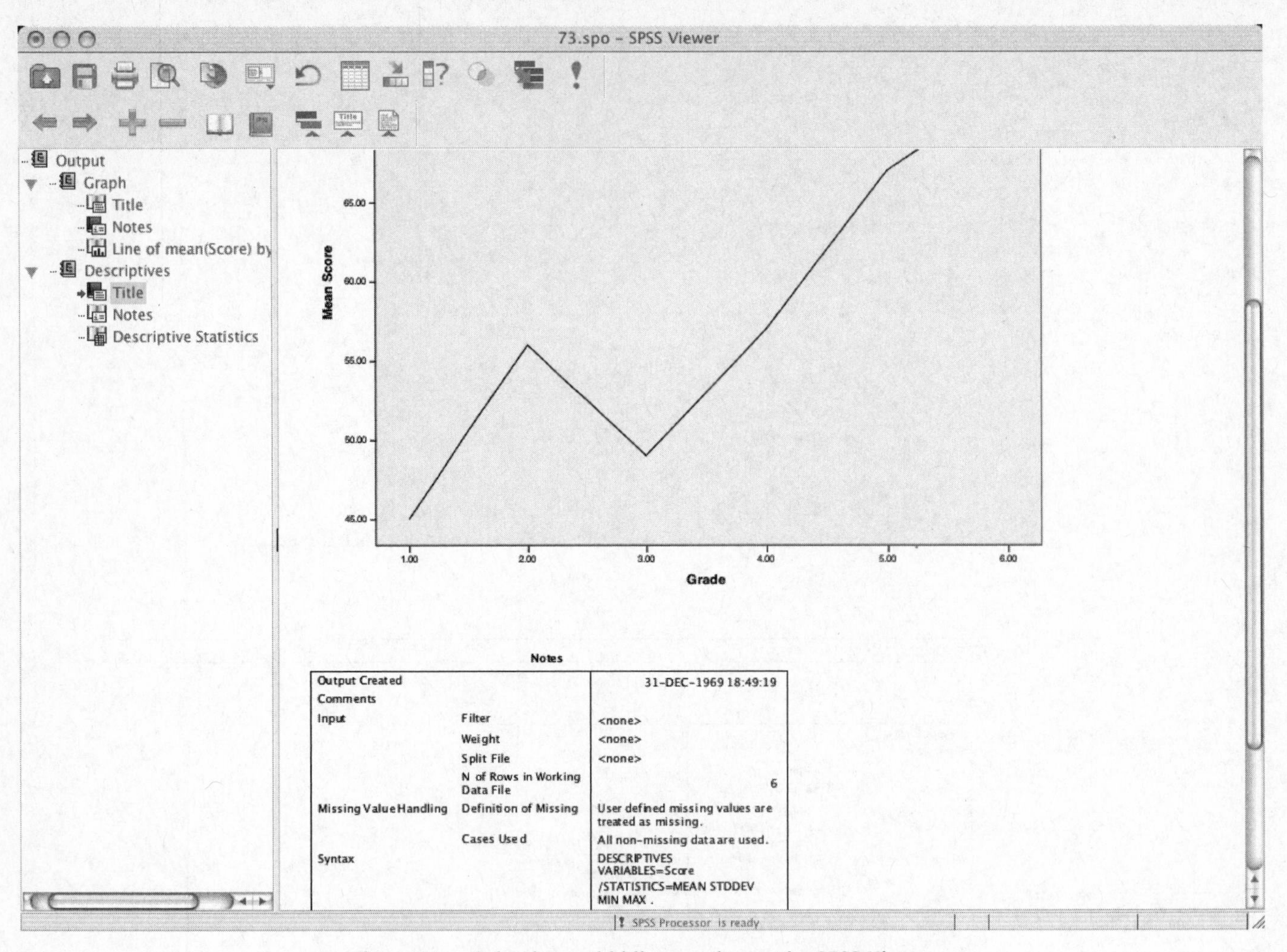

Figure 99m. **Selecting and hiding results on the SPSS Viewer.**

For example, in Figure 99m, you will see that the title of the procedure Descriptives (which appears in Figure 98m on page 119) in the Outline pane was hidden in the Results pane. Although it is not deleted, it no longer appears in the Results pane. But, as you can see in Figure 99m, it still appears in the Outline pane of the Viewer window.

To show a table or chart once it has been hidden, follow these steps:

1. Click on the item icon in the Outline pane. If you click on the item name, then you will simply highlight the name and the Show option on the View menu will not be active.
2. Click **View ⟶ Show**. The item reappears in the **Contents pane**.

You'll want to hide and show output as you decide what you want to appear in a hard copy of the Viewer.

Printing the Contents of the Viewer Window

To print the entire contents of the Viewer, follow these steps:

1. Click **File ⟶ Print**.
2. Click **OK**. All the tables and charts in the Viewer will be printed.

Printing a Selection from the Viewer Window

To print a specific selection from the Viewer window, such as a table or a chart, follow these steps:

1. Click on the table, chart, or both that you want to print in the Viewer window. You can select more than one element by holding down the Shift or Ctrl key when you click each element.
2. Click **File → Print**.
3. Click **Selection → OK**. The element or elements you selected will be printed.

Any element surrounded by a black line is selected and will print. You can select as many elements as appear in the Contents pane to print at once.

> **TIP**
>
> If you select an item in the Results pane and press the Del key, the item will be deleted and can be recovered using the Option + Z key combination.

Deleting Output

To delete output from the Viewer, follow these steps:

1. Click on the output you want deleted, either in the Outline or Contents pane of the Viewer window.
2. Click **Edit → Delete**, or press the **Del** key.

Moving Output

Although the results in the Viewer window appear in the order they were created, you can change the order in which they appear. To change the order, do the following.

In the Outline pane of the Viewer, drag the icon (not the name) representing the output to its new location.

For example, in Figure 100m, you can see that the element titled Descriptive Statistics is moved to the place above the element titled Notes and below Title.

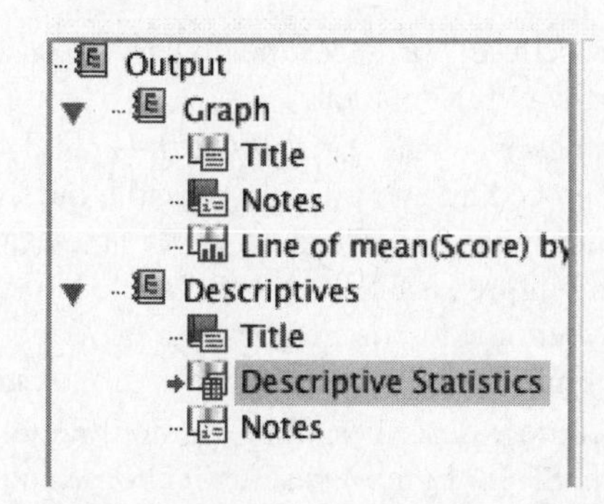

Figure 100m. **Moving output to another location.**

PART II Working with SPSS Procedures

UNIT 5 Creating Variables and Computing Descriptive Statistics

In Part I of the book, you learned how to use the program to perform basic data entry and manipulation, create and edit graphs, and create and work with tables. Part II of the book moves beyond these basic techniques into discussions about many of the statistical procedures that are available in SPSS.

Part II consists of 6 units involving 27 lessons. Each lesson presents a topic—from simple analyses, such as descriptive statistics for quantitative variables and one-sample *t* tests, to more complex analyses, such as two-way analysis of variance, multiple regression, and factor analysis. It should be noted that the terms used by SPSS for various types of analyses might be different from the terms used in many textbooks and classes. For example, the statistical test for the difference between two related means is often called a *matched-samples t test* or a *correlated-samples t test*. SPSS calls it a *paired-samples t test*.

Lessons are divided into several major sections that show you when and how to apply the different statistical procedures and how to analyze and interpret their results. Lessons may not include all nine of these sections.

- Each lesson begins with a section that introduces the statistical procedure and describes how to structure the SPSS data file to conduct the procedure.
- "Applying the [Procedure]" illustrates the research designs and situations associated with the procedure.
- "Understanding the [Procedure]" provides the underlying assumptions of the procedure and additional pertinent information.
- "The Data Set" section describes the variables in the application data set used to illustrate the statistical procedure and presents research questions that will be addressed by the statistical analyses. The data sets for the examples are on the Web at http://www.prenhall.com/greensalkind and are labeled in the general format of "Lesson? Data File?" where the lesson number is substituted for the first question mark and the data set number is substituted for the second question mark.
- "Conducting the [Procedure]" describes all of the steps required to conduct the statistical analysis by using SPSS. In most cases, we choose analysis options that have relatively broad applicability. However, some options may need to be modified to meet the needs of the data analysis. We also present selected SPSS output and discussion of the output for each analysis in this section. Because it is not necessary to view all of the SPSS output to understand the procedure being discussed, we present abridged output for many analyses.
- "Using SPSS Graphs to Display the Results" demonstrates various graphical techniques that might be used in reporting results. Each graph we present has been edited to be consistent with the recommendations in the *Publication Manual of the American Psychological Association*. Therefore, the graph you see in your output may look somewhat different from the graph we present in

the book. We present graphs in this form so you can see how to present graphs for publication.

- The purpose of the next section, either "Writing an APA Results Section" or "Writing an APA Participants Section," is to provide actual models of Results and Participants sections in APA manuscript format. Because of the simplified nature of the research questions and analyses, the APA results sections are much briefer and less complex than those published in research journals.
- "Alternative Analyses" describes related statistical procedures that may be used instead of the particular procedure described in the lesson.
- Homework problems are given in the "Exercise" section. The data needed to complete the exercises at the end of each lesson are in the data files on the Web at http://www.prenhall.com/greensalkind.

Many statistical procedures in Part II are extremely complex, and we cannot provide the in-depth understanding that a textbook or a statistics class can provide. We strongly recommend that you consult textbooks on any statistical procedures you are unfamiliar with before conducting analyses with those procedures.

The steps required to conduct analyses in Part II of our book are similar with SPSS for Windows 15 and SPSS for Macintosh 13, although the dialog boxes for analyses have somewhat different appearances across platforms. In addition, the outputs from analyses are essentially comparable across platforms.

The way we perform analyses do differ across platforms in a few ways. One place they differ is in the selection of multiple variables for your analyses. For SPSS for Windows, you hold down the Ctrl key and click on the variables of interest. In contrast, for SPSS for Macintosh, you hold down the Apple key and click on the variables of interest. Another place where the two platforms differ is in the creation of graphs. With SPSS for Windows 15, you will click on **Graphs**, click on **Legacy Dialogs**, and click on your choice of graphs. In contrast with SPSS for Macintosh 13, you will click on **Graphs** and then click on your choice of graphs. Finally, if you work with SPSS for Macintosh 13, the editing features for graphs are somewhat less flexible than they are with SPSS for Windows 15.

In Part II of our book, we present stepwise instructions, dialog boxes, and output for the Windows platform, but describe differences with the Macintosh platform where we believe them to be important ones. We made this decision because the two platforms function similarly when conducting analyses in Part II of the book. The detailed inclusion of materials from both platforms would have made the book fragmented and would have detracted from the readability of the text. We believe that SPSS for Macintosh will be modified in future releases to make it even more comparable to SPSS for Windows.

When conducting your analyses in Part II you may find some differences between what you see in SPSS and what we have displayed in our book. SPSS allows you to customize the look of SPSS. Conduct the following steps to make the look of your SPSS comparable to the SPSS images presented in Part II of the book.

1. Click **Edit** from the main menu and click **Options**. (SPSS for Macintosh 13: Click **SPSS 13** from the main menu and click **Preferences.)**
2. Under the General tab in the Variable Lists area, click on the **Display Names** option and the **File** option. (SPSS for Macintosh 13: Under the General tab in the Variable Lists area, click on the **Names** option next to Display and on the **File** option next to Sorting.)
3. Under the General tab, click on the box next to **No scientific notation for small numbers in tables** under Output.
4. Click **OK** in the dialog box.
5. When viewing the SPSS Data Editor, click **View** from the main menu and click on **Value labels** so that no check appears next to this option.

Many analyses in the lessons require multiple tests of hypotheses. In these lessons, we frequently use one of three methods to control for Type I errors across the multiple tests. A step-by-step guide for each of these three methods is described in Appendix B.

In Unit 5, the first unit of Part II, we discuss how to create variables and how to compute and interpret descriptive statistics. In the first lesson of Unit 5, we describe how to create new quantitative and qualitative variables from existing variables (e.g., overall scale scores from item scores). In the next lesson, we illustrate how to summarize the results on qualitative variables by using tabular and graphical displays. In the last lesson of the unit, we discuss how to describe the results on a quantitative variable, overall and within levels of a qualitative variable. We also explain how to determine percentile ranks associated with scores on a quantitative variable.

Because the objective of Unit 5 is to understand how to conduct descriptive statistics and excludes inferential techniques, we have developed applications aimed at describing research participants rather than answering research questions. We discuss applications for answering research questions in Units 6 through 10 of Part II.

LESSON

Creating Variables

19

Before conducting the statistical analyses described in later lessons, it is important to know how to create the variables that you might analyze. In this lesson, we will present methods for developing an overall scale from a set of quantitative variables that assess the same dimension. The quantitative variables that are combined together may have the same or different metrics. We will also learn how to create a qualitative variable from one or more quantitative variables.

Quantitative variables have the same metric if they are measured on the same response scale (e.g., items responded to on a 5-point scale with the same verbal anchors). In creating an overall scale from quantitative variables with the same metric, we may need initially to reverse-scale some variables. We reverse-scale variables that are in the opposite direction of the dimension that we are interested in assessing. For example, items that assess self-esteem may be worded to reflect high self-esteem, such as "I feel good about myself," or low self-esteem, such as "I do not think I am a good person." To create a meaningful overall self-esteem scale, we have to reverse-scale the item representing low self-esteem so that higher values on it reflect higher levels of self-esteem. After reverse-scaling variables that are in the opposite direction of the dimension of interest, we sum all variables, those that were reverse-scaled and those that were not. For our self-esteem example, we add together the reverse-scaled low self-esteem item with the high self-esteem item. We discuss in this lesson how to reverse-scale variables and how to combine variables to create overall scores with or without missing data.

A similar strategy is required for creating an overall scale from variables with different metrics. Because the variables have different metrics, we standardize them by converting them to *z* scores before creating an overall scale. For example, before summing across three different personality measures—a self-report measure, a behavioral observation, and a teacher rating—we convert the scores on the three measures to standardized scores. We explain in this lesson how to transform scores on variables with different metrics to *z* scores, how to reverse-scale *z*-scored variables, and how to combine them to create overall scores. Although we create overall scores by computing total scores, the techniques can be easily adapted to develop overall scores by calculating mean scores.

For some analyses, you may want to create a qualitative or categorical variable from one or more quantitative variables. We can divide a single quantitative variable into categories. For example, we might transform a quantitative variable measuring stress to a categorical variable with three categories: low, medium, and high stress. Qualitative variables may also be created from two or more quantitative variables. For example, we could construct a four-category variable from two quantitative variables—performance IQ and verbal IQ. The categories would be low performance and low verbal IQ, low performance and high verbal IQ, high performance and low verbal IQ, and high performance and high verbal IQ.

Applications for Creating Variables

We will demonstrate initially how to create overall scores from variables of particular types:

- Variables that have the same metric and require no reverse-scaling
- Variables that have the same metric and require reverse-scaling for one or more of the variables
- Variables that have different metrics and require no reverse-scaling
- Variables that have different metrics and require reverse-scaling for one or more of the variables
- Variables that have missing data

Next we will show how to create a qualitative variable from

- A single quantitative variable
- Multiple quantitative variables

We demonstrate all seven applications for creating new variables from existing ones with the following example study.

Casey gains the cooperation of 40 women college students to participate in his study. All participants gained at least 10 pounds during their first year in college. He is interested in their eating habits and their feelings about eating during their second year of college. Casey asks the students to complete three scales at the beginning of their second year: the Positive Body Image (PBI) scale, the Positive Attitude toward Eating (ATE) scale, and the Overeating Guilt (OG) scale. In addition, he asks students to make ratings about their hunger on four consecutive days during the first month of their sophomore year. Finally at the end of their sophomore year, they estimate the number of days they skipped lunch or dinner during a typical month of their second college year and the proportion of days that they were on diets. Casey enters the data from the six items of the PBI, the five items of the ATE, the eight items of the OG, the four ratings of daily hunger, and the two eating habit measures. Accordingly, Casey's SPSS data file contains 40 cases and 25 variables.

The Data Set

All seven applications use the data set named *Lesson 19 Data File 1* on the Web at http://www.prenhall.com/greensalkind. Table 7 (on page 127) shows the variables in the data set.

Creating Variables

In this section, we demonstrate how to create scales for seven applications.

Creating an Overall Scale from Variables with the Same Metric and No Reverse-Scaling

Casey wants to compute scores on the Positive Attitude toward Eating (ATE) scale. Higher scores on all five items indicate positive feelings toward eating, and therefore none of the items requires reverse-scaling. Casey creates a total scale score by summing across all five items.

To create an overall ATE scale score, follow these steps:

1. Click **Transform**, and click **Compute Variable**.
2. Type **ate** in the Target Variable text box.

Table 7

Variables in Lesson 19 Data File 1

Variable	Definition
skip	Estimated number of days skipped lunch or dinner during a typical month of sophomore year
diet	Estimated proportion of days on diet during sophomore year
hunger1 to hunger4	Ratings of hunger on four consecutive days during first month of sophomore year. Participants indicated their hunger on a 0-to-100 scale, with 0 = no hunger during entire day and 100 = voracious during entire day. A missing score was assigned if participant failed to submit rating via e-mail at the end of a day.
pbi01 to pbi06	Six-item Positive Body Image (PBI) scale, completed at beginning of sophomore year. Participants respond on a 5-point scale, with 1 = totally disagree to 5 = totally agree. 1. I dislike the size of my waist.* 2. I have a good-looking body. 3. I hate my thighs.* 4. I like the firmness of my abdominal muscles. 5. I dislike the size of my hips.* 6. I like how I look.
ate01 to ate05	Five-item Positive Attitude toward Eating (ATE) scale, completed at beginning of sophomore year. Participants indicate how they feel about eating on a 3-point scale (–1 = negative feeling; 0 = neutral feeling; 1 = positive feeling). 1. Eating desserts 2. Eating at fast food restaurants 3. Eating snacks 4. Eating appetizers 5. Eating breakfast
og01 to og08	Eight-item Overeating Guilt (OG) scale, completed at beginning of sophomore year. Participants indicate agreement with statement on a 5-point scale, with 1 = totally disagree to 5 = totally agree. 1. I feel like I am a weak person. 2. It is not a big deal.* 3. I am not upset with myself.* 4. I hate myself. 5. I feel like I have no self-discipline. 6. I feel absolutely nauseous. 7. I feel ok about myself.* 8. I do not feel guilty.*

*Higher scores on these items represent lower scores on scale.

3. Click on **All** in the Function group box. In the Functions and Special Variables box, scroll down until you find an expression that says **SUM**. Highlight the expression and click ▲ to move it to the Numeric Expression text box.
4. In the Numeric Expression text box, highlight the two question marks and the comma in the SUM(?,?) expression by dragging your mouse across them.
5. Press the Delete key to remove the question marks and the comma.
6. Click **ate01** and click ▶ to move it within the parentheses in the Numeric Expression text box.
7. Type a comma after ate01.
8. Repeat Steps 6 and 7 for **ate02**, **ate03**, **ate04**, and **ate05**. Do not type a comma after ate05. Figure 106 (on page 128) shows the Compute Variable dialog box after completing Step 8.
9. Click **OK**. SPSS will return you to the Data Editor where you can see your new variable, ate.

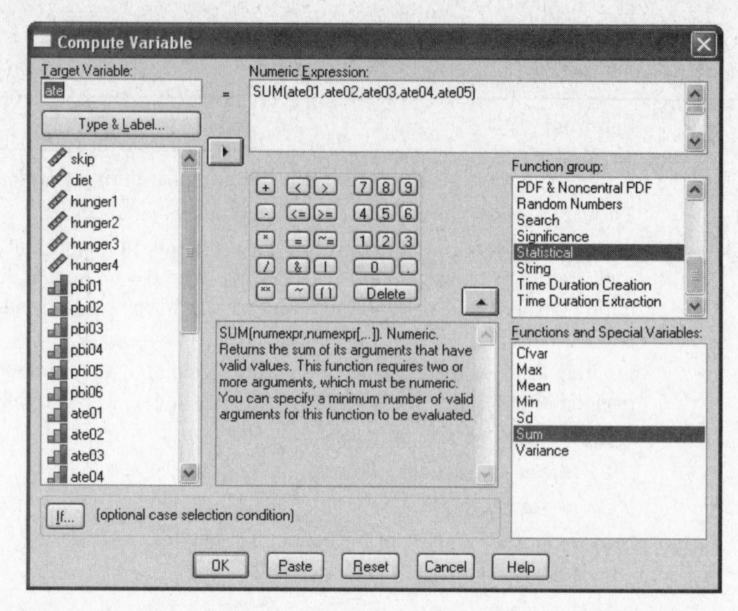

Figure 106. **Compute Variable dialog box for creating ATE scale scores.**

Creating an Overall Scale from Variables with the Same Metric and Reverse-Scaling

Casey wants to compute overall scores on the Positive Body Image (PBI) scale, such that higher PBI scale scores represent a positive body image. Three PBI items (pbi02, pbi04, and pbi06) are worded so that a higher item score represents a positive body image. The other three items (pbi01, pbi03, and pbi05) are worded so that higher item scores indicate a negative body image and, consequently, need to be reverse-scaled before the six PBI items are summed to create overall PBI scale scores.

Reverse-Scaling

To reverse-scale pbi01, pbi03, and pbi05, follow these steps:

1. Click **Transform**, click **Recode**, then click **Into Same Variables**.
2. In the Recode into Same Variables dialog box, hold down the Ctrl key, click **pbi01**, **pbi03**, and **pbi05**, and click ▶ to move them to the Numeric Variables box.
3. Click **Old and New Values**.
4. Within the Old Value area, click **Value** if not chosen. Type **1** in the text box to the right of Value.
5. Within the New Value area, click **Value** if not chosen. Type **5** in the text box next to Value.
6. Click **Add**.
7. Repeat for each of the remaining four values, changing 2 to 4, 3 to 3, 4 to 2, and 5 to 1. Make sure you click **Add** after each new value.
8. Click **Continue**.
9. Click **OK**.

TIP

With SPSS for Windows, select multiple variables by holding down the Ctrl key and then clicking to choose specific variables. With SPSS for Macintosh, follow the same directions, but hold down the Apple key rather than the Ctrl key. (See Step 2 in Reverse-Scaling for an illustration of selecting multiple variables.)

These three items are now scaled in the appropriate direction. You are ready to compute PBI scale scores.

Computing Total Scale Scores

To compute PBI scale scores, follow these steps:

1. Click **Transform**, and click **Compute Variable**.
2. Click **Reset** to clear the dialog box.

3. Type **pbi** in the Target Variable text box.
4. Click on **All** in the Function group box. In the Functions and Special Variables box, scroll down until you find an expression that says SUM. Highlight the expression and click ▲ to move it to the Numeric Expression text box.
5. In the Numeric Expression text box, highlight the two question marks and the comma in the SUM(?,?) expression by dragging your mouse across them.
6. Hit the Delete key to remove the question marks and the comma.
7. Type **pbi01 to pbi06** between the parentheses in the Numeric Expression text box. Figure 107 shows the Compute Variable dialog box after completing Step 7.

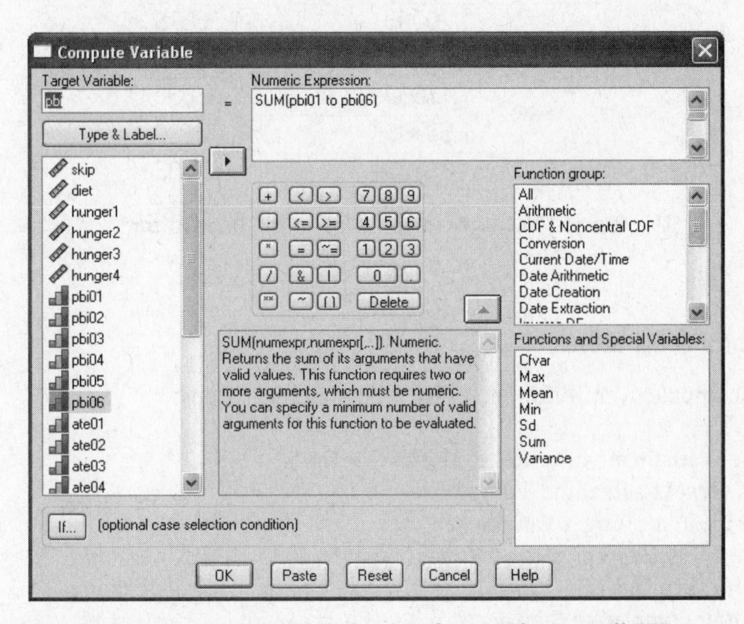

***Figure 107.* Compute Variable dialog box for creating overall PBI scores.**

8. Click **OK**. SPSS will return you to the Data Editor where you can see your new variable, pbi.

Creating an Overall Scale from Variables with Different Metrics and No Reverse-Scaling

Casey wishes to create an index of restrictive eating (IRE) by combining the skip and the diet variables. For the skip variable, students estimated the number of days they skipped lunch or dinner. For the diet variable, students estimated the proportion of days they were on diets. The two variables have different metrics (number of days and proportion of days) and, accordingly, cannot simply be summed together. One solution is for Casey to transform the two variables into z scores to create a comparable metric and then to sum the z scores to create an overall index of restrictive eating.

Creating Standardized Scores

The first step is to standardize the metric on the two restrictive eating variables so that they can be summed together.

1. Click **Analyze**, click **Descriptive Statistics**, then click **Descriptives**.
2. In the Descriptives dialog box, hold down the Ctrl key and click **skip** and **diet** to select them.
3. Click ▶ to move them to the Variable(s) box.

4. Click **Save standardized values as variables.**
5. Click **OK.**
6. Click **Window** and click **Lesson 19 Data File 1** to return to the Data Editor where you can see two new variables, Zskip and Zdiet shown in Figure 108.

Zskip	Zdiet
.20053	-1.50761
2.42861	-.73589
.20053	1.35878
-.69071	-1.06663
-1.13632	.25632
1.53738	-.73589
.86895	.03583

Figure 108. **z-scored variables in the Data Editor.**

Computing Total Scale Scores

Now you can compute overall IRE scores by following these steps:

1. Click **Transform**, and click **Compute Variable**.
2. Click **Reset** to clear the dialog box.
3. Type **ire** in the Target Variable text box.
4. In the Numeric Expression text box, type the following expression: **SUM(Zskip, Zdiet)**. Figure 109 shows the Compute Variable dialog box after completing Step 4.

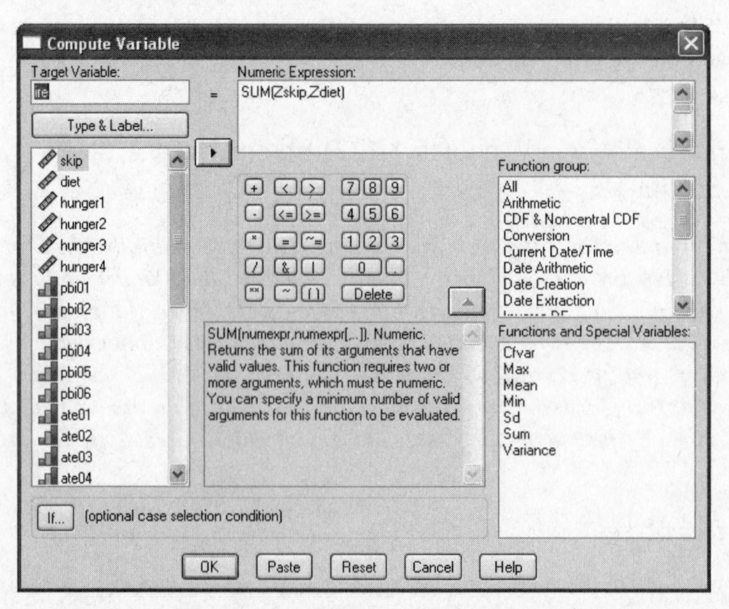

Figure 109. **Compute Variable dialog box for creating IRE scores.**

5. Click **OK**. SPSS will return you to the Data Editor where you can see your new variable, ire.

Creating an Overall Scale from Variables with Different Metrics and Reverse-Scaling

Casey believes that both items from the Positive Attitude toward Eating (ATE) scale and from the Overeating Guilt (OG) scale assess positive emotional reaction to eating. He would like to combine them together to create a positive emotional reaction to food (PERF) index. However, the items from the two scales have different metrics. The ATE items describe eating activities and ask individuals to give their emotional response on a –1 to +1 scale, while the OG items are statements about how they feel after overeating and ask individuals to indicate whether they agree with these statements. Consequently, the item scores from the two scales have to be transformed to z scores so that they have comparable metrics. In addition, higher scores on 4 of the 13 items (items: og01, og04, og05, og06) reflect negative feelings rather than positive feelings and need to be reverse-scaled. Given items have been transformed to z scores, Casey can reverse-scale these four items by multiplying them by −1 (making positive values negative and negative values positive). Now Casey can sum the 13 transformed items to yield PERF index scores.

Creating Standardized Scores

Follow these steps to create standardized scores:

1. Transform the items for the ATE scale (ate01 to ate05) and the Overeating Guilt scale (og01 to og08) to *z* scores. To transform these items, follow the procedure (Steps 1–5) described in the previous section for "Creating Standardized Scores" in "Creating an Overall Scale from Variables with Different Metrics and No Reverse-Scaling." The names for the *z*-scaled items are Zate01 to Zate05 and Zog01 to Zog08.

Reverse-Scaling Standardized Scores

To reverse-scale the *z*-scored OG items Zog01, Zog04, Zog05, and Zog06, follow these steps:

1. Click **Transform**, and click **Compute Variable**.
2. Click **Reset** to clear the dialog box.
3. Type **Zog01** in the Target Variable text box.
4. In the Numeric Expression text box, type the following expression: **–1 * Zog01**. Figure 110 shows the Compute Variable dialog box after completing Step 4.

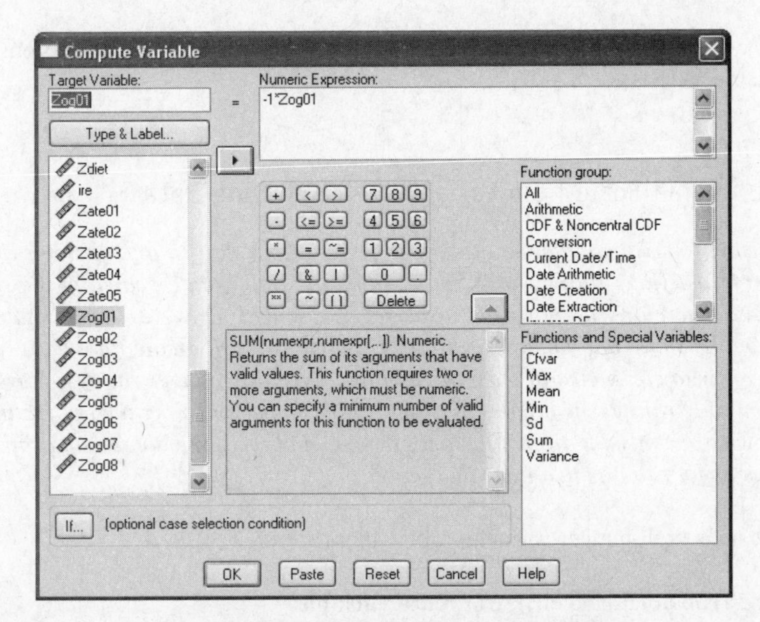

***Figure 110.* Reverse-scaling a *z*-scored variable.**

5. Click **OK**.
6. Click **OK** in response to the Change Existing Variable prompt.
7. Repeat steps 1 through 6 to reverse-scale **Zog04**, **Zog05**, and **Zog06**.

Computing Total Scale Scores

To compute overall PERF index scores, follow these steps:

1. Click **Transform** and click **Compute Variable**.
2. Click **Reset** to clear the dialog box.
3. Type **perf** in the Target Variable text box.
4. In the Numeric Expression text box, type **SUM(Zate01 to Zate05) + SUM(Zog01 to Zog08)**. Figure 111 shows the Compute Variable dialog box after completing Step 4.

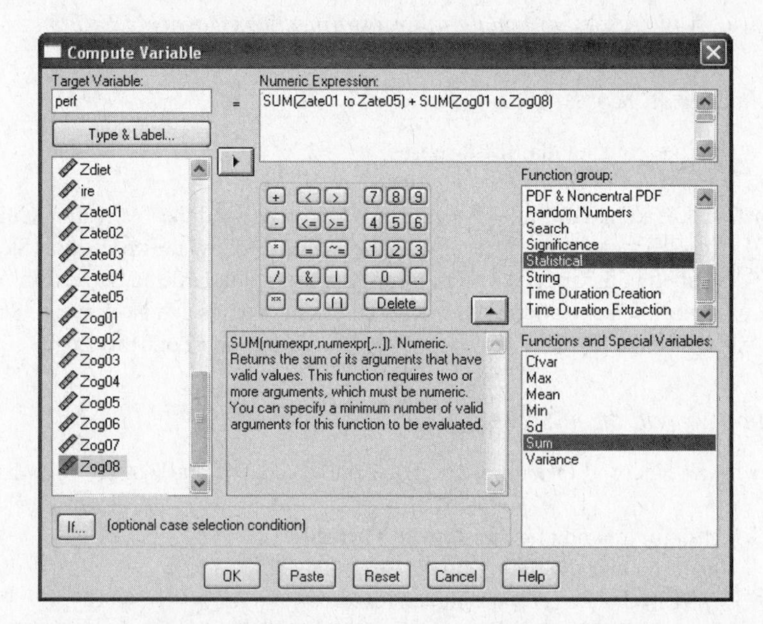

Figure 111. **Compute Variable dialog box for creating PERF index scores.**

5. Click **OK**. SPSS will return you to the Data Editor where you can see your new variable, perf.

Creating an Overall Scale from Variables with Missing Data

Casey wishes to create a total hunger score from his four daily hunger ratings (hunger1 to hunger4). The four ratings have the same metric and are all scaled in the same direction and, therefore, do not need to be reverse-scaled. However, some students are missing one or more daily hunger ratings. Casey decides to create total hunger scores for students who completed at least three of the four daily hunger ratings. For students with no missing ratings, he creates total scores by simply summing the four ratings. For students with one missing rating, he takes the mean of the three nonmissing ratings and multiplies by four to create a total-like score.

To create an overall hunger variable, follow these steps:

1. Click **Transform** and click **Compute Variable**.
2. Click **Reset** to clear the dialog box.

3. Type **hunger** in the Target Variable text box.
4. Click on **All** in the Function group box. In the Functions and Special Variables box, scroll down until you find an expression that says **Mean**. Highlight the expression and click ▲ to move it to the Numeric Expression text box. The first question mark will be highlighted.
5. Click **hunger1** and click ▶ to move it to the Numeric Expression text box.
6. Move your cursor to the other side of the comma.
7. Click **hunger2** and click ▶ to move it to the Numeric Expression text box.
8. Type a comma after **hunger2**.
9. Repeat Steps 7 and 8 for **hunger3 and hunger4**. Do not type a comma after **hunger4**.
10. Delete the extra question mark.
11. Place your cursor between MEAN and the first parentheses and type **.3**. This tells SPSS that you must have at least three nonmissing data scores for it to compute a mean score for an individual.
12. Type **4*** in front of MEAN.3 (hunger1, hunger2, hunger3, hunger4) to multiply the mean score times 4, the number of variables in the parentheses. This multiplication process produces a total-like score. Figure 112 shows the Compute Variable dialog box after completing Step 12.

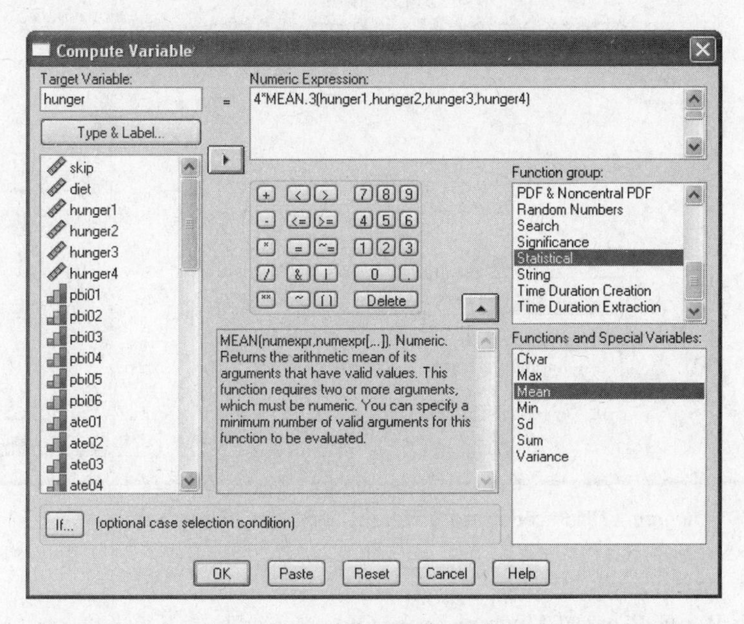

***Figure 112.* Creating an overall hunger variable with missing data.**

13. Click **OK**. SPSS will return you to the Data Editor where you can see your new variable, hunger.

For all cases but one, overall hunger scores were computed. However, for Case 2, there is no overall hunger score because that individual has scores on only two of the daily hunger ratings rather than the three required.

Warning! This method could be used for a wide range of applications. When faced with missing data, a researcher could create total scores by taking the mean of the nonmissing items, given participants responded to at least some number of them, and then multiply the mean by the number of items. For example, a researcher could create a total score for 20 items by taking the mean of the nonmissing items, given participants responded to at least 10 items, and then multiply the mean by 20—e.g., 20*mean.10(item01 to item20). However, we do not recommend general use of this method. The method should be used only if all items are equivalent to each other, the missing values are completely missing at random, and the mean is based on a high percentage of items with nonmissing values.

Creating a Qualitative Variable from One Quantitative Variable

Casey is interested in categorizing students into those who have negative feelings about eating and those who have nonnegative feelings about eating. He assigns labels of negative and nonnegative based on the students' scores on the Positive Attitude toward Eating (ATE) scale. Students who score less than 0 on the ATE are classified as negative eaters (eatcat = −1), and those who score 0 or higher are classified as nonnegative eaters (eatcat = 1).

To create this categorization, follow these steps:

1. Click **Transform**, click **Recode**, then click **Into Different Variables**.
2. Click **ate**, then click ▶ to move it to Input Variable → Output Variable text box.
3. Move your cursor to the Name text box in the Output Variable area, and type **atecat**.
4. Click **Change**. Figure 113 shows the Recode into Different Variables dialog box after completing Step 4.

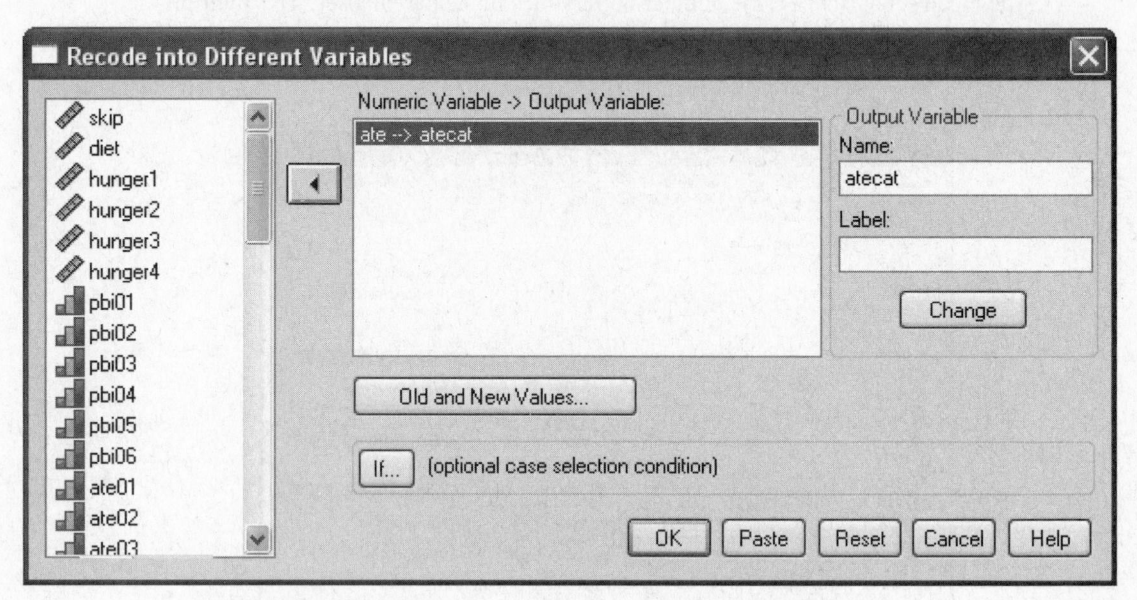

Figure 113. **Recode into Different Variables dialog box.**

5. Click **Old and New Values**.
6. Click **Range: Lowest Through** in the Old Value area.
7. Type **-1** in the text box after Through.
8. Type **-1** in the Value text box in the New Value area.
9. Click **Add**.
10. Click **All other values** in the Old Value area.
11. Type **1** in the Value text box in the New Value area.
12. Click **Add**.
13. Click **Continue**.
14. Click **OK**. SPSS will return you to the Data Editor where you can see your new variable, atecat.

Creating a Qualitative Variable from Two Quantitative Variables

Casey wishes to categorize his sample into four groups based on their Positive Body Image scores (pbi) and Positive Attitude toward Eating scores (ate).

- *Individuals with positive body images and positive attitudes toward eating*
- *Individuals with positive body images and nonpositive attitudes toward eating*

- *Individuals with nonpositive body images and positive attitudes toward eating*
- *Individuals with nonpositive body images and nonpositive attitudes toward eating*

He selects the midpoint value on each scale to differentiate among students who fall into these four categories. For the PBI, the midpoint on the 1-to-5 response scale for any one item is 3 and there are 6 items; therefore, the PBI scale midpoint is 18 (= 3 × 6). Similarly, for the ATE, the midpoint on the response scale for any one item is 0 and there are 5 items; therefore, the ATE scale midpoint is 0 (= 0 × 5). Casey creates a variable called catfood that classifies a student as having

- *a positive body image and a positive attitude toward eating (catfood = 1)*
- *a positive body image and a nonpositive attitude toward eating (catfood = 2)*
- *a nonpositive body image and a positive attitude toward eating (catfood = 3)*
- *a nonpositive body image and a nonpositive attitude toward eating (catfood = 4)*

To create the catfood variable, follow the steps for creating each of the four categories. To create a category for students with positive body images and positive attitudes toward eating, follow these steps:

1. Click **Transform**, and click **Compute Variable**.
2. Click **Reset** to clear the dialog box.
3. Type **catfood** in the Target Variable text box.
4. In the Numeric Expression text box, type **1**.
5. Click **If**.
6. In the Compute Variable: If Cases dialog box, click **Include if case satisfies condition**.
7. In the text box below Include if case satisfies condition, type **pbi > 18 & ate > 0**. Figure 114 shows the Compute Variable: If Cases dialog box with the appropriate commands.

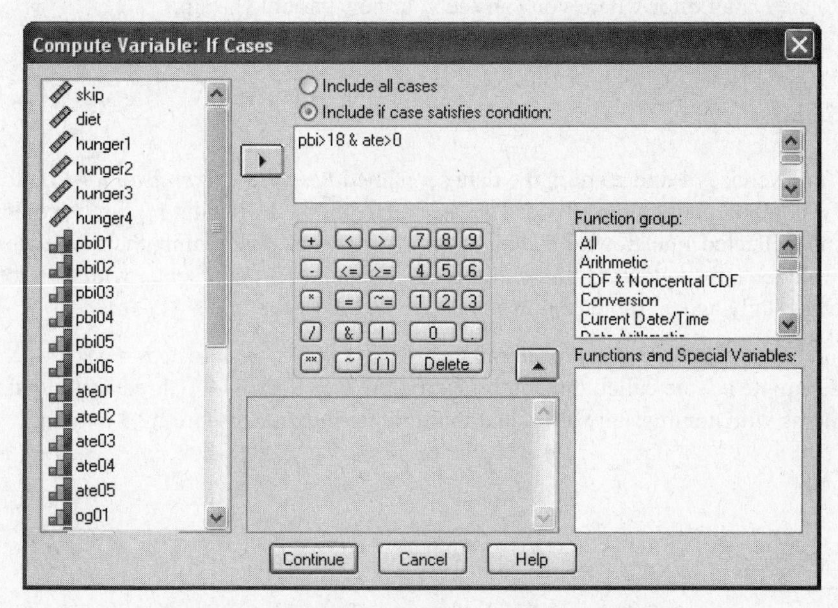

Figure 114. **Compute Variable: If Cases dialog box.**

8. Click **Continue**.
9. Click **OK**.

To create a category for students with positive body images and nonpositive attitudes toward eating, follow these steps:

1. Click **Transform**, and click **Compute Variable**.
2. In the Numeric Expression text box, type **2**.
3. Click **If**.
4. In the text box below Include if case satisfies condition, type **pbi > 18 & ate <= 0**.
5. Click **Continue**.
6. Click **OK**.
7. Click **OK** in response to Change Existing Variable? prompt.

To create a category for students with nonpositive body images and positive attitudes toward eating, follow these steps:

1. Click **Transform**, and click **Compute Variable**.
2. In the Numeric Expression text box, type **3**.
3. Click **If**.
4. In the text box below Include if case satisfies condition, type **pbi <= 18 & ate > 0**.
5. Click **Continue**.
6. Click **OK**.
7. Click **OK** in response to Change Existing Variable? prompt.

To create a category for students with nonpositive body images and nonpositive attitudes toward eating, follow these steps:

1. Click **Transform**, and click **Compute Variable**.
2. In the Numeric Expression text box, type **4**.
3. Click **If**.
4. In the text box below Include if case satisfies condition, type **pbi <= 18 & ate <= 0**.
5. Click **Continue**.
6. Click **OK**.
7. Click **OK** in response to Change Existing Variable? prompt. SPSS will return you to the Data Editor where you can see your new variable group.

Exercises

The data for Exercises 1 and 2 are in the data set named *Lesson 19 Exercise File 1* on the Web at http://www.prenhall.com/greensalkind. The data are from the following research problem.

Marion collected questionnaire data from 12 students concerning their attitudes toward math. The students responded to three attitudinal items on a 5-point scale, with 1 = totally disagree to 5 = totally agree. Table 8 shows the three items.

1. Reverse-scale the values for item3 (1 = 5, 2 = 4, 3 = 3, 4 = 2, 5 = 1).
2. Compute a scale called mathatt by taking the mean across the three attitudinal items with nonmissing values and multiplying that mean score by 3 to get a

Table 8
Math Attitude Items

Variable	Definition
item1	I like my mathematics classes.
item2	I find math to be a positive challenge.
item3	I am fearful of math.

total-like score. Compute this score so that if any case has more than one missing value, mathatt will show a missing value for that case.

The data for the Exercises 3 through 5 are in the data set named *Lesson 19 Exercise File 2* on the Web at http://www.prenhall.com/greensalkind. The data are from the following research problem.

Susan is interested in how much coffee people drink at work. To answer this question, she has 10 individuals record the number of cups of coffee they drink daily at work for four days. She also collects an index of job stress and each person's age. The data file contains the four variables concerning coffee drinking (day1 through day4), the job stress variable (jobstres), and the age variable (age). There are no missing data in this data set.

3. Create a categorical age variable named agecat with three age categories, 20–29, 30–39, and 40 and beyond.
4. Create a categorical job stress variable named jobcat with two categories, 1 (jobstres score less than or equal to 21) and 2 (jobstres score greater than 21).
5. Create a new variable called group with four categories based on low and high amounts of coffee drinking and low and high degree of job stress. Split the sample into low and high amounts of coffee drinking based on the total amount of coffee drinking, such that individuals who drink less than 5 cups of coffee are considered low and those who drink 5 or more cups are considered high.

The data for the Exercises 6 through 8 are in the data set named *Lesson 19 Exercise File 3* on the Web at http://www.prenhall.com/greensalkind. The data are from the following research problem.

Matt is interested in negative peer interaction in preschoolers. He collects data on 20 preschoolers. He has three measures: (1) a teacher rating of prosocial behavior with peers, (2) behavioral observations of the number of aggressive acts against peers per hour, and (3) peer reports of aggression. He wants to combine his three measures into a single measure of negative peer interaction. There are no missing data in this dataset.

6. Transform the scores for the three variables into z scores.
7. Reverse-scale the prosocial rating so that higher values reflect negative peer interaction.
8. Create an overall negative peer interaction score.

LESSON

20 Univariate Descriptive Statistics for Qualitative Variables

Qualitative or categorical variables are those whose values represent mutually exclusive categories with no quantitative meaning, such as gender, race, or religious preference. In this lesson, we focus on methods for describing qualitative variables using the Frequencies procedure. These methods can also be used to describe a variable with ordinal, interval, or ratio data if the variable has a limited number of values.

The Frequencies procedure in SPSS can be used to describe one or more qualitative variables. It allows us to determine the frequencies or percentages for categories of a variable as well as the mode, a measure of central tendency. In addition, Frequencies can be used to depict qualitative data graphically in charts such as bar charts and pie charts.

Applications for Describing Qualitative Variables

We will discuss applications of Frequencies for three types of variables.

- Qualitative variables with two categories
- Qualitative variables with a moderate number of categories
- Qualitative variables with many categories

Qualitative Variables with Two Categories

Linda has collected data on the gender of her research participants. There are 120 participants in her study. She wants to know the percentages for men and women in her sample.

Qualitative Variables with a Moderate Number of Categories

Linda also has collected information on the ethnic status of the participants in her study. Her ethnicity variable consists of five categories: White, Hispanic, Black, Asian, and other. Again she is interested in the percentages or frequencies of research participants in these categories.

Qualitative Variables with Many Categories

Finally, Linda has collected data on the occupations of the participants. The occupational variable consists of 10 categories with each value representing a different job description. She is interested in obtaining the frequencies or percentages of participants in the 10 job categories.

Understanding Descriptive Statistics for Qualitative Variables

Frequency distributions are tabular or graphical presentations of data that show each category for a variable and the frequency of the category's occurrence in the data set. Percentages for each category are often reported instead of, or in addition to, the frequencies.

The mode is the measure of central tendency associated with a qualitative variable. The mode is the most frequently occurring score or category in a data set. Distributions of variables may be unimodal (one category appears more frequently than others), bimodal (two scores or categories occur most frequently), or multimodal (more than two categories occur most frequently).

The Data Set

The data for this lesson involve the three applications described earlier and are in the data file named *Lesson 20 Data File 1* on the Web at http://www.prenhall.com/greensalkind. The variables in the data file are shown in Table 9.

Table 9
Variables in Lesson 20 Data File 1

Variables	Definition
gender	A qualitative variable that indicates the gender of the participants with the following categories: 1 = Men 2 = Women
ethnic	A qualitative variable that indicates the ethnic group of the participants with the following categories: 1 = Black 2 = Hispanic 3 = Asian 4 = White 5 = Other
job	A qualitative variable that indicates the job classification of the participants with the following categories: 1 = Clerical 2 = Management 3 = Maintenance 4 = Academic 5 = Professional 6 = Self-Employed 7 = Medical 8 = Skilled 9 = Military 10 = Unemployed

The Research Question

For descriptive statistics, there is no research question as such. However, Linda is interested in the frequencies or percentages of research participants in each category of her variables.

1. What are the percentages of men and women in the sample?
2. What are the frequencies associated with each ethnic status category?
3. How many participants are in each of the 10 job categories?

Conducting Descriptive Statistics for Qualitative Variables

To compute frequencies for the gender, ethnicity, and occupational variables, follow these steps:

1. Click **Analyze**, click **Descriptive Statistics**, then click **Frequencies**. When you do this, you will see the Frequencies dialog box as shown in Figure 115.

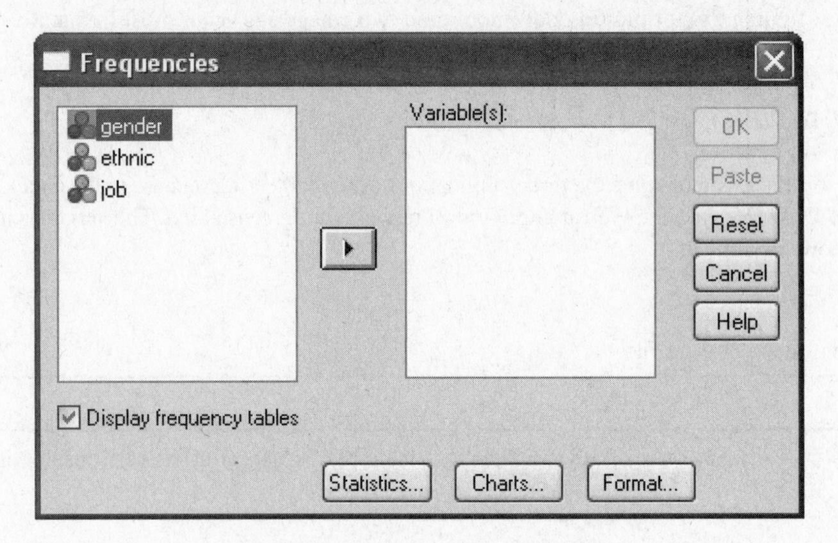

***Figure 115*. The Frequencies dialog box.**

2. Holding down the Ctrl key, click **ethnic**, **gender**, and **job** to select them, then click ▶ to place them in the Variable(s) box.
3. Click **Statistics**. You will see the Frequencies: Statistics dialog box as shown in Figure 116.

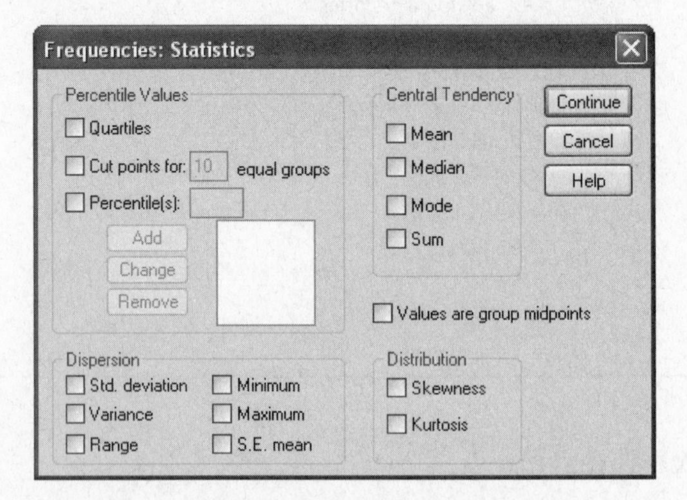

***Figure 116*. The Frequencies: Statistics dialog box.**

4. Under the Central Tendency area, click **Mode**.
5. Click **Continue**.
6. Click **OK**.

Statistics

		Gender of Participant	Ethnic Status	Job Category
N	Valid	120	120	120
	Missing	0	0	0
Mode		2	4	7

Frequency Table

Gender of Participant

		Frequency	Percent	Valid Percent	Cumulative Percent
Valid	Men	58	48.3	48.3	48.3
	Women	62	51.7	51.7	100.0
	Total	120	100.0	100.0	

Ethnic Status

		Frequency	Percent	Valid Percent	Cumulative Percent
Valid	Black	20	16.7	16.7	16.7
	Hispanic	24	20.0	20.0	36.7
	Asian	17	14.2	14.2	50.8
	White	51	42.5	42.5	93.3
	Other	8	6.7	6.7	100.0
	Total	120	100.0	100.0	

Job Category

		Frequency	Percent	Valid Percent	Cumulative Percent
Valid	Clerical	13	10.8	10.8	10.8
	Management	5	4.2	4.2	15.0
	Maintenance	7	5.8	5.8	20.8
	Academic	15	12.5	12.5	33.3
	Professional	13	10.8	10.8	44.2
	Self-Employed	13	10.8	10.8	55.0
	Medical	20	16.7	16.7	71.7
	Skilled	13	10.8	10.8	82.5
	Military	12	10.0	10.0	92.5
	Unemployed	9	7.5	7.5	100.0
	Total	120	100.0	100.0	

***Figure 117.* Summary statistics for the frequencies analysis.**

Selected SPSS Output for Frequencies

The output for the analysis is shown in Figure 117. The output initially presents the mode for each of the three variables. The tables labeled Gender of Participant, Ethnic Status, and Job Category show the frequencies (counts) for these variables. In addition, the tables show the percentage of participants in each category. Because the variables are qualitative, the column labeled Cumulative Percent is not meaningful.

Using SPSS Graphs to Display the Results

Bar charts are often used to display the results of categorical or qualitative variables. Bar charts allow for the visual representation of the frequencies in each category and can be used to display results of variables with a relatively large number of categories. An alternative to the bar chart is the pie chart. Pie charts also present the frequencies or percentages in each category but in a form that can be edited to highlight specific categories. Pie charts are more useful for displaying results of variables that have relatively few categories, in that pie charts become cluttered and difficult to read if variables have many categories.

Creating a Bar Chart

To create a bar chart, follow these steps:

1. Click **Analyze**, click **Descriptive Statistics**, then click **Frequencies**.
2. Click **Reset** to clear the dialog box.

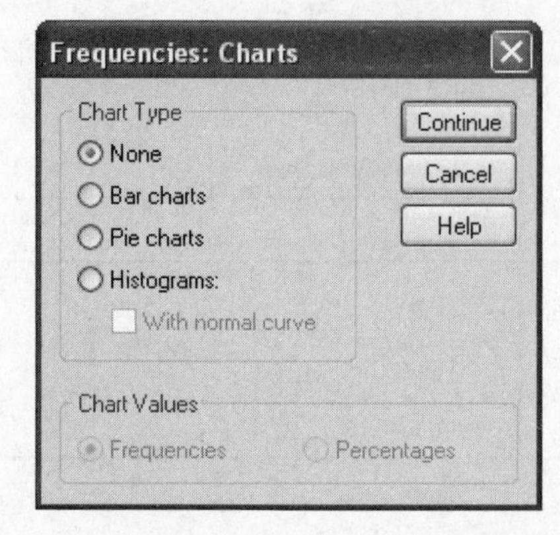

Figure 118. **The Frequencies: Charts dialog box.**

3. Click **job**, then click ▶ to place it in the Variable(s) box.
4. Click **Charts**. You will see the Frequencies: Charts dialog box as shown in Figure 118.
5. Click **Bar charts**.
6. Click **Continue**.
7. Click **OK**.

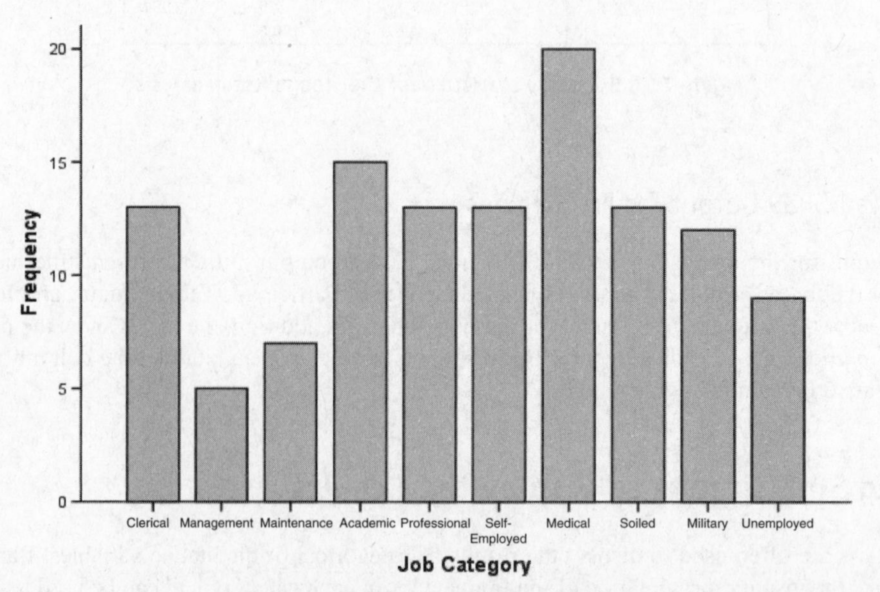

Figure 119. **Bar chart of job categories.**

To edit the graph to make it appear as it does in Figure 119, conduct the following steps:

1. Double-click on the chart to select it for editing, and maximize the chart window.
2. Click inside the axes of the graph but not inside the bars.
3. Click on **Edit** and click on **Properties**.
4. Click on the **Fill and Border** tab of the Properties box if not currently selected.
5. Click on box labeled **Fill** and then click on the white rectangle in the color chart on the right.
6. Click on box labeled **Border** and then click on the white rectangle with the red diagonal line in the color chart on the right.
7. Click **Apply** in the Properties dialog box.
8. Click once on the label of the *Y*-axis. The graph should now appear as it does in our figure.

Creating a Pie Chart

To create a pie chart for the ethnicity variable, follow these steps:

1. Click **Graphs**, click on **Legacy Dialogs**, and click **Pie**. With SPSS for a Macintosh, click **Graphs** and click **Pie**.
2. Click **Summaries for groups of cases**.
3. Click **Define**. You will see the Define Pie: Summaries for Groups of Cases dialog box in Figure 120.
4. Highlight **ethnic** and click ▶ to move it to the Define Slices by box.

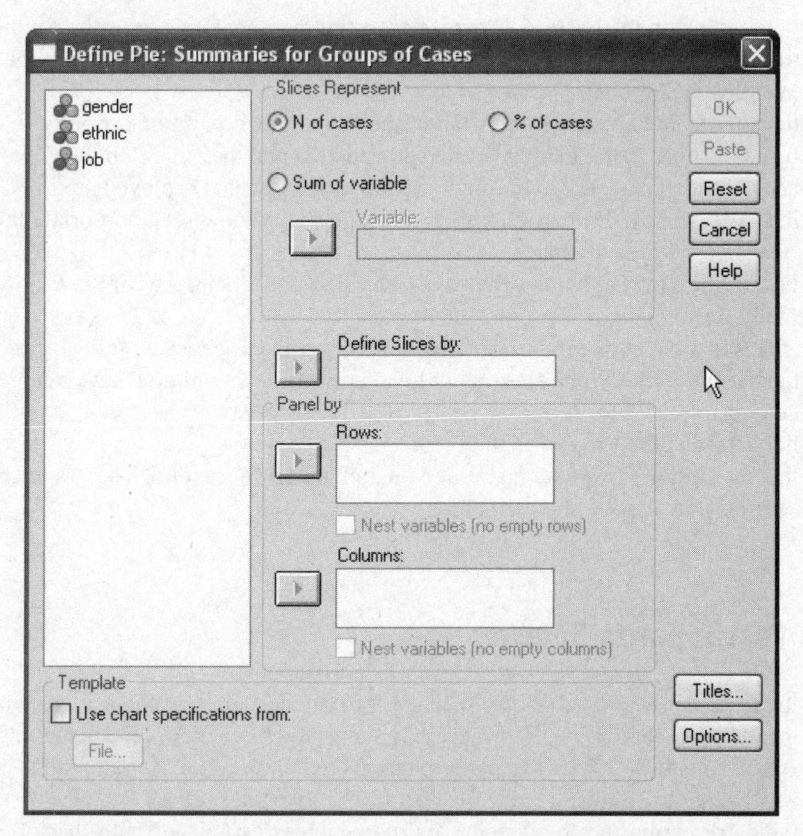

Figure 120. **The Define Pie: Summaries for Groups of Cases dialog box.**

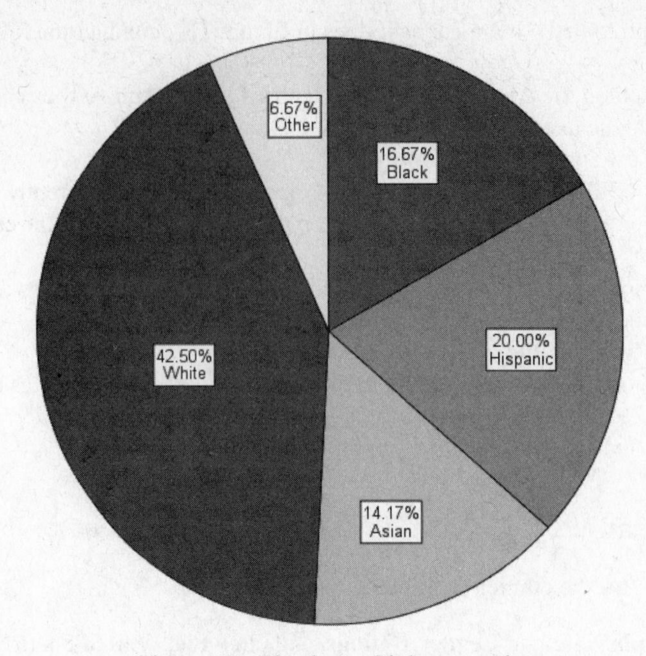

***Figure 121.* Pie chart of job categories.**

5. Click **% of cases** in the Slice Represent area.
6. Click **OK**.

To obtain Figure 121, you have to eliminate the legend and add labels to the slices of your pie chart. To do this, conduct the following steps:

1. Double-click on the chart to select it for editing, and maximize the chart window.
2. Click on **Elements** and click on **Show Data Labels**. You should now see the Properties dialog box.
3. Click on the **Data Value Labels** tab of the Properties box if not currently selected.
4. Click on **ethnic** in the Labels Not Displayed area and then click on the green upward pointing arrow to move Ethnic Status to the Labels Displayed area.
5. Click **Apply** in the Properties dialog box. You should now see appropriate labels in the slices of the pie chart.
6. Click on the **Text Style** tab (**Text** tab with SPSS for Macintosh) of the Properties dialog box.
7. In the font area, click on the **Size** drop-down menu and choose a font size of 10.
8. Click **Apply** in the Properties dialog box. The labels should now have a larger font size.
9. Click **Close** in the Properties dialog box.
10. Click on **Options** from the main menu. Click on **Hide Legend**. The graph should now appear as it does in our figure.

An APA Participants Section

The sample consisted of 120 participants, approximately half of them were women (n = 62), while the rest were men (n = 58). Figure 121 shows the frequencies and percentages for ethnic status. Approximately 42% of the participants were Whites, and the remaining participants were relatively evenly distributed across Blacks (17%), Hispanics (20%), Asians (14%), and others (8%). Table 10 reports the frequencies and percentages associated with the job categories. The most frequently occurring job category was the medical profession, and the least common job category was management.

Table 10

Frequencies and Percentages of Job Categories

Job categories	Frequency	Percentage
Medical	20	16.7
Academic	15	12.5
Self-Employed	13	10.8
Skilled labor	13	10.8
Clerical	13	10.8
Professional	13	10.8
Military	12	10.0
Unemployed	9	7.5
Maintenance	7	5.8
Management	5	4.2

Exercises

Exercises 1 through 4 are based on the following research problem. The data for these exercises can be found in the data file named *Lesson 20 Exercise File 1* on the Web at http://www.prenhall.com/greensalkind.

Ann wants to describe the demographic characteristics of a sample of 25 individuals who completed a large-scale survey. She has demographic data on the participants' gender (two categories), educational level (four categories), marital status (three categories), and community population size (eight categories).

1. Conduct a frequency analysis on the gender and marital status variables. From the output, identify the following:
   a. Percent of men
   b. Mode for marital status
   c. Frequency of divorced people in the sample
2. Create a frequency table to summarize the data on the educational level variable.
3. Create a bar chart to summarize the data from the community population variable.
4. Write a Participants section describing the participants in Ann's sample.

Exercises 5 through 7 are based on the following research problem. The data for these exercises can be found in the data file named *Lesson 20 Exercise File 2* on the Web at http://www.prenhall.com/greensalkind.

Julie asks 50 men and 50 women to indicate what type of books they typically read for pleasure. She codes the responses into 10 categories: drama, mysteries, romance, historical nonfiction, travel books, children's books, poetry, autobiographies, political science, and local interest books. She also asks the participants how many books they read in a month. She categorizes their responses into four categories: nonreaders (no books a month), light readers (1–2 books a month), moderate readers (3–5 books a month), and heavy readers (more than 5 books a month). Julie's SPSS data file contains two variables: book, a 10-category variable for type of books read, and reader, a four-category variable indicating the number of books read per month.

5. Create a table to summarize the types of books that people report reading.
6. Create a pie chart to describe how many books per month Julie's sample reads.
7. Write a Participants section describing your results on the three variables.

LESSON

21 Univariate Descriptive Statistics for Quantitative Variables

The goal of univariate descriptive statistics is to portray accurately and succinctly data from a variable. When there are only a few values for a quantitative variable (e.g., a variable with a 4-point response scale), we can describe its results very accurately by presenting all of its values and their relative frequencies. The relative frequency of a value is the number of times it occurs divided by the total frequency across all values. If a quantitative variable has many values, however, it would be impractical to present all values of the variable and their relative frequencies. Rather, it would be easier to present grouped values and their relative frequencies. The relative frequencies for values or grouped values can be presented in a table or a graph.

Graphical and tabular presentations are not always possible in journals because of space limitations. Thus, data are often summarized by using statistical indices. If it were important for readers of an article to understand the distribution of a variable, it would be desirable, nevertheless, to present its distribution with the use of a graph. In general, descriptive statistical indices are presented in the text of a manuscript if there are relatively few variables. If there are a large number of variables to be described, statistical indices for these variables can be presented in a table.

In this lesson, we will focus on two SPSS procedures, Explore and Descriptives, to describe the results of quantitative variables by means of statistical indices, including measures of central tendency, variability, skewness, and kurtosis. We also will demonstrate how to create charts that visually represent the distributions of variables, including stem-and-leaf plots, boxplots, error bar charts, and histograms. Finally, we will illustrate how to convert raw scores on a variable to z scores and percentile ranks. Percentile ranks will be computed under conditions when we can and cannot assume that raw scores are normally distributed.

Applications for Describing Quantitative Variables

We present applications for describing the results for two types of variables:

- Quantitative variables with few values
- Quantitative variables with many values

We present two additional applications:

- Describing the results for quantitative variables within levels of a qualitative variable
- Computing z scores and percentile ranks for quantitative variables

Quantitative Variables with Few Values

Joanne has 90 college students complete a four-item Likert scale that assesses study motivation in school. Each item has scores that range from 1 (not at all like me) to 5 (very much like me). Joanne is interested in describing the students' responses. Because there are only a few values for each variable, Joanne will create relative frequency tables to summarize her data.

Quantitative Variables with Many Values

Joanne also collected information on each student's college GPA and high school GPA. Joanne is interested in describing the typical high school and college GPAs of her research participants. To summarize these variables, she will compute various statistical indices to look at the distributions. She also will create graphs to visually display the distributions.

Quantitative Variables within Levels of a Qualitative Variable

Joanne collected information about each student's gender (a qualitative variable). She wants to describe the typical high school and college GPAs (quantitative variables) for men and women separately. To summarize these GPA variables, she will compute statistical indices to assess the characteristics of the GPA distributions for men and for women. Alternatively, she could create graphs that visually display differences in the distributions for men and women students.

z Scores and Percentile Ranks on Quantitative Variables

Joanne is interested in determining where GPA scores for particular individuals lie in relation to all GPA scores. She will convert her GPA scores to z scores and then to percentile ranks to assess the relative position of these GPA scores. Transforming the z scores to percentile ranks assumes that the z scores (and the GPA scores) are normally distributed. In contrast, she could convert her GPA scores directly to percentile ranks and would not have to assume that the GPA scores are normally distributed.

Understanding Descriptive Statistics for Quantitative Variables

Descriptive statistics involves summarizing distributions of scores by developing tabular or graphical presentations and computing descriptive statistical indices. You can also describe individual scores within a distribution by converting these scores to percentile ranks.

A frequency distribution gives the frequencies associated with all values of a variable. A simple frequency distribution table can be used to delineate the results of a quantitative variable when it has only a few values. For a variable with many levels, the values are grouped together into intervals of values to create a grouped frequency table. There is no specific interval length that would be appropriate for all grouped frequency tables, but the interval length should be wide enough to summarize the results adequately, yet narrow enough to present the distribution accurately. The current version of SPSS does not provide an easy way to generate a grouped frequency table.

SPSS does allow us to display graphically grouped frequency distributions by using histograms. Stem-and-leaf plots also are used to display a distribution of a quantitative variable graphically. Each score on a variable is divided into two parts, the stem and the leaf. The stem gives the leading digits and the leaf shows the trailing digits. For example, for a score of 40, the stem of 4 represents the 10s place, and the leaf of 0 represents the units place.

The distribution of a quantitative variable can also be summarized with the use of statistical indices. Measures of central tendency include the mean, the arithmetic average of a set of scores; the median, the middle value of ranked scores; and the mode, the most frequently occurring score. Indices of variability can also be computed to describe a distribution's variability around a measure of central tendency or, more generally, the spread of scores on a variable. The most common measures of variability are the variance and its square-rooted form, the standard deviation. Skewness reflects the degree to which a variable's scores fall at one end or the other end of the variable's scale. If the majority of the scores fall at the high end of the scale, with relatively few scores at the low end, the distribution is negatively skewed. Conversely, if most scores fall at the low end of the scale, with relatively few scores at the high end, the distribution is positively skewed. The kurtosis of a variable reflects the thickness of the tail regions (relative frequency of scores in both extremes) of a distribution.

Measures of central tendency and variability can also be displayed graphically. With error bar charts, means and standard deviations can be displayed for each quantitative variable. An alternative chart that presents a richer picture is a boxplot. With a boxplot, the line in the middle of a box represents the median score on that variable. The length of the box in the boxplot is the interquartile range. In other words, the bottom and top of the box represent the 25th and 75th percentiles, respectively. Outliers are scores that fall between 1.5 and 3 box lengths, while extreme scores are scores that fall more than 3 box lengths from the lower or upper edge of the box. The whiskers, or the lines at the top and the bottom of the plot that originate from the box, represent the smallest and largest values that are not outliers or extreme scores.

We can also describe where individual scores fall with respect to all scores in a distribution. A score may be converted to a percentile rank, the percent of individuals who have scores that fall at or below the score of interest. Percentile ranks can be computed by using one of two methods. One method requires that raw scores first be converted to z scores and then to percentile ranks based on the normal distribution. This method assumes that the raw scores are normally distributed. The second method requires the computation of percentile ranks directly from raw scores and makes no assumptions about the distribution of raw scores.

The Data Set

The data set for this lesson includes the high school and college GPA data described in the four examples. They are in the data file named *Lesson 21 Data File 1* on the Web at http://www.prenhall.com/greensalkind. The variables are presented in Table 11.

Table 11

Variables in Lesson 21 Data File 1

Variables	Definition
hsgpa	High school GPA scores
collgpa	College GPA scores
gender	1 = Men
	2 = Women

Conducting Descriptive Statistics for Quantitative Variables

Now we will compute statistical indices for a single quantitative variable and for a quantitative variable within levels of a qualitative variable, and convert a set of scores to z scores and percentile ranks.

Computing Descriptive Statistics for Quantitative Variables

To compute descriptive statistics for high school and college GPA scores across the entire sample, follow these steps:

1. Click **Analyze**, click **Descriptive Statistics**, then click **Explore**. You will see the Explore dialog box in Figure 122.

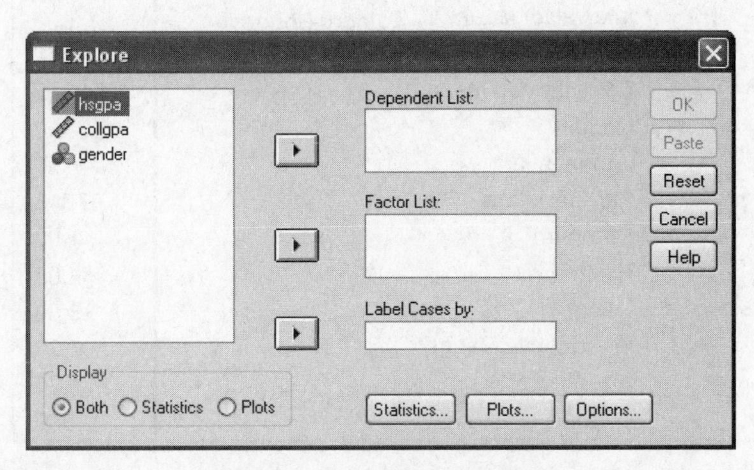

Figure 122. **The Explore dialog box.**

2. Holding down the Ctrl key, click **hsgpa** and **collgpa**, then click ▶ to place them in the Dependent List box.
3. In the Display box, click **Statistics**.
4. Click the **Statistics** button. You will see the Explore: Statistics dialog box as shown in Figure 123.

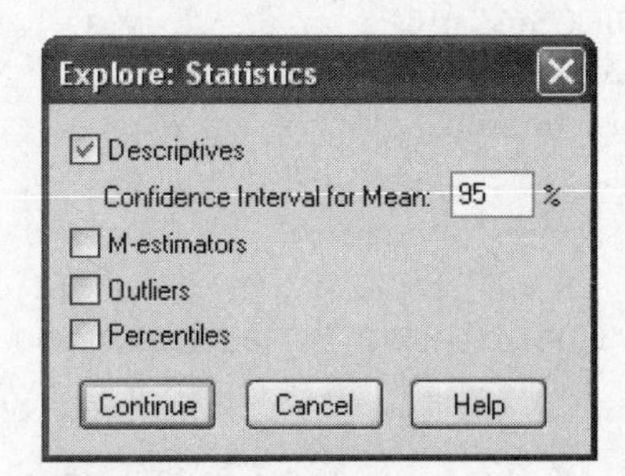

Figure 123. **The Explore: Statistics dialog box.**

5. Be sure **Descriptives** is checked.
6. Click **Continue**.
7. Click **OK**.

Selected SPSS Output for Quantitative Variables

Your output should have descriptive statistics for both high school and college GPA scores, although we show you only the descriptive statistics for high school GPA scores in Figure 124.

Descriptives

			Statistic	Std. Error
High School GPA Scores	Mean		2.5379	.09175
	95% Confidence Interval for Mean	Lower Bound	2.3556	
		Upper Bound	2.7202	
	5% Trimmed Mean		2.5372	
	Median		2.5350	
	Variance		.758	
	Std. Deviation		.87046	
	Minimum		1.00	
	Maximum		4.00	
	Range		3.00	
	Interquartile Range		1.26	
	Skewness		.125	.254
	Kurtosis		-.889	.503

***Figure 124*. Descriptive statistics for high school GPA scores.**

Computing Descriptive Statistics for Quantitative Variables within Levels of a Qualitative Variable

To compute descriptive statistics for college and high school GPA scores separately for men and women, follow these steps:

1. Click **Analyze**, click **Descriptive Statistics**, then click **Explore**.
2. Click **Reset** to clear the dialog box.
3. Holding down the Ctrl key, click **collgpa** and **hsgpa** to highlight them. Then click ▶ to place them in the Dependent List box.
4. Click **gender**, then click ▶ to place it in the Factor List box.
5. In the Display box, click **Statistics**.
6. Click the **Statistics** button.
7. Be sure Descriptives is checked.
8. Click **Continue**.
9. Click **OK**.

Selected SPSS Output for Quantitative Variables within Levels of a Qualitative Variable

We present only the descriptive statistics for the high school GPA scores for men and for women in Figure 125.

Converting Scores to *z* Scores and Percentile Ranks Not Assuming Normality

To convert the college GPA scores to percentile ranks without assuming they are normally distributed, follow these steps:

1. Click **Transform** and click **Rank Cases**. You will see the Rank Cases dialog box in Figure 126.

Descriptives

	Gender			Statistic	Std. Error
High School GPA Scores	Men	Mean		2.5276	.13596
		95% Confidence Interval for Mean	Lower Bound	2.2538	
			Upper Bound	2.8015	
		5% Trimmed Mean		2.5263	
		Median		2.3800	
		Variance		.850	
		Std. Deviation		.92214	
		Minimum		1.00	
		Maximum		4.00	
		Range		3.00	
		Interquartile Range		1.57	
		Skewness		.296	.350
		Kurtosis		-.996	.688
	Women	Mean		2.5486	.12415
		95% Confidence Interval for Mean	Lower Bound	2.2983	
			Upper Bound	2.7990	
		5% Trimmed Mean		2.5483	
		Median		2.6200	
		Variance		.678	
		Std. Deviation		.82349	
		Minimum		1.06	
		Maximum		4.00	
		Range		2.94	
		Interquartile Range		1.12	
		Skewness		-.113	.357
		Kurtosis		-.710	.702

Figure 125. **Descriptive statistics for high school GPA scores for men and for women.**

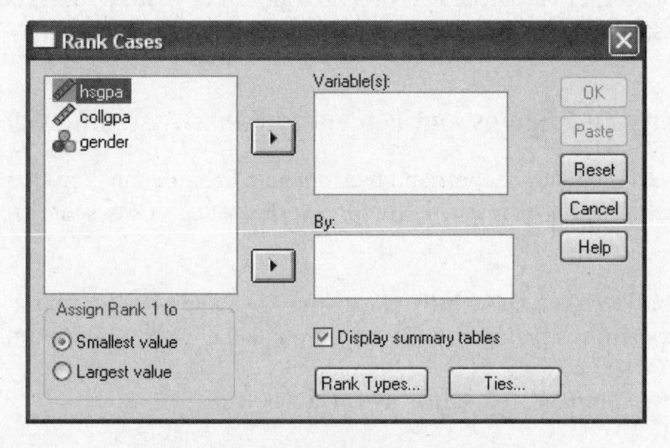

Figure 126. **The Rank Cases dialog box.**

2. Click **collgpa**, then click ▶ to place it in the **Variable(s)** text box.
3. Click **Rank Types**. Be sure **Rank** is checked.
4. Click **Continue**.
5. Click **Ties**. You will see the Rank Cases: Ties dialog box.
6. Click **High** as shown in Figure 127 (on page 152).

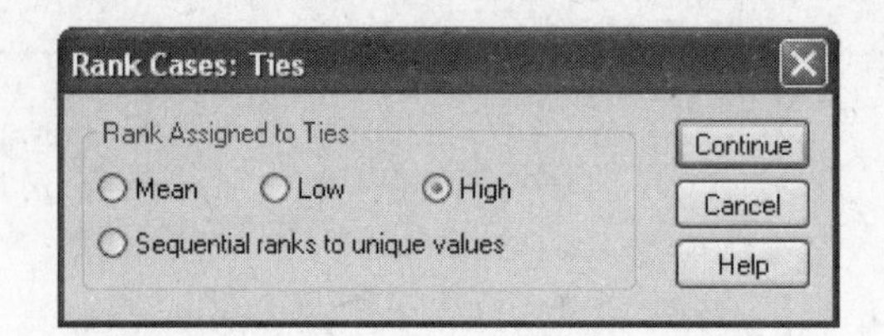

Figure 127. **Rank Cases: Ties dialog box.**

7. Click **Continue**.
8. Click **OK** and return to the Data Editor.
9. Click **Transform**, and click **Compute Variable**.
10. In the Target Variable text box, type **perrank**.
11. In the Numeric Expression text box, type **((Rcollgpa)/ 90) * 100**. The divisor in the expression is equal to the number of cases in the data set.
12. Click **OK**.

Selected SPSS Output for Converting Scores Not Assuming Normality

The percentile ranks of the college GPA scores for the first 10 cases are shown in Figure 128 in the column labeled perrank.

	hsgpa	collgpa	gender	Rcollgpa	perrank
1	4.00	1.75	1	6.000	6.67
2	1.87	1.48	1	2.000	2.22
3	3.85	2.65	2	59.000	65.56
4	1.31	2.57	2	47.000	52.22
5	2.62	2.93	2	75.000	83.33
6	1.91	2.59	2	50.000	55.56
7	3.00	2.61	2	52.000	57.78
8	2.95	2.67	2	62.000	68.89
9	2.92	2.64	2	58.000	64.44
10	2.74	3.03	2	80.000	88.89

Figure 128. **Percentile ranks for college GPA of first 10 cases, not assuming that GPA is normally distributed.**

Converting Scores to *z* Scores and Percentile Ranks Assuming Normality

We will now convert scores to *z* scores and then create percentile ranks from *z* scores, assuming that the scores are normally distributed. To convert the college GPA scores to *z* scores, follow these steps:

1. Click **Analyze**, click **Descriptive Statistics**, then click **Descriptives**.
2. In the Descriptives dialog box, click **collgpa**, and click ▶ to move it to the Variable(s) text box.
3. Click **Save standardized values as variables**.
4. Click **OK**.

To create percentile ranks for the *z*-scored college GPA scores, return to the Data Editor and follow these steps:

1. Click **Transform**, and click **Compute Variable**.
2. Type **gparank** in the Target Variable text box.
3. Click on **All** in Function groups box. In the Functions and Special Variables box, scroll down until you find an expression that says **Cdf.Normal**.
4. Highlight the expression, and click ▲ to move it to the Numeric Expression text box.

5. Click **Zcollgpa**, and click ▶ to place it in the Numeric Expression text box in place of the first question mark (quant).
6. Replace the second question mark (mean) with **0** and the third question mark (stddev) with **1**. Type *** 100** after Cdf.Normal(Zcollgpa,0,1). Figure 129 shows the numeric expression used to create the percentile ranks.
7. Click **OK**.
8. Click **Window**, and click **Lesson 21 Data File 1** to return to the Data Editor where you can see your new variable, gparank.

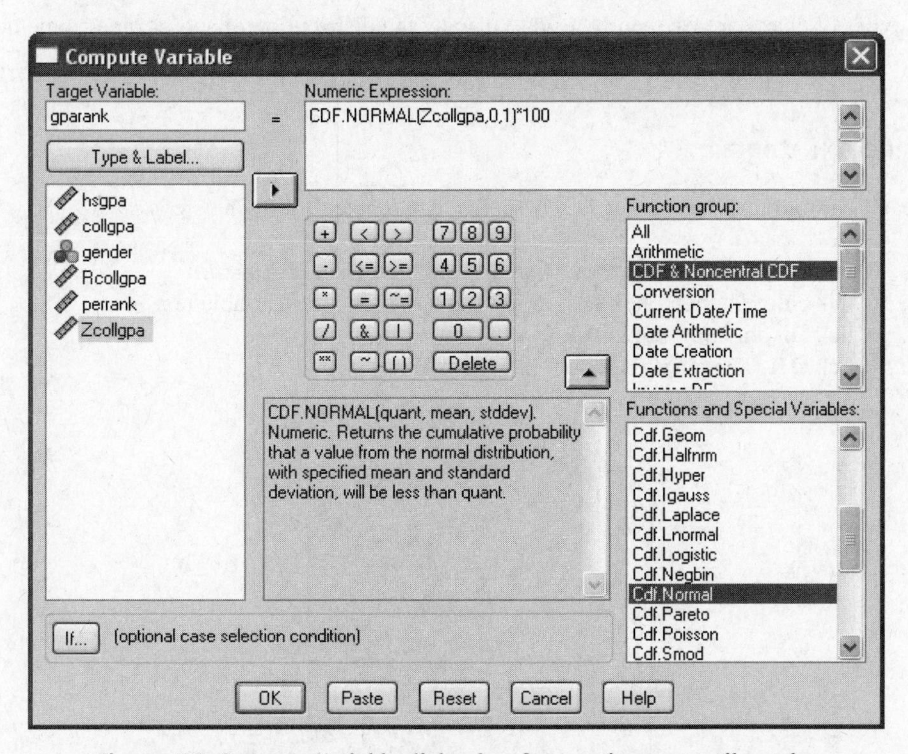

***Figure 129*. Compute Variable dialog box for creating percentile ranks.**

The percentile ranks obtained when we made no assumptions about the distribution are not necessarily the same percentile ranks we obtained when we assumed the scores were normally distributed. Table 12 shows the scores and percentile ranks for the first 10 cases in the data set, assuming and not assuming a normal distribution.

Table 12

Percentile Ranks for First 10 College GPA Scores, Not Assuming and Assuming Normality

College GPA	Percentile ranks not assuming normality	Percentile ranks assuming normality
1.75	07	05
1.48	02	01
2.65	66	64
2.57	52	57
2.93	83	84
2.59	56	59
2.61	58	61
2.67	69	66
2.64	64	63
3.03	89	89

You can see that the percentile ranks associated with various college GPAs are not identical. Variables with distributions that are substantially nonnormal will have greater discrepancies between percentile ranks computed using the two methods, and variables with relatively normal distributions will yield similar percentile ranks.

Using SPSS Graphs to Display the Results

Three types of charts are frequently used to display the distribution of scores on a single quantitative variable: the histogram, the stem-and-leaf plot, and the boxplot.

Creating a Histogram

To create a histogram of the college GPA scores, conduct the following steps:

1. Click **Graphs**, click **Legacy Dialogs**, and then click **Histogram**.
2. Click **collgpa**, then click ▶ to move the variable to the Variable text box.
3. Click **Display normal curve**.
4. Click **OK**. The graph is shown in Figure 130.

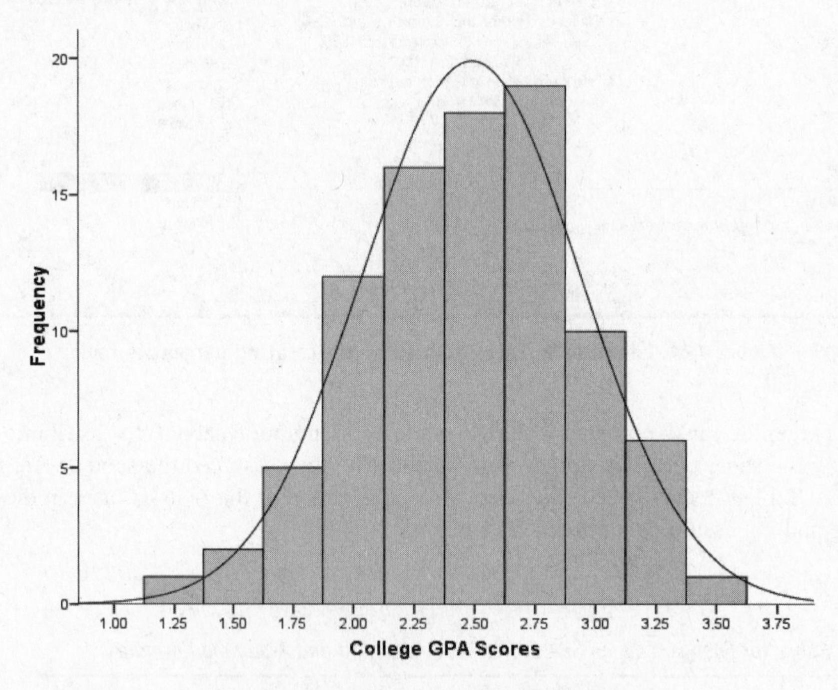

Figure 130. **A histogram of college GPAs.**

The histogram in Figure 130 has been edited. We deleted the mean, standard deviation, and *N*; changed the scale of the *X*-axis; and deleted the inner frame.

In the social sciences, many researchers think of their quantitative variables as being normally distributed. However, the variables are almost never normally distributed. By imposing the normal curve on the histogram, we can assess visually the degree to which college GPA scores approximate a normal distribution.

Creating a Stem-and-Leaf Plot

A stem-and-leaf plot, like the histogram, effectively communicates the distribution of a variable that has many possible values.

To create a stem-and-leaf plot for the high school GPA scores, follow these steps:

1. Click **Analyze**, click **Descriptive Statistics**, then click **Explore**.
2. Click **Reset** to clear the dialog box.
3. In the Explore dialog box, click **hsgpa**, then click ▶ to move it to the Dependent List box.
4. In the Display box, click **Plots**.
5. Click the **Plots** button. You will see the Explore: Plots dialog box in Figure 131.

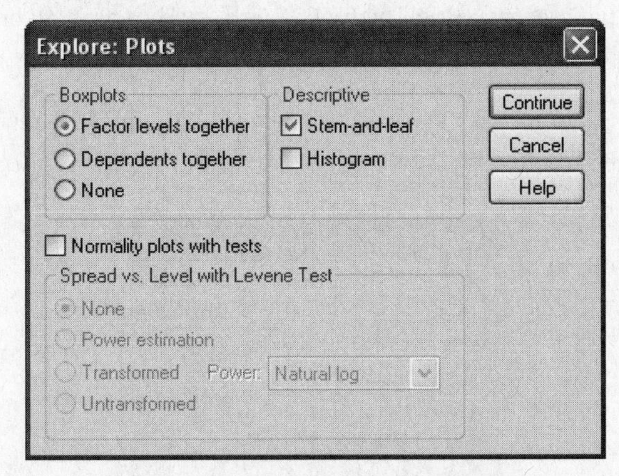

Figure 131. **Explore: Plots dialog box.**

6. Click the **Stem-and-leaf** box if it does not contain a check.
7. Under Boxplots, click **None**.
8. Click **Continue**.
9. Click **OK**. You can see the stem-and-leaf chart in Figure 132.

High School GPA Scores

```
High School GPA Scores Stem-and-Leaf Plot

 Frequency    Stem &  Leaf

    14.00        1 .  00111122333334
    11.00        1 .  56677888999
    15.00        2 .  000000111222234
    22.00        2 .  5555555566666667789999
    11.00        3 .  00001111223
     9.00        3 .  555778889
     8.00        4 .  00000000

 Stem width:       1.00
 Each leaf:        1 case(s)
```

Figure 132. **A stem-and-leaf plot of high school GPA scores.**

The stems in the stem-and-leaf plot are the leading digits of each score, and the leaves are the trailing digits of each score. In Figure 132, the first row of the stem-and-leaf plot finds the following digits: 1• 00111122333334. The numeral displayed before the dot is the stem, and the numerals after the dot are the leaves. At the bottom of the plot, SPSS reports the stem width to be 1.0, indicating that the value of the stems represents the units place. Attaching the first leaf of 0 to the stem of 1 produces the score of 1.0; attaching the leaf of 1 to the stem of 1 yields scores of 1.1, and so on. Consequently, the scores displayed in the first row are two 1.0s, four 1.1s, two 1.2s, five 1.3s, and one 1.4. Looking at the data, you can see that there are two values (1.00 and 1.06) summarized as 1.0s and four scores (1.12, 1.14, 1.17, and 1.19) summarized as 1.1s, and so on.

Creating a Boxplot for Different Levels of a Qualitative Variable

An alternative graph that presents a rich picture of data is a boxplot. To create a boxplot of the distributions of the college GPA scores by gender, follow these steps:

1. Click **Graphs**, click **Legacy Dialogs**, and then click **Boxplot**. You will see the Boxplot dialog box in Figure 133.

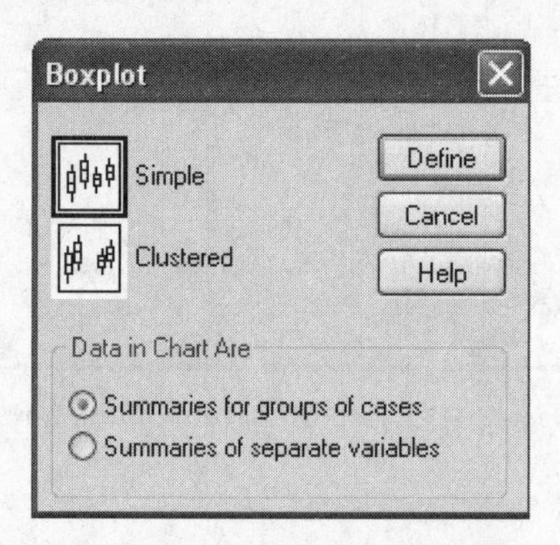

Figure 133. **The Boxplot dialog box.**

2. Click **Simple**, then click **Summaries for groups of cases**.
3. Click **Define**. You will see the Define Simple Boxplot: Summaries dialog box in Figure 134.

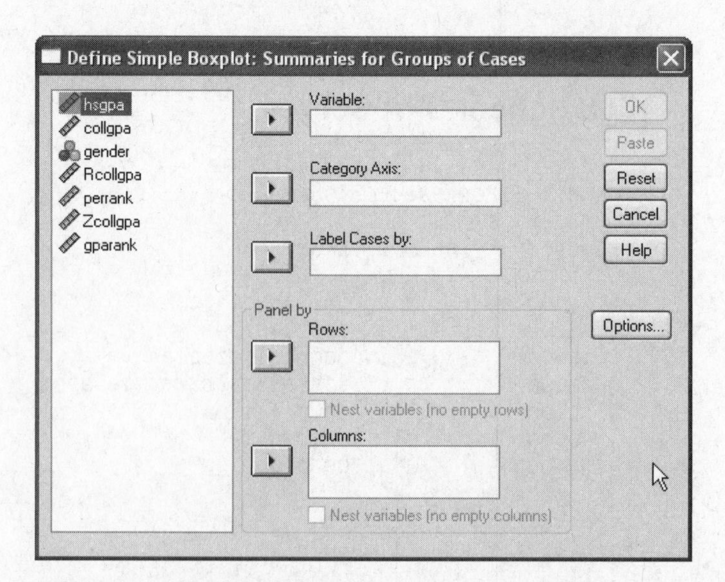

Figure 134. **The Define Simple Boxplot: Summaries dialog box.**

4. In the Define Simple Boxplot: Summaries dialog box, click **collgpa**, and click the top ▶ to move it to the Variable box.
5. Click **gender**, and click the middle ▶ to move it to the Category Axis box.
6. Click **OK**. The edited graph appears as shown in Figure 135.

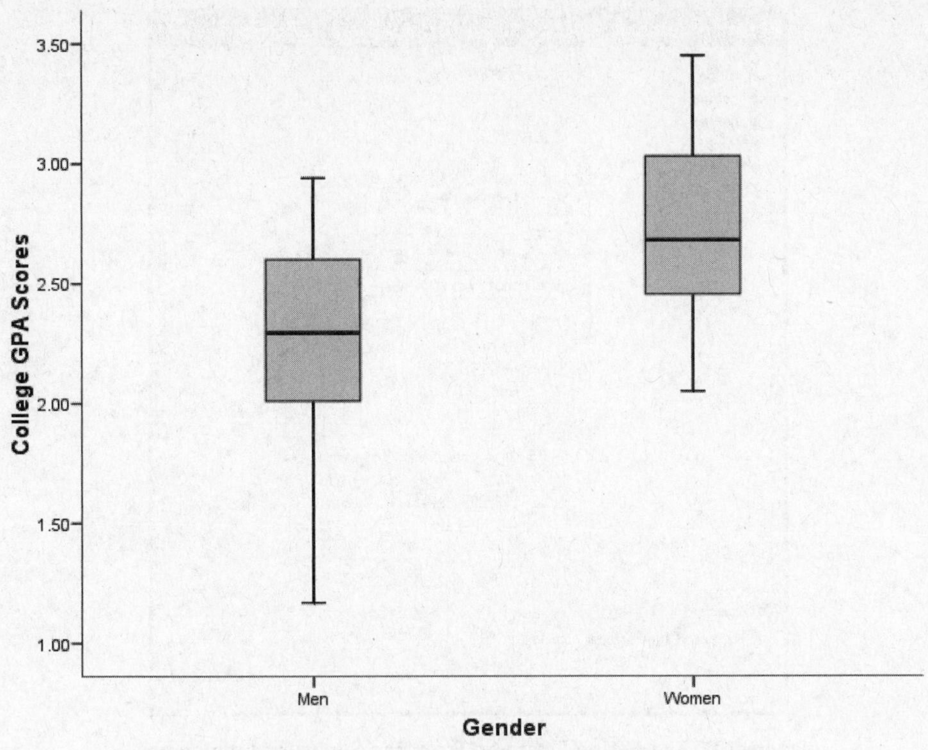

Figure 135. **Distributions of college GPA for men and women.**

Creating Error Bar Charts

A simple error bar chart involves displaying the means and standard deviations for different variables. To create an error bar chart for high school and college GPA scores by gender, follow these steps:

1. Click **Graphs**, click **Legacy Dialogs**, and then click **Error Bar**. You will see the Error Bar dialog box in Figure 136.

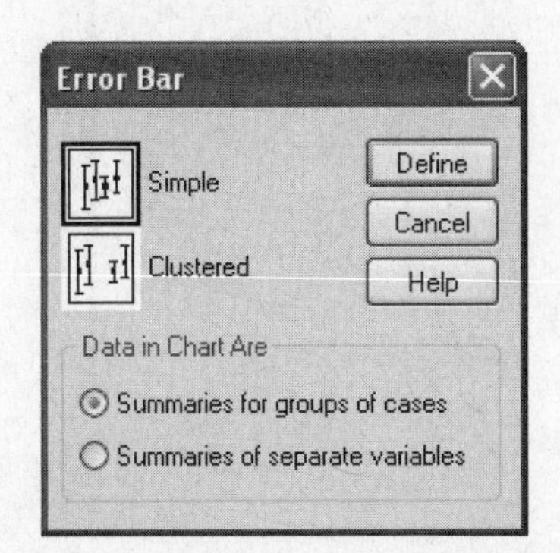

Figure 136. **The Error Bar dialog box.**

2. Click **Clustered**, then click **Summaries of Separate Variables**.
3. Click **Define**. You will see the Define Clustered Error Bar: Summaries . . . dialog box in Figure 137 (on page 158).
4. Holding down the Ctrl key, click **hsgpa** and **collgpa**. Click the top ▶ to move them to the Variables box.
5. Click **gender**, and click the second ▶ to move it to the Category Axis box.
6. Click on **standard deviation** in the Bars Represent drop-down menu.

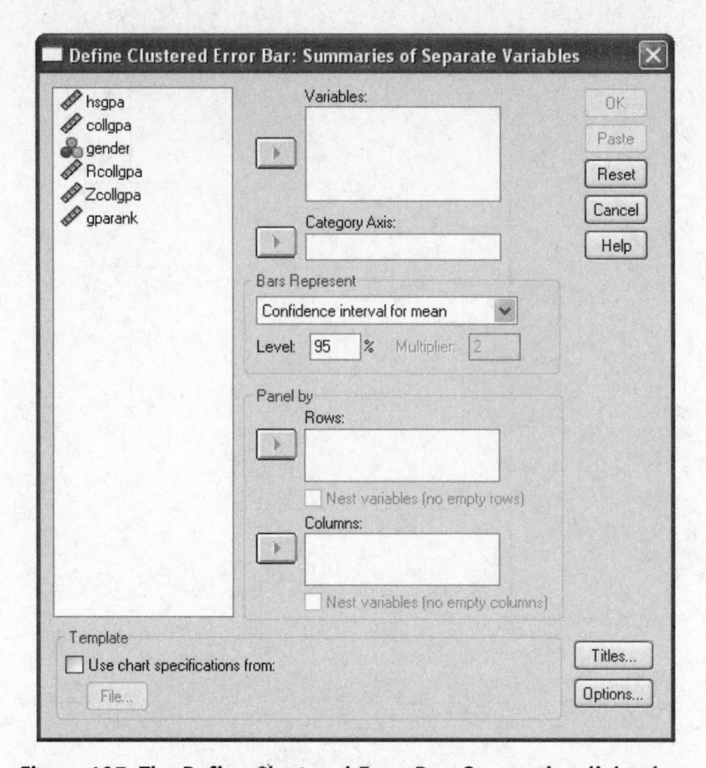

Figure 137. **The Define Clustered Error Bar: Summaries dialog box.**

7. The Multiplier box will now be available. Set the Multiplier at **2**.
8. Click **OK**. The edited chart is shown in Figure 138.

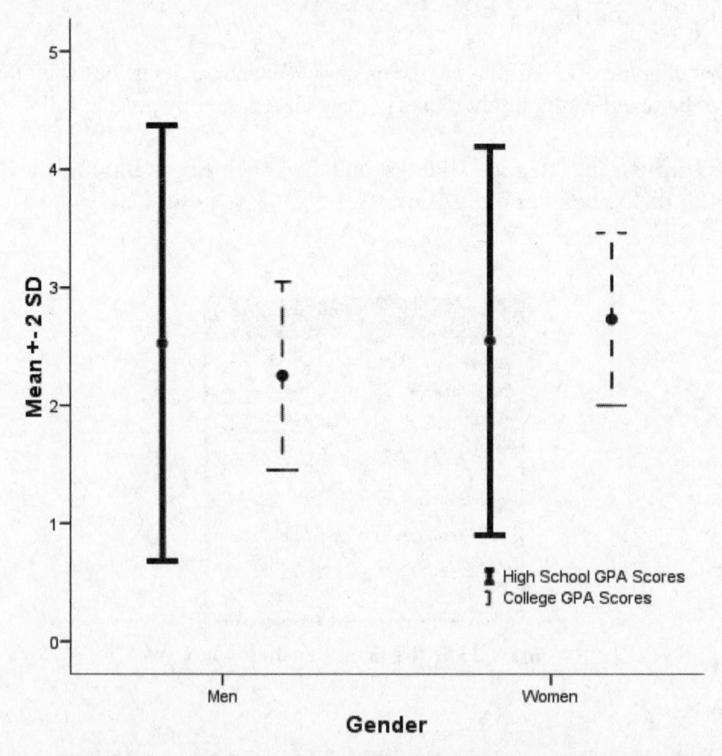

Figure 138. **Distributions of high school and college GPA scores for men and women.**

Other options could be chosen in the Bars Represent and Multiplier boxes. We chose standard deviation rather than standard error because we are interested in describing the data, not making inferences about it. Also, by setting the multiplier to 2, we can look at two standard deviations around each mean. We used two standard deviations because the error bars would encompass approximately 95% of the scores if they were normally distributed.

An APA Participants Section

The participants were 46 men and 44 women college students. Table 13 presents the means and standard deviations of high school and college GPAs by gender. The means and standard deviations of high school GPAs were relatively similar for men and for women. However, on the average, women had higher college GPAs than men did.

Table 13

Means and Standard Deviations for College and High School GPA Scores of Each Gender

	Men		Women	
	M	*SD*	*M*	*SD*
High school GPA	2.53	.92	2.55	.82
College GPA	2.25	.40	2.73	.37

We did not present results for the percentile rank analyses in the preceding Participants section because they are not typically reported in manuscripts.

Creating Figures in APA Format

Figures (i.e., graphs) can provide a visual representation of complex data or statistical results. Here are some guidelines for creating figures in APA format. See the *Publication Manual of the American Psychological Association* (2001) for more detailed information.

1. Label each figure with consecutive Arabic numbers (e.g., *Figure 1.*).
   - Italicize Figure, the number, follow by a period (e.g., *Figure 1.*, not Figure 1.).
   - Refer to each figure in the text.
   - Refer to figures in the text by their numbers (nonitalicized in text).
2. Figure captions should be brief and explain the figure's content.
   - Capitalize the first letter of the first word of a figure caption.
   - Do not italicize the caption.
   - End the caption with a period.
3. The presentation of figure numbers and captions depends on whether the manuscript containing the figures is in submission form (e.g., submitted manuscript to a journal) or in final form (e.g., in a dissertation or published journal).
   - For manuscripts in submitted form, no labels appear on individual figures. In its place, a page labeled Figure Captions precedes all figures. This page includes the figure number and caption of the first figure on the first line, the figure number and caption of the second figure on the second line, and so on.
   - For manuscripts in final form, the figure number followed by the figure caption is included at the bottom of individual figures.
4. Use the following guidelines for labeling the axes of graphs and legends.
   - Center the labels on the axes of the graphs.
   - Capitalize all important words on the axis labels and in the title of legends.
   - Capitalize the first word of legend entries.
5. All statistical symbols in figures should be italicized except subscripts and Greek symbols.

6. Content footnotes amplify information in figures and should communicate a single idea.
7. In the text of the manuscript, draw the reader's attention to the highlights of the information presented in the figure. Do not reiterate in the text all the information in the figure.

Creating Tables in APA Format

Besides figures, tables also provide a summary of results. We offer some guidelines for creating tables in APA format, but the *Publication Manual of the American Psychological Association* (2001) should be consulted for more detailed information.

1. Label each table with consecutive Arabic numbers (e.g., Table 1).
   - Do not italicize the table number (e.g., Table 1, not *Table 1*).
   - Refer to each table in the text.
   - Refer to tables in the text by their numbers.
2. Table titles should be brief and explain the table's contents.
   - Capitalize the first letter of each important word in the title.
   - Italicize table titles.
3. The table number comes first and the table title comes next on the following line.
4. Headings in tables provide labels for the information in the body of the table.
   - Provide a brief heading for each column.
   - Headings identify information below them, not across from them.
   - Capitalize the first word of headings.
   - Abbreviations may be used for statistical indices (e.g., *M* for mean and *SD* for standard deviation).
   - All statistical symbols should be italicized except subscripts and Greek symbols.
   - Set off a heading with a horizontal line if subheadings are below it.
   - Do not use vertical lines in headings.
5. The body of the table contains the data.
   - Do not use vertical lines to differentiate columns in a table's body.
   - Horizontal lines can be used in the body of more complex tables to divide it into sections. Provide a title in the center at the top of each demarcated section.
   - Decimalized values are generally reported to two decimal places.
   - A zero does not precede a value less than 1 if the statistic cannot be greater than 1 (e.g., a correlation coefficient).
   - Report *p* values to two decimal places, as a rule of thumb (e.g., $p < .01$).
6. Notes may be general, specific, or probability notes and are presented at the bottom of tables.
   - General notes provide information about the table as a whole and are designated by the word *Note*.
   - Specific notes refer to a specific column, row, or individual entry in the table and are designated with superscripted lowercase letters.
   - Probability notes indicate the level of significance and are designated with asterisks (e.g., $^{*}p < .05$ or $^{**}p < .01$).
7. In the text of the manuscript, draw the reader's attention to the highlights of the information presented in the table. Do not reiterate in the text all the information in the body of the table.

Exercises

The data for the Exercises 1 through 5 are in the data set named *Lesson 21 Exercise File 1* on the Web at http://www.prenhall.com/greensalkind. The data are from the following research problem.

David collects anxiety scores from 15 college students who visit the university health center during finals week.

1. Compute descriptive statistics on the anxiety scores. From the output, identify the following:
   a. Skewness
   b. Mean
   c. Standard deviation
   d. Kurtosis
2. Compute percentile ranks on the anxiety scores assuming that the distribution of scores is normal. What are the scores associated with the percentile ranks of 12, 27, 38, 73, and 88?
3. Compute percentile ranks on the anxiety scores not assuming that the distribution of scores is normal.
4. Create a histogram to show the distribution of the anxiety scores. Edit the graph so that most of the normal curve is visible.
5. Based on the histogram and the descriptive statistics, which percentile rank method should you use?

The data for the Exercises 6 through 8 are in the data set named *Lesson 21 Exercise File 2* on the Web at http://www.prenhall.com/greensalkind. The data are from the following research problem.

Michelle collects questionnaire data from 40 college students to assess whether they have a positive attitude toward political campaigns. She asks them to rate five statements about campaigns on a 5-point Likert scale (1 = disagree to 5 = agree). She also has information on respondents' political party affiliation (politic with 1 = Republican and 2 = Democrat). Table 14 shows the items on the scale.

Table 14
Political Campaign Attitude Items

Variables	Definition
att1	Use of inappropriate tactics in campaign strategies.
att2	Candidates address the issues.
att3	Accurate presentation of candidate's political agenda.
att4	Focus on issues relevant to the average citizen.
att5	Honesty in making campaign promises.

6. Compute total attitude scores from the scores for the five attitudinal items. The total attitude scores should reflect whether students have a general positive attitude toward campaigning.
7. Compute means on the total attitude scores for the two political parties.
8. Create boxplots showing the distributions of the total attitude scores for the two political parties.

PART II

UNIT 6 t Test Procedures

Unit 6 examines techniques that use a t test to assess hypotheses involving a single mean or differences between two means. The t test can be applied to address research questions for designs that involve a single sample, paired samples, or two independent samples.

Lesson 22 introduces a one-sample t test that evaluates whether a mean for a population is equal to a hypothesized test value based on the research hypothesis. Lesson 23 presents a paired-samples t test that assesses whether the mean difference between paired observations is significantly different from zero. Then, Lesson 24 discusses an independent-samples t test that evaluates whether means for two independent groups are significantly different from each other.

LESSON

22

One-Sample t Test

A one-sample t test evaluates whether the mean on a test variable is significantly different from a constant, called a test value by SPSS. Each case must have a score on one variable, the test variable.

A major decision in using the one-sample t test is choosing the test value. This test value typically represents a neutral point. Individuals who score higher than the neutral point are given one label, and those who score lower are given a different label. Those who fall exactly on the neutral point are given neither label.

Applications of the One-Sample t Test

Various applications of the one-sample t test are distinguished by the choice of the test value. The test value may be the

- midpoint on the test variable
- average value of the test variable based on past research
- chance level of performance on the test variable

The following are examples for each of the three types of applications.

Test Value as the Midpoint on the Test Variable

Dan wants to determine whether students have a negative or a positive view toward current world leaders. He develops the World Leader Scale (WLS), which includes the names of 15 world leaders. He asks 100 students to identify the country that each leader governs and to rate the leader on a scale from 0 to 10, where 0 = a totally worthless leader, 5 = neither a worthless nor a worthy leader, and 10 = a totally worthy leader. A WLS score for a student is the mean rating for the leaders whose countries were correctly identified. Dan conducts a one-sample t *test on the 100 WLS scores to determine if the population mean is significantly different from 5. The test value of 5 is chosen because a value less than 5 implies a negative view of world leaders, a score greater than 5 implies a positive view of world leaders, and a value of 5 implies neither a negative nor a positive view of world leaders. Dan's data file for analyzing the results contains 100 cases, one for each student, and one test variable, the WLS scores.*

Test Value as the Average Value of the Test Variable Based on Past Research

Janet is interested in determining if male adolescents who do not engage in sports are more depressed than the average male adolescent. To test this hypothesis, she obtains Kansas University Depression Inventory (KUDI) scores from 30 adolescent boys who have indicated on a questionnaire that they do not engage in sports outside of school gym classes. The KUDI is a depression measure that was standardized to have a mean of 50 for male adolescents. Janet conducts a one-sample t *test on the 30 KUDI scores to determine if the population mean is different from 50. The test value of 50 is chosen because a value less than 50 implies less depression than a typical male adolescent, a value greater than 50 implies more depression than a typical male adolescent, and a value of 50 implies neither less nor more depression than a typical male adolescent. Janet's SPSS data file for analyzing the results contains 30 cases, one for each adolescent, and one test variable, the KUDI scores.*

Test Value as the Chance Level of Performance on the Test Variable

Lena believes that the Matching Figures Test (MFT), a visual identification task, is too difficult for children who are younger than five years old. The MFT consists of 24 items. For each item, a child is shown a picture for two seconds. Five seconds later, the child is asked to select the same picture from a set of three pictures. Lena collects MFT scores on 75 four-year-olds and conducts a one-sample t *test on these scores to determine if the population mean is different from chance level. Because the likelihood of a child getting any item correct is one in three, and a child has 24 opportunities to obtain a correct answer, chance level on the MFT is 8 (1/3 times 24). A score less than the test value of 8 implies that 4-year-olds do worse than chance; a score greater than 8 implies that 4-year-olds do better than chance; and a value of 8 implies a score neither below nor above chance level. Lena's SPSS data file for analyzing the results contains 75 cases, one for each child, and one test variable, the MFT scores.*

Understanding the One-Sample *t* Test

We consider next the assumptions underlying the one-sample *t* test and an associated effect size statistic.

Assumptions Underlying the One-Sample *t* Test

There are two assumptions underlying a one-sample *t* test.

Assumption 1: The Test Variable Is Normally Distributed in the Population

In many applications with a moderate or larger sample size, the one-sample *t* test may yield reasonably accurate *p* values even when the normality assumption is violated. A commonly accepted value for a moderate sample size is 30. Larger sample sizes may be required to produce relatively valid *p* values if the population distribution is substantially nonnormal. In addition, the power of this test may be reduced considerably if the population distribution is nonnormal and, more specifically, thick-tailed or heavily skewed. See Wilcox (2001) for an extended discussion of assumptions.

Assumption 2: The Cases Represent a Random Sample from the Population, and the Scores on the Test Variable Are Independent of Each Other

The one-sample *t* test will yield an inaccurate *p* value if the assumption of independence is violated.

Effect Size Statistics for the One-Sample *t* Test

SPSS supplies all the information necessary to compute an effect size, *d*, given by:

$$d = \frac{Mean\ Difference}{SD}$$

where the mean difference and standard deviation are reported in the SPSS output. We can also compute d from the t value by using the equation

$$d = \frac{t}{\sqrt{N}}$$

where N is the total sample size.

d evaluates the degree that the mean scores on the test variable differ from the test value in standard deviation units. If d equals 0, the mean of the scores is equal to the test value. As d deviates from 0, the effect size becomes larger. Potentially, d can range in value from negative infinity to positive infinity. What is a small versus a large d is dependent on the area of investigation. However, d values of .2, .5, and .8, regardless of sign, are by convention interpreted as small, medium, and large effect sizes, respectively.

The Data Set

The data set used in this lesson is named *Lesson 22 Data File 1* on the Web at http://www.prenhall.com/greensalkind. It represents data from the depression example presented earlier. The single variable in the data set is shown in Table 15.

Table 15
Variables in Lesson 22 Data File 1

Variables	Definition
kudi	Values on the KUDI represent depression scores. Based on previous research, we know that young male adolescents have a mean of 50 on the KUDI. Accordingly, 50 is the test value.

The Research Question

The research question should reflect the difference between the mean of the test variable and the test value: Are male adolescents who do not engage in sports more or less depressed than the average male adolescent?

Conducting a One-Sample *t* Test

To conduct a one-sample t test, follow these steps:

1. Click **Analyze**, click **Compare Means**, then click **One-Sample T Test**. You will see the One-Sample T Test dialog box as shown in Figure 139 (on page 166).
2. Click **kudi**, then click ▶ to move it to the Test Variable(s) box.
3. Type **50** in the Test Value text box.
4. Click **OK**.

Selected SPSS Output for the One-Sample *t* Test

The results of the one-sample t test are shown in Figure 140 (on page 166). SPSS computes a mean and a standard deviation for the test variable. It also reports the mean difference value, which is the difference between the mean of the test variable and the hypothesized value. The difference between the mean KUDI value of 54.63 and the test value of 50 is 4.63. This difference indicates that male adolescents in the sample on the average were more depressed than the average adolescent.

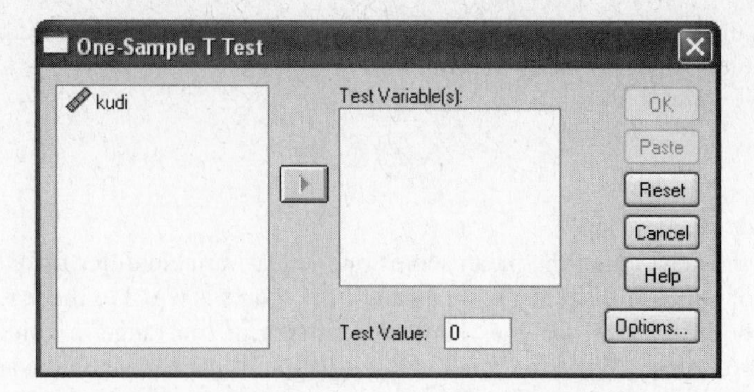

***Figure 139.* The One-Sample T Test dialog box.**

One-Sample Statistics

	N	Mean	Std. Deviation	Std. Error Mean
KUDI Depression Scale	30	54.63	10.327	1.886

One-Sample Test

	Test Value = 50					
					95% Confidence Interval of the Difference	
	t	df	Sig. (2-tailed)	Mean Difference	Lower	Upper
KUDI Depression Scale	2.457	29	.020	4.633	.78	8.49

***Figure 140.* The results of the one-sample *t* test.**

To determine whether the test was significant, examine the table labeled One-Sample Test. The test is significant, $t(29) = 2.46$, $p = .02$. The p value is located in the column labeled Sig. Because the p value is less than .05, we reject the null hypothesis that the population mean is equal to 50 at the .05 level.

Using SPSS Graphs to Display the Results

Although graphs are sometimes not included in the Results section of a manuscript because of space limitations, graphs should be presented when possible since they convey a rich understanding of the data. Two graphs that can display the data for a one-sample *t* test are the histogram and the stem-and-leaf plot. Lesson 21 presents steps for creating both types of graphs. Figure 141 is a histogram showing the distribution of KUDI scores for the 30 cases.

An APA Results Section

A one-sample *t* test was conducted on the KUDI scores to evaluate whether their mean was significantly different from 50, the accepted mean for male adolescents in general. The sample mean of 54.63 ($SD = 10.33$) was significantly different from 50, $t(29) = 2.46$, $p = .02$. The 95% confidence interval for the KUDI mean ranged from 50.78 to 58.49. The effect size d of .45 indicates a medium effect. Figure 141 shows the distribution of KUDI scores. The results support the conclusion that young male adolescents who do not engage in sports outside of school are somewhat more depressed than average.

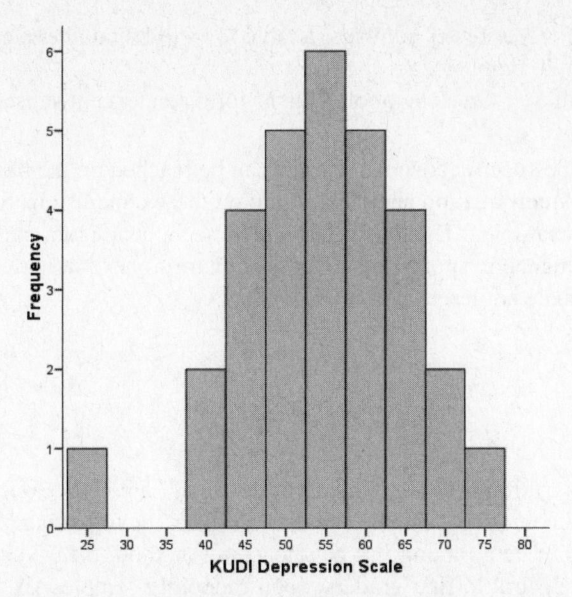

***Figure 141.* Distribution of KUDI scores for adolescent boys who do not participate in sports.**

Writing an APA Results Section

Here are guidelines for writing Results sections for statistical methods that require no follow-up procedures, such as the one-sample t test, correlations, the Mann-Whitney U test, or the binomial test.

1. Describe the test, the variables, and the purpose of the test. For example, "A one-sample t test was conducted on the KUDI scores to evaluate whether their mean was significantly different from 50, the accepted mean for male adolescents in general."
2. Report the results of the statistical test. For example, "With alpha set at .05, the one-sample t test was significantly different from 50, $t(29) = 2.46$, $p = .02$. The effect size d of .45 indicates a medium effect."
   - Discuss the assumptions of a test if necessary to describe why it was chosen or, more generally, why the test was valid. It is unnecessary to present routinely the assumptions of tests.
   - As a rule of thumb, report statistics to two decimal places.
   - State the alpha level chosen for a statistical test and whether the test is significant or not. The alpha level might be specified for individual tests when reported. Alternatively, the alpha level for all presented statistical tests might be stated in the Methods section or at the beginning of the Results section.
   - Report the test value, degrees of freedom, and significance level. When SPSS reports a p value of .000, we should indicate in the Results section that "$p < .01$."
   - Report a confidence interval when possible. A statistical test allows us to make a decision about whether we can or cannot reject a null hypothesis, while a confidence interval allows us to reach the same statistical decision, but also provides an interval estimate of the statistic of interest (e.g., mean, mean difference, or correlation). For example, "The 95% confidence interval for the KUDI mean ranged from 50.78 to 58.49, and therefore the hypothesis that the population KUDI mean is 50 was rejected at the .05 alpha level."
   - Report a statistic that allows the reader to make a judgment about the magnitude of the effect, such as a d statistic for a one-sample t test.
3. Report relevant descriptive statistics, such as the mean and the standard deviation for a one-sample t test.
   - For a simple analysis, the descriptive statistics can be reported in the text as $M = 54.63$, $SD = 10.33$.
   - APA Publication Manual (2001) offers the following interesting guideline for choosing among various reporting methods when presenting results: Use a

sentence for 3 or fewer numbers, a table for 4 to 20 numbers, and a graph for more than 20 numbers.

- Italicize all non-Greek symbols, like *M* for mean, except subscripts and superscripts.

4. Summarize the specific conclusions that can be reached on the basis of the analyses, but save interpretation and elaboration on these conclusions for a Discussion section. For example, "The results support the conclusion that male adolescents who do not engage in sports outside of school are somewhat more depressed than the average male adolescent."

Exercises

The data for Exercises 1 through 4 are in the data set named *Lesson 22 Exercise File 1* on the Web at http://www.prenhall.com/greensalkind. The data are from the following research problem.

John is interested in determining if a new teaching method, the Involvement Technique, is effective in teaching algebra to first graders. John randomly samples six first graders from all first graders within the Lawrence City School System and individually teaches them algebra with the new method. Next, the pupils complete an eight-item algebra test. Each item describes a problem and presents four possible answers to the problem. The scores on each item are 1 or 0 where 1 indicates a correct response, and 0 indicates a wrong response. The SPSS data file contains six cases, each with eight item scores for the algebra test.

1. Compute total scores for the algebra test from the item scores. A one-sample *t* test will be computed on the total scores.
2. What is the test value for this problem?
3. Conduct a one-sample *t* test on the total scores. On the output, identify the following:
   a. Mean algebra score
   b. *t* test value
   c. *p* value
4. Given the results of the children's performance on the test, what should John conclude? Write a Results section based on your analyses.

The data for Exercises 5 and 6 are in the data set named *Lesson 22 Exercise File 2* on the Web at http://www.prenhall.com/greensalkind. The data are from the following research problem.

As part of a larger study, Dana collected data from 20 college students on their emotional responses to classical music. Students listened to two 30-second segments from "The Collection from the Best of Classical Music." After listening to a segment, students rated it on a scale from 1 to 10, with 1 indicating "Makes me very sad" and 10 indicating "Makes me very happy." Dana computes a total score (hap_sad) for each student by summing the student's two ratings. Dana conducts a one-sample *t* test to evaluate whether classical music makes students sad or happy.

5. Conduct the analyses on these data, and write a Results section for it.
6. If you have not presented a graph in your Results section, create a histogram for the hap_sad scores. Label the graph following APA guidelines.

LESSON 23

Paired-Samples *t* Test

Each case must have scores on two variables for a paired-samples *t* test. The paired-samples *t* test procedure evaluates whether the mean of the difference between these two variables is significantly different from zero. It is applicable to two types of studies, repeated-measures and matched-subjects designs.

For a repeated-measures design, a participant is assessed on two occasions or under two conditions on one measure. In the SPSS data file created to conduct a paired-samples *t* test, each participant has scores on two variables. The first variable represents the first score on the measure, and the second variable represents the second score. The primary question of interest is whether the mean difference between the scores on the two occasions (or under the two conditions) is significantly different from zero.

For a matched-subjects design, participants are paired, and each participant in a pair is assessed once on a measure. Each pair of participants is a case in the SPSS data file and has scores on two variables, the score obtained by one participant under one condition and the score obtained by the other participant under the other condition. The primary question for the matched-subjects design is whether the mean difference in scores between the two conditions differs significantly from zero.

Applications of the Paired-Samples *t* Test

We discuss four applications of the paired-samples *t* test based on the design of a study.

- Repeated-measures designs with an intervention
- Repeated-measures designs with no intervention
- Matched-subjects designs with an intervention
- Matched-subjects designs with no intervention

Repeated-Measures Designs with an Intervention

Moe wants to assess the effectiveness of a leadership workshop for 60 middle managers. The 60 managers are rated by their immediate superiors on the Leadership Rating Form (LRF), before the workshop and after the workshop. Moe's SPSS data file contains 60 cases, one for each manager, and two variables, LRF before the workshop and LRF after the workshop.

Repeated-Measures Designs with No Intervention

Woody is interested in determining if workers are more concerned with job security or pay. He gains the cooperation of 30 individuals who work in different settings and asks each employee to rate his or her concern about salary level and job security on a scale from 1 to 10 (1 = no concern and 10 = ultimate concern). Woody's SPSS data file contains 30 cases, one for each employee, and two variables, ratings about concern for salary level and ratings about concern for job security.

Matched-Subjects Designs with an Intervention

Marilyn is interested in determining if individuals who are exposed to depressed individuals become sad. She administers the Mood Measure of Sadness (MMS) to 40 students and rank-orders their scores. She then pairs students with similar MMS scores. Students within a matched pair have more similar MMS scores than two students from different pairs. Marilyn then randomly assigns the two students within each matched pair to one of two conditions: one student within each pair is exposed to a depressed individual, and the other person is exposed to an individual who shows no signs of depression. MMS scores (which she calls post-MMS scores) are obtained on the students after they are exposed to a depressed or nondepressed individual. Marilyn's SPSS data file contains 20 cases, one for each matched pair of students, and two variables, the post-MMS scores for the students exposed to the depressed person and the post-MMS scores for the students exposed to the nondepressed individual.

Matched-Subjects Design with No Intervention

Kristy is interested in investigating whether husbands and wives with infertility problems feel equally anxious. She obtains the cooperation of 24 infertile couples and then administers the Infertility Anxiety Measure (IAM) to both the husbands and the wives. Kristy's SPSS data file contains 24 cases, one for each husband–wife pair, and two variables, the IAM scores for the husbands and the IAM scores for the wives. Note that in this study, the participants come to the study as couples or matched pairs, and that the conditions under which the experimenter obtains the IAM scores are the two levels of gender, men and women.

Understanding the Paired-Samples *t* Test

We next discuss the assumptions and associated effect size statistics for the paired-samples *t* test.

Assumptions Underlying the Paired-Samples *t* Test

Assumption 1: Difference Scores Are Normally Distributed in the Population

Difference scores are created by subtracting one of the variables from the other. In many applications with a moderate or larger sample size, the paired-samples *t* test may yield reasonably accurate *p* values even when the difference scores are not normally distributed in the population. A commonly accepted value for a moderate sample size is 30 pairs of scores. Larger sample sizes may be required to produce relatively valid *p* values if the population distribution is substantially nonnormal. In addition, the power of this test may be reduced considerably if the population distribution is nonnormal and, more specifically, thick-tailed or heavily skewed. See Wilcox (2001) for an extended discussion of assumptions.

Assumption 2: The Cases Represent a Random Sample from the Population, and the Difference Scores Are Independent of Each Other

The paired-samples *t* test yields inaccurate *p* values if the assumption of independence is violated.

Effect Size Statistics for the Paired-Samples *t* Test

SPSS supplies all the information necessary to compute two types of effect size indices, *d* and η^2 (eta square).

The d statistic may be computed by using the equation

$$d = \frac{Mean}{SD}$$

where the Mean and the SD (standard deviation) are reported in the output under Paired Differences. d also can be computed from the reported values for t and N (the number of pairs) as follows:

$$d = \frac{t}{\sqrt{N}}$$

The d statistic evaluates the degree that the mean of the difference scores deviates from 0 in standard deviation units. If d equals 0, the mean of the difference scores is equal to zero. As d diverges from 0, the effect size becomes larger. The value of d can range from negative infinity to positive infinity. What is a small versus a large d is dependent on the area of investigation. However, d values of .2, .5, and .8, regardless of sign, are, by convention, interpreted as small, medium, and large effect sizes, respectively.

An eta square (η^2) may be computed as an alternative to d. η^2 ranges in value from 0 to 1. An η^2 of 0 indicates that the mean of the difference scores is equal to zero. In contrast, a value of 1 indicates that the difference scores in the sample are all the same nonzero value (i.e., perfect replication). η^2 can be computed as follows:

$$\eta^2 = \frac{N\ Mean^2}{N\ Mean^2 + (N - 1)\ SD^2}$$

Alternatively, it can be computed as

$$\eta^2 = \frac{t^2}{t^2 + N - 1}$$

The Data Set

The data set in this lesson illustrates the paired-samples t test that Woody used to study workers' concerns about job security and pay. The data are in the file named *Lesson 23 Data File 1* on the Web at http://www.prenhall.com/greensalkind, and this file contains the variables shown in Table 16.

Table 16
Variables in Lesson 23 Data File 1

Variables	Definition
pay	1 to 10 rating, with higher scores indicating greater concern for salary level
security	1 to 10 rating, with higher scores indicating greater concern for job security

The Research Question

Research questions can be asked about either mean differences or relationships between variables.

1. Mean differences: Are employees more concerned on the average about job security or salary level?
2. Relationship between variables: Is degree of concern related to the job characteristics of pay and security?

Conducting a Paired-Samples *t* Test

To conduct a paired-samples *t* test, follow these steps:

1. Click **Analyze**, click **Compare Means**, then click **Paired-Samples T Test**. When you do this, you will see the Paired-Samples T Test dialog box shown in Figure 142.

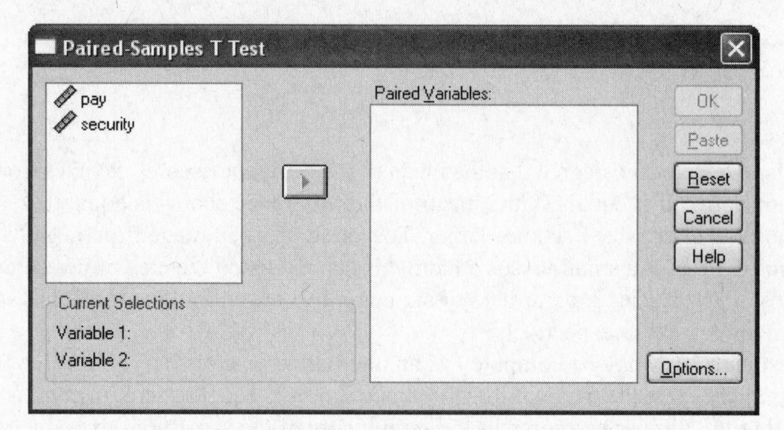

***Figure 142.* The Paired-Samples T Test dialog box.**

2. Click **pay**. Pay now appears as Variable 1 in Current Selections.
3. Click **security**. Security appears as Variable 2 in Current Selections.
4. Now click on ▶ and pay - security appears in the Paired Variables box.
5. Click **OK**.

Selected SPSS Output for the Paired-Samples *t* Test

Partial results of the analysis are shown in Figure 143. The mean difference between pay and security (1.17) is equal to the mean for pay minus the mean for security (5.67 − 4.50) and is shown under Paired Differences. The difference between the two means is computed based on the ordering of the two variables in the Paired Variables box. The variable that is listed second is subtracted from the one that is listed first.

Paired-Samples Statistics

		Mean	N	Std. Deviation	Std. Error Mean
Pair 1	concern for pay	5.67	30	1.493	.273
	concern for security	4.50	30	1.834	.335

Paired Samples Correlations

		N	Correlation	Sig.
Pair 1	concern for pay & concern for security	30	.088	.643

Paired-Samples Test

		Paired Differences					t	df	Sig. (2-tailed)
					95% Confidence Interval of the Difference				
		Mean	Std. Deviation	Std. Error Mean	Lower	Upper			
Pair 1	concern for pay - concern for security	1.167	2.260	.413	.323	2.011	2.827	29	.008

***Figure 143.* Partial results of the paired-samples *t* test.**

To determine whether the test was significant, examine the table labeled Paired-Samples Test. The test is significant, $t(29) = 2.83, p < .01$. The p value is located in the column labeled Sig. Because the p value is less than .05, we reject the null hypothesis that the population mean difference is equal to 0 at the .05 level.

Using SPSS Graphs to Display the Results

Graphs are useful for illustrating not only the differences between means but also the overlap of the two distributions. For the paired-samples *t* test, error bar charts or boxplots of the paired variables could be used. A histogram of the difference scores could also be used. In Figure 144, we present the results of the pay and security ratings with boxplots. To create these boxplots, follow these steps:

1. Click **Graphs**, click **Legacy Dialogs**, and then click **Boxplot**.
2. Click **Simple**, then click **Summaries of separate variables**. Click **Define** in the Boxplots dialog box.
3. Holding down the Ctrl key, click **pay** and **security**, and click ▶ to move them to the Boxes Represent.
4. Click **OK**. The edited graph is shown in Figure 144.

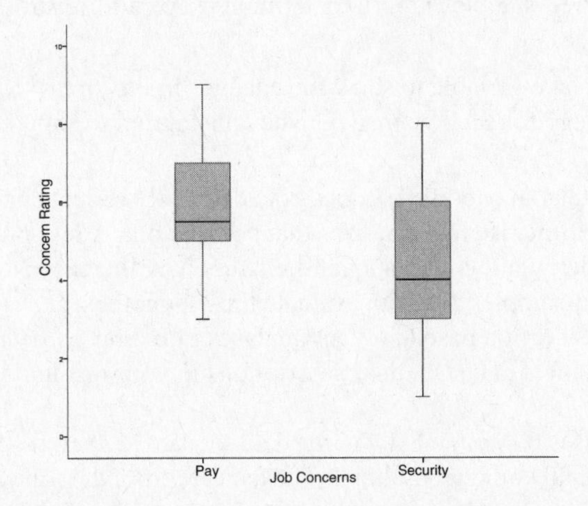

***Figure 144.* Boxplots of pay and security ratings.**

An APA Results Section

A paired-samples *t* test was conducted to evaluate whether employees were more concerned with pay or job security. The results indicated that the mean concern for pay (M = 5.67, SD = 1.49) was significantly greater than the mean concern for security (M = 4.50, SD = 1.83), $t(29) = 2.83, p < .01$. The standardized effect size index, d, was .52, with considerable overlap in the distributions for the 10-point Likert ratings of pay and security, as shown in Figure 144. The 95% confidence interval for the mean difference between the two ratings was .32 to 2.01.

Alternative Analyses

Data for a paired-samples *t* test can be analyzed by using a repeated-measures analysis of variance (see Lesson 29). The paired-samples *t* test and the repeated-measures ANOVA yield identical results in that the p values associated with the two tests are always the same. The advantage of the repeated-measures ANOVA over the paired-samples *t* test is that the former computes the effect size statistic, η^2 (labeled Eta Squared).

Data for a paired-samples *t* test also can be analyzed with the use of nonparametric procedures. (See Lesson 44.) Nonparametric procedures may be considered if the population distribution of the difference scores is not normal.

Exercises

The data for Exercises 1 through 5 are in the data set named *Lesson 23 Exercise File 1* on the Web at http://www.prenhall.com/greensalkind. The data are from the following research problem.

Mike, a developmental psychologist, wants to know if overall life stress increases or decreases as working women grow older. He obtains scores on the ILS (Index of Life Stress) from a group of 100 working women when they are 40 years of age. He is able to obtain a second ILS from 45 of these women at age 60 to assess whether they felt significantly more or less stress as they grew older. The ILS consists of two scores, interpersonal life stress and occupational life stress. The two scores combine to form an index of overall life stress. The SPSS data file contains 45 cases, one for each woman, and four variables, their interpersonal and occupational life stress at age 40 and their interpersonal and occupational life stress at age 60.

1. Compute scores to obtain a total Index of Life Stress (ILS) at age 40 and age 60.
2. Compute a paired-samples *t* test to determine if overall life stress increases or decreases with age.
3. Create a difference variable to show the changes in life stress from 40 years of age to 60 years of age for each woman. Create a histogram to show these changes graphically.
4. Mike decides that an overall ILS does not adequately reflect changes in women's life stress over time. He hypothesizes that occupational stress probably declines as women get older, while interpersonal life stress may increase or stay the same. Conduct paired-samples *t* tests to evaluate these hypotheses.
5. Write a Results section based on your analyses in Exercises 1 through 4. Be sure to include graphical and statistical descriptions in your results.

The data for Exercises 6 through 8 are in the data set named *Lesson 23 Exercise File 2* on the Web at http://www.prenhall.com/greensalkind. The data are from the following research problem.

Kristy is interested in investigating whether husbands and wives who are having infertility problems feel equally anxious. She obtains the cooperation of 24 infertile couples. She then administers the Infertility Anxiety Measure (IAM) to both the husbands and the wives. Her SPSS data file contains 24 cases, one for each husband–wife pair, and two variables, the IAM scores for the husbands and the IAM scores for the wives.

6. Conduct a paired-samples *t* test on these data. On your output, identify the following:
   a. mean IAM score for husbands
   b. mean IAM score for wives
   c. *t* test value
   d. *p* value
7. Write a Results section in APA style based on your output.
8. If you did not include a graph in your Results section, create a boxplot to show the differences between husbands' and wives' IAM scores.

LESSON 24

Independent-Samples *t* Test

The independent-samples *t* test evaluates the difference between the means of two independent groups. With an independent-samples *t* test, each case must have scores on two variables, the grouping variable and the test variable. The grouping variable divides cases into two mutually exclusive groups or categories, such as men or women for the grouping variable gender, while the test variable describes each case on some quantitative dimension such as verbal comprehension. The *t* test evaluates whether the mean value of the test variable for one group differs significantly from the mean value of the test variable for the second group.

Applications of the Independent-Samples *t* Test

Independent-samples *t* tests are used to analyze data from different types of studies.

- Experimental studies
- Quasi-experimental studies
- Field studies

We illustrate next the three types of studies.

Experimental Study

Bart is interested in determining if individuals talk more when they are nervous. To evaluate his hypothesis, he obtains 30 volunteers from undergraduate classes to participate in his experiment. He randomly assigns 15 students to each of two groups. Next, he schedules separate appointments for every student. When students in the first group arrive for their appointments, they are told that in 10 minutes they will be asked to describe how friendly they are by completing a true-false personality measure. In contrast, students in the second group are told that in 10 minutes they will be asked a set of questions by a panel of measurement specialists to determine how friendly they are. In both groups, the conversations between the students and the experimenter are recorded for 10 minutes. During this 10-minute period, the experimenter makes prescribed comments if no conversation occurs within a 1-minute time interval and responds to student questions with a brief response such as yes or no. The percent of time that each student talks during the 10-minute period is recorded. Bart's SPSS data file includes a grouping variable and a test variable for 30 cases. The grouping variable distinguishes between the low-stress group, whose members anticipate completing a personality measure, and the high-stress group, whose members anticipate questions from a panel of measurement specialists. The test or dependent variable is the percent of time that each student talks during the 10-minute period.

Quasi-Experimental Study

Mike is interested in assessing whether college students are more likely to argue about their test results with men professors or with women professors. He restricts the sample to professors who teach subjects commonly taught by both men and women, such as history, social work, psychology, education, English, and foreign languages. Mike obtains the permission of 20 men and 20 women professors to observe their classes the day they return graded midterm exams. In each class, Mike records the proportion of students who complain about some aspect of the midterm and how it was graded. Mike's SPSS data file includes both a grouping variable and a test variable for 40 cases. The grouping variable distinguishes between men and women professors, while the test variable is the proportion of students who complained about the midterm.

Field Study

Elaine is interested in determining if men prisoners who are classified as depressed eat less of their meals than those prisoners who are not classified as depressed. Of the 500 prisoners who volunteer for the study, 50 prisoners are classified as depressed. Elaine then randomly samples 50 prisoners from the nondepressed prisoners for use as a comparison or control group. She then monitors for a week the amount of each meal eaten by the 100 participants. To determine the amount of each meal eaten, she weighs their plates before and after each meal. At the end of the week, she computes the proportion of food that was eaten by each prisoner during the week. Elaine's SPSS data file includes both a grouping and a test variable for 100 cases. The grouping variable for the t *test distinguishes prisoners who are depressed and not depressed. The test variable is the proportion of food eaten by each inmate.*

Understanding the Independent-Samples *t* Test

We discuss next the assumptions and associated effect size statistics for the independent-samples *t* test.

Assumptions Underlying the Independent-Samples *t* Test

Assumption 1: The Test Variable is Normally Distributed in Each of the Two Populations (as Defined by the Grouping Variable)

In many applications with a moderate or larger sample size, the independent-samples *t* test may yield reasonably accurate *p* values even when the normality assumption is violated. In some applications with nonnormal populations, a sample size of 15 cases per group might be sufficiently large to yield fairly accurate *p* values. Larger sample sizes may be required to produce relatively valid *p* values if the population distributions are substantially nonnormal. In addition, the power of this test may be reduced considerably if the population distributions are nonnormal and, more specifically, thick-tailed or heavily skewed. See Wilcox (2001) for an extended discussion of assumptions.

Assumption 2: The Variances of the Normally Distributed Test Variable for the Populations are Equal

To the extent that this assumption is violated and the sample sizes differ for the two populations, the *p* value from the independent-samples *t* test should not be trusted. The independent-samples *t* test procedure, however, computes an approximate *t* test that does not assume that the population variances are equal in addition to the traditional *t* test that assumes equal population variances.

Assumption 3: The Cases Represent a Random Sample from the Population, and the Scores on the Test Variable are Independent of Each Other

If the independence assumption is violated, the *p* value should not be trusted.

Effect Size Statistics for the Independent-Samples *t* Test

SPSS supplies all the information necessary to compute two effect size indices, *d* and η^2 (eta square).

The *d* statistic may be computed by using the equation

$$d = \frac{Mean\ Difference}{SD_{\text{pooled}}}$$

where the mean difference is reported in the SPSS output, but the pooled standard deviation (SD_{pooled}) is not. SD_{pooled} can be calculated from the reported standard deviations for the two groups by using the equation

$$SD_{\text{pooled}} = \sqrt{\frac{(N_1 - 1)SD_1^2 + (N_2 - 1)SD_2^2}{N_1 + N_2 - 2}}$$

However, it is easier to compute *d* from the following equation:

$$d = t\sqrt{\frac{N_1 + N_2}{N_1 N_2}}$$

d can range in value from negative infinity to positive infinity. The value of 0 for *d* indicates that there are no differences in the means. As *d* diverges from 0, the effect size becomes larger. What is a small versus a large *d* is dependent on the area of investigation. However, *d* values of .2, .5, and .8, regardless of sign, are, by convention, interpreted as small, medium, and large effect sizes, respectively.

Eta square, η^2, may be computed as an alternative to *d*. An η^2 ranges in value from 0 to 1. It is interpreted as the proportion of variance of the test variable that is a function of the grouping variable. A value of 0 indicates that the difference in the mean scores is equal to 0, while a value of 1 indicates that the sample means differ, but the test scores do not differ within each group (i.e., perfect replication). You can compute η^2 with the following equation:

$$\eta^2 = \frac{\frac{N_1 N_2}{N_1 + N_2}(Mean\ Difference)^2}{\frac{N_1 N_2}{N_1 + N_2}(Mean\ Difference)^2 + (N_1 + N_2 - 2)SD_{\text{pooled}}^2}$$

Alternatively, we can say

$$\eta^2 = \frac{t^2}{t^2 + (N_1 + N_2 - 2)}$$

What is a small versus a large η^2 is dependent on the area of investigation. However, η^2 of .01, .06, and .14 are, by convention, interpreted as small, medium, and large effect sizes, respectively.

The Data Set

The data set used to illustrate the independent-samples *t* test is named *Lesson 24 Data File 1* on the Web at http://www.prenhall.com/greensalkind. The file presents data from the example about anxiety and talking that we gave earlier. The variables in the data set are described in Table 17 (on page 178).

The Research Question

The research question can be stated to reflect mean differences or relationships between variables.

1. Mean differences: Does the average amount of talking differ under low-stress versus high-stress conditions?
2. Relationship between variables: Is percent of time that students talk related to level of stress?

Table 17
Variables in Lesson 24 Data File 1

Variables	Definition
stress	Stress is the grouping or independent variable. The stress variable distinguishes between two treatment conditions. If stress = 1, then a student was in the low-stress condition and anticipated completing a personality measure. If stress = 2, then a student was in the high-stress condition and anticipated questions from a panel of measurement specialists.
talk	Talk is the test or dependent variable. It is defined as the percent of time that each student talked during the 10-minute test period.

Conducting an Independent-Samples *t* Test

To conduct an independent-samples *t* test, follow these steps:

1. Click **Analyze**, click **Compare Means**, then click **Independent-Samples T Test**. You will see the Independent-Samples T Test dialog box shown in Figure 145.

TIP

The independent-samples *t* test procedure also allows you to define a grouping variable by using a cut point on a quantitative variable in the Define Groups dialog box. All individuals at or above the cut point are in one group, and all those below the cut point are in the second group. You should be selective about using this option. An independent-samples *t* test with a dichotomized variable (created using the cut-point option) generally has less power than a test of a correlation with a non-dichotomized variable. However, an independent-samples *t* test using a dichotomized quantitative variable might be conducted for non-statistical reasons if the groups that are defined by the cut point have substantive meaning beyond that of the quantitative scores themselves. For example, 70 has traditionally been used as the cut point between normality and abnormality on the MMPI, a measure of psychopathology.

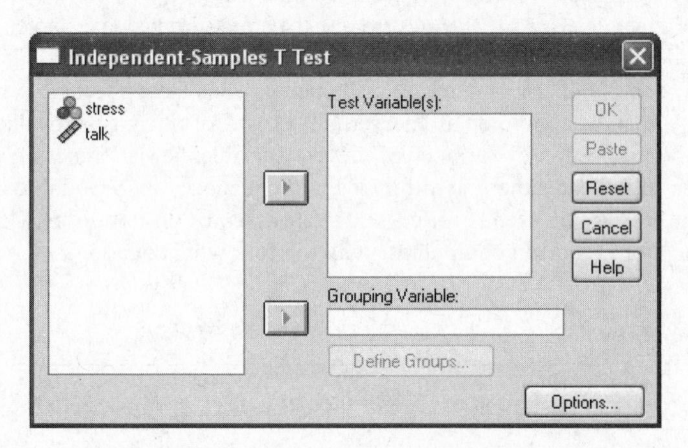

Figure 145. **The Independent-Samples *t* Test dialog box.**

2. Click **talk**, then click ▶ to move it to the Test Variable(s) area.
3. Click **stress**, then click ▶ to move it to the Grouping Variable area.
4. Click **Define Groups**. You will see the Define Groups dialog box shown in Figure 146.

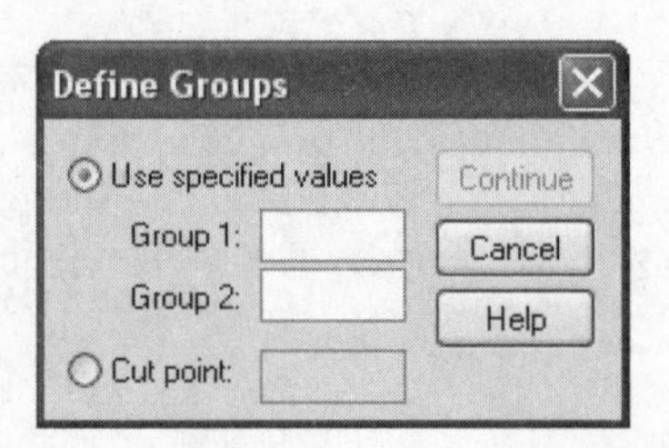

Figure 146. **The Define Groups dialog box.**

5. In the Group 1 box, type **1**.
6. In the Group 2 box, type **2**.
7. Click **Continue**.
8. Click **OK**.

Selected SPSS Output for the Independent-Samples *t* Test

You can see the results of the analysis in Figure 147. The mean difference is obtained by subtracting the mean for Group 2 from the mean for Group 1, as Groups 1 and 2 were defined in the Define Groups dialog box. The sign of the mean difference dictates the sign of the *t* value. The results should be interpreted accordingly. For our example, students in the low-stress group were defined as a 1 on the grouping variable, and students in the high-stress group were defined as a 2. The mean difference on the test variable talk, 23.13, was obtained by subtracting the mean (22.07) for Group 2 (high stress) students from the mean (45.20) for Group 1 (low stress) students. The positive *t* value indicates that the mean amount of talking for the students in the low-stress group is significantly greater than the mean for the students in the high-stress group.

Group Statistics

	stress	N	Mean	Std. Deviation	Std. Error Mean
Percent Time Talking	Low Stress	15	45.20	24.969	6.447
	High Stress	15	22.07	27.136	7.006

Independent Samples Test

		Levene's Test for Equality of Variances		t-test for Equality of Means						
									95% Confidence Interval of the Difference	
		F	Sig.	t	df	Sig. (2-tailed)	Mean Difference	Std. Error Difference	Lower	Upper
Percent Time Talking	Equal variances assumed	.023	.881	2.430	28	.022	23.133	9.521	3.630	42.637
	Equal variances not assumed			2.430	27.808	.022	23.133	9.521	3.624	42.643

***Figure 147.* Partial results of the independent-samples *t* test.**

Levene's test evaluates the assumption that the population variances for the two groups are equal. If the test is significant, we may conclude that the equality-of-variance assumption is violated. If the variances for the two groups are different and the sample sizes are unequal, you should not report the standard *t* value, but rather the *t* value that does not assume equal variances. An argument can be made for always reporting the *t* value for unequal variances and, thereby, avoiding the homogeneity-of-variance assumption.

In our example, the variances are very similar and, consequently, the standard *t* test, $t(28) = 2.43$, $p = .02$, and the *t* test for unequal variances, $t(27.81) = 2.43$, $p = .02$, yield comparable results. As shown in this example, the degrees of freedom for the *t* test for unequal variances may not be an integer.

Using SPSS Graphs to Display the Results

A variety of graphical displays can help the reader understand the results of the independent-samples *t* test analysis. Common examples of graphs used to display mean differences and distributions of variables across multiple groups are error bar graphs, which display the means and standard deviations for each group, and boxplots. (See Lesson 21 for a discussion of the boxplot.) Figure 148 is an error bar graph that shows the means and standard deviations of the percent of time spent talking in each of the stress groups.

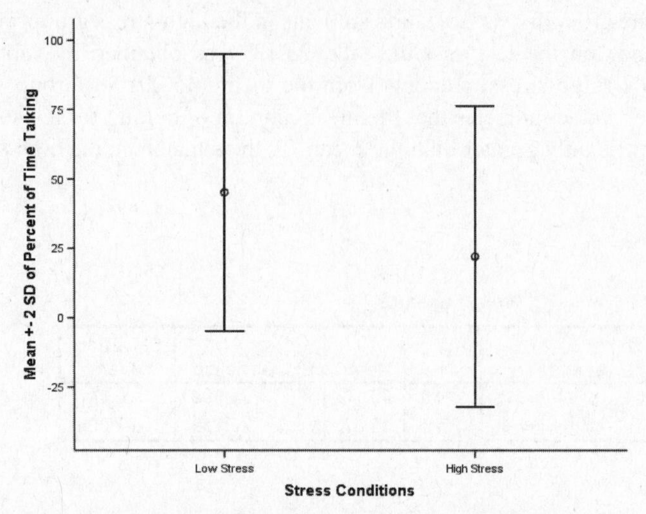

Figure 148. **Error bars (two standard deviations above and below the mean) for the percent of time talking scores for each stress group.**

An APA Results Section

An independent-samples *t* test was conducted to evaluate the hypothesis that students talk more under a high-stress condition (anticipating questions from a panel of measurement specialists) as opposed to a low-stress condition (anticipating taking a personality measure). The test was significant, $t(28) = 2.43$, $p = .02$, but the results were counter to the research hypothesis. Students in the high-stress condition ($M = 22.07$, $SD = 27.14$) on the average talked less than those in the low-stress condition ($M = 45.20$, $SD = 24.97$). The 95% confidence interval for the difference in means was quite wide, ranging from 3.63 to 42.64. The eta square index indicated that 17% of the variance of the talk variable was accounted for by whether a student was assigned to a low-stress or a high-stress condition. Figure 148 shows the distributions for the two groups.

Alternative Analyses

The type of data described in this lesson can also be analyzed by using one-way analysis of variance (see One-Way ANOVA, Lesson 25). The independent-samples *t* test assuming equal variances and the one-way ANOVA yield identical results in that the *p* values are the same. The advantage of the independent-samples *t* test procedure over one-way ANOVA using the General Linear Model—Univariate procedure is that the *t* test procedure calculates a *t* test that does not require the population variances to be equal. On the other hand, an effect size statistic, labeled eta square, is computed by using methods described in Lesson 25. Eta square is not available in the independent-samples *t* test procedure.

Data for an independent-samples *t* test can also be analyzed by using nonparametric procedures. (See Lesson 42.) If the assumptions of the independent-samples *t* test are met, the parametric test is a more powerful analysis than nonparametric alternatives. If the normality assumption for an independent-samples t test is not met, the nonparametric alternative may be more powerful in some circumstances.

Exercises

The data for Exercises 1 through 5 are in the data set named *Lesson 24 Exercise File 1* on the Web at http://www.prenhall.com/greensalkind. The data are from the following research problem.

Billie wishes to test the hypothesis that overweight individuals tend to eat faster than normal-weight individuals. To test this hypothesis, she has two assistants sit in a McDonald's restaurant and identify individuals who order the Big Mac special (Big Mac, large fries, and large Coke) for lunch. The Big Mackers, as they are affectionately called by the assistants, are classified as overweight, normal-weight, or neither overweight nor normal weight. The assistants identify 10 overweight and 30 normal-weight Big Mackers. The assistants record the amount of time it takes for the individuals to complete their Big Mac special meals. The SPSS data file contains two variables, a grouping variable with two levels, overweight (= 1) and normal weight (= 2), and time in seconds to eat the meal.

1. Compute an independent-samples t test on these data. Report the t values and the p values assuming equal population variances and not assuming equal population variances.
2. On the output, identify the following:
   a. Mean eating time for overweight individuals
   b. Standard deviation for normal weight individuals
   c. Results for the test evaluating homogeneity of variances
3. Compute an effect size that describes the magnitude of the relationship between weight and the speed of eating Big Mac meals.
4. Write a Results section based on your analyses.
5. If you did not include a graph in your Results section, create a graph in APA format that shows the differences between the two groups.

The data for Exercises 6 through 10 are in the data set named *Lesson 24 Exercise File 2* on the Web at http://www.prenhall.com/greensalkind. The data are from the following research problem.

Fred wishes to determine whether children in regular education are hurt or helped academically by the inclusion of special education students in their classes. He has gained the cooperation of the seventh-grade teachers in two similar junior high schools to participate in his study. One school recently integrated special education students into their regular classroom, while the other one has not. The teachers in both schools supply Fred the students' raw scores on a standardized achievement test at the beginning of the year and at the end of the year. The SPSS data file contains 40 cases and three variables: (1) a factor called integrat with two levels, regular education students who are not in integrated classrooms (integrat = 1) and regular education students who are in the integrated classrooms (integrat = 2), (2) a pretest, raw scores on the standardized achievement test obtained at the beginning of the year, and (3) a posttest, raw scores on the same achievement measure obtained at the end of the year.

6. Create a dependent variable by computing a difference score between the pretest and posttest achievement scores.
7. Conduct an independent-samples t test to evaluate the effectiveness of integration on academic achievement.
8. What did the Levene's test evaluate?
9. Which t ratio is the appropriate result? Why?
10. Write a Results section based on your analyses. What would you conclude?

PART II

UNIT 7 Univariate and Multivariate Analysis-of-Variance Techniques

Unit 7 covers univariate and multivariate analysis-of-variance (ANOVA) techniques. These procedures assess the relationship of one or more factors with a dependent variable (univariate ANOVA) or with multiple dependent variables (MANOVA). The factors are either between-subjects or within-subjects factors. A between-subjects factor divides research participants into different groups such as gender or multiple treatments. A within-subjects factor has multiple levels, and each participant is observed on a dependent variable across those levels. An example of a within-subjects factor might be trials, with each subject having scores on the dependent variable for all trials.

Lesson 25 deals with one-way ANOVA, which allows us to analyze mean differences between two or more groups on a between-subjects factor. Lesson 26 describes two-way ANOVA, a procedure that relates two between-subjects factors to a dependent variable. Lesson 27 considers analysis of covariance (ANCOVA), an approach that evaluates differences between two or more groups on a dependent variable, statistically controlling for differences on one or more covariates. Lesson 28 presents one-way MANOVA, which evaluates the relationship between a single between-subjects factor and two or more dependent variables. Lesson 29 deals with one-way repeated-measures ANOVA. This technique allows us to analyze mean differences between two or more levels of a within-subjects factor. Lesson 30 describes two-way repeated-measures ANOVA, which relates two within-subjects factors to a dependent variable.

SPSS also can be used to analyze data from other analysis-of-variance designs. For example, analyses can be conducted that have three between-subjects factors or one between-subjects factor and one within-subjects factor. These analyses use the same procedures but require specifying slightly different choices in the dialog boxes. We briefly discuss methods used to conduct ANOVAs for designs with more than two between-subjects factors in Lesson 26.

LESSON

25

One-Way Analysis of Variance

For a one-way analysis of variance (one-way ANOVA), each individual or case must have scores on two variables: a factor and a dependent variable. The factor divides individuals into two or more groups or levels, while the dependent variable differentiates individuals on a quantitative dimension. The ANOVA *F* test evaluates whether the group means on the dependent variable differ significantly from each other. Each case in an SPSS data file used to conduct a one-way ANOVA contains a factor that divides participants into groups and one quantitative dependent variable.

Applications of One-Way ANOVA

A one-way ANOVA can be used to analyze data from different types of studies.

- Experimental studies
- Quasi-experimental studies
- Field studies

We will illustrate two of the three types of studies.

Experimental Study

Dana wishes to assess whether vitamin C is effective in the treatment of colds. To evaluate her hypothesis, she decides to conduct a 2-year experimental study. She obtains 30 volunteers from undergraduate classes to participate. She randomly assigns an equal number of students to three groups: placebo (group 1), low doses of vitamin C (group 2), and high doses of vitamin C (group 3). In the first and second years of the study, students in all three groups are monitored to assess the number of days that they have cold symptoms. During the first year, students do not take any pills. In the second year of the study, students take pills that contain one of the following: no active ingredients (group 1), low doses of vitamin C (group 2), or high doses of vitamin C (group 3). Dana's SPSS data file includes 30 cases and two variables: a factor distinguishing among the three treatment groups and a dependent variable, the difference in the number of days with cold symptoms in the first year versus the number of days with cold symptoms in the second year.

Quasi-Experimental Study

Karin believes that students with behavior problems react best to teachers who have a humanistic philosophy and have firm control of their classrooms. In a large city school system, she classifies 40 high school teachers of students with behavior problems into three categories: humanists with firm control (group 1, n = 9); strict disciplinarians (group 2, n = 21), and keepers of the behavior problems (group 3, n = 10). Karin is interested in assessing whether students who are taught by one type of teacher as opposed to another type of teacher are more likely to stay out of difficulty in school. By reviewing the records of all students in these 40 teachers' classes, she determines the number of times the students of each teacher were sent to the main office for disciplinary action during the second half of the academic year. All classes had 30 students in them. Karin's SPSS data file has 40 cases with a factor distinguishing among the three types of teachers and a dependent variable, the number of disciplinary actions for students in a class.

Understanding One-Way ANOVA

An overall analysis-of-variance test is conducted to assess whether means on a dependent variable are significantly different among groups. If the overall ANOVA is significant and a factor has more than two levels, follow-up tests are usually conducted. These follow-up tests frequently involve comparisons between pairs of group means. For example, if a factor has three levels, three pairwise comparisons might be conducted to compare the means of groups 1 and 2, the means of groups 1 and 3, and the means of groups 2 and 3. SPSS calls these follow-up tests post hoc multiple comparisons.

Assumptions Underlying One-Way ANOVA

Assumption 1: The Dependent Variable Is Normally Distributed for Each of the Populations as Defined by the Different Levels of the Factor

In many applications with a moderate or larger sample size, a one-way ANOVA may yield reasonably accurate p values even when the normality assumption is violated. In some applications with nonnormal populations, a sample size of 15 cases per group might be sufficiently large to yield fairly accurate p values. Larger sample sizes may be required to produce relatively valid p values if the population distributions are substantially nonnormal. In addition, the power of the ANOVA test may be reduced considerably if the population distributions are nonnormal and, more specifically, thick-tailed or heavily skewed. See Wilcox (2001) for an extended discussion of assumptions.

Assumption 2: The Variances of the Dependent Variable Are the Same for All Populations

To the extent that this assumption is violated and the sample sizes differ among groups, the resulting p value for the overall F test is untrustworthy. Under these conditions, it is preferable to use statistics that do not assume equality of population variances, such as the Browne-Forsythe or the Welch statistics. They are accessible in SPSS for Windows by selecting Analysis, Compare Means, One-Way ANOVA, and Options.

For post hoc tests, the validity of the results is questionable if the population variances differ regardless of whether the sample sizes are equal or unequal. If the variances are different, it is appropriate to choose one of the four methods available for conducting post hoc multiple comparison tests that do not assume that the population variances are equal. These four methods are available within the Univariate procedure of the General Linear Model or the Compare Means procedure within One-Way ANOVA. Of the four methods, no one method is superior in all circumstances (Hotchberg and Tamhane, 1987). We have chosen to use the Dunnett's C procedure in instances where the variances are unequal.

Assumption 3: The Cases Represent Random Samples from the Populations and the Scores on the Test Variable Are Independent of Each Other

The ANOVA F yields inaccurate p values if the independence assumption is violated.

Effect Size Statistics for One-Way ANOVA

The General Linear Model procedure computes an effect size index, η^2 (eta square). Eta square ranges in value from 0 to 1. An η^2 value of 0 indicates that there are no differences in the mean scores among groups. A value of 1 indicates that there are differences between at least two of the means on the dependent variable and that there are no differences on the dependent variable scores within each of the groups (i.e., perfect replication). In general, η^2 is interpreted as the proportion of variance of the dependent variable that is related to the factor. What is a small versus a large η^2 is dependent on the area of investigation. However, η^2 of .01, .06, and .14 are, by convention, interpreted as small, medium, and large effect sizes, respectively.

The Data Set

The data set used to illustrate a one-way ANOVA is named *Lesson 25 Data File 1* on the Web at http://www.prenhall.com/greensalkind. It represents data from the vitamin C example described earlier. The variables in the data set are presented in Table 18.

Table 18
Variables in Lesson 25 Data File 1

Variables	Definition
group	1 = Placebo
	2 = Low doses of vitamin C
	3 = High doses of vitamin C
diff	Number of days with cold symptoms in the second year minus the number of days with cold symptoms in the first year

The Research Question

The research question can be stated to reflect mean differences or relationships between variables.

1. Mean differences. Does the mean change in the number of days of cold symptoms differ among the three experimental populations: those who take placebo, those who take low doses of vitamin C, and those who take high doses of vitamin C?
2. Relationship between variables. Is there a relationship in the population between the amount of vitamin C taken and the change in the number of days that individuals show cold symptoms?

Conducting a One-Way ANOVA

To conduct an overall one-way ANOVA and pairwise comparisons among the three vitamin C treatment means, follow these steps:

1. Click **Analyze**, click **General Linear Model**, then click **Univariate**. You'll see the Univariate dialog box in Figure 149.
2. Click **diff**, then click ▶ to move it to the Dependent Variable box.
3. Click **group**, then click ▶ to move it to the Fixed Factor(s) box.
4. Click **Options**. You'll see the Univariate: Options dialog box in Figure 150.
5. Click **group** in the Factor(s) and Factor Interactions box, then click ▶ to make it appear in the Display Means for box.
6. Click **Homogeneity tests**, **Estimates of effect size**, and **Descriptive statistics** in the Display box.
7. Click **Continue**.
8. Click **Post Hoc**. You'll see the Univariate: Post Hoc Multiple Comparisons for Observed Means dialog box in Figure 151.
9. In the Factor(s) box, click **group**, and click ▶ to make it appear in the Post Hoc Tests for box.

TIP

One-Way ANOVA assumes equality of population variances. If this assumption is violated, use the Browne-Forsythe or the Welch statistic available within One-Way ANOVA in Compare Means. For further discussion, see the section labeled "Assumptions Underlying One-Way ANOVA."

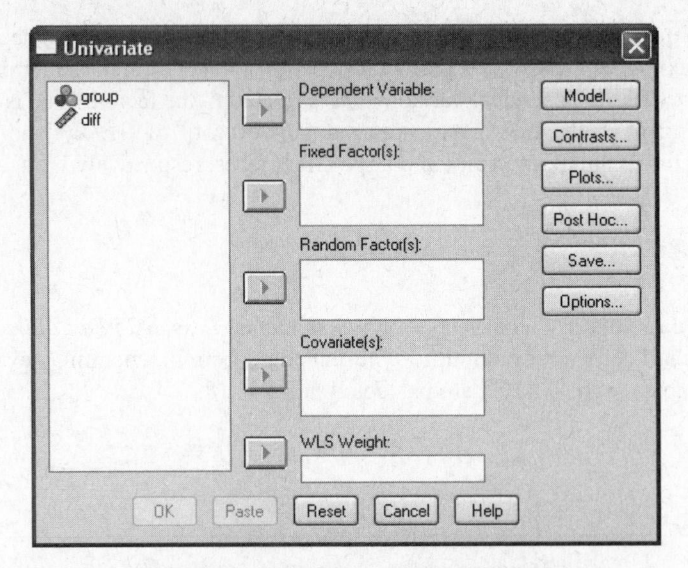

Figure 149. **The Univariate dialog box.**

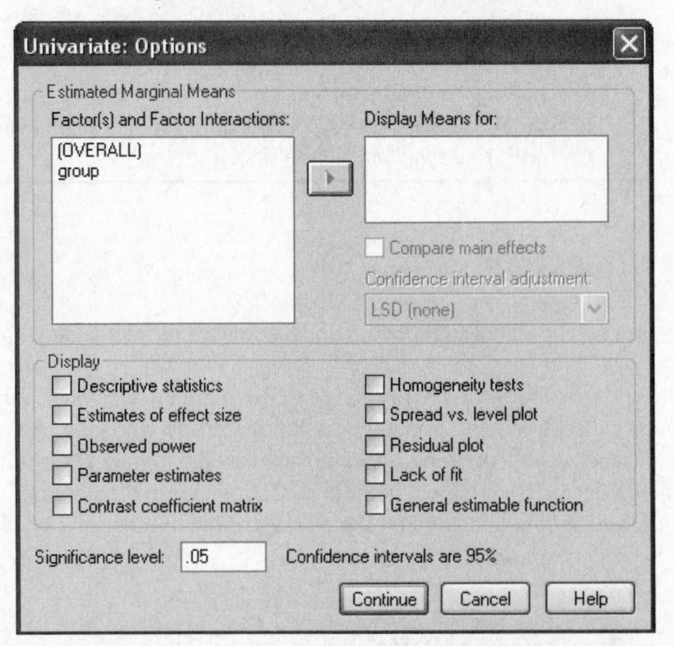

Figure 150. **The Univariate: Options dialog box.**

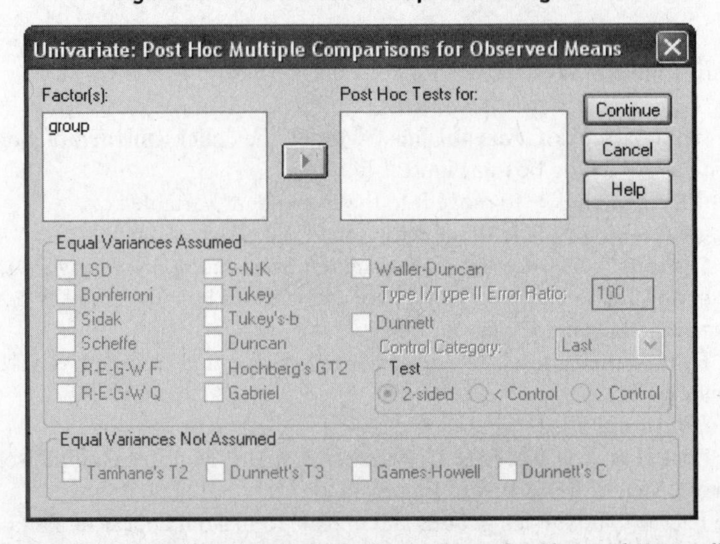

Figure 151. **The Univariate: Post Hoc Multiple Comparisons for Observed Means dialog box.**

10. In the Equal Variances Assumed box, click **Tukey** and **R-E-G-W-Q**. Other post hoc tests assuming equal variances are equally valid.
11. In the Equal Variances Not Assumed box, click **Dunnett's *C***. Other post hoc tests not assuming equal variances are equally valid.
12. Click **Continue**.
13. Click **OK**.

Selected SPSS Output for One-Way ANOVA

The results of analyses are shown in Figure 152, including the means and standard deviations, the homogeneity-of-variance test, and the one-way ANOVA *F* test. To determine whether the overall ANOVA was significant, examine the table labeled Tests of Between-Subjects Effects. The test is significant, $F(2,27) = 4.84$, $p = .02$. The *p* value is located in the column labeled Sig. Because the *p* value is less than .05, we reject the null hypothesis that there are no differences among the groups. The η^2 (labeled Partial Eta Squared in the output) of .26 indicates a strong relationship between the vitamin C factor and the change in the number of days with cold symptoms.

Descriptive Statistics

Dependent Variable: Difference in Days with Colds

Vitamin C Treatment	Mean	Std. Deviation	N
Placebo	3.50	4.143	10
Low Vitamin C Dose	-2.10	4.067	10
High Vitamin C Dose	-2.00	5.477	10
Total	-.20	5.182	30

Levene's Test of Equality of Error Variances[a]

Dependent Variable: Difference in Days with Colds

F	df1	df2	Sig.
1.343	2	27	.278

Tests the null hypothesis that the error variance of the dependent variable is equal across groups.

a. Design: Intercept+group

Tests of Between-Subjects Effects

Dependent Variable: Difference in Days with Colds

Source	Type III Sum of Squares	df	Mean Square	F	Sig.	Partial Eta Squared
Corrected Model	205.400[a]	2	102.700	4.836	.016	.264
Intercept	1.200	1	1.200	.057	.814	.002
group	205.400	2	102.700	4.836	.016	.264
Error	573.400	27	21.237			
Total	780.000	30				
Corrected Total	778.800	29				

a. R Squared = .264 (Adjusted R Squared = .209)

Figure 152. **The results of a one-way analysis of variance.**

Because the overall *F* test was significant, follow-up tests were conducted to evaluate pairwise differences among the means. A decision has to be made whether to use a post hoc procedure that assumes equal variances (Tukey or R-E-G-W-Q) or one that does not assume equal variances (Dunnett's *C*) to control for Type I error across the multiple pairwise comparisons. In our example, the standard deviations range from 4.07 to 5.48 and the variances (the standard deviations squared) range from 16.54 to 30.00, indicating that the variances are somewhat, but not drastically, different from each other. The test of homogeneity of variance was nonsignificant, $p = .28$. Because there may be a lack of power associated with the test due to the small sample size, the result of the homogeneity test does not necessarily imply that there are no differences in the population variances. Therefore, the prudent choice for these data would be to ignore the results of the Tukey and R-E-G-W-Q tests and to use the results of the Dunnett's *C* test, a multiple comparison procedure that does not require the population variances to be equal.

Selected SPSS Output for Post Hoc Comparisons

The results of the post hoc comparisons are shown in Figure 153. Using the Dunnett's *C* test, groups 1 and 2 differed significantly from one another. In the table labeled Multiple Comparisons, the stars (*) in the Mean Difference column indicate which pairwise comparisons are significant.

Post Hoc Tests

Vitamin C Treatment

Multiple Comparisons

Dependent Variable: Difference in Days with Colds

	(I) Vitamin C Treatment	(J) Vitamin C Treatment	Mean Difference (I-J)	Std. Error	Sig.	95% Confidence Interval Lower Bound	Upper Bound
Tukey HSD	Placebo	Low Vitamin C Dose	5.60*	2.061	.030	.49	10.71
		High Vitamin C Dose	5.50*	2.061	.033	.39	10.61
	Low Vitamin C Dose	Placebo	-5.60*	2.061	.030	-10.71	-.49
		High Vitamin C Dose	-.10	2.061	.999	-5.21	5.01
	High Vitamin C Dose	Placebo	-5.50*	2.061	.033	-10.61	-.39
		Low Vitamin C Dose	.10	2.061	.999	-5.01	5.21
Dunnett C	Placebo	Low Vitamin C Dose	5.60*	1.836		.47	10.73
		High Vitamin C Dose	5.50	2.172		-.56	11.56
	Low Vitamin C Dose	Placebo	-5.60*	1.836		-10.73	-.47
		High Vitamin C Dose	-.10	2.157		-6.12	5.92
	High Vitamin C Dose	Placebo	-5.50	2.172		-11.56	.56
		Low Vitamin C Dose	.10	2.157		-5.92	6.12

Based on observed means.

*. The mean difference is significant at the .05 level.

Homogeneous Subsets

Difference in Days with Colds

	Vitamin C Treatment	N	Subset 1	2
Tukey HSD[a,b]	Low Vitamin C Dose	10	-2.10	
	High Vitamin C Dose	10	-2.00	
	Placebo	10		3.50
	Sig.		.999	1.000
Ryan-Einot-Gabriel-Welsch Range[b]	Low Vitamin C Dose	10	-2.10	
	High Vitamin C Dose	10	-2.00	
	Placebo	10		3.50
	Sig.		.962	1.000

Means for groups in homogeneous subsets are displayed.
Based on Type III Sum of Squares
The error term is Mean Square(Error) = 21.237.

a. Uses Harmonic Mean Sample Size = 10.000.

b. Alpha = .05.

Figure 153. **The results of post hoc pairwise comparisons.**

Using SPSS Graphs to Display the Results

The same graphical methods that were presented for the independent-samples *t* test (Lesson 24) can be used for one-way ANOVA as well. ANOVA results may be depicted using boxplots to show the distributions of the dependent variable across the groups.

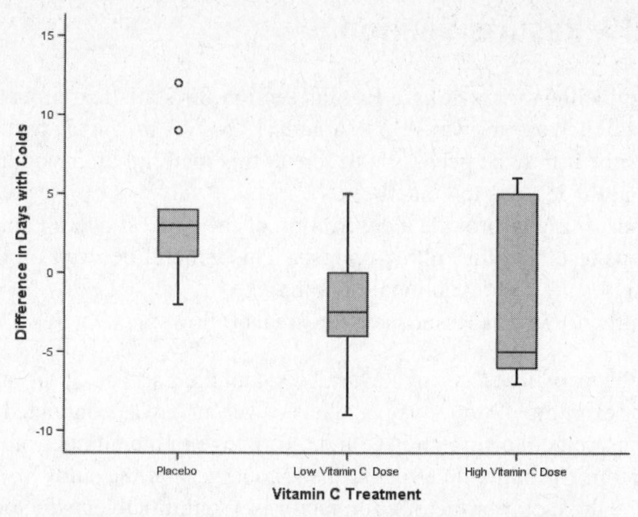

Figure 154. **Changes in number of days with cold symptoms for placebo, low-dose, and high-dose vitamin C treatment groups.**

We created a boxplot to show the distributions of the changes in number of days with cold symptoms for the three vitamin C treatment groups. Figure 154 shows the distributions of the three groups.

An APA Results Section

A one-way analysis of variance was conducted to evaluate the relationship between vitamin C and the change in the number of days with cold symptoms from the first year to the second year of the study. The independent variable, the vitamin C factor, included three levels: placebo, low doses of vitamin C, and high doses of vitamin C. The dependent variable was the change in the number of days of cold symptoms from the first year to the second year. The ANOVA was significant, $F(2, 27) = 4.84$, $p = .02$. The strength of relationship between the vitamin C treatment and the change in the number of days with cold symptoms, as assessed by η^2, was strong, with the vitamin C factor accounting for 26% of the variance of the dependent variable.

Follow-up tests were conducted to evaluate pairwise differences among the means. Because the variances among the three groups ranged from 16.54 to 30.00, we chose not to assume that the variances were homogeneous and conducted post hoc comparisons with the use of the Dunnett's *C* test, a test that does not assume equal variances among the three groups. There was a significant difference in the means between the group that received a low dose of vitamin C and the placebo group, but no significant differences between the two vitamin C groups and between the high dose and placebo groups. The group that received a low dose of vitamin C showed a greater decrease in number of days with cold symptoms in comparison to the placebo group. The 95% confidence intervals for the pairwise differences, as well as the means and standard deviations for the three vitamin C groups, are reported in Table 19.

Table 19

95% Confidence Intervals of Pairwise Differences in Mean Changes in Number of Days of Cold Symptoms

Vitamin C group	*M*	*SD*	Placebo	Low dose
Placebo	3.50	4.14		
Low dose	−2.10	4.07	.47 to 10.73*	
High dose	−2.00	5.48	−.56 to 11.56	−6.12 to 5.92

Note: An asterisk indicates that the 95% confidence interval does not contain zero, and therefore the difference in means is significant at the .05 significance using Dunnett's *C* procedure.

Writing an APA Results Section

We present some guidelines for writing a Results section for statistical procedures that may require follow-up tests, such as one-way ANOVA and MANOVA in Unit 7, or the Friedman test in Unit 10. Consequently, it may be necessary to reread this material after you have read the other lessons in Unit 7 and the lessons in Unit 10.

Some researchers initially provide a description of the general overall analytic strategy that includes the omnibus tests and the follow-up tests. This general description is necessary to the degree that the analyses are unconventional or complex.

The steps required to write a Results section are as follows:

1. Describe the statistical test(s), the variables, and the purpose of the statistical test(s). For example, "A one-way analysis of variance was conducted to evaluate the relationship between vitamin C and the change in the number of days with cold symptoms from the first year to the second year of the study."
   - Describe the factor or factors. If a factor is a within-subjects factor, be sure to label it as such. Otherwise the reader may assume that it is a between-subjects factor. If a multifactorial design has one or more within-subjects factors, describe each factor as a between-subjects or a within-subjects factor.
   - Indicate the number of levels of each factor. It may also be informative to the reader to have a description of each level if the levels are different treatments. However, it is not necessary to report the number of levels and what the levels are for factors with obvious levels such as gender.
   - Describe what the dependent variable(s) are.
2. Report the results of the overall test(s).
   - Describe any decisions about which test was chosen based on assumptions. For example, for a one-way within-subjects ANOVA, justify the choice of using a traditional univariate test instead of a multivariate test.
   - Report the test value and significance level (for the one-way ANOVA, $F(2, 27) = 4.84$, $p = .02$). For p values of .000, report them as $p < .01$. For multifactor designs, report the statistic for each of the main and interaction effects. Tell the reader whether the test(s) are significant or not.
   - Report statistics that allow the reader to make a judgment about the magnitude of the effect for each overall test (e.g., for the one-way ANOVA, $\eta^2 = .45$).
   - Italicize all non-Greek symbols except subscripts and superscripts.
3. Report the descriptive statistics. Refer the reader to a table or figure that presents the relevant descriptive statistics (e.g., means and standard deviations for ANOVA designs). A table or figure may not be necessary for simpler designs, such as a one-way ANOVA with three groups. For these simple designs, the descriptive statistics may be presented in the text.
4. Describe and summarize the general conclusions of the analysis. For example, "The results of the one-way ANOVA supported the hypothesis that different types of vitamin C treatment had a differential effect on the reduction of cold symptoms in individuals."
5. Report the results of the follow-up tests.
   - Describe the procedures used to conduct the follow-up tests. Explain any decisions you made about choice of tests based on their assumptions.
   - Report the method used to control for Type I error across the multiple tests.
   - Summarize the results of the follow-up procedures. It may be useful to present the results of the significance tests among pairwise comparisons with a table of means and standard deviations. When possible, report confidence intervals for pairwise comparisons.

- Describe and summarize the general conclusions of the follow-up analyses. Make sure to include in your description the directionality of the test. For example, the mean for one treatment group is higher or lower than the mean for another group.

6. Report the distributions of the dependent variable for levels of the factor(s) in a graph, if space is available. The graph should be inserted in the text where appropriate. For example, if the graph pertains to assumptions, you would insert it in the section where assumptions are discussed. Likewise, if the graph reflects the means and standard deviations, it should be presented with the discussion of the descriptive statistics.

Alternative Analyses

Data for a one-way ANOVA can also be analyzed by using the Kruskal-Wallis nonparametric procedure (see K independent-samples tests in Lesson 43). If the assumptions of a one-way ANOVA are met, ANOVA is more powerful than the nonparametric alternatives. However, if the assumptions are not met, nonparametric alternatives may be more powerful.

Exercises

The data for Exercises 1 through 3 are in the data set named *Lesson 25 Exercise File 1* on the Web at http://www.prenhall.com/greensalkind. The data are from the following research problem.

Marvin is interested in whether blonds, brunets, and redheads differ with respect to their extrovertedness. He randomly samples 18 men from his local college campus: six blonds, six brunets, and six redheads. He then administers a measure of social extroversion to each individual.

1. Conduct a one-way ANOVA to investigate the relationship between hair color and social extroversion. Be sure to conduct appropriate post hoc tests. On the output, identify the following:
   a. F ratio for the group effect
   b. Sums of squares for the hair color effect
   c. Mean for redheads
   d. p value for the hair color effect
2. What is the effect size for the overall effect of hair color on extroversion?
3. Create a boxplot to display the differences among the distributions for the three hair color groups.

The data for Exercises 4 through 6 are in the data set named *Lesson 25 Exercise File 2* on the Web at http://www.prenhall.com/greensalkind. The data are from Karin's research problem, described earlier in this lesson, involving students with behavior problems and different teaching styles. Karin's SPSS data file has 40 cases with a factor distinguishing among the three types of teachers and a dependent variable, the number of disciplinary actions for students in a class. Conduct a one-way ANOVA to answer Karin's question.

4. What does the Levene's test tell us about our data?
5. Which follow-up tests would you choose and why?
6. Write a Results section based on the analyses you have conducted.

LESSON

26 Two-Way Analysis of Variance

With a two-way analysis of variance (two-way ANOVA), each participant must have scores on three variables: two factors and a dependent variable. Each factor divides cases into two or more levels, while the dependent variable describes cases on a quantitative dimension. *F* tests are performed on the main effects for the two factors and the interaction between the two factors. Follow-up tests may be conducted to assess specific hypotheses if main effect tests, interaction tests, or both are significant.

Higher-way ANOVA is not the focus of this lesson, but is briefly discussed. With higher-way ANOVA, there are more than two factors in a design. Therefore, each participant must have scores on at least four variables: the three or more factors and a dependent variable. *F* tests are performed on multiple effects. For example, with a three-way ANOVA, tests are conducted on the three main effects for the factors, the 3 two-way interactions between pairs of factors, and the 1 three-way interaction among the 3 factors. Depending on which main and interaction tests yield significance, a variety of follow-up tests may be performed.

Applications of Two-Way ANOVA

We can analyze data from different types of studies by using two-way ANOVA.

- Experimental studies
- Quasi-experimental studies
- Field studies

We illustrate an experimental and a field study.

An Experimental Study

Ethel is interested in two methods of note-taking strategies and the effect of these methods on the overall GPAs of college freshmen. She believes that men would benefit most from Method 1, while women would benefit most from Method 2. After obtaining 30 men and 30 women volunteers in freshmen orientation, she randomly assigns 10 women and 10 men to Method 1, 10 women and 10 men to Method 2, and 10 women and 10 men to a control condition. During the first month of the spring semester, individuals in the two note-taking method groups receive daily instruction on the particular note-taking method to which they were assigned. The control group receives no note-taking instruction. Fall and spring GPAs for all participants are recorded. One factor for this study is note-taking method with three levels, and the second factor is gender with two levels. The design for this study may be described as a 3 × 2 ANOVA (the number of levels of note-taking method by the number of levels of gender). Ethel's SPSS data file has 60 cases and three variables: a factor distinguishing among the three note-taking method groups, a second factor differentiating men from women, and a dependent variable, the students' spring semester GPA minus their fall semester GPA.

A Field Study

Ted is interested in evaluating whether the gender of a client and the gender of a therapist affect the outcome of therapy. Specifically, he is interested in testing the gender-matching hypothesis that client–therapist pairs of the same gender produce the most positive outcomes. Ted has access to a large database and selects a sample of 40 client–therapist pairs so that there are an equal number of each type of pair: 10 pairs each of female client with female therapist, female client with male therapist, male client with female therapist, and male client with male therapist. All 40 clients are being treated for difficulty with coping with daily stresses. Each of the 40 therapists indicates that he or she has an eclectic approach to therapy. Finally, the length of therapy for any one pair is never fewer than five sessions.

There are two factors in this study, gender of the client with two levels and gender of the therapist with two levels. The dependent variable is the amount of client improvement over the course of therapy. The design for this study may be described as a 2 × 2 ANOVA (the number of levels of client gender by the number of levels of therapist gender). Ted's SPSS data file has 40 cases and three variables: a factor classifying the gender of the client, a second factor classifying the gender of the therapist, and a dependent variable, client improvement.

Understanding Two-Way ANOVA

The initial tests conducted in a two-way ANOVA are the omnibus or overall tests of the main and interaction effects. These omnibus tests evaluate the following hypotheses:

- First Main Effect: Are the population means on the dependent variable the same across levels of the first factor averaging across levels of the second factor?
- Second Main Effect: Are the population means on the dependent variable the same across levels of the second factor averaging across levels of the first factor?
- Interaction Effect: Are the differences in the population means on the dependent variable among levels of the first factor the same across levels of the second factor?

If one or more of the overall effects are significant, various follow-up tests can be conducted. The choice of which follow-up procedure to conduct depends on which effects are significant.

If the interaction effect is significant, follow-up tests can be conducted to evaluate simple main effects, interaction comparisons, or both. The choice among tests depends on which best address the research questions.

- Depending on the research questions, a number of simple main effect tests can be evaluated: differences in population means among levels of the first factor for each level of the second factor, differences in means among levels of the second factor for each level of the first factor, or both. If any of these simple main effect tests is significant and involves more than two means, additional tests are conducted typically to assess pairwise differences among the means.
- Interaction contrasts may be conducted in place of or in addition to simple main effects. The simplest interaction contrasts involve four means and are referred to as tetrad contrasts. Tetrad contrasts evaluate whether differences in population means between two levels of one factor are the same across two levels of a second factor.

If the interaction effect is not significant, the focus switches to the main effects. If a main effect for a factor with more than two levels is significant, then follow-up tests can be conducted. These tests evaluate whether there are differences in the means among the levels of this factor averaged across levels of the other factor. These follow-up tests most often involve comparing means for pairs of levels of the factor associated with the significant main effect.

Assumptions Underlying Two-Way ANOVA

Assumption 1: The Dependent Variable Is Normally Distributed for Each of the Populations

The cells of the design (i.e., the combinations of levels of the two factors) define the different populations. For example, for a 3×2 ANOVA, there are six cells ($3 \times 2 = 6$) and, consequently, the assumption requires the population distributions on the dependent variable to be normally distributed for all six cells. In many applications with a moderate or larger sample size, a two-way ANOVA may yield reasonably accurate *p* values even when the normality assumption is violated. In some applications with nonnormal populations, a sample size of 15 cases per group might be sufficiently large to yield fairly accurate *p* values. Larger sample sizes may be required to produce relatively valid *p* values if the population distributions are substantially nonnormal. In addition, the power of the ANOVA tests may be reduced considerably if the population distributions are nonnormal and, more specifically, thick-tailed or heavily skewed. See Wilcox (2001) for an extended discussion of assumptions.

Assumption 2: The Population Variances of the Dependent Variable Are the Same for All Cells

To the extent that this assumption is violated and the sample sizes differ for the cells, the *p* values from the overall two-way ANOVA are not trustworthy. In addition, the results of the follow-up tests that require equal variances should be mistrusted if the population variances differ.

Assumption 3: The Cases Represent Random Samples from the Populations, and the Scores on the Dependent Variable Are Independent of Each Other

The two-way ANOVA yields inaccurate *p* values if the independence assumption is violated.

Effect Size Statistics for Two-Way ANOVA

The General Linear Model procedure computes an effect size index, labeled partial eta squared. It may be computed for a main or interaction source with the use of the following equation:

$$Partial\ \eta^2_{Main\ or\ Interaction\ Source} = \frac{Sum\ of\ Squares_{Main\ or\ Interaction\ Source}}{Sum\ of\ Squares_{Main\ or\ Interaction\ Source} + Sum\ of\ Squares_{Error}}$$

Partial η^2 ranges in value from 0 to 1. A partial η^2 is interpreted as the proportion of variance of the dependent variable that is related to a particular main or interaction source, excluding the other main and interaction sources. It is unclear what are small, medium, and large values for partial η^2. What is a small versus a large η^2 is dependent on the area of investigation. In all likelihood, the conventional cutoffs of .01, .06, and .14 for small, medium, and large η^2 are too large for partial η^2.

The Data Set

We will illustrate how to conduct a two-way ANOVA by using our first example, a study evaluating the effects of different note-taking methods on GPA for men and women. We developed two data files: *Lesson 26 Data File 1* and *Lesson 26 Data File 2*, which can be found on the Web at http://www.prenhall.com/greensalkind. The two data sets were constructed to yield different results for tests of main and interaction effects so that we could illustrate the variety of follow-up tests associated with a two-way ANOVA. With the first data set, the main effects are significant, while the interaction effect is not. These data allow us to show how to conduct pairwise comparisons among the means for levels of a significant main-effect factor. With the second data set, the interaction effect is significant. These data help us illustrate how to conduct simple main effects and interaction contrasts following a significant interaction. The variables in the two data sets are the same and are described in Table 20.

Table 20
Variables in Lesson 26 Data Files 1 and 2

Variable	Definition
gender	1 = Men
	2 = Women
method	1 = Note-Taking Method 1
	2 = Note-Taking Method 2
	3 = Control
gpaimpr	Spring GPA − Fall GPA

The Research Question

As described earlier, Ethel conducted the study to assess primarily an interaction hypothesis. The hypothesis was that men will find the Method 1 note-taking approach more helpful in learning classroom information, and women will find Method 2 more beneficial. In addition, a method main effect was expected: The mean on the dependent variable for Method 1 and 2 should be higher than the mean for the control condition, averaging across gender. On the other hand, we have no hypothesis associated with the gender main effect in that the description contained no statements indicating that women will do better or worse than men averaging across note-taking methods.

Of course, the data do not always turn out as expected. Consequently, it might be useful to describe what questions the main and the interaction effects address. The questions are phrased in terms of differences among means, although they could be rewritten so that they address relationships between the factors and the dependent variable.

1. Method main effect: Do the means on change in GPA differ among Method 1, Method 2, and control conditions? The means for the three method conditions are averaged across men and women students.
2. Gender main effect: Do the means on change in GPA differ for men and women students? The means for men and women students are averaged across the three method conditions.
3. Method × Gender interaction effect: Do the differences in the means on change in GPA among the three method conditions vary as a function of gender?

Conducting a Two-Way ANOVA

We conduct two-way ANOVAs on two data sets: *Lesson 26 Data File 1* and *Lesson 26 Data File 2* on the Web at http://www.prenhall.com/greensalkind. After conducting these analyses, we will use the first data set to illustrate how to conduct follow-up tests when the initial analyses yield one or more significant main effects, but no significant interaction effects. Then, we will use the second data set to illustrate how to conduct follow-up tests when the initial analyses yield a significant interaction effect.

Conducting Tests of Main and Interactions Effects

To conduct a two-way ANOVA on the two data sets, follow these steps:

1. Click **Analyze**, click **General Linear Model**, then click **Univariate**. You will see the Univariate dialog box, as shown in Figure 155 (on page 196).
2. Click **gpaimpr**, then click ▶ to move it to the Dependent Variable box.

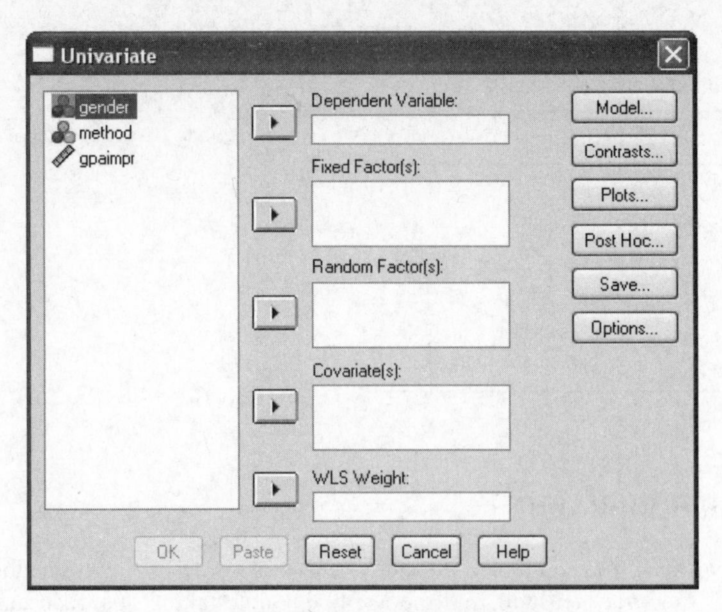

***Figure 155.* The Univariate dialog box.**

3. Holding down the Ctrl key, click **gender** and **method**. Click ▶ to move them to the Fixed Factor(s) box.
4. Click **Options**.
5. In the Univariate: Options dialog box, hold down the Ctrl key, and click **gender**, **method**, and **gender*method** in Factor(s) and Factor Interactions box.
6. Click ▶ to move them to the Display Means for box.
7. In the Display box, click **Homogeneity tests**, **Descriptive statistics**, and **Estimates of effect size**.
8. Click **Continue**.
9. Click **OK**.

Selected SPSS Output for Two-Way ANOVA Tests for Lesson 26 Data File 1

Figure 156 shows the means and standard deviations for each cell and the results of the omnibus two-way ANOVA test for *Lesson 26 Data File 1*. From these analyses, we can see that both the gender main effect and the method main effect are significant and that the interaction between gender and method is not significant. In addition, an examination of the means indicates that the ordering of the means for the three method groups is the same for both genders. Based on these results, we chose to follow up our overall tests with pairwise comparisons among the three method means averaged across gender. Because gender has only two levels, men and women, the main effect associated with gender tells us that men and women differ significantly from one another, and no further analyses are needed to evaluate the gender main effect. To see how to conduct follow-up tests to this two-way ANOVA, see "Conducting Follow-Up Analyses to a Significant Main Effect."

TIP

How would we conduct analyses if we had more than two factors? No problem! For example, with a three-way ANOVA, include all three factors in the Fixed Factor(s) box in the GLM-Univariate dialog box. In the Univariate: Options dialog box, include all three factors and their interactions in the Display Means for box.

Selected SPSS Output for Two-Way ANOVA Tests for Lesson 26 Data File 2

Figure 157 shows the cell means and standard deviations and the results of the omnibus two-way ANOVA F test for *Lesson 26 Data File 2*. From these analyses we can see that both the method main effect and the method-by-gender interaction effect are significant. If the interaction effect is significant, we generally do not interpret main effects, but we do wish to understand why the interaction was significant. We can follow up a significant interaction by conducting simple main effect tests, interaction comparisons, or both. To see how to conduct follow-up tests to this two-way ANOVA, see "Conducting Follow-Up Analyses to a Significant Interaction."

Descriptive Statistics

Dependent Variable: Change in GPA

Gender	Note-Taking methods	Mean	Std. Deviation	N
Men	Method 1	.3350	.22858	10
	Method 2	.6400	.17764	10
	Control	.1650	.14916	10
	Total	.3800	.26993	30
Women	Method 1	.1700	.18288	10
	Method 2	.3050	.19214	10
	Control	.1050	.14615	10
	Total	.1933	.18880	30
Total	Method 1	.2525	.21853	20
	Method 2	.4725	.24893	20
	Control	.1350	.14699	20
	Total	.2867	.24938	60

Tests of Between-Subjects Effects

Dependent Variable: Change in GPA

Source	Type III Sum of Squares	df	Mean Square	F	Sig.	Partial Eta Squared
Corrected Model	1.889[a]	5	.378	11.463	.000	.515
Intercept	4.931	1	4.931	149.582	.000	.735
gender	.523	1	.523	15.856	.000	.227
method	1.174	2	.587	17.809	.000	.397
gender * method	.193	2	.096	2.921	.062	.098
Error	1.780	54	.033			
Total	8.600	60				
Corrected Total	3.669	59				

a. R Squared = .515 (Adjusted R Squared = .470)

Figure 156. **Partial results of the two-way ANOVA omnibus tests for Lesson 26 Data File 1.**

Descriptive Statistics

Dependent Variable: Change in GPA

Gender	Note-Taking methods	Mean	Std. Deviation	N
Men	Method 1	.3350	.22858	10
	Method 2	.3050	.19214	10
	Control	.1650	.14916	10
	Total	.2683	.20064	30
Women	Method 1	.1700	.18288	10
	Method 2	.6400	.17764	10
	Control	.1050	.14615	10
	Total	.3050	.29254	30
Total	Method 1	.2525	.21853	20
	Method 2	.4725	.24893	20
	Control	.1350	.14699	20
	Total	.2867	.24938	60

Tests of Between-Subjects Effects

Dependent Variable: Change in GPA

Source	Type III Sum of Squares	df	Mean Square	F	Sig.	Partial Eta Squared
Corrected Model	1.889[a]	5	.378	11.463	.000	.515
Intercept	4.931	1	4.931	149.582	.000	.735
gender	.020	1	.020	.612	.438	.011
method	1.174	2	.587	17.809	.000	.397
gender * method	.695	2	.348	10.543	.000	.281
Error	1.780	54	.033			
Total	8.600	60				
Corrected Total	3.669	59				

a. R Squared = .515 (Adjusted R Squared = .470)

Figure 157. **Partial results of the two-way ANOVA omnibus tests for Lesson 26 Data File 2.**

Conducting Follow-up Analyses to a Significant Main Effect

We have conducted a two-way ANOVA on the data in *Lesson 26 Data File 1* and found significant main effects, but no significant interaction. Because the main effect for method involves more than two levels, we conduct follow-up tests to evaluate pairwise differences among the three methods, averaging across gender. To conduct these pairwise comparisons, follow these steps:

1. Click **Analyze**, click **General Linear Model**, then click **Univariate**.
2. The appropriate options should be selected in the Univariate dialog box. If they are not, conduct Steps 2 through 8 described earlier for our overall analysis.
3. Click **Post Hoc**.
4. In the Factor(s) box, click **method**, and then click ▶ to move it to the Post Hoc Tests for box.
5. In the Equal Variances Assumed box, click **R-E-G-W-Q** and **Tukey**. Other post hoc options assuming equal variances are also valid.
6. In the Equal Variances Not Assumed box, click **Dunnett's *C***. Other post hoc options not assuming equal variances are also valid.
7. Click **Continue**.
8. Click **OK**.

Selected SPSS Output for Pairwise Comparisons of Main Effect Means

Figure 158 shows the results of the pairwise comparisons, assuming equal variances and not assuming equal variances. With the Tukey and Dunnett's *C* tests, only the comparison between Method 1 and the control condition was not significant. For the R-E-G-W-Q test, all three comparisons were significant.

Conducting Follow-up Analyses to a Significant Interaction

We previously conducted a two-way ANOVA on the data in *Lesson 26 Data File 2* and found a significant interaction. We may conduct two types of follow-up tests: simple main effects and interaction comparisons. We first discuss conducting tests of simple main effects.

Simple Main Effect Tests Following a Significant Interaction

Simple main effect tests evaluate differences in population means among levels of one factor for each level of another factor. Conducting simple main effect tests involves two decisions.

1. Which Simple Main Effects Should Be Analyzed?

We must decide whether to examine the simple main effects for one of the factors or for both factors. Which simple main effects to explore should depend on the research hypotheses. For our example, we could explore two simple main effects:

- Method simple main effects involve differences in means among methods for men and differences in means among methods for women
- Gender simple main effects involves differences in means between men and women for Method 1, differences in means between men and women for Method 2, and differences in means between men and women for the control group.

We suspect that most researchers would be more interested in the method simple main effects. However, both could be of interest.

Post Hoc Tests

Note-Taking methods

Multiple Comparisons

Dependent Variable: Change in GPA

	(I) Note-Taking methods	(J) Note-Taking methods	Mean Difference (I-J)	Std. Error	Sig.	95% Confidence Interval Lower Bound	Upper Bound
Tukey HSD	Method 1	Method 2	-.2200*	.05741	.001	-.3584	-.0816
		Control	.1175	.05741	.111	-.0209	.2559
	Method 2	Method 1	.2200*	.05741	.001	.0816	.3584
		Control	.3375*	.05741	.000	.1991	.4759
	Control	Method 1	-.1175	.05741	.111	-.2559	.0209
		Method 2	-.3375*	.05741	.000	-.4759	-.1991
Dunnett C	Method 1	Method 2	-.2200*	.07407		-.4082	-.0318
		Control	.1175	.05889		-.0321	.2671
	Method 2	Method 1	.2200*	.07407		.0318	.4082
		Control	.3375*	.06464		.1733	.5017
	Control	Method 1	-.1175	.05889		-.2671	.0321
		Method 2	-.3375*	.06464		-.5017	-.1733

Based on observed means.

*. The mean difference is significant at the .05 level.

Homogeneous Subsets

Change in GPA

	Note-Taking methods	N	Subset 1	2	3
Tukey HSD[a,b]	Control	20	.1350		
	Method 1	20	.2525		
	Method 2	20		.4725	
	Sig.		.111	1.000	
Ryan-Einot-Gabriel-Welsch Range[b]	Control	20	.1350		
	Method 1	20		.2525	
	Method 2	20			.4725
	Sig.		1.000	1.000	1.000

Means for groups in homogeneous subsets are displayed.
Based on Type III Sum of Squares
The error term is Mean Square(Error) = .033.

a. Uses Harmonic Mean Sample Size = 20.000.

b. Alpha = .05.

Figure 158. **Partial results of pairwise comparisons of method main effect means.**

2. What Error Term Should Be Used to Conduct the Simple Main Effects?

We must decide whether to use the same error term for each simple main effect analysis and assume homogeneity of error variances or to use different error terms for each simple main effect test and not assume homogeneity of error variances. If the error variances differ across simple main effects, it would be preferable to conduct the simple main effect tests using different error variances.

Because of space limitations, we describe only how to conduct simple main effects assuming homogeneity of variances. We will describe two sets of steps. The first set of steps is for conducting simple main effect analyses, while second set of steps is for conducting pairwise comparisons of individual means if a simple main effect is significant.

Conducting Simple Main Effects Analyses

We cannot conduct simple main effects analyses by choosing options within dialog boxes. Instead we must paste syntax created using dialog boxes and then make revisions to the estimated means statement (denoted as /EMMEANS). We present the required revisions for examining differences between men and women for each of the three methods as well as evaluating differences among the methods for each gender.

1. Click **Analyze**, click **General Linear Model**, then click **Univariate**.
2. Conduct Steps 2 through 8 as described previously in the section "Conducting Tests of Main and Interaction Effects."
3. Click **Paste**. The Syntax Editor should include the following statements:

```
UNIANOVA
gpaimpr BY gender method
/METHOD = SSTYPE(3)
/INTERCEPT = INCLUDE
/EMMEANS = TABLES(gender)
/EMMEANS = TABLES(method)
/EMMEANS = TABLES(gender*method)
/PRINT = DESCRIPTIVE ETASQ HOMOGENEITY
/CRITERIA = ALPHA(.05)
/DESIGN = gender method gender*method.
```

4. Copy the /EMMEANS = TABLES(gender*method) statement so it appears twice in the syntax:

```
/EMMEANS = TABLES(gender*method)
/EMMEANS = TABLES(gender*method)
```

5. Revise these statements so they appear as follows:

```
/EMMEANS = TABLES(gender*method) COMPARE(gender) ADJ(LSD)
/EMMEANS = TABLES(gender*method) COMPARE(method) ADJ(LSD)
```

   The first statement instructs SPSS to compare men and women for each method (i.e., gender simple main effects), while the second statement instructs SPSS to compare methods for each gender (i.e., method simple main effects). The syntax generated by conducting Steps 1 through 5 is shown in Figure 159.

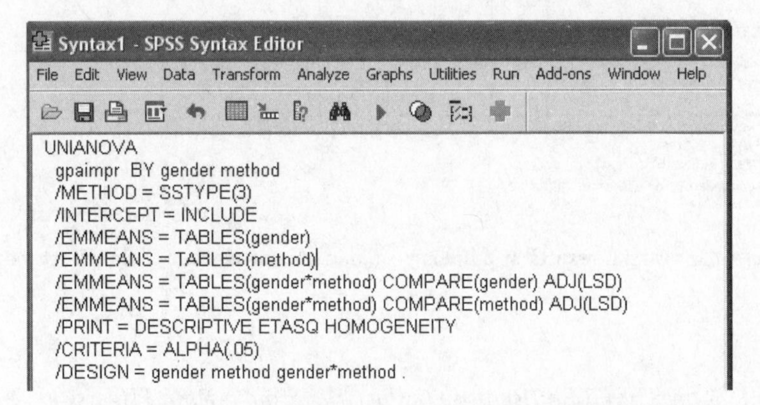

***Figure 159*. Syntax for conducting simple main effects analyses.**

6. Highlight the syntax, click **Run**, and then click **Selection**.

Selected SPSS Output for Simple Main Effects

The output for the simple main effects for method within gender is shown in Figure 160. Although we conducted all possible simple main effects, we are probably interested in evaluating only the simple main effects for method within gender. Accordingly, we restrict our output and discussion to these two simple main effect tests.

The ouput for the method simple main effects for each gender are labeled "4. Gender * Note-Taking methods." In the Univariate Tests table, tests are presented to evaluate differences among methods for men, $F(2, 54) = 2.50$, $p = .09$, partial $\eta^2 = .08$, and differences among methods for women, $F(2, 54) = 25.86$, $p < .01$, partial $\eta^2 = .49$. These results indicate significant differences among methods for women only, regardless as to whether one controlled for Type I error rate across the two tests using the Bonferroni approach or not (i.e., testing each F test at the .025 or the .05 level).

Pairwise Comparisons

Dependent Variable: Change in GPA

Gender	(I) Note-Taking methods	(J) Note-Taking methods	Mean Difference (I-J)	Std. Error	Sig.[a]	95% Confidence Interval for Difference[a] Lower Bound	Upper Bound
Men	Method 1	Method 2	.030	.081	.713	-.133	.193
		Control	.170*	.081	.041	.007	.333
	Method 2	Method 1	-.030	.081	.713	-.193	.133
		Control	.140	.081	.090	-.023	.303
	Control	Method 1	-.170*	.081	.041	-.333	-.007
		Method 2	-.140	.081	.090	-.303	.023
Women	Method 1	Method 2	-.470*	.081	.000	-.633	-.307
		Control	.065	.081	.427	-.098	.228
	Method 2	Method 1	.470*	.081	.000	.307	.633
		Control	.535*	.081	.000	.372	.698
	Control	Method 1	-.065	.081	.427	-.228	.098
		Method 2	-.535*	.081	.000	-.698	-.372

Based on estimated marginal means

*. The mean difference is significant at the .05 level.

a. Adjustment for multiple comparisons: Least Significant Difference (equivalent to no adjustments).

Univariate Tests

Dependent Variable: Change in GPA

Gender		Sum of Squares	df	Mean Square	F	Sig.	Partial Eta Squared
Men	Contrast	.165	2	.082	2.498	.092	.085
	Error	1.780	54	.033			
Women	Contrast	1.705	2	.852	25.855	.000	.489
	Error	1.780	54	.033			

Each F tests the simple effects of Note-Taking methods within each level combination of the other effects shown. These tests are based on the linearly independent pairwise comparisons among the estimated marginal means.

Figure 160. **Results for simple main effects analyses.**

Because the method factor has three levels, it is necessary to conduct follow-up tests for the significant simple main effects for method within women. These follow-up tests are in the section labeled Women in the Pairwise Comparisons table. The largest difference in means was .54 between Methods 2 and Control, while the next largest difference in means was −.47 between Methods 1 and 2. The p values for tests of these differences were both less than .01. The smallest difference was .06 between Method 1 and Control, which had an associated p value of .43. Assuming that the omnibus F tests for the simple main effects for method within each gender (those reported previously) were tested at the .025 level, each of the pairwise differences could be to tested at the .025 level based on the LSD (Appendix B). Consequently, the only pairwise difference that was not statistically significant was between Method 1 and Control.

Conducting Interaction Comparisons Following a Significant Interaction

As an alternative (or perhaps in addition) to simple main effect analyses, interaction comparisons can be conducted as follow-up tests to a significant interaction. The tetrad contrast, the simplest interaction comparison associated with a two-way interaction, involves four cell means. In particular, tetrad contrasts evaluate whether the differences in population means between two levels of one factor are equal across two levels of the second factor. (Note: The terms of *contrast* and *comparison* are used interchangeably.)

For example, Ethel's hypothesis that men benefit most from Method 1, while women benefit most from Method 2 can be addressed by a specific interaction comparison involving four cell means. This comparison would evaluate the null hypothesis that the difference in population means between Method 1 and Method 2 for men is equal to the difference in population means between Method 1 and Method 2 for women is equal to zero: $(\mu_{\text{men,method1}} - \mu_{\text{men,method2}}) - (\mu_{\text{women,method1}} - \mu_{\text{women,method2}}) = 0$. Ethel's research hypothesis is that this null hypothesis is incorrect, and that the difference in means between methods is larger for men than it is for women (i.e., the difference is not equal to 0, but is a positive value). Two additional tetrad contrasts exist for our example:

$$(\mu_{\text{men,method1}} - \mu_{\text{men,control}}) - (\mu_{\text{women,method1}} - \mu_{\text{women,control}}) = 0 \text{ and}$$
$$(\mu_{\text{men,method2}} - \mu_{\text{men,control}}) - (\mu_{\text{women,method2}} - \mu_{\text{women,control}}) = 0$$

We conduct these three tetrad contrasts by specifying **L** matrices using the /lmatrix command. After showing how to conduct the three contrasts, we offer a very brief explanation about the use of **L** matrices.

1. Return to the Syntax Editor. Delete all lines in the Syntax window.
2. Click **Analyze**, click **General Linear Model**, then click **Univariate**. The choices that we made previously should still be selected within the Univariate dialog box.
3. Click **Paste**. The first three lines in the Syntax Editor are shown:
   UNIANOVA
   gpaimpr BY gender method
   /METHOD = SSTYPE(3)
4. After the third line, type in the following statements to evaluate whether the difference in means between Method 1 and Method 2 for men minus the difference in means between Method 1 and Method 2 for women is equal to zero:
   /lmatrix
   '(Method 1 vs Method 2) for men vs (Method 1 vs Method 2) for women'
   gender*method 1 -1 0 -1 1 0
5. Next, type in the following statements to evaluate whether the difference in means between Method 1 and control for men minus the difference in means between Method 1 and control for women is equal to zero:
   /lmatrix
   '(Method 1 vs Control) for men vs (Method 1 vs Control) for women'
   gender*method 1 0 -1 -1 0 1
6. Next, type in the following statements to evaluate whether the difference in means between Method 2 and control for men minus the difference in means between Method 2 and control for women is equal to zero:
   /lmatrix
   '(Method 2 vs Control) for men vs (Method 2 vs Control) for women'
   gender*method 0 1 -1 0 -1 1
7. Type a period (.) at the end of the last line.
8. Highlight all lines in the Syntax window, click **Run**, and then click **Selection**.

Selected SPSS Output for Interaction Comparisons

Figure 161 shows only the results for the tetrad contrast created in Step 4 above and labeled Custom Hypothesis #1 on the output. The Contrast Estimate of .500 listed in the first table for Custom Hypothesis Tests #1 represents the difference in means between Method 1 and Method 2 for men (.030 = .035 − .305) minus the difference in means between Method 1 and Method 2 for women (−.470 = .170 − .640). The test of this tetrad contrast is presented in the table labeled Test Results, $F(1, 54) = 18.96, p < .01$.

We do not present the tables for the second and third tetrad contrast, but you should see the results for both in your output. The Contrast Estimate of .105 for Custom Hypothesis Tests #2 represents the difference in means between Method 1 and control for men (.170 = .335 − .165) minus the difference in means between Method 1 and control for women (.065 = .170 − .105). The test of this tetrad contrast is presented in the table labeled Test Results, $F(1, 54) = .84, p = .36$.

The Contrast Estimate of −.395 listed in the table for Custom Hypothesis Tests #3 represents the difference in means between Method 2 and Control for men (.140 = .305 − .165) minus the difference in means between Method 2 and Control for women (.535 = .640 − .105). The test of this tetrad contrast is presented in the table labeled Test Results, $F(1, 54) = 11.83, p < .01$.

Custom Hypothesis Tests #1

Contrast Results (K Matrix)[a]

Contrast			Dependent Variable: Change in GPA
L1	Contrast Estimate		.500
	Hypothesized Value		0
	Difference (Estimate - Hypothesized)		.500
	Std. Error		.115
	Sig.		.000
	95% Confidence Interval for Difference	Lower Bound	.270
		Upper Bound	.730

a. Based on the user-specified contrast coefficients (L') matrix: (Method 1 vs Method 2) for men vs (Method 1 vs Method 2) for women

Test Results

Dependent Variable: Change in GPA

Source	Sum of Squares	df	Mean Square	F	Sig.
Contrast	.625	1	.625	18.961	.000
Error	1.780	54	.033		

Figure 161. **Results of interaction comparison of Method 1 and Method 2 for men versus women.**

The F tests for the three tetrad contrasts evaluate mean differences without controlling for Type I error across the three tests. Some method, such as the Holm's approach, may be used to control for Type I error across the three interaction comparisons.

Explaining the /lmatrix Commands for Tetrad Contrasts

Customized hypotheses can be tested using the /lmatrix command. When using the /lmatrix command, we must define hypotheses by specifying coefficients for main and interaction effects (typically denoted as α, β, and $\alpha\beta$), even though we typically conceptualize our research hypotheses in terms of means. Given this and other complexities associated with **L** matrices and the /lmatrix command, we offer a very restricted discussion of them. We refer the reader to Green, Marquis, Hershberger, Thompson, and McCollum (1999) for a general discussion of **L** matrices with GLM analyses.

Fortunately it is relatively straightforward to test tetrad contrasts in a two-way ANOVA using **L** matrices. For these tetrad comparisons, the coefficients for the effects are the same as the coefficients for the means. Below we specify a six-step approach for defining the coefficients for the /lmatrix command. We illustrate the stepwise approach for the tetrad comparison evaluating whether the difference in population means between Method 1 and Method 2 for men minus the difference in population means between Method 1 and Method 2 for women is equal to zero.

STEP 1. Specify the null hypothesis for the tetrad contrast of interest. We suspect that the null hypothesis is most frequently specified in the following form: the mean difference between two levels of the first factor for a particular level of a second factor is equal to the mean difference between the same two levels of the first factor for a different level of the second factor. (The designations of *first* and *second* are arbitrary.)

EXAMPLE: The tetrad contrast could be stated as follows: The difference in population means between Method 1 and Method 2 for men is the same as the difference in population means between Method 1 and Method 2 for women. Mathematically,

$$\mu_{\text{men,method1}} - \mu_{\text{men,method2}} = \mu_{\text{women,method1}} - \mu_{\text{women,method2}}$$

STEP 2. Respecify the hypothesis so that all means are on the left side of the equation.

EXAMPLE: $\mu_{\text{men,method1}} - \mu_{\text{men,method2}} - \mu_{\text{women,method1}} + \mu_{\text{women,method2}} = 0$

STEP 3. Rewrite the left side of the equation so that the means are summed together, with coefficients in parentheses next to each mean.

EXAMPLE: $(+1)\mu_{\text{men,method1}} + (-1)\mu_{\text{men,method2}} + (-1)\mu_{\text{women,method1}} + (+1)\mu_{\text{women,method2}} = 0$

STEP 4. Create a two-way layout of coefficients for means. The rows and columns of the layout are defined by the order of the factors on the second statement of syntax (i.e., the line after UNIANOVA). The row factor is the first factor specified immediately following the word *BY* on this statement, while the column factor is the second factor after the word *BY*. The ordering of the rows (or columns) is determined by the values of the levels of the factor, with the lowest number representing the first row (or column).

EXAMPLE: The second statement of syntax is 'gpaimpr BY gender method.' Thus, gender is the row factor, and method is the column factor. For gender, the first and second rows are men (value of 1) and women (value of 2), respectively. For method, the first, second, and third columns are Method 1 (value of 1), Method 2 (value of 2), and control (value of 3), respectively. The resulting layout is as follows:

	Method 1	Method 2	Control
Men			
Women			

STEP 5. Place the coefficient (in parentheses) for each mean, as written in Step 3, in the appropriate cell of the two-way layout.

EXAMPLE:

	Method 1	Method 2	Control
Men	1	−1	0
Women	−1	1	0

STEP 6. Translate the two-way layout of coefficients into SPSS syntax.

- The first line specifies /lmatrix.
- The second line labels the tested hypothesis in single quotes; the wording of the label is determined by the user.
- The third line specifies the interaction effect followed by the coefficients. The interaction is the names of the factors (as specified in the second statement of the syntax), with an asterisk between them. The coefficients are ordered such that the coefficients in the first row (from left to right) are presented first, then the second row from left to right, and so on. At least one space must exist between coefficients.

EXAMPLE:

```
/lmatrix
'(Method 1 vs Method 2) for men vs (Method 1 vs Method 2) for women'
gender*method 1 -1 0 -1 1 0.
```

A period should be placed at the end of the last line if the /lmatrix statements represent the last tetrad comparison evaluated by the syntax.

Using SPSS Graphs to Display Results

A graph displaying a boxplot for each cell of a two-way ANOVA can make a significant interaction more understandable. To create a boxplot for *Lesson 26 Data File 2*, conduct the following steps:

1. Click **Graphs**, click **Legacy Dialogs**, and then click **Boxplot**.
2. Click **Clustered** and **Summaries for groups of cases** in the Boxplot dialog box.

3. Click **Define**.
4. In the Define dialog box, click **gpaimpr**, and click ▶ to move it to the Variable box.
5. Click **method**, and click ▶ to move it to the Category Axis box.
6. Click **gender**, and click ▶ to move it to the Define Clusters by box.
7. Click **OK**. The edited graph is shown in Figure 162.

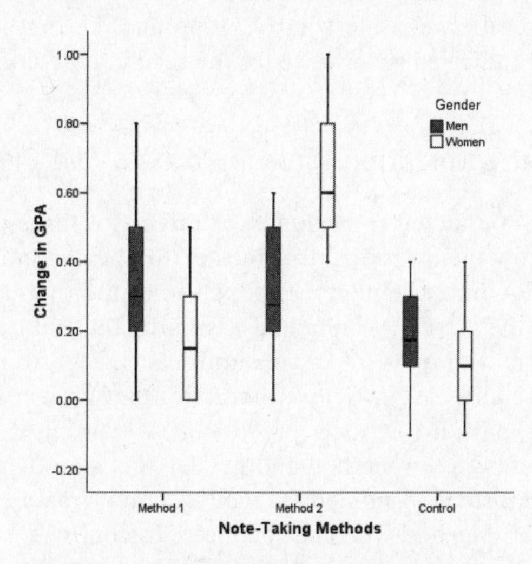

Figure 162. **Boxplots of GPA improvement by note-taking method for men and women.**

Two APA Results Sections

We present two results sections: one for a significant main effect and nonsignificant interaction effect and one for a significant interaction effect.

Results for Significant Main Effect and a Nonsignificant Interaction (*Lesson 26 Data File 1*)

A 3 × 2 ANOVA was conducted to evaluate the effects of three note-taking conditions and gender on GPA improvement from fall semester to spring semester. The means and standard deviations for GPA improvement as a function of the two factors are presented in Table 21. The ANOVA indicated no significant interaction between note taking and gender, $F(2, 54) = 2.92$, $p = .06$, partial $\eta^2 = .10$, but significant main effects for note taking, $F(2, 54) = 17.81$, $p < .01$, partial $\eta^2 = .40$, and gender, $F(1, 54) = 15.86$, $p < .01$, partial $\eta^2 = .23$. The gender main effect indicated that men tended to have greater improvements in GPA than women, but it was not the focus of this study.

Table 21

Means and Standard Deviations for GPA Improvement

Gender	Method	Mean	*SD*
Men	Method 1	.34	.23
	Method 2	.64	.18
	Control	.17	.15
Women	Method 1	.17	.18
	Method 2	.31	.19
	Control	.11	.15

The primary purpose of the study was to determine which method of note taking is more effective. Follow-up analyses to the main effect for note taking examined this issue. The follow-up tests consisted of all pairwise comparisons among the three types of note taking methods. The Tukey HSD procedure was used to control for Type I error across the pairwise comparisons. The results of this analysis indicate that the group using the second note-taking method improved GPA significantly more than either the first method group or the control group. There was no significant difference between the control group and the first method of note-taking. Overall, the 3×2 ANOVA indicates superiority for the second note-taking method.

Results for a Significant Interaction (*Lesson 26 Data File 2*)

A 3×2 ANOVA was conducted to evaluate the effects of three note-taking conditions and gender on GPA improvement from fall semester to spring semester. The means and standard deviations for GPA improvement as a function of the two factors are presented in Table 22. The results for the ANOVA indicated a significant main effect for note taking, $F(2, 54) = 17.81$, $p < .01$, partial $\eta^2 = .40$, a nonsignificant effect for gender, $F(1, 54) = .61$, $p = .44$, partial $\eta^2 = .01$, and a significant interaction between note taking and gender, $F(2, 54) = 10.54$, $p < .01$, partial $\eta^2 = .28$.

Because the interaction between method and gender was significant, we chose to ignore the method main effect and instead examined the method simple main effects—that is, the differences among methods for men and women separately. To control for Type I error across the two simple main effects, we set alpha for each at .025. There were no significant differences between note-taking conditions for men, $F(2, 54) = 2.50$, $p = .09$, but there were significant differences for women, $F(2, 54) = 25.86$, $p < .01$.

Follow-up tests were conducted to evaluate the three pairwise differences among the means for women, with alpha set at .008 (.025/3 = .008) to control for Type I error over the three pairwise comparisons. The second note-taking method had significantly higher GPA improvements than either the first note-taking method or the control group for the women students. There were no significant differences between the control group and the first method of note taking.

One final analysis was conducted to address the hypothesis that men would benefit most from note-taking Method 1, while women would benefit most from note-taking Method 2. The difference in means between Method 1 and Method 2 for men minus the difference in means between Method 1 and Method 2 for women was .50, $F(1, 54) = 18.96$, $p < .01$. The results of this comparison supported the research hypothesis.

A Word of Caution: Additional Complexities Occur with Unequal Sample Sizes across Cells

In our examples, the sample sizes were the same for all cells. The analyses become much more complex, however, if the number of cases differ across cells. We will discuss briefly some of the issues with ANOVAs with unequal cell frequencies. However, our discussion is

Table 22

Means and Standard Deviations for GPA Improvement

Gender	Method	Mean	*SD*
Men	Method 1	.34	.23
	Method 2	.31	.19
	Control	.17	.15
Women	Method 1	.17	.18
	Method 2	.64	.15
	Control	.11	.15

very rudimentary and includes just enough information to make it clear that much more needs to be said before this topic area is clearly understood. For a more extensive discussion of this topic, analysis-of-variance textbooks like Maxwell and Delaney (2000) should be consulted.

The means associated with the main effects (e.g., the mean for males and the mean for females for the gender main effect) can be calculated several ways if the number of cases are unequal across the different cells of a two-way ANOVA. For example, you might compute weighted means or unweighted means. In the SPSS output, the weighted means are presented in the Descriptive Statistics table in the row labeled Total, while the unweighted means are given in the table labeled Estimated Marginal Means. If the sample sizes for the cells are equal, the two types of means yield the same results.

To help us understand the difference between these two types of means, let's examine a new data set for our note taking example, one that has three cases in three cells (males in Method 1 and females in Method 2 and control) and two cases in remaining three cells. The data for this example are shown in Table 23. For these data, we can compute unweighted means for method levels or gender levels by simply averaging cell means. Thus, the unweighted mean for Method 1 is .230 and is computed by averaging the cell mean for males in Method 1 of .300 and the cell mean for females in Method 1 of .160: $.230 = (.300 + .160)/2$. Alternatively, we could calculate a weighted mean for each level of method or gender by weighting each cell mean by its frequency and dividing by the sum of the cell frequencies. For Method 1, the weighted mean is .244 and is computed by weighting the cell mean for males in Method 1 of .300 by its frequency of 3 and weighting the cell mean for females in Method 1 of .160 by its frequency of 2 and dividing the result by the sum of the cell frequencies of 5: $.244 = [(3)\ .300 + (2)\ .160]\ /\ (3 + 2)$.

We may choose to analyze either unweighted or weighted means (or, potentially, other types of means). We should choose the type of mean that best addresses our research question. For example, unweighted means are preferable if a study is designed to have equal sample sizes, the unequal sample sizes are due to missing data, and the data are missing completely at random. Under these conditions, the unweighted means should produce similar means to those that would have been obtained if the original design had been completed with no missing data. We'd guess that most researchers choose unweighted means, wisely or unwisely.

F tests that evaluate differences in unweighted means are likely to yield different results from F tests that evaluate differences in weighted means. By default, SPSS computes F tests that evaluate unweighted population means. To conduct F tests that

Table 23

Cell Means (Raw Data in Parentheses) and Unweighted and Weighted Means for Gender and Methods

	Note-taking method				
Gender	Method 1	Method 2	Control	Unweighted means for each gender	Weighted means for each gender
Males	.30 (.20, .34, .36)	.18 (.16, .20)	.12 (.10, .14)	.200	.214
Females	.16 (.14, .18)	.22 (.24, .26, .16)	.16 (.15, .17, .16)	.180	.183
Unweighted means for each method	.230	.200	.140		
Weighted means for each method	.244	.204	.144		

evaluate another type of mean, we must select an option other than Type III sums of squares in the drop-down menu in the Univariate: Model dialog box. The unweighted means (estimated marginal means) should be reported if the default option (Type III sums of squares) is selected.

Exercises

Exercises 1 through 4 are based on the following research problem. The data are in the file labeled *Exercise File 1* on the Web at http://www.prenhall.com/greensalkind.

An experimenter wanted to investigate simultaneously the effects of two types of reinforcement schedules and three types of reinforcers on the arithmetic problem-solving performance of second-grade students. A sample of 66 second graders was identified, and 11 were randomly assigned to each of the six combinations of reinforcement schedules and reinforcers. Students studied arithmetic problem solving under these six conditions for three weeks, and then took a test on the material they studied. The SPSS data file includes 66 cases and three variables: a factor distinguishing between the two types of reinforcement schedules (random or spaced), a second factor distinguishing among three types of reinforcers (token, money, or food), and the dependent variable, an arithmetic problem-solving test.

Conduct a two-way ANOVA to analyze these data.

1. Identify the following information from the output:
   a. F value for the schedule main effect
   b. Mean GPA dependent variable score for the cell associated with the random schedule and the money reinforcer
   c. Effect size for the interaction effect
   d. p value for the reinforcer type main effect
2. What follow-up tests would you conduct? Why?
3. Write a Results section based on your output.
4. Create a boxplot to visually represent the interaction between schedule type and reinforcer type on test scores.

Exercises 5 through 8 are based on the following research problem. The data set for this problem is named *Lesson 26 Exercise File 2* on the Web at http://www.prenhall.com/greensalkind.

Vicki was interested in how much time fathers of children with a disability play with their children who are disabled. To address this question, she found 60 fathers in six categories: (a) fathers with a male child with no physical or mental disability, (b) fathers with a female child with no physical or mental disability, (c) fathers with a male physically disabled child, (d) fathers with a female physically disabled child, (e) fathers with a male mentally retarded child, and (f) fathers with a female mentally retarded child. She asked fathers to record how many minutes per day they spent playing with their child for five days. The SPSS data file contains 60 cases and three variables, a factor indicating the disability status of the child, a factor indicating the child's gender, and a play time score averaged across the five days.

5. Conduct a two-way ANOVA to evaluate differences among the groups, according to gender and disability status of the child, in the amount of time fathers spent playing with their children.
6. Which follow-up procedures should you use? Why?
7. Write a Results section reporting the results of the analysis.
8. Create a boxplot to show the distributions for the six groups.

LESSON 27

One-Way Analysis of Covariance

A one-way analysis of covariance (ANCOVA) evaluates the null hypothesis that population means on the dependent variable are equal across levels of a factor, adjusting for differences on the covariate, or, more simply stated, the population adjusted means are equal across groups. An assumption underlying ANCOVA is that the slopes relating the covariate to the dependent variable are the same for all groups (i.e., the homogeneity-of-slopes assumption). If this assumption is violated, then between-group differences in adjusted means are not interpretable. In practice, the homogeneity-of-slopes assumption should be empirically evaluated prior to conducting an ANCOVA, and ANCOVA should not be conducted if the results indicate that this assumption has been violated. In place of ANCOVA, analyses should be carried out to assess simple main effects, that is, differences in group means on the dependent variable for particular levels of the covariate.

With a one-way analysis of covariance, each individual or case must have scores on three variables: a factor or independent variable, a covariate, and a dependent variable. The factor divides individuals into two or more groups or levels, while the covariate and the dependent variable differentiate individuals on quantitative dimensions.

Applications of the One-Way ANCOVA

One-way ANCOVA is used to analyze data from several types of studies.

- Studies with a pretest and random assignment of subjects to factor levels
- Studies with a pretest and assignment to factor levels based on the pretest
- Studies with a pretest, matching based on the pretest, and random assignment to factor levels
- Studies with potential confounding

We will next illustrate each of these four types of studies and then briefly discuss them. Maxwell and Delaney (2000) discuss more extensively the use of ANCOVA with these four applications.

Studies with a Pretest and Random Assignment to Factor Levels

Dana wishes to assess whether vitamin C is effective in the treatment of colds. To evaluate her hypothesis, she decides to conduct a two-year experimental study. She obtains 30 volunteers from undergraduate classes to participate. She randomly assigns an equal number of students to three groups: placebo (group 1), low doses of vitamin C (group 2), and high doses of vitamin C (group 3). In the first and second years of the study, students in all three groups are monitored to assess the number of days that they have cold symptoms.

During the first year, students do not take any pills. In the second year of the study, students take pills that contain one of the following: no active ingredient (group 1), a low dose of vitamin C (group 2), or a high dose of vitamin C (group 3). Dana's SPSS data file includes 30 cases and three variables: a factor distinguishing among the three medication groups; a covariate, the number of days with cold symptoms in the first year before any treatment; and a dependent variable, the number of days with cold symptoms in the second year when they are taking pills.

Studies with a Pretest and Assignment to Factor Levels Based on the Pretest

Jean wishes to assess whether individuals who are depressed about some event would benefit from intensively writing about their emotions. To evaluate this hypothesis, she obtains a sample of 60 runners who have had physical injuries that prevent them from running. She administers a measure of depression to these runners. The individuals who have the highest scores on the premeasure of depression (those who are most depressed) are assigned to the writing condition, while those who have lower scores are assigned to the attention placebo group. The runners in both groups meet with Jean's research assistant twice, four hours per occasion. The runners in the writing group are asked to write about how they feel about their injuries, while the runners in the attention placebo group are asked to write a short story involving adventure. Runners in both groups then retake the depression measure. Jean's SPSS data file includes 60 cases and three variables: a factor distinguishing among the two treatment groups, a covariate that assesses depression before treatment, and a dependent variable that assesses depression after treatment.

Studies with a Pretest, Matching Based on the Pretest, and Random Assignment to Factor Levels

Jim decides to conduct a study to learn how much time basketball players need to spend at the foul line preparing themselves to shoot free throws to be effective. Jim asks boys from his physical education classes to volunteer for his free-throw shooting study. Sixty boys consent to participate. These boys initially shoot 50 free throws, and their accuracy is recorded. Next, Jim rank-orders the accuracy scores from lowest to highest and places the boys into 20 groups of three based on these scores. In other words, the three boys who have the three lowest accuracy scores are in one group, the three boys who have the next three lowest accuracy scores are in another group, and so on. Within each of these 20 triads, one boy is randomly assigned to treatment Group 1, another boy to treatment Group 2, and the third boy to treatment Group 3. Over the next two weeks, boys are taught a specific method to shoot free throws. The boys in Group 1 are taught to shoot approximately 2 seconds after going to the free throw line; the boys in Group 2 are taught to shoot approximately 5 seconds after going to the free throw line; and the boys in Group 3 are taught to shoot approximately 8 seconds after going to the free throw line. Finally, all boys again must shoot 50 free throws, and their scores are recorded. Jim's SPSS data file has 60 cases and three variables: a factor distinguishing among the three treatment groups, a covariate that assesses their accuracy in shooting free throws before treatment, and a dependent variable that assesses their accuracy in shooting free throws after treatment.

Studies with Potential Confounding

Larry works with 60 boys who have behavior problems in the classroom. Recently, 20 of these boys participated in a special summer program. The other 40 boys did not participate because their parents did not want them to participate or failed to return the permission slip before the deadline. After the summer program, Larry had the teachers of all 60 boys complete a rating form indicating the degree that they found the boys to be behavioral problems in their classrooms. Larry wanted to compare the two groups of boys on the behavior ratings to assess the effectiveness of the summer program. Because he felt that the boys who participated in the summer program might come from families with a higher socioeconomic status (SES), he also computed an SES index for each of the 60 boys. Larry's SPSS data file includes 60 cases and three variables: a factor distinguishing between the two groups (boys who participated in the summer program and boys who did not), a covariate SES, and the behavior ratings completed by the teachers.

Understanding One-Way ANCOVA

Prior to conducting an ANCOVA, it is necessary to evaluate empirically the homogeneity-of-slopes assumption. The results of this empirical analysis leads researchers to proceed in one of two ways:

- If the results support the homogeneity-of-slopes assumption, researchers may conduct an ANCOVA. The ANCOVA F test evaluates whether the population means on the dependent variable, adjusted for differences on the covariate, differ across levels of a factor. If a factor has more than two levels and the F is significant, follow-up tests should be conducted to assess differences between adjusted means for the groups. These follow-up tests most often involve comparisons of pairs of adjusted means. For example, if a factor has three levels, three pairwise comparisons among adjusted means can be conducted: Group 1 versus Group 2, Group 1 versus Group 3, and Group 2 versus Group 3.
- If the results fail to support the homogeneity-of-slopes assumption, researchers should not conduct an ANCOVA. The implication of finding the slopes to be heterogeneous is that the mean differences between groups vary as a function of the covariate score. Accordingly, follow-up tests are required to assess mean differences between groups for particular scores on the covariate. These tests are referred to as simple main effects tests. If mean differences exist for any particular covariate score and the factor has more than two levels, additional analyses may be conducted to assess pairwise differences among groups (i.e., levels of the factor) at that covariate score.

Adequacy of One-Way ANCOVA for Each Application

The adequacy of ANCOVA to adjust the dependent variable scores for covariate differences depends on the type of study.

Studies with a Pretest and Random Assignment to Factor Levels

ANCOVA may be applied to data in which (1) all cases are measured initially on a pretest, (2) cases are randomly assigned to different groups, (3) groups receive different treatments, and (4) all cases subsequently are measured on a posttest. The pretest and the posttest could be the same measure. Alternatively, the pretest and posttest could be different measures and could even assess different constructs. Assuming that the ANCOVA assumptions are met, the one-way ANCOVA should adequately adjust dependent variable scores for initial covariate differences among groups for this design.

Studies with a Pretest and Assignment to Factor Levels Based on the Pretest

ANCOVA may be applied to data in which (1) all cases are measured on a pretest, (2) cases are assigned to different treatment groups based on their pretest scores, (3) groups receive different treatments, and (4) all cases are measured on a posttest. The pretest and the posttest could be the same measure, or they could be different measures that assess the same or different dimensions.

If the ANCOVA assumptions are met, ANCOVA should adequately adjust the dependent variable scores for initial covariate differences among groups for this design.

Studies with a Pretest, Matching Based on the Pretest, and Random Assignment to Factor Levels

ANCOVA may be applied to data in which (1) all cases are measured on a pretest, (2) cases are assigned to different groups based on their pretest scores, (3) cases are randomly assigned to levels of the factor within each of the groups that were formed based on the pretest, (4) cases in different levels of the factor receive different treatments, and (5) all cases are measured on a posttest. Similar to the previous application, the pretest and the posttest could be the same measure or different measures assessing the same or different constructs. Assuming that the ANCOVA assumptions are met, the one-way ANCOVA should adequately adjust dependent variable scores for initial covariate differences among groups.

Studies with Potential Confounding

For studies with potential confounding, cases are in different groups, but are neither randomly assigned to groups nor assigned to groups based on their pretest scores. The difficulty with these studies is that the groups may differ due to variables other than the factor and the covariate. Because these studies are potentially confounded, conclusions about the group differences are difficult to reach. The results of an ANCOVA can be misleading for studies with this design.

Assumptions Underlying a One-Way ANCOVA

Assumption 1: The Dependent Variable Is Normally Distributed in the Population for Any Specific Value of the Covariate and for Any One Level of a Factor

This assumption describes multiple conditional distributions of the dependent variable, one for every combination of values of the covariate and levels of the factor, and requires them all to be normally distributed. To the extent that population distributions are not normal and sample sizes are small, p values may be invalid. In addition, the power of ANCOVA tests may be reduced considerably if the population distributions are nonnormal and, more specifically, thick-tailed or heavily skewed. See Wilcox (2001) for an extended discussion of assumptions.

Assumption 2: The Variances of the Dependent Variable for the Conditional Distributions Described in Assumption 1 Are Equal

To the extent that this assumption is violated and the group sample sizes differ, the validity of the results of the one-way ANCOVA analysis should be questioned. Even with equal sample sizes, the results of the standard post hoc tests should be mistrusted if the population variances differ.

Assumption 3: The Cases Represent a Random Sample from the Population, and the Scores on the Dependent Variable Are Independent of Each Other

The test will yield inaccurate results if the independence assumption is violated.

Assumption 4: The Covariate Is Linearly Related to the Dependent Variable within All Levels of the Factor, and the Weights or Slopes Relating the Covariate to the Dependent Variable Are Equal across All Levels of the Factor

The latter part of this assumption is the homogeneity-of-slopes assumption. To the extent that homogeneity of slopes does not hold, the results of ANCOVA are likely to be misinterpreted in that the differences on the dependent variable between groups vary as a function of the covariate. Later in this lesson, we will discuss a method for evaluating the homogeneity-of-slopes assumption.

In addition, we will present simple main effect tests that may be conducted if this assumption does not hold. These tests evaluate differences between groups on the dependent variable for particular values of the covariate.

Effect Size Statistics for One-Way ANCOVA

The General Linear Model procedure computes an effect size index, the partial η^2. Although partial η^2s are computed for the factor and the covariate, via the formula

$$\text{Partial } \eta^2_{\text{Factor or Covariate Source}} = \frac{\text{Sum of Squares}_{\text{Factor or Covariate Source}}}{\text{Sum of Squares}_{\text{Factor or Covariate Source}} + \text{Sum of Squares}_{\text{Error}}}$$

the partial η^2 of primary interest is the one associated with the factor.

The partial η^2 ranges in value from 0 to 1. The partial η^2 for the factor is interpreted as the proportion of variance of the dependent variable related to the factor, holding constant (partialling out) the covariate. It is unclear what are small, medium, and large values for partial η^2; however, conventional cutoffs are .01, .06, and .14, respectively.

The Data Set

The data set used to illustrate this procedure is named *Lesson 27 Data File 1* on the Web at http://www.prenhall.com/greensalkind and represents data from the vitamin C example described earlier in this lesson. The variables in the data set are shown in Table 24.

Table 24
Variables in Lesson 27 Data File 1

Variable	Definition
group	1 = Placebo
	2 = Low doses of vitamin C
	3 = High doses of vitamin C
predays	Number of days with cold symptoms in the first year
days	Number of days with cold symptoms in the second year

The Research Question

The research question can be stated to reflect mean differences or relationships between variables.

1. Mean differences: Does the number of days of cold symptoms differ for those who take a placebo, those who take low doses of vitamin C, and those who take high doses of vitamin C assuming no prior differences in the number of days of cold symptoms among groups?
2. Relationship between variables: Is there a relationship between how much vitamin C is taken and the number of days that an individual shows cold symptoms, holding constant the number of days with cold symptoms in the year prior to treatment?

The research question would have to be modified if the slopes were found to be heterogeneous among groups. For example, the first research question might be rephrased as follows: Does the number of days of cold symptoms differ for students who take a placebo, students who take low doses of vitamin C, and students who take high doses of vitamin C if the students had a specific number of days of cold symptoms in the year prior to these treatments (e.g., 9 days of cold symptoms)?

Conducting a One-Way ANCOVA and Related Analyses

In this section, we will illustrate a number of analyses:

- **Test of Homogeneity of Slopes** We initially demonstrate how to test the hypothesis of the homogeneity of slopes, which is an assumption of ANCOVA. This analysis represents the first step in the testing process when conducting an ANCOVA.
- **Analysis of Covariance** Next, we show how to conduct an ANCOVA. ANCOVA evaluates differences in adjusted means. This step presumes that the initial analysis indicated that the slopes appear homogeneous.
- **Simple Main Effects** Finally, we decribe how to assess group differences on the dependent variable for particular levels of the covariate (i.e., simple main effects). This step assumes that the initial analysis indicated that the slopes appear heterogeneous.

Conducting a Test of the Homogeneity-of-Slopes Assumption

To conduct the test of the homogeneity-of-slopes assumption, follow these steps:

1. Click **Analyze**, click **General Linear Model**, then click **Univariate**.
2. Click **days**, then click ▶ to move it to the Dependent Variable box.
3. Click **group**, then click ▶ to move it to the Fixed Factor(s) box.
4. Click **predays**, then click ▶ to move it to the Covariate(s) box.
5. Click on **Options**.
6. In the Factor(s) and Factor Interactions box, click **group**.
7. Click ▶ to move it to the Display Means for box.
8. Select **Descriptive statistics**, **Estimates of effect size**, and **Homogeneity tests** in Display box.
9. Click **Continue**.
10. Click **Model**.
11. Click **Custom** under Specify Model.
12. Click **group(F)** under Factors & Covariates and click ▶ to make it appear in the Model box.
13. Click **predays(C)** under Factors & Covariates and click ▶ to make it appear in the Model box.
14. Holding down the Ctrl key, click **group(F)** and **predays(C)** in the Factors & Covariates box. Check to see that the default option Interaction is specified in the drop-down menu in the Build Terms box. If it is not, select it.
15. Click ▶ and group*predays should now appear in the Model box.
16. Click **Continue**.
17. Click **OK**.

Selected SPSS Output for Test of the Homogeneity-of-Slopes Assumption

Selected results of the analysis are shown in Figure 163.

Before conducting an ANCOVA, the homogeneity-of-slopes assumption should be tested. The null hypothesis for this test is that the population slopes are homogeneous. The null hypothesis can also be conceptualized as the population effect of the interaction between the covariate and the factor in predicting the dependent variable is zero. A significant interaction between the covariate and the factor suggests that population slopes differ or that the differences on the dependent variable among groups vary as a function of the covariate. If the interaction is significant, ANCOVA should not be conducted. Instead, the differences between groups on the dependent variables should be assessed at particular levels of the covariate (i.e., simple main effects).

The interaction source is labeled group*predays. The interaction is not significant, $F(2, 24) = 1.47$, $p = .25$, although the partial η^2 of .11 is of moderate size. The results of the partial η^2 indicate that in the sample the mean differences in posttreatment days of cold symptoms

Tests of Between-Subjects Effects

Dependent Variable: Days with Colds: Post

Source	Type III Sum of Squares	df	Mean Square	F	Sig.	Partial Eta Squared
Corrected Model	351.594[a]	5	70.319	5.917	.001	.552
Intercept	186.218	1	186.218	15.670	.001	.395
group	10.114	2	5.057	.426	.658	.034
predays	156.618	1	156.618	13.179	.001	.354
group * predays	35.039	2	17.519	1.474	.249	.109
Error	285.206	24	11.884			
Total	2960.000	30				
Corrected Total	636.800	29				

a. R Squared = .552 (Adjusted R Squared = .459)

***Figure 163*. Selected results of the test of homogeneity of slopes.**

among the vitamin C groups varied to some extent as a function of the number of pretreatment days of cold symptoms. However, given the interaction test was nonsignificant, the data are not sufficiently convincing that we can reach this conclusion in the population. Under these conditions, researchers may take one of two approaches. Based on the nonsignicant test results, an ANCOVA could be conducted assuming homogeneity of slopes and, if the ANCOVA is significant, follow-up tests could be computed to assess differences in adjusted means. Alternatively, based on the results of the partial η^2, simple main effects tests could be conducted that allow for heterogeneity of slopes. We prefer the second strategy in that the homogeneity assumption is obviated.

We next proceed to the ANCOVA assuming homogeneity of slopes. However, we later consider methods that allow for heterogeneity of slopes for this problem.

Conducting One-Way ANCOVA

To conduct a one-way ANCOVA, follow these steps:

1. Click **Analyze**, click **General Linear Model**, then click **Univariate**.
2. If you have not exited SPSS, the appropriate options should still be selected. If not, conduct Steps 2 through 9 in the preceding homogeneity-of-slopes test.
3. Click **Model**.
4. Click **Full Factorial**.
5. Click **Continue**.
6. Click on **Options**.
7. Click on the box next to **Compare main effects**.
8. Click **Continue**.
9. Click **OK**.

Selected SPSS Results for the Main Effect and the Covariate

The results are shown in Figure 164. The group source (labeled GROUP on the SPSS output) evaluates the null hypothesis that the population adjusted means are equal. The results of the analysis indicate that this hypothesis should be rejected, $F(2, 26) = 6.45$, $p < .01$, and the partial η^2 of .33 suggests a strong relationship between treatment and posttreatment days with cold symptoms, controlling for pretreatment days with cold symptoms. The test assesses the differences among the adjusted means for the three groups, which are reported in the output as the Estimated Marginal Means (i.e., 12.01, 7.71, and 6.67). It should be noted that differences among the adjusted means are not the same as differences among the means on the dependent measure (i.e., 11.60, 8.40, and 6.40) in that the three treatment groups had differing number of pretreatment days with colds.

Descriptive Statistics

Dependent Variable: Days with Colds: Post

Vitamin C Treatment	Mean	Std. Deviation	N
Placebo	11.60	5.358	10
Low Vitamin C Dose	8.40	3.836	10
High Vitamin C Dose	6.40	3.471	10
Total	8.80	4.686	30

Tests of Between-Subjects Effects

Dependent Variable: Days with Colds: Post

Source	Type III Sum of Squares	df	Mean Square	F	Sig.	Partial Eta Squared
Corrected Model	316.555[a]	3	105.518	8.567	.000	.497
Intercept	172.111	1	172.111	13.973	.001	.350
predays	178.955	1	178.955	14.529	.001	.358
group	158.903	2	79.452	6.450	.005	.332
Error	320.245	26	12.317			
Total	2960.000	30				
Corrected Total	636.800	29				

a. R Squared = .497 (Adjusted R Squared = .439)

Estimates

Dependent Variable: Days with Colds: Post

Vitamin C Treatment	Mean	Std. Error	95% Confidence Interval Lower Bound	95% Confidence Interval Upper Bound
Placebo	12.011[a]	1.115	9.719	14.303
Low Vitamin C Dose	7.715[a]	1.124	5.404	10.026
High Vitamin C Dose	6.674[a]	1.112	4.388	8.960

a. Covariates appearing in the model are evaluated at the following values: Days with Colds: Prior = 9.00.

Pairwise Comparisons

Dependent Variable: Days with Colds: Post

(I) Vitamin C Treatment	(J) Vitamin C Treatment	Mean Difference (I-J)	Std. Error	Sig.[a]	95% Confidence Interval for Difference[a] Lower Bound	95% Confidence Interval for Difference[a] Upper Bound
Placebo	Low Vitamin C Dose	4.296*	1.596	.012	1.016	7.576
	High Vitamin C Dose	5.337*	1.570	.002	2.110	8.564
Low Vitamin C Dose	Placebo	-4.296*	1.596	.012	-7.576	-1.016
	High Vitamin C Dose	1.041	1.590	.518	-2.227	4.308
High Vitamin C Dose	Placebo	-5.337*	1.570	.002	-8.564	-2.110
	Low Vitamin C Dose	-1.041	1.590	.518	-4.308	2.227

Based on estimated marginal means

*. The mean difference is significant at the .05 level.

a. Adjustment for multiple comparisons: Least Significant Difference (equivalent to no adjustments).

***Figure 164.* Results of the one-way ANCOVA.**

The covariate is included in the analysis to control for differences on this variable and is not the focus of the analysis. Consequently, the results for the covariate are frequently not reported in a Results section. Nevertheless, the results are available as part of the output. The test of the covariate evaluates the relationship between the covariate and the dependent variable within groups (i.e., controlling for the factor). In the current example, this relationship is significant, $F(1, 26) = 14.53$, $p < .01$, with the covariate accounting for about 36% (i.e., the partial η^2 of .36) of variance of the posttreatment days with cold symptoms, controlling for the treatment factor.

The results of the pairwise comparisons are shown in the bottom portion of Figure 164. The differences in adjusted means are 4.30 (12.01 − 7.71) between the placebo and the low-dose vitamin C groups, 5.34 (12.01 − 6.67) between the placebo and high-dose vitamin C groups, and −1.04 (7.71 − 6.67) between the low-and high-dose vitamin C groups. Based on the LSD (Appendix B), the first two pairwise differences were significant, $p = .01$ and $p < .01$, whereas the last pairwise difference was nonsignificant, $p = .52$.

Conducting Tests of Simple Group Main Effects for Particular Values of the Covariate

If the slopes are heterogeneous in the population, ANCOVA is inappropriate. Under these conditions, analyses are required that assess mean differences between groups on the dependent variable for particular levels of the covariate—that is, simple group main effects. To conduct these analyses, researchers must choose levels on the covariate. We recommend selecting at least three levels representing low, medium, and high values. One empirical approach to determine levels would be to choose one standard deviation below the mean, the mean, and one standard deviation above the mean on the covariate. For our data, the mean and standard deviation on predays, ignoring groups, are 9.00 and 5.55, respectively. Accordingly, low, medium, and high values are 3.45, 9.00, and 14.55. To assess differences between groups at these levels of the covariate, follow these steps:

1. Click **Analyze**, click **General Linear Model**, then click **Univariate**.
2. Conduct Steps 2 through 16 as described in Conducting a Test of the Homogeneity-of-Slopes Assumption.
3. Click on **Options**.
4. Click on the box next to **Compare main effects.**
5. Click **Continue**.
6. Click on **Paste**.
7. Copy the /EMMEANS statement so it appears three times in the syntax:
   /EMMEANS = TABLES(group) WITH(predays=MEAN) COMPARE ADJ(LSD)
   /EMMEANS = TABLES(group) WITH(predays=MEAN) COMPARE ADJ(LSD)
   /EMMEANS = TABLES(group) WITH(predays=MEAN) COMPARE ADJ(LSD)
8. Substitute 3.45 for MEAN in the first /EMMEANS statement and 14.55 for MEAN in the third /EMMEANS statement. The syntax should appear as it does in Figure 165.
9. Highlight the syntax, click **Run**, and then click **Selection**.

```
UNIANOVA
 days BY group WITH predays
 /METHOD = SSTYPE(3)
 /INTERCEPT = INCLUDE
 /EMMEANS = TABLES(group) WITH(predays=3.45) COMPARE ADJ(LSD)
 /EMMEANS = TABLES(group) WITH(predays=MEAN) COMPARE ADJ(LSD)
 /EMMEANS = TABLES(group) WITH(predays=14.55) COMPARE ADJ(LSD)
 /CRITERIA = ALPHA(.05)
 /DESIGN = group predays group*predays .
```

Figure 165. **Syntax for conducting tests of simple group main effects.**

Selected SPSS Output for Simple Group Main Effects

Simple main effect tests were conducted at low, medium, and high values on the covariate. Accordingly, we required a p value of .017 (.05/3) for significance. If any one simple main effect was significant, pairwise comparisons were evaluated at the same level (i.e., .017) as the simple main effects test following the LSD procedure.

The results of the tests for the simple group main effects when the pretreatment days of cold symptoms is equal to 3.45 are shown in Figure 166. For individuals with a pretreatment score of 3.45, the means for the posttreatment days of cold symptoms are estimated to be 8.51, 5.06, and 5.63 for the placebo, low-dose, and high-dose groups, respectively. As indicated in the table labeled Univariate Tests, we cannot conclude that the corresponding population means differ, $F(2,24) = 1.48$, $p = .25$, partial η^2 of .11. Given nonsignificance among means, we did not examine the pairwise differences among groups.

Estimates

Dependent Variable: Days with Colds: Post

			95% Confidence Interval	
Vitamin C Treatment	Mean	Std. Error	Lower Bound	Upper Bound
Placebo	8.508[a]	1.416	5.586	11.430
Low Vitamin C Dose	5.064[a]	1.762	1.428	8.699
High Vitamin C Dose	5.631[a]	1.559	2.414	8.849

a. Covariates appearing in the model are evaluated at the following values: Days with Colds: Prior = 3.

Pairwise Comparisons

Dependent Variable: Days with Colds: Post

					95% Confidence Interval for Difference[a]	
(I) Vitamin C Treatment	(J) Vitamin C Treatment	Mean Difference (I-J)	Std. Error	Sig.[a]	Lower Bound	Upper Bound
Placebo	Low Vitamin C Dose	3.444	2.260	.141	-1.220	8.109
	High Vitamin C Dose	2.877	2.106	.185	-1.470	7.223
Low Vitamin C Dose	Placebo	-3.444	2.260	.141	-8.109	1.220
	High Vitamin C Dose	-.568	2.352	.811	-5.423	4.287
High Vitamin C Dose	Placebo	-2.877	2.106	.185	-7.223	1.470
	Low Vitamin C Dose	.568	2.352	.811	-4.287	5.423

Based on estimated marginal means

a. Adjustment for multiple comparisons: Least Significant Difference (equivalent to no adjustments).

Univariate Tests

Dependent Variable: Days with Colds: Post

	Sum of Squares	df	Mean Square	F	Sig.	Partial Eta Squared
Contrast	35.178	2	17.589	1.480	.248	.110
Error	285.206	24	11.884			

The F tests the effect of Vitamin C Treatment. This test is based on the linearly independent pairwise comparisons among the estimated marginal means.

Figure 166. **Results of tests of simple group main effects when the covariate is equal to 3.45.**

The results of the tests of simple main effects are not shown for medium and high values on the covariate. For individuals with a pretreatment score of 9.00, the means for posttreatment days of cold symptoms are estimated to be 12.20, 7.69, and 6.49 for the placebo, low-dose, and high-dose groups, respectively. As indicated in the Univariate Tests table, the differences among the means on posttreatment measure when the pretreatment measure is 9.00 differ significantly, $F(2, 24) = 7.43$, $p < .01$, partial η^2 of .38. Pairwise comparisons indicate differences in the population between the placebo and low-dose vitamin C groups, $p < .01$, between the placebo and high-dose vitamin C groups, $p < .01$, but not between the low- and high-dose vitamin C groups, $p = .45$.

For individuals with high scores on the covariate (i.e., 14.55), the means for posttreatment days of cold symptoms are estimated to be 15.89, 10.32, and 7.36 for the placebo, low-dose, and high-dose groups, respectively. As indicated in the Univariate Tests table, the differences among the means on days when the pretest is 9.00 differ significantly, $F(2, 24) = 6.58$, $p < .01$, partial η^2 of .35. Pairwise comparisons indicate differences in the population between the placebo and low-dose vitamin C group, $p = .016$, between the placebo and high- dose vitamin C group, $p < .01$, but not between the low- and high-dose vitamin C groups, $p = .19$.

Using SPSS Graphs to Display the Results

A scatterplot may help in the interpretation of results when the slopes are heterogeneous. This graph shows differences in slopes as well as differences between estimated group means on the dependent variable for low, medium, and high values on the covariate. To obtain the graph, create a simple scatterplot with the covariate on the *X*-axis, the dependent variable on the *Y*-axis, and set markers by the factor. In editing the graph, choose Fit Line at Subgroups within Elements (main menu) and choose vertical ("*X* Axis") reference lines within Options (main menu) to portray low, medium, and high values on the covariate. The option *X*-Axis reference line must be chosen three times, once for each line. Each time, type the value in the Position box.

The scatterplot for our example is presented in Figure 167. An estimated group mean on the dependent variable for a low, medium, or high covariate score can be determined by following the appropriate vertical line from the covariate axis to the regression line for the group of interest and reading the corresponding dependent variable score at that point. For example, for individuals with a low number of pretreatment day with cold symptoms, the estimated mean on post-treatment days with cold symptoms is approximately 5 if given low doses of vitamin C and only slightly higher if given high doses of vitamin C. We are able to describe the results in the sample more easily based on our scatterplot. For example, in comparison with placebo, high doses of vitamin C demonstrated, on average, greater effectiveness to the extent that individuals had a greater number of days with cold symptoms prior to treatment.

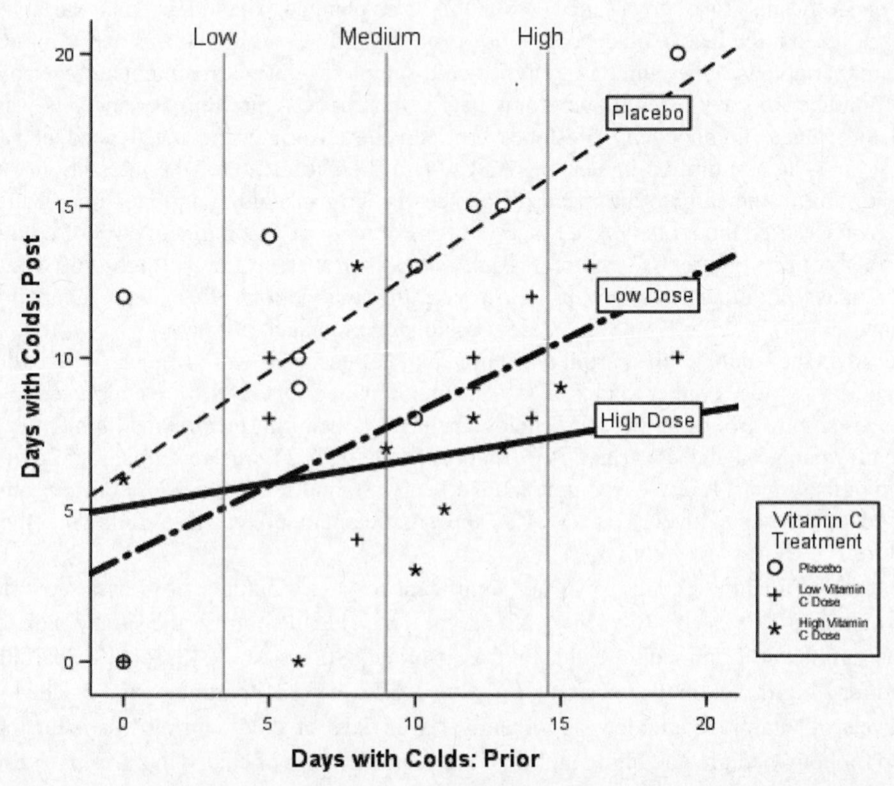

Figure 167. **Graph showing the differences between groups for three levels of the covariate.**

An APA Results Section

We present two results sections, one in which homogeneity of slope is assumed and one in which it is not.

Results Assuming Homogeneity of Slopes

A one-way analysis of covariance (ANCOVA) was conducted. The independent variable, vitamin C, included three levels: placebo, low doses of vitamin C, and high doses of vitamin C. The dependent variable was the number of days of cold symptoms during treatment and the covariate was the number of days of cold symptoms before treatment. A preliminary analysis evaluating the homogeneity-of-slopes assumption indicated that the relationship between the covariate and the dependent variable did not differ significantly as a function of the independent variable, $F(2, 24) = 1.47$, $MSE = 11.88$, $p = .25$, partial $\eta^2 = .11$. The ANCOVA was significant, $F(2, 26) = 6.45$, $MSE = 12.32$, $p < .01$. The strength of relationship between the vitamin C factor and dependent variable was very strong, as assessed by a partial η^2, with the vitamin C factor accounting for 33% of the variance of the dependent variable, holding constant the number of days with pretreatment cold symptoms.

The means of the number of days with cold symptoms adjusted for initial differences were ordered as expected across the three vitamin C groups. The placebo group had the largest adjusted mean ($M = 12.01$), the low-dose vitamin C group had a smaller adjusted mean ($M = 7.71$), and the high-dose vitamin C group had the smallest adjusted mean ($M = 6.67$). Follow-up tests were conducted to evaluate pairwise differences among these adjusted means. Based on the LSD procedure, the adjusted means for both vitamin C groups differed significantly from the placebo group, but the adjusted means for the two vitamin C groups did not differ significantly.

Results Not Assuming Homogeneity of Slopes

A one-way analysis of covariance (ANCOVA) was planned. The independent variable, vitamin C, included three levels: placebo, low doses of vitamin C, and high doses of vitamin C. The dependent variable was the number of days of cold symptoms during treatment and the covariate was the number of days of cold symptoms before treatment. A preliminary analysis was conducted to evaluate homogeneity of slopes between the covariate and the dependent variable across groups, an assumption underlying ANCOVA. The partial η^2 for the interaction was .11, indicating that in the sample the mean differences in days of cold symptoms among the vitamin C groups varied moderately as a function of the number of pretreatment days of cold symptoms. As shown in Figure 167, the regression line was less steep for the high-dose vitamin C group versus the other two groups. However, the interaction effect was nonsignificant, $F(2, 24) = 1.47$, $MSE = 11.88$, $p = .25$, possibly due to a lack of power.

Based on the results of the partial η^2, simple main effects tests were conducted that allow for heterogeneity of slopes rather than ANCOVA. Simple main effect tests were conducted to assess differences among groups at low (1 *SD* below the mean), medium (mean), and high (1 *SD* above the mean) values on the covariate. A p value of .017 (.05/3) was required for significance for each of these tests. If any one simple main effect was significant, pairwise comparisons were evaluated at the same level (i.e., .017) as the simple main effects test, following the LSD procedure.

The simple main effects test was not significant for a low number of pretreatment days of cold symptoms, $F(2, 24) = 1.48$, $p = .25$, partial η^2 of .11. In contrast, the simple main effects test was significant for a medium value on the covariate, $F(2, 24) = 7.43$, $p < .01$, partial η^2 of .38, and for a high level on the covariate, $F(2, 24) = 6.58$, $p < .01$, partial η^2 of .35. The low and high levels of vitamin C yielded significantly fewer days of colds during treatment than the placebo for both medium and high number of pretreatment days of cold symptoms. However, the differences between the two doses of vitamin C were not significant.

Alternative Analyses

SPSS offers no procedures for conducting nonparametric alternatives to ANCOVA. See Huitema (1980) for a discussion of nonparametric alternatives.

As for parametric methods, researchers sometimes use alternative methods such as ANOVA on change scores. If subjects are randomly assigned to groups or assigned based on pretest scores, ANCOVA is in most cases more powerful than these alternatives. If it is unclear how subjects are assigned to groups, none of the methods including ANCOVA are likely to control adequately for prior differences. (See Maxwell and Delaney, 2000, for a discussion of this topic.)

Exercises

The data for Exercises 1 through 5 are in the data set named *Lesson 27 Exercise File 1* on the Web at http://www.prenhall.com/greensalkind. The data are from the following research problem.

Sam is interested in the relationship between college professors' academic discipline and their actual ability to fix a car, holding constant mechanical aptitude. Five professors were randomly selected from mechanical engineering, psychology, and philosophy departments at a major university. Each professor completed a mechanical aptitude scale. Scores on this measure have a mean of 100 and a standard deviation of 15. The professors were then rated on how well they performed four automotive maintenance tasks: changing oil, changing the points and plugs, adjusting the carburetor, and setting the timing on a 1985 Pontiac. Ratings were based on the degree of success in completing a task and the amount of time needed to complete it. Lower scores reflect more efficiency at completing the automotive maintenance tasks.

1. Transform the scores on the four mechanical task ratings by z-scoring them and then summing the z scores so that Sam has a single measure of professors' mechanical performance efficiency.
2. Evaluate whether the relationship between mechanical aptitude and mechanical performance efficiency is the same for all three types of professors (homogeneity-of-slopes assumption). What should Sam conclude about the homogeneity-of-slopes assumption? Report the appropriate statistics from the output to justify your conclusion.
3. Conduct the standard ANCOVA on these data. From the output, identify the following:
   a. p value associated with the effect due to professor type
   b. Effect size associated with the effect due to professor type
   c. F value associated with the covariate, mechanical aptitude
4. Conduct the appropriate post hoc tests.
5. Write a Results section based on your analyses.

The data for Exercises 6 through 8 are in the data set named *Lesson 27 Exercise File 2* on the Web at http://www.prenhall.com/greensalkind. The data are from the following research problem.

Marilyn was interested in the effects of journal therapy on depression. She randomly assigned 60 individuals who had been clinically diagnosed as having severe depression to three treatment groups. Individuals in the first group wrote intensively about their feelings in journals for an hour each day for three weeks (journal therapy) and received regular counseling. The second group received only journal therapy, and the third group received only regular counseling. All 60 participants were given a depression subscale before and after treatment.

6. Evaluate the homogeneity-of-slopes assumption. What should Marilyn conclude about this assumption? Report the appropriate statistics from the output to justify your conclusion.
7. Conduct an ANCOVA on these data. What conclusions should Marilyn draw?
8. Should Marilyn conduct post hoc tests? Why or why not? Report the appropriate statistics from the output to justify your conclusion.

LESSON

28 One-Way Multivariate Analysis of Variance

Multivariate analysis of variance (MANOVA) is a multivariate extension of analysis of variance. As with ANOVA, the independent variables for a MANOVA are factors, and each factor has two or more levels. Unlike ANOVA, MANOVA includes multiple dependent variables rather than a single dependent variable. MANOVA evaluates whether the population means on a set of dependent variables vary across levels of a factor or factors. Here we will discuss only a MANOVA with a single factor, that is, a one-way MANOVA. Each case in the SPSS data file for a one-way MANOVA contains a factor distinguishing participants into groups and two or more quantitative dependent variables.

Applications of One-Way MANOVA

One-way MANOVAs can analyze data from different types of studies.

- Experimental studies
- Quasi-experimental studies
- Field studies

We give examples of an experimental study and a quasi-experimental study.

An Experimental Study

Dan wished to examine the effects of various study strategies on learning. Thirty students from different sections of an introductory psychology course were randomly assigned to one of three study conditions: the think condition, the write condition, or the talk condition. Students attended one general lecture after which they were placed into study rooms based on their assigned condition. Students in all rooms received the same set of study questions, but each room received different instructions about how to study. The write group was instructed to write responses to each question, the think group was instructed to think about answers to the questions, and the talk group was instructed to develop a talk that they could deliver centering on the answers to the questions. At the completion of the study session, all students took a quiz consisting of recall, application, analysis, and synthesis questions. Dan's SPSS data file includes 30 cases and five variables. The variables include a factor differentiating the three types of study groups (write, think, and talk groups) and the four dependent variables, scores on the recall, application, analysis, and synthesis questions.

A Quasi-Experimental Study

Wanda wanted to determine whether assurances of confidentiality or anonymity have an effect on how students respond to personality and psychopathology questionnaires. She had previously administered the same three questionnaires—the Beck Depression Inventory (BDI), the 10 clinical scales of the Minnesota Multiphasic Personality Inventory (MMPI), and the Eating Disorders Scale (EDS)—to 479 students at the beginning of three different studies. In one study, she told the students that they would complete the questionnaires anonymously. In a second study, she told the students that the results would be confidential, but she requested that they put their names on their questionnaires. In the third study, she told the students that the results would be confidential, but that she would contact them for possible referral to mental health counselors if their results warranted it. Wanda's SPSS data file has 479 cases and 13 variables, a factor differentiating among the instructional sets for the three studies (anonymity, confidentiality, and confidentiality with referral) and the 12 dependent variables (the BDI, the 10 MMPI scales, and the EDS).

Understanding One-Way MANOVA

A one-way MANOVA tests the hypothesis that the population means for the dependent variables are the same for all levels of a factor, that is, across all groups. If the population means of the dependent variables are equal for all groups, the population means for any linear combination of these dependent variables are also equal for all groups. Consequently, a one-way MANOVA evaluates a hypothesis that includes not only equality among group means on the dependent variables, but also equality among group means on linear combinations of these dependent variables.

SPSS reports a number of statistics to evaluate the MANOVA hypothesis, labeled Wilks's lambda, Pillai's trace, Hotelling's trace, and Roy's largest root. (Please note that the APA style manual prefers Wilks' lambda to Wilks' lambda.)

Each statistic evaluates a multivariate hypothesis that the population means on the multiple dependent variables are equal across groups. We will use Wilks's lambda because it is frequently reported in the social science literature. Pillai's trace is a reasonable alternative to Wilks's lambda.

If the one-way MANOVA is significant, follow-up analyses can assess whether there are differences among groups on the population means for certain dependent variables and for particular linear combinations of dependent variables. A popular follow-up approach is to conduct multiple ANOVAs, one for each dependent variable, and to control for Type I error across these multiple tests by using one of the Bonferroni approaches. If any of these ANOVAs yield significance and the factor contains more than two levels, additional follow-up tests are performed. These tests typically involve post hoc pairwise comparisons among levels of the factor, although they may involve more complex comparisons. We will illustrate follow-up tests using this strategy.

Some people have criticized the strategy of conducting follow-up ANOVAs after obtaining a significant MANOVA because the individual ANOVAs do not take into account the multivariate nature of MANOVA. Conducting follow-up ANOVAs ignores the fact that the MANOVA hypothesis includes subhypotheses about linear combinations of dependent variables. Of course, if we have particular linear combinations of variables of interest, we can evaluate these linear combinations by using ANOVA in addition to, or in place of, the ANOVAs conducted on the individual dependent variables. For example, if two of the dependent variables for a MANOVA measure the same construct of introversion, then we may wish to represent them by transforming the variables to z scores, adding them together, and evaluating the resulting combined scores by using ANOVA. This ANOVA could be performed in addition to the ANOVAs on the remaining dependent variables.

If we have no clue as to what linear combinations of dependent variables to evaluate, we may choose to conduct follow-up analyses to a significant MANOVA with the use of discriminant analysis. Discriminant analysis (see Lesson 35) yields one or more uncorrelated linear combinations of dependent variables that maximize differences among the groups. These linear combinations are empirically determined and may not be interpretable.

Assumptions Underlying One-Way MANOVA

Assumption 1: The Dependent Variables Are Multivariately Normally Distributed for Each Population, with the Different Populations Being Defined by the Levels of the Factor

If the dependent variables are multivariately normally distributed, each variable is normally distributed, ignoring the other variables and each variable is normally distributed at every combination of values of the other variables. It is difficult to imagine that this assumption could be met. To the extent that population distributions are not multivariate normal and sample sizes are small, the p values may be invalid. In addition, the power of this test may be reduced considerably if the population distributions are not multivariate normal and, more specifically, thick-tailed or heavily skewed.

Assumption 2: The Population Variances and Covariances among the Dependent Variables Are the Same across All Levels of the Factor

To the extent that the sample sizes are disparate and the variances and covariances are unequal, MANOVA yields invalid results. SPSS allows us to test the assumption of homogeneity of the variance-covariance matrices with Box's M statistic. The F test from Box's M statistics should be interpreted cautiously in that a significant result may be due to violation of the multivariate normality assumption for the Box's M test, and a nonsignificant result may be due to a lack of power.

Assumption 3: The Participants Are Randomly Sampled, and the Score on a Variable for any One Participant Is Independent from the Scores on this Variable for All Other Participants

MANOVA should not be conducted if the independence assumption is violated.

Effect Size Statistics for a One-Way MANOVA

The multivariate General Linear Model procedure computes a multivariate effect size index. The multivariate effect size associated with Wilks's lambda (Λ) is the multivariate eta square:

$$\textit{Multivariate}\ \eta^2 = 1 - \Lambda^{\frac{1}{s}}$$

Here, s is equal to the number of levels of the factor minus 1 or the number of dependent variables, whichever is smaller. This statistic should be interpreted similar to a univariate eta square and ranges in value from 0 to 1. A 0 indicates no relationship between the factor and the dependent variable, while a 1 indicates the strongest possible relationship. It is unclear what should be considered a small, medium, and large effect size for this statistic.

The Data Set

The data set used to illustrate a one-way MANOVA is named *Lesson 28 Data File 1* on the Web at http://www.prenhall.com/greensalkind. It represents data from the experimental study described earlier in the lesson. That research study involved evaluating differences in learning among three study strategies. It included four dependent variables, while here we include only two dependent variables (scores on recall and application questions). The variables in the data set are shown in Table 25.

Table 25
Variables in Lesson 28 Data File 1

Variable	Definition
group	The factor in this study is group. There are three levels to the factor: think (= 1), write (= 2), and talk (= 3).
recall	Recall is the total correct score on recall questions and ranges from 0 to 8.
applicat	Applicat is the total correct score on application questions and ranges from 0 to 8.

The Research Question

We could ask the research question to reflect mean differences or relationships between variables.

1. Mean differences: Are the population means for the scores on recall and application tests (or linear combinations of these test scores) the same or different for students in the three study groups?
2. Relationships between variables: Is there a relationship between type of study strategy used and performance on recall and application test items?

Conducting a One-Way MANOVA

Here, we will outline steps to conduct a one-way MANOVA, the univariate ANOVAs following a significant overall test, and the post hoc pairwise comparisons. Although these tests are typically sequential, we will illustrate how to conduct all three tests in a single step.

To conduct a one-way MANOVA, follow-up univariate tests, and post hoc comparisons, follow these steps:

1. Click **Analyze**, click **General Linear Model**, then click **Multivariate**. You'll see the Multivariate dialog box shown in Figure 168.

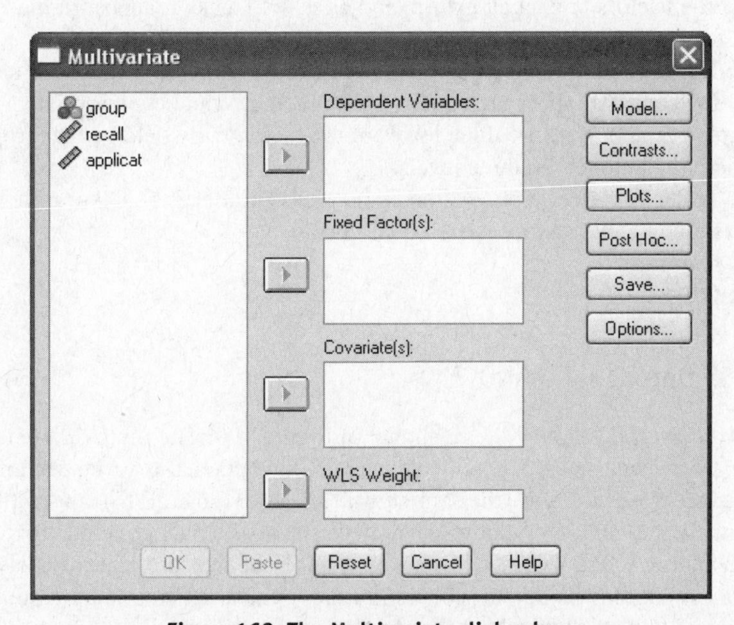

Figure 168. **The Multivariate dialog box.**

2. Click **recall**, hold down the Ctrl key and click **applicat**, and then click ▶ to move the two variables to the Dependent Variables box.
3. Click **group**, and then click ▶ to move the variable to the Fixed Factor(s) box.
4. Click **Options**. You'll see the Multivariate: Options dialog box shown in Figure 169.

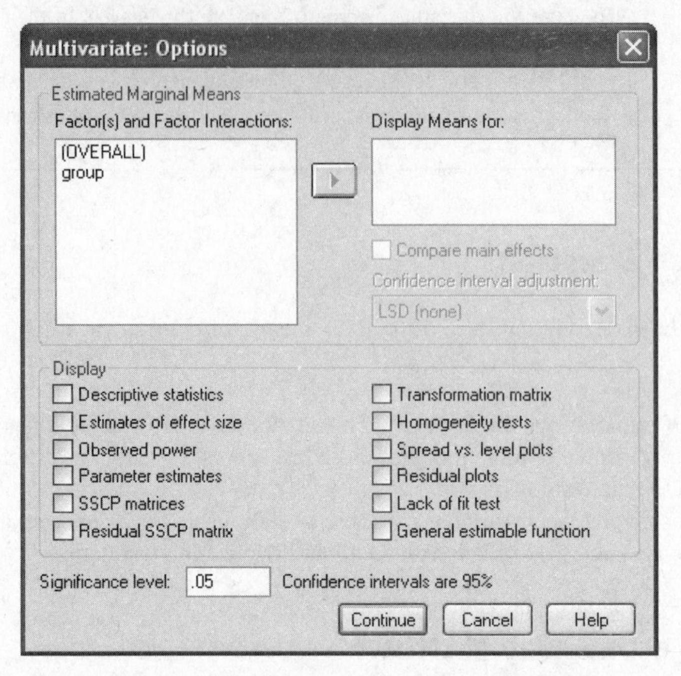

***Figure 169.* The Multivariate: Options dialog box.**

5. Click **group** in the Factor(s) and Factor Interactions box, then click ▶ to make it appear in the Display Means for box.
6. Click **Descriptive Statistics**, **Estimates of effect size**, and **Homogeneity tests** in the Display box.
7. Change the Significance Level from .05 to .025. (.025 was chosen by dividing .05 by the number of dependent variables. We will talk about choosing the significance level when we discuss output for pairwise comparisons.)
8. Click **Post Hoc**.
9. Under the Factors box, click **group**, and click ▶ to make it appear in the Post Hoc Tests for box.
10. In the Equal Variances Assumed box, click **Bonferroni**. Other post hoc tests assuming equal variances are equally valid. (LSD would be appropriate for our example because there are three levels of the factor, but would be inappropriate if the factor had more than three levels.)
11. In the Equal Variances Not Assumed box, click **Dunnett's *C***. Other post hoc tests not assuming equal variances are equally valid.
12. Click **Continue**.
13. Click **OK**.

Selected SPSS Output for MANOVA

The results of the overall MANOVA are shown in Figure 170. The multivariate test for homogeneity of dispersion matrices, Box's Test, evaluates whether the variances and covariance among the dependent variables are the same for all levels of a factor. If the F test for Box's test is significant, the homogeneity hypothesis is rejected, and we may conclude that there are differences in the matrices. The results of Box's test should be interpreted cautiously, however, in that a significant result may be due to violation of the multivariate normality assumption for this test and a nonsignificant result may be due to small sample size and a lack of power. In this example, the test for homogeneity of dispersion matrices is nonsignificant, $F(6, 18169) = 1.04, p = .40$.

Descriptive Statistics

	Study Strategy Groups	Mean	Std. Deviation	N
Recall Exam	Think	3.30	.675	10
	Write	5.80	1.033	10
	Talk	4.20	1.135	10
	Total	4.43	1.406	30
Application Exam	Think	3.20	1.229	10
	Write	5.00	1.764	10
	Talk	4.40	1.174	10
	Total	4.20	1.562	30

Box's Test of Equality of Covariance Matrices[a]

Box's M	6.980
F	1.039
df1	6
df2	18168.923
Sig.	.398

Tests the null hypothesis that the observed covariance matrices of the dependent variables are equal across groups.

a. Design: Intercept+group

Multivariate Tests[c]

Effect		Value	F	Hypothesis df	Error df	Sig.	Partial Eta Squared
Intercept	Pillai's Trace	.962	326.035[a]	2.000	26.000	.000	.962
	Wilks' Lambda	.038	326.035[a]	2.000	26.000	.000	.962
	Hotelling's Trace	25.080	326.035[a]	2.000	26.000	.000	.962
	Roy's Largest Root	25.080	326.035[a]	2.000	26.000	.000	.962
group	Pillai's Trace	.602	5.811	4.000	54.000	.001	.301
	Wilks' Lambda	.421	7.028[a]	4.000	52.000	.000	.351
	Hotelling's Trace	1.318	8.240	4.000	50.000	.000	.397
	Roy's Largest Root	1.275	17.215[b]	2.000	27.000	.000	.560

a. Exact statistic

b. The statistic is an upper bound on F that yields a lower bound on the significance level.

c. Design: Intercept+group

Figure 170. **The results of the one-way MANOVA.**

Also reported on the output and of primary concern are the results of the MANOVA test. In this example, the Wilks's Λ of .42 is significant, $F(4, 52) = 7.03$, $p < .01$, indicating that we can reject the hypothesis that the population means on the dependent variables are the same for the three teaching strategies. The multivariate $\eta^2 = .35$ indicates 35% of multivariate variance of the dependent variables is associated with the group factor.

Selected SPSS Output for the Univariate ANOVAs

The results of the univariate ANOVAs are shown in Figure 171 (on page 228).

If there are no missing data, the ANOVAs that are printed as part of the MANOVA output are identical to those obtained by conducting individual one-way ANOVAs. However, if data are missing, the univariate ANOVAs on the MANOVA output will differ from the univariate ANOVAs obtained by conducting separate ANOVAs using the Univariate General Linear Model procedure. The MANOVA procedure excludes data on all dependent variables for an individual if a score is missing on any one dependent variable. In contrast, data are excluded by the Univariate General Linear Model procedure only if a score is missing on the particular dependent variable being analyzed. To be consistent with the MANOVA results, follow-up analyses of variance should be reported for individuals who have scores on all of the dependent variables. These are the ANOVAs that are produced by the Multivariate General Linear Model procedure.

Tests of Between-Subjects Effects

Source	Dependent Variable	Type III Sum of Squares	df	Mean Square	F	Sig.	Partial Eta Squared
Corrected Model	Recall Exam	32.067[a]	2	16.033	17.111	.000	.559
	Application Exam	16.800[b]	2	8.400	4.200	.026	.237
Intercept	Recall Exam	589.633	1	589.633	629.253	.000	.959
	Application Exam	529.200	1	529.200	264.600	.000	.907
group	Recall Exam	32.067	2	16.033	17.111	.000	.559
	Application Exam	16.800	2	8.400	4.200	.026	.237
Error	Recall Exam	25.300	27	.937			
	Application Exam	54.000	27	2.000			
Total	Recall Exam	647.000	30				
	Application Exam	600.000	30				
Corrected Total	Recall Exam	57.367	29				
	Application Exam	70.800	29				

a. R Squared = .559 (Adjusted R Squared = .526)

b. R Squared = .237 (Adjusted R Squared = .181)

Figure 171. **The results of the univariate ANOVAs.**

The p values for the ANOVAs on the MANOVA output do not take into account that multiple ANOVAs have been conducted. Methods may be used to control for Type I error across the multiple ANOVAs. For our application, we chose to control for Type I error using a traditional Bonferroni procedure and test each ANOVA at the .025 level (.05 divided by the number of ANOVAs conducted). Consequently, the univariate ANOVA for the recall scores was significant, $F(2, 27) = 17.11, p < .01$, while the univariate ANOVA for the application scores was nonsignificant, $F(2, 27) = 4.20, p = .026$. The ANOVA for the application scores was nonsignificant because .026 exceeded the required level of .025.

TIP

The LSD procedure is a powerful method to control for Type I errors across all pairwise comparisons if a factor has three levels. See Appendix B for details about the LSD procedure.

Selected SPSS Output for Follow-up Pairwise Comparisons

The results of the pairwise comparisons are shown in Figure 172. We had previously controlled for Type I error across the two univariate ANOVAs by testing each at the .025 level. To be consistent with this decision, we also need to control the probability of committing one or more Type I errors across the multiple pairwise comparisons for a dependent variable at the .025 level. We are able to maintain this familywise error rate across comparisons for a dependent variable by selecting .025 for the Significance Level in the Multivariate: Options dialog box.

Because the ANOVA for the application scores was nonsignificant, we examined only the pairwise comparisons for the recall scores. Although more powerful methods are available, the Bonferroni approach was used to control for Type I error across the pairwise comparisons for recall. With the Bonferroni method, each comparison is tested at the alpha level for the ANOVA divided by the number of comparisons; for our example, $.025/3 = .008$. Two of the three comparisons were significant, the comparisons associated with the thinking and writing groups and with the writing and talking groups. It should be noted that the same two comparisons were significant with Dunnett's C method.

Using SPSS Graphs to Display the Results

Graphical methods, such as boxplots, can be used in the Results section to allow the reader to evaluate the differences among groups. The boxplots that display the distributions for the multiple dependent variables of a MANOVA are created using a method that is slightly different from the ones used for an independent-samples t test or one-way ANOVA. To create the boxplot shown in Figure 173 (on page 230), conduct the following steps:

1. Click **Graphs**, click **Legacy Dialogs**, and then click **Boxplot**.
2. Click **Clustered** and **Summaries for Separate Variables** in the Boxplot dialog box.

Multiple Comparisons

Dependent Variable		(I) Study Strategy Groups	(J) Study Strategy Groups	Mean Difference (I-J)	Std. Error	Sig.	95% Confidence Interval	
							Lower Bound	Upper Bound
Recall Exam	Bonferroni	Think	Write	-2.50*	.433	.000	-3.60	-1.40
			Talk	-.90	.433	.142	-2.00	.20
		Write	Think	2.50*	.433	.000	1.40	3.60
			Talk	1.60*	.433	.003	.50	2.70
		Talk	Think	.90	.433	.142	-.20	2.00
			Write	-1.60*	.433	.003	-2.70	-.50
	Dunnett C	Think	Write	-2.50*	.390		-3.59	-1.41
			Talk	-.90	.418		-2.07	.27
		Write	Think	2.50*	.390		1.41	3.59
			Talk	1.60*	.485		.24	2.96
		Talk	Think	.90	.418		-.27	2.07
			Write	-1.60*	.485		-2.96	-.24
Application Exam	Bonferroni	Think	Write	-1.80*	.632	.025	-3.41	-.19
			Talk	-1.20	.632	.206	-2.81	.41
		Write	Think	1.80*	.632	.025	.19	3.41
			Talk	.60	.632	1.000	-1.01	2.21
		Talk	Think	1.20	.632	.206	-.41	2.81
			Write	-.60	.632	1.000	-2.21	1.01
	Dunnett C	Think	Write	-1.80	.680		-3.70	.10
			Talk	-1.20	.537		-2.70	.30
		Write	Think	1.80	.680		-.10	3.70
			Talk	.60	.670		-1.27	2.47
		Talk	Think	1.20	.537		-.30	2.70
			Write	-.60	.670		-2.47	1.27

Based on observed means.

*. The mean difference is significant at the .05 level.

Figure 172. **The results of the post hoc comparison procedures.**

3. Click **Define**.
4. Click **group**, then click ▶ to move it to the Category Axis box.
5. Holding down the Ctrl key, click **recall** and **applicat**, then click ▶ to move them to the Boxes Represent box.
6. Click **OK**. Make appropriate edits to obtain Figure 173 (on page 230).

An APA Results Section

A one-way multivariate analysis of variance (MANOVA) was conducted to determine the effect of the three types of study strategies (thinking, writing, and talking) on the two dependent variables, the recall and the application test scores. Significant differences were found among the three study strategies on the dependent measures, Wilks's $\Lambda = .42$, $F(4, 52) = 7.03$, $p < .01$. The multivariate η^2 based on Wilks's Λ was quite strong, .35. Table 26 (on page 230) contains the means and the standard deviations on the dependent variables for the three groups.

Analyses of variances (ANOVA) on the dependent variables were conducted as follow-up tests to the MANOVA. Using the Bonferroni method, each ANOVA was tested at the .025 level. The ANOVA on the recall scores was significant, $F(2, 27) = 17.11$, $p < .01$, $\eta^2 = .56$, while the ANOVA on the application scores was nonsignificant, $F(2, 27) = 4.20$, $p = .026$, $\eta^2 = .24$.

Post hoc analyses to the univariate ANOVA for the recall scores consisted of conducting pairwise comparisons to find which study strategy affected performance most strongly. Each pairwise comparison was tested at the .025 divided by 3 or .008 level. The writing group produced significantly superior performance on the recall questions in comparison with either of the other two groups. (See Table 26.) The thinking and talking groups were not significantly different from each other.

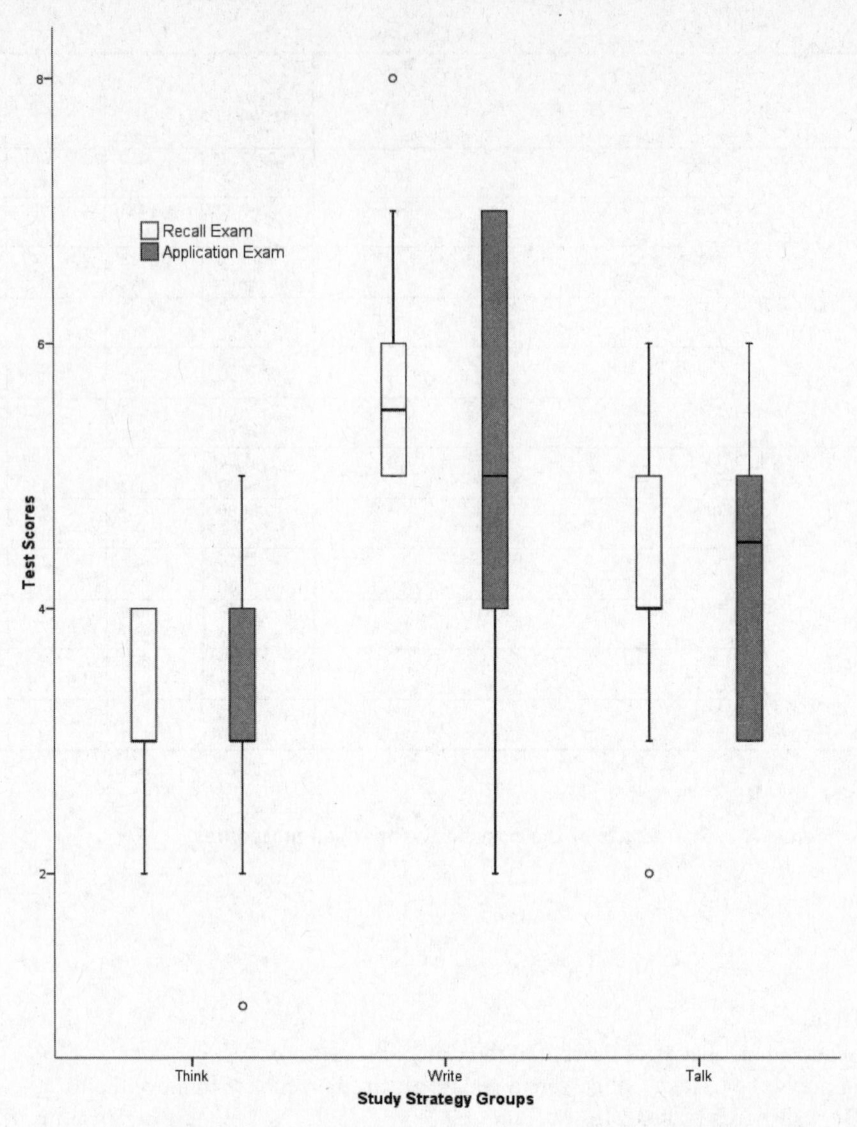

Figure 173. **Distributions of application and recall scores for the three study strategy groups.**

Table 26

Means and Standard Deviations on the Dependent Variables for the Three Groups

	Recall		Application	
Study strategy	*M*	*SD*	*M*	*SD*
Thinking	3.30	0.67	3.20	1.23
Writing	5.80	1.03	5.00	1.76
Talking	4.20	1.14	4.40	1.17

Exercises

The data for Exercises 1 through 4 are in the data set named *Lesson 28 Exercise File 1* on the Web at http://www.prenhall.com/greensalkind. The data are from the following research problem.

Dana wished to evaluate the effects of conflict-resolution training on the social skills of elementary school children. Fourth-grade students in a school district were randomly assigned to one of three groups. Two groups received training in conflict resolution from either a cognitive-based approach (group = 1) or a behavior-based approach (group = 2). The third group did not receive formal instruction in conflict resolution (group = 3). To measure the effects of the treatment, a social problem-solving inventory (SPSI) was administered to the children, and rating scales were completed by the teachers (rate_t) and by the parents (rate_p).

1. Conduct a MANOVA to evaluate the effects of the training programs. Identify the following values on the output:
   a. Wilks's lambda
   b. *F* statistic associated with Wilks's lambda
   c. Multivariate η^2 associated with Wilks's lambda
   d. Univariate *F* test for the teacher rating
   e. Mean parent rating for group 2
2. Conduct the appropriate follow-up tests.
3. Write a Results section reporting the results of her analysis.
4. Create a boxplot to display graphically Dana's results.

The data for Exercises 5 and 6 are in the data set named *Lesson 28 Exercise File 2* on the Web at http://www.prenhall.com/greensalkind. The data are from the following research problem.

Vicki was interested in whether fathers of children with a disability feel competent to parent their children who are disabled. To address this question, she found 60 fathers with a child in one of three categories: (1) fathers with a child with no disabilities, (2) fathers with a physically disabled child, and (3) fathers with a mentally retarded child. She administered the Perceived Parenting Competence Scale (PPCS), a self-report measure. It has three subscales: (1) instrumental competence, which assesses caretaking responsibilities (instrum); (2) emotional parenting competence, which evaluates the quality of emotional support provided to the child (emot); and (3) play competence, which measures the quality of the recreational time spent with the child (play). Vicki's SPSS data file contains 60 cases and four variables, a factor indicating the disability status of the child and scores on the three PPCS subscales.

5. Conduct a one-way MANOVA to evaluate differences among the disability status groups on the three perceived competence variables. Is the multivariate test significant?
6. Write a Results section.

LESSON

29

One-Way Repeated-Measures Analysis of Variance

With one-way repeated-measures designs, each subject or case in a study is exposed to all levels of a qualitative variable and measured on a quantitative variable during each exposure. The qualitative variable is referred to as a repeated-measures factor or a within-subjects factor. The quantitative variable is called the dependent variable.

To conduct a repeated-measures ANOVA in SPSS, we do not specify the repeated-measures factor and the dependent variable in the SPSS data file. Instead, the SPSS data file contains several quantitative variables. The number of quantitative variables is equal to the number of levels of the within-subjects factor. The scores on any one of these quantitative variables are the scores on the dependent variable for a single level of the within-subjects factor.

Although we do not define the within-subjects factor in the SPSS data file, we specify it in the dialog box for the General Linear Model Repeated-Measures procedure. To define the factor, we give a name to the within-subjects factor, specify the number of levels of this factor, and indicate the quantitative variables in the data set associated with the levels of the within-subjects factor.

Applications of One-Way Repeated Measures ANOVA

One-way within-subjects ANOVAs can analyze data from different types of studies.

- Experimental studies
- Quasi-experimental studies
- Field studies
- Longitudinal studies

We illustrate a longitudinal study and an experimental study.

Longitudinal Study

Janice is interested in determining whether the average man wants to express his worries to his wife more (or less) the longer they are married. Janice develops the Desire to Express Worry (DEW) scale. She has men take the DEW when they initially get married and then at their 5th, 10th, and 15th wedding anniversaries. She decides to conduct a one-way within-subjects ANOVA that involves only those 30 men who made it to their 15th wedding anniversary. The within-subjects factor is time with four levels (0, 5, 10, and 15 years of marriage), and the dependent variable is the DEW scores at the four times. Janice's SPSS data file has 30 cases and four variables. The four variables are the initial DEW score and the DEW scores after 5, 10, and 15 years of marriage.

Experimental Study

Shep is interested in determining if teachers feel greater stress when coping with problems associated with students, parents, or administrators. To make this comparison, he develops nine scenarios representing problems that high school teachers are likely to encounter when they teach. Three scenarios represent problems with students, three represent problems with parents, and three represent problems with administrators. The scenarios are randomly ordered and presented to 100 teachers. The teachers are to read each scenario, think of a similar problem they have had, and indicate on a stress measure how stressed they felt as they coped with their problem. Three scores are computed by summing the stress scores for the scenarios of the same type (i.e., scenarios describing problems with students, parents, or administrators). An ANOVA is conducted with type of stressor as the within-subjects factor with three levels (students, parents, or administrators) and the stress measure as the dependent variable. For the analysis, Shep's SPSS data file includes 100 cases and three variables. The three variables are the student stress scores, the parent stress scores, and the administrator stress scores.

Understanding One-Way Repeated-Measures ANOVA

In many studies using a one-way repeated-measures design, the levels of a within-subject factor represent multiple observations on a scale over time or under different conditions. However, for some studies, levels of a within-subject factor may represent scores from different scales, and the focus may be on evaluating differences in means among these scales. In such a setting, the scales must be commensurable for the ANOVA significance tests to be meaningful. That is, the scales must measure individuals on the same metric, and the difference scores between scales must be interpretable.

In some studies, individuals are matched on one or more variables so that individuals within a set are similar on a matching variable(s), while individuals not in the same set are dissimilar. The number of individuals within a set is equal to the number of levels of a factor. The individuals within a set are then observed under the various levels of this factor. The matching process for these designs is likely to produce correlated responses on the dependent variable like those of repeated-measures designs. Consequently, the data from these studies can be analyzed as if the factor is a within-subjects factor. For further discussion of these designs, consult textbooks on experimental design and analysis of variance.

Let's look at how to conduct the analyses for one-way repeated-measures designs. We address tests to evaluate the overall hypothesis as well as post hoc tests.

SPSS conducts a standard univariate F test if the within-subjects factor has only two levels. Three types of tests are conducted if the within-subjects factor has more than two levels: the standard univariate F test, alternative univariate tests, and multivariate tests. All three types of tests evaluate the same hypothesis—the population means are equal for all levels of the factor. The choice of what test to report should be made prior to viewing the results.

The standard univariate ANOVA F test is not recommended when the within-subjects factor has more than two levels because one of its assumptions, the sphericity assumption, is commonly violated, and the ANOVA F test yields inaccurate p values to the extent that this assumption is violated. (We'll talk later about assumptions underlying one-way repeated-measures ANOVA.) The alternative univariate tests take into account violations of the sphericity assumption. These tests employ the same calculated F statistic as the standard univariate test, but its associated p value potentially differs. In determining the p value, an epsilon statistic is calculated based on the sample data to assess the degree that the sphericity assumption is violated. The numerator and denominator degrees of freedom of the standard test are multiplied by epsilon to obtain a corrected set of degrees of freedom for the tabled F value and to determine its p value.

The multivariate test does not require the assumption of sphericity. Difference scores are computed by comparing scores from different levels of the within-subjects factor. For example, for a within-subjects factor with three levels, difference scores might be computed between the first and second level and between the second and third level. The multivariate test then would evaluate whether the population means for these two sets of difference scores are simultaneously equal to zero. This test evaluates not only the means associated with these two sets of difference scores, but also evaluates whether the mean of the difference scores between the first and third levels of the factor is equal to zero as well as linear combinations of these difference scores.

It should be noted that the SPSS Repeated Measures procedure computes the difference scores used in the analysis for us. However, these difference scores do not become a part of our data file and, therefore, we may or may not be aware that the multivariate test is conducted on these difference scores.

Applied statisticians tend to prefer the multivariate test to the standard or the alternative univariate test because the multivariate test and follow-up tests have a close conceptual link to each other. If the initial hypothesis that the means are equal is rejected and there are more than two means, then follow-up tests are conducted to determine which of the means differs significantly from each other. Although more complex comparisons can be performed, most researchers choose to conduct pairwise comparisons. These comparisons may be evaluated by using paired-samples t test, and a Bonferroni approach, such as the Holm's sequential procedure, can be used to control for Type I error across the multiple pairwise tests.

Standard Univariate Assumptions

Assumption 1: The Dependent Variable Is Normally Distributed in the Population for Each Level of the Within-Subjects Factor

In many applications with a moderate or larger sample size, the one-way repeated-measures ANOVA may yield reasonably accurate p values, even when the normality assumption is violated. A commonly accepted value for a moderate sample size is 30 subjects. Larger sample sizes may be required to produce relatively valid p values if the population distribution is substantially nonnormal. In addition, the power of this test may be reduced considerably if the population distribution is nonnormal and, more specifically, thick-tailed or heavily skewed. See Wilcox (2001) for an extended discussion of assumptions.

Assumption 2: The Population Variance of Difference Scores Computed between Any Two Levels of a Within-Subjects Factor Is the Same Value Regardless of Which Two Levels Are Chosen

This assumption is sometimes referred to as the sphericity assumption or as the homogeneity-of-variance-of-differences assumption. The sphericity assumption is meaningful only if there are more than two levels of a within-subjects factor.

If this assumption is violated, the p value associated with the standard within-subjects ANOVA cannot be trusted. However, other methods do not require the sphericity assumption. Two approaches are alternative univariate methods that correct the degrees of freedom to take into account violation of this assumption, as well as the multivariate approach that does not require the sphericity assumption.

Assumption 3: The Cases Represent a Random Sample from the Population, and There Is No Dependency in the Scores between Participants

The only type of dependency that should exist among dependent variable scores is the dependency introduced by having the same individuals produce multiple scores. Even this type of dependency introduced by the within-subjects factor is limited for the standard univariate test and must conform to the sphericity assumption. The results for a one-way, within-subjects ANOVA should not be trusted if the scores between individuals are related.

Multivariate Assumptions

The multivariate test is conducted on difference scores, and, therefore, the assumptions underlying the multivariate test concern these difference scores. The number of variables with difference scores is equal to the number of levels of the within-subjects factor minus 1. Although the difference-score variables may be computed in a number of ways, we will compute them by subtracting the scores associated with one level of the within-subjects factor from the scores for an adjacent level of the within-subjects factor. In our first example involving changes in desire to express worry, our within-subjects factor has four levels and, therefore, we would compute three difference scores. These three variables would be obtained by subtracting (1) the DEW scores at 5 years of marriage from the DEW scores at 0 years of marriage, (2) the DEW scores at 10 years of marriage from the DEW scores at 5 years of marriage, and (3) the DEW scores at 15 years of marriage from the DEW scores at 10 years of marriage.

Assumption 1: The Difference Scores Are Multivariately Normally Distributed in the Population

If the difference scores are multivariately normally distributed, each difference score is normally distributed, ignoring the other difference scores. Also, each difference score is normally distributed at every combination of values of the other difference scores. To the extent that population distributions are not multivariate normal and sample sizes are small, the p values may be invalid. In addition, the power of this test may be reduced considerably if the population distributions are not multivariate normal and, more specifically, thick-tailed or heavily skewed.

Assumption 2: The Individual Cases Represent a Random Sample from the Population, and the Difference Scores for Any One Subject Are Independent from the Scores for Any Other Subject

The test should not be used if the independence assumption is violated.

Effect Size Statistics for One-Way Repeated-Measures ANOVA

The effect size reported for the standard univariate approach is a partial eta square and may be calculated using the following equation:

$$\textit{Partial } \eta^2_{factor} = \frac{\textit{Sum of Squares}_{factor}}{\textit{Sum of Squares}_{factor} + \textit{Sum of Squares}_{Error}}$$

The effect size for the multivariate test associated with Wilks's lambda (Λ) is the multivariate eta square:

$$\textit{Multivariate } \eta^2 = 1 - \Lambda$$

Both of these statistics range from 0 to 1. A 0 indicates no relation between the repeated-measures factor and the dependent variable, while a 1 indicates the strongest possible relationship.

The Data Set

The data set used to illustrate a one-way repeated-measures ANOVA is named *Lesson 29 Data File 1* on the Web at http://www.prenhall.com/greensalkind and represents data coming from the DEW example. The definitions for the variables in the data set are given in Table 27 on page 236.

The Research Question

The research question can be stated to reflect mean differences or relationships between variables.

1. Mean differences: Do population means on the Desire to Express Worry scale vary with how long men are married?
2. Relationship between two variables: Is there a relationship between length of marriage and how much men want to express their worries?

Table 27
Variables in Lesson 29 Data File 1

Variable	Definition
time1	DEW scores after 0 years of marriage
time2	DEW scores after 5 years of marriage
time3	DEW scores after 10 years of marriage
time4	DEW scores after 15 years of marriage

Conducting a One-Way Repeated-Measures ANOVA

We next illustrate how to conduct a one-way repeated-measures ANOVA. The following steps produce results to assess overall differences in means among levels of a within-subjects factor (univariate and multivariate), pairwise comparisons among levels of a within-subjects factor, and polynomial contrasts.

1. Click **Analyze**, click **General Linear Model**, then click **Repeated Measures**. You will see the Repeated Measures Define Factor(s) dialog box as shown in Figure 174.

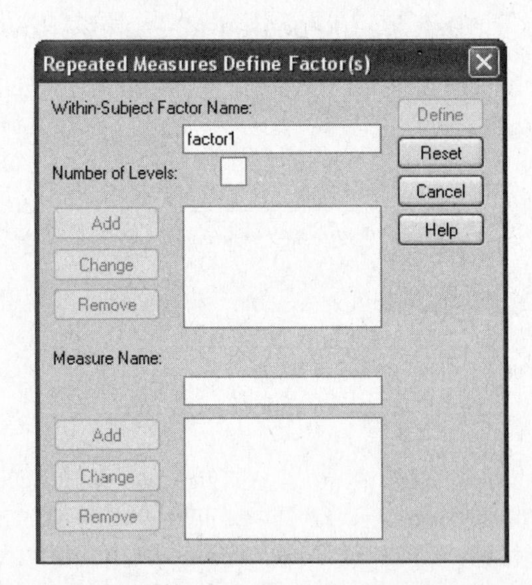

***Figure 174.* The Repeated Measures Define Factor(s) dialog box.**

2. In the Within-Subject Factor Name text box, type **time** in place of factor1.
3. Type **4** in the Number of Levels box.
4. Click **Add**.
5. Click **Define**. You will see the Repeated Measures dialog box, as shown in Figure 175.
6. Holding down the Ctrl key, click on **time1**, **time2**, **time3**, and **time4**, then click ▶ to move them to the Within-Subjects Variables box.
7. Click **Options**.
8. Highlight time in the Factor(s) and Factor Interactions box. Click ▶ to make it appear in the Display Means for box.
9. Click **Compare main effects**.
10. Click **Estimates of effect size** and **Descriptive statistics** in the Display box.

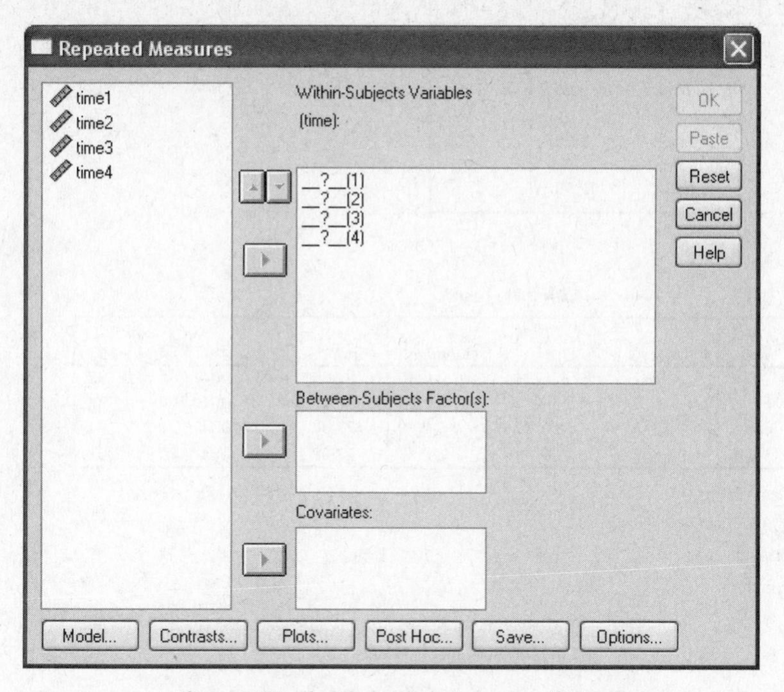

***Figure 175.* The Repeated Measures dialog box.**

11. Click **Continue**.
12. Click **OK**.

Selected SPSS Output for One-Way Repeated-Measures ANOVA

SPSS for a one-way within-subjects ANOVA produces a lot of output: descriptive statistics, statistics for evaluating the sphericity assumption, univariate ANOVA results, ANOVA results using multivariate tests, and more. We describe the output in multiple sections, beginning with the overall results based on the univariate and multivariate tests. The results for these tests are shown in Figure 176 (on page 238).

The standard univariate ANOVA (labeled Sphericity Assumed) indicates a significant time effect, $F(3, 87) = 7.66$, $p < .01$. The alternative univariate tests (i.e., Greenhouse-Geisser and Huynh-Feldt) yield the same F value, but correct the degrees of freedom of the F as a function of the degree to which the data indicate that the sphericity assumption is violated. The p value is greater for these alternative tests, but all three tests are significant at the traditional .05 level.

We will concentrate our attention on interpreting the ANOVA results by using the multivariate tests and, therefore, avoid the controversy surrounding the sphericity assumption. As we have previously discussed, sphericity is not an assumption for the multivariate tests. For a one-way within-subjects ANOVA, all multivariate tests (Pillai's trace, Wilks's lambda, and so on) must yield the same result. We recommend reporting Wilks's Λ because researchers are most likely to be familiar with it. For our example, the multivariate tests indicate a significant time effect, Wilks's $\Lambda = .62$, $F(3, 27) = 5.57$, $p < .01$.

Selected SPSS Output for Pairwise Comparisons

The results of the paiwise comparisons are shown in Figure 177 (on page 238). Six unique pairwise comparisons were conducted among the means for time 1, time 2, time 3, and time 4. Four of the six pairwise comparisons are significant, controlling for familywise error rate across the six tests at the .05 level, using the Holm's sequential Bonferroni procedure.

Descriptive Statistics

	Mean	Std. Deviation	N
DEW Scores at the Beginning of Marriage	65.80	9.227	30
DEW Scores After 5 Years of Marriage	65.43	10.686	30
DEW Scores After 10 Years of Marriage	63.10	10.685	30
DEW Scores After 15 Years of Marriage	61.93	12.567	30

Multivariate Tests[b]

Effect		Value	F	Hypothesis df	Error df	Sig.	Partial Eta Squared
time	Pillai's Trace	.382	5.567[a]	3.000	27.000	.004	.382
	Wilks' Lambda	.618	5.567[a]	3.000	27.000	.004	.382
	Hotelling's Trace	.619	5.567[a]	3.000	27.000	.004	.382
	Roy's Largest Root	.619	5.567[a]	3.000	27.000	.004	.382

a. Exact statistic

b.

Design: Intercept
Within Subjects Design: time

Tests of Within-Subjects Effects

Measure: MEASURE_1

Source		Type III Sum of Squares	df	Mean Square	F	Sig.	Partial Eta Squared
time	Sphericity Assumed	310.733	3	103.578	7.664	.000	.209
	Greenhouse-Geisser	310.733	2.094	148.410	7.664	.001	.209
	Huynh-Feldt	310.733	2.260	137.483	7.664	.001	.209
	Lower-bound	310.733	1.000	310.733	7.664	.010	.209
Error(time)	Sphericity Assumed	1175.767	87	13.515			
	Greenhouse-Geisser	1175.767	60.719	19.364			
	Huynh-Feldt	1175.767	65.545	17.938			
	Lower-bound	1175.767	29.000	40.544			

Figure 176. **The results of the one-way repeated-measures analysis.**

Pairwise Comparisons

Measure: MEASURE_1

(I) time	(J) time	Mean Difference (I-J)	Std. Error	Sig.[a]	95% Confidence Interval for Difference[a]	
					Lower Bound	Upper Bound
1	2	.367	.898	.686	-1.469	2.202
	3	2.700*	.931	.007	.796	4.604
	4	3.867*	1.288	.005	1.232	6.501
2	1	-.367	.898	.686	-2.202	1.469
	3	2.333*	.620	.001	1.065	3.602
	4	3.500*	1.029	.002	1.395	5.605
3	1	-2.700*	.931	.007	-4.604	-.796
	2	-2.333*	.620	.001	-3.602	-1.065
	4	1.167	.794	.152	-.457	2.790
4	1	-3.867*	1.288	.005	-6.501	-1.232
	2	-3.500*	1.029	.002	-5.605	-1.395
	3	-1.167	.794	.152	-2.790	.457

Based on estimated marginal means

*. The mean difference is significant at the .05 level.

a. Adjustment for multiple comparisons: Least Significant Difference (equivalent to no adjustments).

Figure 177. **The results of the pairwise comparisons.**

The smallest p value is for the comparison of time 2 and time 3, and its p value of .001 is less than $\alpha = .05/6 = .0083$; therefore, the difference between the means for these two times is significant. The next smallest p value is for the comparison of time 2 and time 4, and its p value of .002 is less than $\alpha = .05/5 = .010$; therefore, this comparison is also significant. The next smallest p value is for the comparison of time 1 and time 4 and its p value of .005 is less than $\alpha = .05/4 = .0125$; therefore, this comparison is also significant. The p value of .007 for the comparison of time 1 and time 3 is smaller than $\alpha = .05/3 = .0167$; therefore, this comparison is also significant. The next comparison is not significant at the .025, and therefore none of the remaining comparisons are significant.

It should be noted that the pairwise comparisons generated by the repeated-measures ANOVA could also be obtained using paired-samples t tests. To conduct the paired-samples t tests, follow these steps:

1. Click **Analyze**, click **Compare Means**, then click **Paired-Samples *t*-Test**.
2. Click **time1**. Time1 now appears as Variable 1 in Current Selections.
3. Click **time2**. Time2 appears as Variable 2 in Current Selections.
4. Now click ►, and time1–time2 appears in the Paired Variables box.
5. Repeat steps 2 through 4 for each pairwise comparison (time1–time3, time1–time4, time2–time3, time2–time4, time3–time4)
6. Click **OK**.

Selected SPSS Output for Polynomial Contrasts

If levels of a within-subjects factor represent quantitative scores that are equally spaced, in most cases, it is more appropriate to conduct polynomial contrasts than paired-samples t tests. For our example, the levels of the within-subjects factor are 0, 5, 10, and 15 years. These levels are equally spaced with five years between adjacent levels. Consequently, polynomial contrasts are the more appropriate follow-up analysis. SPSS automatically produces the results for the polynomial contrasts unless you click on the **Contrast** button and change the contrast in the Repeated Measures dialog box. This dialog box is shown in Figure 178.

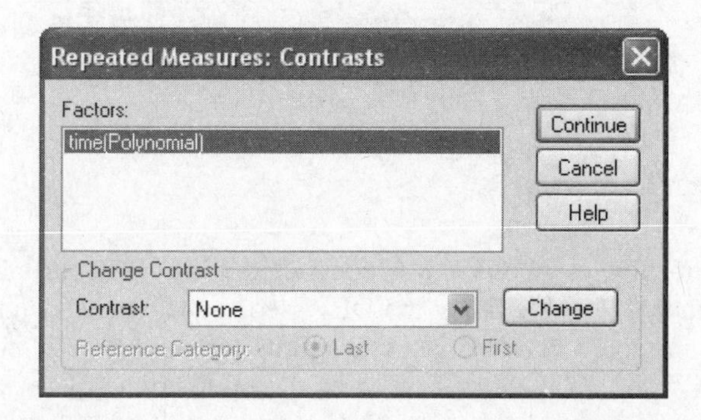

Figure 178. **The Repeated Measures: Contrasts dialog box.**

The results for the polynomial contrasts are shown in Figure 179 (on page 240). There is a significant linear effect, $F(1, 29) = 11.56, p < .01$. Higher-order polynomial contrasts (quadratic and cubic) were nonsignificant. By inspecting the means in the Descriptive Statistics table in Figure 176, it is evident that the linear effect is due to the DEW means decreasing over time. It should be noted that there was little change in means from 0 to 5 years; therefore, the significant linear trend was due to changes after 5 years of marriage.

Tests of Within-Subjects Contrasts

Measure: MEASURE_1

Source	time	Type III Sum of Squares	df	Mean Square	F	Sig.	Partial Eta Squared
time	Linear	291.207	1	291.207	11.564	.002	.285
	Quadratic	4.800	1	4.800	.486	.491	.016
	Cubic	14.727	1	14.727	2.690	.112	.085
Error(time)	Linear	730.293	29	25.183			
	Quadratic	286.700	29	9.886			
	Cubic	158.773	29	5.475			

Figure 179. **The results of the polynomial contrasts.**

Using SPSS Graphs to Display the Results

Boxplots are often used to display the results of a one-way repeated measures ANOVA. For our example, we created boxplots to show the distributions of the DEW scores for the different time periods. We produced the boxplot shown in Figure 180 by using the steps described in Lesson 23 (on page 173) and editing the resulting figure.

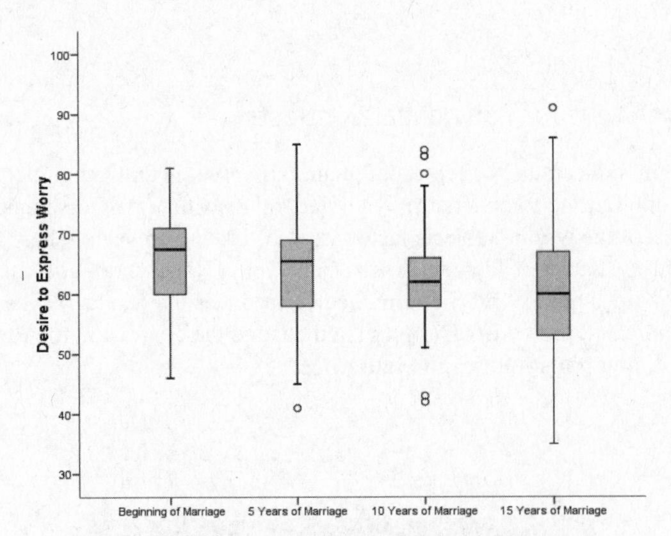

Figure 180. **Distributions of DEW scores across time.**

An APA Results Section

A one-way within-subjects ANOVA was conducted with the factor being number of years married and the dependent variable being the DEW scores. The means and standard deviations for DEW scores are presented in Table 28. The results for the ANOVA indicated a significant time effect, Wilks's $\Lambda = .62$, $F(3, 27) = 5.57$, $p < .01$, multivariate $\eta^2 = .38$.

Table 28

Means and Standard Deviations for DEW Scores

Number of years married	*M*	*SD*
0 years	65.80	9.23
5 years	65.43	10.69
10 years	63.10	10.68
15 years	61.93	12.57

Follow-up polynomial contrasts indicated a significant linear effect with means decreasing over time, $F(1, 29) = 11.56, p < .01$, partial $\eta^2 = .29$. Higher-order polynomial contrasts were nonsignificant. It should be noted that there was little change in means from 0 to 5 years; therefore, the significant trend was due to changes after 5 years of marriage. These results suggest that men are more eager to express worry to their wives early in their marriage, and this desire decreases after 5 years of marriage.

Exercises

Exercises 1 through 4 are based on Shep's study, described earlier, about whether teachers feel greater stress when coping with problems associated with students, parents, or administrators. The data for these exercises are named *Lesson 29 Exercise File 1* on the Web at http://www.prenhall.com/greensalkind. The data file contains the stress scores for the three types of problems.

1. Conduct a one-way repeated-measures ANOVA on these data. Identify the following information from the output:
   a. Multivariate F value based on the Wilks's lambda
   b. p value for Wilks's lambda
   c. Mean of stress scores when coping with problems associated with students
2. Conduct the appropriate follow-up tests.
3. Write a Results section.
4. Create an error bar chart to show the distributions of the stress scores.

Exercises 5 through 7 are based on the following research problem. The data for these examples are in the data file named *Lesson 29 Exercise File 2* on the Web at http://www.prenhall.com/greensalkind.

Mary, a developmental psychologist, is interested in how children's self-esteem develops over time. She measures 25 children at ages 5, 7, 9, 11, and 13 using the Self-Esteem Descriptor (SED). The data file contains 25 cases and five variables, the self-esteem scores for each child at each age period (sed1 through sed5).

5. Conduct a repeated-measures ANOVA to see if self-esteem changes significantly over time.
6. What follow-up tests should you conduct? How would you control for Type I error across your follow-up tests?
7. Write a Results section based on your analyses. Be sure to include a graph.

LESSON

30 Two-Way Repeated-Measures Analysis of Variance

With a two-way repeated-measures analysis of variance (ANOVA), each subject or case in a study is exposed to all combinations of levels of two qualitative variables and measured on a quantitative variable for each combination. The qualitative variables are referred to as repeated-measures factors or within-subjects factors. The quantitative variable is called the dependent variable.

Similar to one-way repeated-measures ANOVA (Lesson 29), we do not create an SPSS data file that includes two within-subject (repeated-measures) factors and a dependent variable to conduct a two-way repeated-measures ANOVA. Instead, the data file contains a number of quantitative variables. The number of variables equals the number of levels of one within-subjects factor times the number of levels of the other within-subjects factor. The scores on any one of these quantitative variables in the SPSS data file are the scores on the dependent variable for a level of one within-subjects factor in combination with a level of the other within-subjects factor.

Although we do not define the within-subjects factors in the SPSS data set, we specify them in the dialog boxes for the General Linear Model Repeated-Measures procedure. In these boxes, we name the within-subjects factors, specify the number of levels for these factors, and indicate the quantitative variables in the data set that are associated with all combinations of levels of the two within-subjects factors.

Applications of Two-Way Repeated-Measures ANOVA

We can analyze data from different types of studies by using a two-way within-subjects ANOVA.

- Experimental studies
- Quasi-experimental studies
- Field studies
- Longitudinal studies

Let's look at examples of two types of studies—an experimental study with a single scale and a field study with multiple scales—that can be analyzed using two-way repeated-measures analyses.

Experimental Study with a Single Scale

Dan wishes to evaluate various coping strategies for pain. He obtains eight student volunteers to come to the lab on two consecutive days for his pain research. On both days, the students plunge their hands into freezing cold water for 90 seconds. They rate how painful the experience is on a scale from 1 (= Not painful at all) to 50 (= Excruciatingly painful) at the end of 30 seconds, 60 seconds, and 90 seconds. The eight students receive avoidance instructions one day and concentration-on-pain instructions the other day. The avoidance instructions tell the students to think about a pleasant time that they have had and to try not to think about what was occurring in the lab. The concentration-on-pain instructions tell the students to focus on the sensations in their hands. Four students are given the avoidance instructions on the first day and the concentration-on-pain instructions on the second day. The other four students are given the instructions in the opposite order. A preliminary analysis indicates that the ordering of instructions had no effect. The within-subjects factors are the following: coping treatment with two levels (avoidance and concentration) and time with three levels (30, 60, and 90 seconds). The dependent variable is rating of pain on a 50-point scale. Dan's SPSS data file includes eight cases and six variables. Three of the variables are the scores at the end of 30, 60, and 90 seconds for the avoidance instructions, and the other three variables are the scores at the end of 30, 60, and 90 seconds for the concentration-on-pain instructions.

Field Study with Multiple Scales

Jean wants to know whether students tend to emphasize one type of study strategy over others, regardless of how information is presented in class. To examine this question, she has three guest instructors come to make presentations to her class of 30 students. One instructor stresses general principles, a second emphasizes applications, and a third facts. Jean tells the class that each guest lecturer has submitted questions to be included on the exam. The day of the exam, the students complete a study questionnaire three times: once for how they prepared for the exam for the principle-oriented instructor, once for the applications-oriented instructor, and once for the facts-oriented instructor. The questionnaire yielded three scale scores for each administration: principle-level studying, application-level studying, and surface-level studying. The scales were normed previously to have a mean of 50 and a standard deviation of 10. For this study, one of the within-subjects factors is type of presentation with three levels (principles-oriented, application-oriented, and facts-oriented), while the other within-subjects factor is study strategy with three levels (principle-level studying, application-level studying, and surface-level studying). The dependent variable is study strategy emphasis. The second factor can be treated as a within-subjects factor because the three scales have a common mean and standard deviation, and differences between scales are meaningful (i.e., a person who scores higher on principle-level studying versus application-level studying puts, in comparison with other persons, a greater emphasis on studying principles than applications). Jean's SPSS data file has 30 cases and nine variables. The nine variables represent the student scores on the three types of scales for each of the three types of presentations.

Understanding Two-Way Repeated-Measures ANOVA

Two-way within-subjects ANOVAs are conducted to analyze data from a variety of studies. In many of these studies, the levels of within-subject factors represent multiple observations on a scale over time or under different conditions. However, for some studies, levels of a within-subject factor may represent scores associated with different scales, and the focus may be on evaluating differences in means among these scales. For the significance tests in the ANOVA to be meaningful for these studies, the scales must be commensurable; that is, the scales must measure individuals on the same metric, and the difference scores between scales must be interpretable.

The factors may also be treated as within-subjects factor for a study if individuals are matched on one or more variables to create matched sets of individuals and if the number of combinations of levels for the two factors is equal to the number of individuals within each matched set. Each individual within a set is then observed under one combination of levels for the two factors. Because of the matching process, the scores on the dependent variable are likely to be correlated, therefore, the factors are treated as within-subjects factors.

Now let's look at how to conduct the analyses for a two-way repeated-measures design. We address tests to evaluate the overall hypotheses for the main and interaction effects as well as post hoc tests.

The output for a two-way repeated-measures ANOVA includes tests of main effects for the two factors as well as their interaction. If a main effect or interaction effect has a single degree of freedom (in the numerator), SPSS produces only a standard univariate test. If a main or interaction effect has more than a single degree of freedom, SPSS can evaluate the same hypothesis with multiple methods. These tests include a traditional univariate test, alternative (degrees-of-freedom corrected) univariate tests, and a multivariate test. The first method is not recommended except in the single degree of freedom case. The choice between the last two methods should be made prior to examining the SPSS output.

The standard univariate ANOVA *F* test is not recommended for tests with more than a single degree of freedom because one of its assumptions, the sphericity assumption, is commonly violated. This *F* test yields inaccurate *p* values to the extent that this assumption is violated. Alternative univariate tests take into account violations of the sphericity assumption. These tests employ the same calculated *F* statistic as the standard univariate test, but their associated *p* values potentially differ. In determining these *p* values, epsilon statistics are computed that assess the degree that the sphericity assumption is violated based on the sample data. The numerator and denominator degrees of freedom of the standard test can then be multiplied by epsilon to obtain a corrected set of degrees of freedom for the tabled *F* value and to determine its *p* value.

The multivariate test employs an entirely different approach and does not require sphericity. Difference scores are computed by comparing scores from different levels of within-subjects factor or factors. It should be noted that the SPSS Repeated Measures procedure computes the difference scores used in conducting this multivariate test. However, these difference scores do not become a part of our data file; therefore, we may or may not be aware that the multivariate test is conducted on these difference scores. As discussed in Lesson 29, applied statisticians tend to prefer the multivariate test to the standard test or the alternative univariate tests. When we conduct a two-way repeated-measures procedure later in this lesson, we do not assume sphericity, and we conduct the multivariate tests.

Regardless of which method is used to initially evaluate the main effects and interaction effect, follow-up tests may be required if one or more of the overall tests is significant. If the interaction hypothesis is rejected, simple main effects or interaction comparison tests frequently are conducted as follow-up tests. If the interaction is not significant, then follow-up tests are conducted for a significant main effect if the factor associated with the main effect has more than two levels.

Several types of difference scores could potentially be evaluated in conducting follow-up tests for the main and interaction effects. We will discuss briefly three types of difference scores to help understand the methods, the sphericity assumption, and the hypotheses evaluated by the multivariate tests.

One type of difference score is computed for each of the main effects and for the interaction. We will use the first example, the pain research, and its data as shown in Table 29, to illustrate the three types of difference scores.

Table 29
Data for Pain Example

	Concentration on pain			Avoidance		
Students	Time 1	Time 2	Time 3	Time 1	Time 2	Time 3
Mary	14	14	25	4	10	30
Jane	16	17	25	1	14	28
Jill	18	17	19	6	13	23
Jean	4	6	22	3	12	24
Corey	16	14	15	8	13	20
Pam	5	11	20	1	10	24
Jennifer	2	6	10	1	12	17
Jersey	21	27	40	8	12	42

Difference Scores Associated with the Coping Main Effect

The coping main effect evaluates differences between the two coping methods averaging across the levels of the time factor. The following steps yield the relevant difference scores for this main effect:

1. Compute a mean score across times 1, 2, and 3 for the avoidance condition for each case.
2. Compute a mean score across times 1, 2, and 3 for the concentration-on-pain condition for each case.
3. Subtract the mean score for Step 1 from the mean score for Step 2 for each case.

The computational results of these steps are shown in Table 30.

Table 30

Difference Scores for the Coping Main Effect

Students	Mean for concentration on pain	Mean for avoidance	Difference
Mary	17.67	14.67	3.00
Jane	19.33	14.33	5.00
Jill	18.00	14.00	4.00
Jean	10.67	13.00	−2.33
Corey	15.00	13.67	1.33
Pam	12.00	11.67	.33
Jennifer	6.00	10.00	−4.00
Jersey	29.33	20.67	8.67

The single set of difference scores in the last column of Table 30 implies a single hypothesis is evaluated by the coping main effect: the population mean of difference scores between the two coping methods (averaged across time) is equal to zero. Identical p values are produced by the coping main effect for our two-way ANOVA and a paired-samples t test on the scores produced in Steps 1 and 2. Because there is only one difference variable associated with the coping main effect, it is not necessary to conduct a multivariate test and follow-up tests.

Difference Scores Associated with the Time Main Effect

The time main effect assesses differences on the scale scores among the three time levels averaging across coping methods. Three pairwise differences for this main effect are relevant and may be computed using the following steps:

1. Compute a mean score across the two coping methods for each of the three levels of time.
2. Compute difference scores between times 1 and 2, between times 1 and 3, and between times 2 and 3.

The computational results of these steps are shown in Table 31 (on page 246).

The multivariate test computed by SPSS for the time main effect assesses whether the population means for these difference scores between time points (as well as more complex difference scores) are simultaneously equal to zero. If the test for this main effect is significant and the interaction is not significant, three follow-up paired-samples t tests are conducted to evaluate which of the mean differences (last three columns of Table 31) are significantly different from zero.

Table 31

Difference Scores for Time Main Effect

Students	Mean for time 1 (T_1)	Mean for time 2 (T_2)	Mean for time 3 (T_3)	$T_2 - T_1$	$T_3 - T_1$	$T_3 - T_2$
Mary	9.0	12.0	27.5	3.0	18.5	15.5
Jane	8.5	15.5	26.5	7.0	18.0	11.0
Jill	12.0	15.0	21.0	3.0	9.0	6.0
Jean	3.5	9.0	23.0	5.5	19.5	14.0
Corey	12.0	13.5	17.5	1.5	5.5	4.0
Pam	3.0	10.5	22.0	7.5	19.0	11.5
Jennifer	1.5	9.0	13.5	7.5	12.0	4.5
Jersey	14.5	19.5	41.0	5.0	26.5	21.5

Difference Scores for the Interaction Effect

The interaction effect assesses whether differences on the scale scores between coping methods differ for the three time levels. Three difference variables for this interaction effect are relevant, and they may be computed using the following steps:

1. Compute a difference score between the two coping methods for time 1 ($A_1 - C_1$), time 2 ($A_2 - C_2$), and time 3 ($A_3 - C_3$).
2. Compute the three pairwise differences among the difference variables created in Step 1.

The computational results of these steps are shown in Table 32.

Table 32

Difference Scores for Interaction Effects

Students	$A_1 - C_1$	$A_2 - C_2$	$A_3 - C_3$	$(A_1 - C_1) - (A_2 - C_2)$	$(A_1 - C_1) - (A_3 - C_3)$	$(A_2 - C_2) - (A_3 - C_3)$
Mary	−10	−4	5	−6	−15	−9
Jane	−15	−3	3	−12	−18	−6
Jill	−12	−4	4	−8	−16	−8
Jean	−1	6	2	−7	−3	4
Corey	−8	−1	5	−7	−13	−6
Pam	−4	−1	4	−3	−8	−5
Jennifer	−1	6	7	−7	−8	−1
Jersey	−13	−15	2	2	−15	−17

The multivariate test computed by SPSS for the interaction effect assesses whether the population means for the difference-of-differences variables created in Step 2 (and more complex difference-of-differences variable) are equal to zero. If the test for this interaction effect is significant, follow-up interaction comparison tests, simple main effects tests, or both may be conducted. The interaction comparisons evaluate whether the means of the three difference-of-differences variables are significantly different from zero and are conducted using a paired-samples *t* test. On the other hand, simple main effect tests can be conducted to evaluate the simple main effects for time within coping or the simple main effects for coping within time. The simple main effects for time within coping evaluates whether mean differences in scores among times within each level of coping strategy are equal to zero. Alternatively, the simple main effects for coping within time evaluates the mean differences in scores between coping methods for each level of time.

Assumptions Underlying a Two-Way Repeated-Measures ANOVA

The assumptions differ for the univariate and multivariate tests. The multivariate test is conducted on difference scores; and, therefore, the assumptions underlying the multivariate test concern the difference scores for the test.

Univariate Assumption 1: The Dependent Variable Is Normally Distributed in the Population for Each Combination of Levels of the Within-Subjects Factors

In many applications with a moderate or larger sample size, the two-way repeated-measures ANOVA may yield reasonably accurate *p* values even when the normality assumption is violated. A commonly accepted value for a moderate sample size is 30 subjects. Larger sample sizes may be required to produce relatively valid *p* values if the population distribution is substantially nonnormal. In addition, the power of this test may be reduced considerably if the population distribution is nonnnormal and, more specifically, thick-tailed or heavily skewed. See Wilcox (2001) for an extended discussion of assumptions.

Univariate Assumption 2: The Population Variances of the Difference Variables Are Equal

For each main and interaction effect, there is a separate set of difference score variances. The relevant variances for a main effect are associated with the pairwise differences among the levels of the main-effects factor. (See the last three columns of Table 31 to see sample quantities of these variables for the time main effect.) The relevant variances for an interaction effect are associated with the difference-of-differences variables. (See the last three columns of Table 32 to see sample quantities of these variables for the interaction effect.)

This assumption is sometimes referred to as the sphericity assumption or as the homogeneity-of-variance-of-differences assumption. The sphericity assumption is meaningful only if the main effect or interaction effect has more than one degree of freedom. If the assumption is violated, the *p* value associated with the standard univariate tests cannot be trusted. The alternative univariate and multivariate approaches do not require the assumption of sphericity.

Univariate Assumption 3: The Individuals Represent a Random Sample from the Population, and Scores Associated with Different Individuals Are Not Related

The only type of dependency that should exist among dependent variable scores is the dependency introduced by having the same individuals produce multiple scores. Even this type of dependency introduced by the within-subjects factor is limited and must conform to the sphericity assumption for the standard univariate tests. The results of the two-way within-subjects ANOVA should not be trusted if this assumption is violated.

Multivariate Assumption 1: The Difference Scores Are Multivariately Normally Distributed in the Population

If the difference scores are multivariately normally distributed, each difference score is normally distributed ignoring the other difference scores, and each difference score is normally distributed at every combination of values of the other difference scores. To the extent that population distributions are not multivariate normal and sample sizes are small, the *p* values may be invalid. In addition, the power of this test may be reduced considerably if the population distributions are nonnormal and, more specifically, thick-tailed or heavily skewed.

Mulitivariate Assumption 2: The Individuals Represent a Random Sample from the Population, and the Difference Scores for Any One Individual Are Independent from the Scores for Any Other Individual

If this assumption is violated, the test should not be used.

Effect Size Statistics for Two-Way Repeated Measures ANOVA

The effect size reported for the standard or the alternative univariate approach is a partial eta square and may be calculated using the following equation:

$$Partial\ \eta^2_{Main\ or\ Interaction} = \frac{Sum\ of\ Squares_{Main\ or\ Interaction}}{Sum\ of\ Squares_{Main\ or\ Interaction} + Sum\ of\ Squares_{Error}}$$

The effect size reported for the multivariate approach is the multivariate eta square,

$$Multivariate\ \eta^2_{Main\ or\ Interaction} = 1 - \Lambda_{Main\ or\ Interaction}$$

where the Wilks's lambda (Λ) is the multivariate statistic for a main or interaction source. Both of these statistics range from 0 to 1. A 0 indicates no relationship between a repeated-measures source and the dependent variable, while a 1 indicates the strongest possible relationship.

The Data Set

The data set used to illustrate a two-way repeated-measures analysis of variance is named *Lesson 30 Data File 1* on the Web at http://www.prenhall.com/greensalkind. The data are based on our first example, the coping-with-pain study. Table 33 presents the definitions of the variables in the data set.

Table 33
Variables in Lesson 30 Data File 1

Variable	Definition
conc1	Rating of pain on a 50-point scale at time 1 using the concentration-on-pain method
conc2	Rating of pain on a 50-point scale at time 2 using the concentration-on-pain method
conc3	Rating of pain on a 50-point scale at time 3 using the concentration-on-pain method
avoid1	Rating of pain on a 50-point scale at time 1 using the avoidance method
avoid2	Rating of pain on a 50-point scale at time 2 using the avoidance method
avoid3	Rating of pain on a 50-point scale at time 3 using the avoidance method

The Research Question

The research questions for the coping main effect, the time main effect, and the interaction between coping and time are as follows:

1. Coping main effect: Do students, on average, report different degrees of pain when using the concentration-on-pain and the avoidance approaches?
2. Time main effect: Do students, on average, report different degrees of pain if their hands are in frigid water for 30 seconds, 60 seconds, and 90 seconds?
3. Interaction between coping and time: Do the differences in means for self-reported pain between the concentration-on-pain and avoidance conditions vary depending upon whether the report was made at 30 seconds, 60 seconds, or 90 seconds?

Conducting a Two-Way Repeated-Measures ANOVA

We will conduct a number of analyses to answer the research questions. The analyses include an overall ANOVA for a 3 × 2 within-subjects design, follow-up tests associated with the time main effect, and follow-up tests associated with the interaction effects (i.e., simple main effect tests and interaction comparisons). In Lesson 26 we discussed selecting appropriate follow-up tests for a two-way ANOVA.

Conducting Tests of Main and Interaction Effects

The following steps evaluate the main and interaction effects (univariate and multivariate) for a two-way repeated-measures ANOVA. The steps also let us obtain the means and standard deviations of the quantitative variables for all combinations of levels of the two within-subjects factors. To conduct these analyses, carry out the following steps:

1. Click **Analyze**, click **General Linear Model**, then click **Repeated Measures**.
2. In the Within-Subject Factor Name text box, type **coping** in place of factor1.
3. Type **2** in the Number of Levels text box.
4. Click **Add**.
5. In the Within-Subject Factor Name text box, type **time**.
6. Type **3** in the Number of Levels text box.
7. Click **Add**.
8. Click **Define**.
9. Holding down the Ctrl key, click **avoid1**, **avoid2**, **avoid 3**, **conc1**, **conc2**, and **conc3**. Click ▶ to move them to the Within-Subjects Variables box.
10. Click **Options**.
11. Click **coping** in the Factor(s) and Factor Interactions box, and click ▶ to have it appear in the Display Means for box.
12. Click **time** in the Factor(s) and Factor Interactions box, and click ▶ to have it appear in the Display Means for box.
13. Click **coping*time** in the Factor(s) and Factor Interactions box, and click ▶ to have it appear in the Display Means for box.
14. Click **Compare main effects**.
15. Click **Descriptives** and **Estimates of effect size**.
16. Click **Continue**.
17. Click **OK**.

Selected SPSS Output for Two-Way Repeated-Measures ANOVA

SPSS produces a lot of output for a two-way within-subjects ANOVA: descriptive statistics, statistics for evaluating the sphericity assumption, traditional and alternative univariate ANOVA results, ANOVA results using multivariate tests and more. Only the multivariate tests for the main and interaction sources and the cell means are shown in Figure 181 (on page 250).

We will concentrate our attention on interpreting the ANOVA results using the multivariate tests and, therefore, avoid the controversy surrounding the sphericity assumption. The multivariate tests indicate a nonsignificant coping main effect, Wilks's $\Lambda = .78$, $F(1, 7) = 1.92$, $p = .21$, a significant time main effect, Wilks's $\Lambda = .11$, $F(2, 6) = 24.59$, $p < .01$, and a significant coping-by-time interaction effect, Wilks's $\Lambda = .11$, $F(2, 6) = 23.43$, $p < .01$. Note that the multivariate result for the coping main effect is identical to the univariate result because the coping main effect has only two levels.

Because the interaction is significant, we would normally conduct follow-up simple main effect tests, interaction comparisons, or both. We will discuss how to conduct these tests shortly. The time main effect was also significant, but normally we would not conduct follow-up tests to this main effect in the presence of a significant interaction effect. However, if there

Descriptive Statistics

	Mean	Std. Deviation	N
avoid1	4.00	3.024	8
avoid2	12.00	1.414	8
avoid3	26.00	7.653	8
conc1	12.00	7.231	8
conc2	14.00	6.803	8
conc3	22.00	8.848	8

Multivariate Tests[b]

Effect		Value	F	Hypothesis df	Error df	Sig.	Partial Eta Squared
coping	Pillai's Trace	.216	1.924[a]	1.000	7.000	.208	.216
	Wilks' Lambda	.784	1.924[a]	1.000	7.000	.208	.216
	Hotelling's Trace	.275	1.924[a]	1.000	7.000	.208	.216
	Roy's Largest Root	.275	1.924[a]	1.000	7.000	.208	.216
time	Pillai's Trace	.891	24.589[a]	2.000	6.000	.001	.891
	Wilks' Lambda	.109	24.589[a]	2.000	6.000	.001	.891
	Hotelling's Trace	8.196	24.589[a]	2.000	6.000	.001	.891
	Roy's Largest Root	8.196	24.589[a]	2.000	6.000	.001	.891
coping * time	Pillai's Trace	.886	23.432[a]	2.000	6.000	.001	.886
	Wilks' Lambda	.114	23.432[a]	2.000	6.000	.001	.886
	Hotelling's Trace	7.811	23.432[a]	2.000	6.000	.001	.886
	Roy's Largest Root	7.811	23.432[a]	2.000	6.000	.001	.886

a. Exact statistic

b. Design: Intercept
Within Subjects Design: coping+time+coping*time

Figure 181. **The cell means and multivariate results of the two-way repeated-measures analysis.**

were no significant interaction effect, we would want to conduct follow-up tests to understand why the main effect test was significant. Next, we will describe how to conduct pairwise comparison tests for the time main effect just to demonstrate how these tests would be done.

Conducting Pairwise Comparisons Following a Significant Main Effect

Pairwise comparisons are most easily conducted by clicking on **Compare main effects** in the Repeated Measures: Options dialog box (Step 14 in the step-by-step instructions). It may be instructional to know that identical results can be obtained by conducting paired-samples *t* tests. Briefly, the latter tests are obtained by first computing time1 as the mean of conc1 and avoid1, time2 as the mean of conc2 and avoid2, and time3 as the mean of conc3 and avoid3. Then, in the **Paired-Samples T Test** dialog box, creating three pairings of time1, time2, time3 in the Paired Variables box.

Selected SPSS Output for Pairwise Comparisons for Time Main Effect

The results of the analysis are shown in Figure 182. All three pairwise comparisons among the means for time 1, time 2, and time 3 are significant controlling for Type I error across the three tests at the .05 level by using the Holm's sequential Bonferroni procedure.

The smallest *p* value is for the comparison of time 1 and time 3 of −16, and its reported *p* value of .000 is less than the of .05/3 = .0167; therefore, the difference between the means for

Pairwise Comparisons

Measure: MEASURE_1

(I) time	(J) time	Mean Difference (I-J)	Std. Error	Sig.[a]	95% Confidence Interval for Difference[a]	
					Lower Bound	Upper Bound
1	2	-5.000*	.813	.000	-6.922	-3.078
	3	-16.000*	2.379	.000	-21.626	-10.374
2	1	5.000*	.813	.000	3.078	6.922
	3	-11.000*	2.138	.001	-16.056	-5.944
3	1	16.000*	2.379	.000	10.374	21.626
	2	11.000*	2.138	.001	5.944	16.056

Based on estimated marginal means

*. The mean difference is significant at the .05 level.

a. Adjustment for multiple comparisons: Least Significant Difference (equivalent to no adjustments).

Figure 182. **The results of the time main effect pairwise comparisons.**

these two times is significant. The next smallest p value is for the comparison of time 1 and time 2 of -11, and its reported p value of .000 is less than the α of $.05/2 = .025$ and, therefore, this comparison is also significant. Finally, the largest p value of .001 for the comparison of time 2 and time 3 of -5 is smaller than an α of $.05/1 = .05$; therefore, this comparison is significant too.

Conducting Simple Main Effect Analyses Following a Significant Interaction

Next we demonstrate how to conduct simple main effect analyses for a two-way within-subjects design following a significant interaction. We show how to conduct these analyses when the simple main effect of interest has only two levels, and paired-samples t tests can be used. To evaluate differences in the two coping approaches within time, follow these steps:

1. Click **Analyze**, click **Compare Means**, then click **Paired-Samples T Test**.
2. Click **Reset** to clear the dialog box.
3. Click **avoid1**. Avoid1 now appears as Variable 1 in Current Selections.
4. Click **conc1**. Conc1 appears as Variable 2 in Current Selections.
5. Now click on ▶, and avoid1–conc1 appears in the Paired Variables box.
6. Repeat Steps 3 through 5 for avoid2–conc2 and avoid3–conc3.
7. Click **OK**.

Selected SPSS Output for Simple Main Effects

We conducted paired-samples t tests to evaluate the differences between the two coping methods for each time period. The results of these analyses are shown in Figure 183 (on page 252).

We could have also conducted the time simple main effects for each coping method. Because there are three levels of time, we would have performed a separate one-way within-subjects ANOVA to evaluate the time factor for each coping method (concentration on pain and avoidance). If a one-way ANOVA yielded significance, we would have conducted pairwise comparisons among the three times. However, we chose to ignore differences among time levels for each coping method because they were of less interest.

Two of the three pairwise comparisons are significant controlling for familywise error rate across the three tests at the .05 level using the Holm's sequential Bonferroni procedure. The comparison with the smallest p value evaluates the mean difference between coping methods for time 3, $t(7) = 6.69$, $p < .01$, which is less than $\alpha = .05/3 = .0167$ and significant. The comparison with the next smallest p value evaluates the mean difference between coping methods for time 1, $t(7) = -4.15$, $p < .01$, which also is less than $\alpha = .05/2 = .025$

Paired-Samples Test

		Paired Differences					t	df	Sig. (2-tailed)
					95% Confidence Interval of the Difference				
		Mean	Std. Deviation	Std. Error Mean	Lower	Upper			
Pair 1	avoid1 - conc1	-8.000	5.451	1.927	-12.557	-3.443	-4.151	7	.004
Pair 2	avoid2 - conc2	-2.000	6.633	2.345	-7.546	3.546	-.853	7	.422
Pair 3	avoid3 - conc3	4.000	1.690	.598	2.587	5.413	6.693	7	.000

Figure 183. **The results of the simple main effects analyses of coping within time.**

and significant. Finally, the last comparison evaluates the mean difference between coping methods for time 2, $t(7) = -.85$, $p = .42$, which is not less than $\alpha = .05/1 = .05$ and, therefore, is not significant.

Conducting Interaction Comparisons Following a Significant Interaction

An alternative way to follow up a significant interaction is to conduct interaction comparisons. Conducting interaction comparisons for a two-way within-subjects design requires computing difference scores and conducting paired-samples *t* tests. To conduct interaction comparisons for our example, follow these steps:

1. Compute differences. Use the Transform → Compute Variable option to compute the three difference variables (a_c) by subtracting the concentration (conc) variables from the avoidance (avoid) variables for each of the three time periods. More specifically, a1_c1 would be avoid1 – conc1, a2_c2 would be avoid2 – conc2, and a3_c3 would be avoid3 – conc3.
2. Click **Analyze**, click **Compare Means**, then click **Paired-Samples T Test**.
3. Click **Reset** to clear the dialog box.
4. Click **a1_c1**. a1_c1 now appears as Variable 1 in Current Selections.
5. Click **a2_c2**. a2_c2 appears as Variable 2 in Current Selections.
6. Now click on ▶ and a1_c1 – a2_c2 appears in the Paired Variables: box.
7. Repeat steps 4 through 6 for a1_c1 – a3_c3 and a2_c2 – a3_c3.
8. Click **OK**.

Selected SPSS Output for Interaction Comparisons

The results of the analysis are shown in Figure 184. Tetrad comparisons involving four means evaluate whether the mean differences between the two coping methods are the same between any two time periods. All three tetrad comparisons are significant controlling for familywise

Paired-Samples Test

		Paired Differences					t	df	Sig. (2-tailed)
					95% Confidence Interval of the Difference				
		Mean	Std. Deviation	Std. Error Mean	Lower	Upper			
Pair 1	a1_c1 - a2_c2	-6.00000	4.07080	1.43925	-9.40328	-2.59672	-4.169	7	.004
Pair 2	a1_c1 - a3_c3	-12.00000	5.12696	1.81265	-16.28625	-7.71375	-6.620	7	.000
Pair 3	a2_c2 - a3_c3	-6.00000	6.09449	2.15473	-11.09512	-.90488	-2.785	7	.027

Figure 184. **The results of the interaction comparison analyses.**

error rate across the three tests at the .05 level, using the Holm's sequential Bonferroni procedure. The comparison with the smallest p value evaluates whether the mean difference between coping methods is the same for time 1 and time 3, $t(7) = -6.62$, $p < .01$, which is less than $\alpha = .05/3 = .0167$. The comparison with the next smallest p value evaluates whether the mean difference between coping methods is the same for time 1 and time 2, $t(7) = -4.17$, $p < .01$, which is less than $\alpha = .05/2 = .025$. Finally, the last comparison evaluates whether the mean difference between coping methods is the same for time 2 and time 3, $t(7) = -2.78$, $p = .03$, which is less than $\alpha = .05/1 = .05$.

Using SPSS Graphs to Display the Results

The General Linear Model-Repeated Measures procedure allows you to create line graphs of the factor means to visually represent the interactions and main effects associated with the factors. However, these graphs are limited in that they do not display the variability of the scores around the means. To create a profile plot depicting the interaction between coping and time, follow these steps:

1. Click **Analyze**, click **General Linear Model**, then click **Repeated Measures**. If you have not exited SPSS, the appropriate options should be selected, then go to Step 2. If they are not, conduct Steps 2 through 9 of "Conducting the Overall Analysis" presented earlier in this lesson and skip to Step 3.
2. In the Repeated Measures: Define Factor(s) dialog box, click **Define**.
3. Click **Plots**. You will see the Repeated Measures: Profile Plots dialog box in Figure 185.
4. In the Repeated Measures: Profile Plots dialog box, click **coping**, and click ▶ to have it appear in the Horizontal Axis box.
5. Click **time**, and click ▶ to have it appear in the Separate Lines box.
6. Click **Add**.
7. Click **Continue**.
8. Click **OK**.

The edited profile plot is shown in Figure 186 (on page 254).

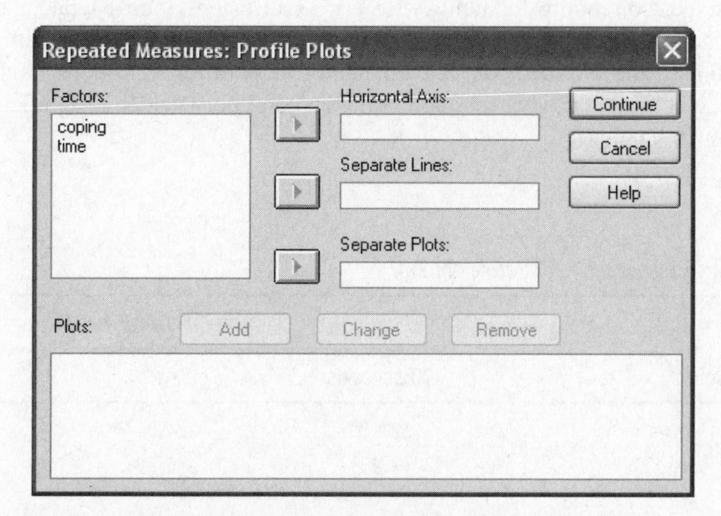

Figure 185. **The Repeated Measures: Profile Plots dialog box.**

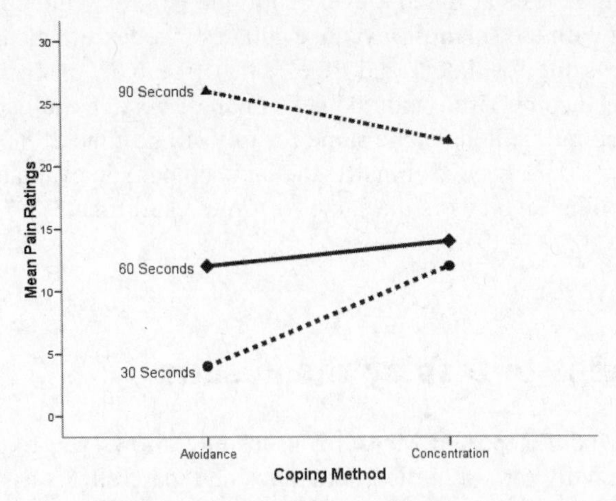

Figure 186. **Profile plot of coping method by time interaction.**

An APA Results Section

A two-way within-subjects analysis of variance was conducted to evaluate the effect of coping strategies and length of time under stress on pain ratings. The dependent variable was a pain rating of 1 to 50. The within-subjects factors were coping methods with two levels (concentration-on-pain and avoidance) and time with three levels (30, 60, and 90 seconds). The Time main effect and Coping × Time interaction effect were tested using the multivariate criterion of Wilks's lambda (Λ). The Time main effect was significant, $\Lambda = .11$, $F(2, 6) = 24.59$, $p < .01$, as well as the Coping × Time interaction effect, $\Lambda = .11$, $F(2, 6) = 23.43$, $p < .01$. The univariate test associated with the Coping main effect was nonsignificant, $\Lambda = .78$, $F(1, 7) = 1.92$, $p = .21$.

Three paired-samples *t* tests were conducted to follow up the significant interaction. We controlled for familywise error rate across these tests by using Holm's sequential Bonferroni approach. Differences in mean ratings of pain between the two coping methods were significantly different between times 1 and 2, $t(7) = -4.17$, $p < .01$, between times 2 and 3, $t(7) = -2.78$, $p = .03$, and between times 1 and 3, $t(7) = -6.62$, $p < .01$. Table 34 shows that although the difference in mean pain ratings was lower for the avoidance method at time 1, the difference decreased at time 2, and reversed itself at time 3. These results support the hypothesis that concentration-on-pain methods are more effective than avoidance methods the longer individuals are exposed to a stressor.

Finally, three paired-samples *t* tests were computed to assess differences between methods at each time period, controlling for familywise error rate using Holm's sequential Bonferroni approach. At time 1, the avoidance condition yielded a significantly lower mean rating, $t(7) = -4.15$, $p < .01$. At time 2, the two methods were not significantly different from each other, $t(7) = -.85$, $p = .42$, while at time 3, the concentration-on-pain condition yielded a significantly lower mean rating, $t(7) = 6.69$, $p < .01$.

Table 34
Means (Standard Deviations) for Ratings of Pain

	Time in frigid water		
Method of coping	30 seconds	60 seconds	90 seconds
Concentration on pain	12.00 (7.23)	14.00 (6.80)	22.00 (8.85)
Avoidance	4.00 (3.02)	12.00 (1.41)	26.00 (7.65)

Exercises

The data for Exercises 1 through 5 are in the data file named *Lesson 30 Exercise File 1* on the Web at http://www.prenhall.com/greensalkind. The data are from the following research problem.

John is interested in the development of visual attention in infants and whether it differs for novel and familiar stimuli. He collects measures of visual attention for novel and familiar stimuli from 60 infants at four time periods: 3 months, 4 months, 6 months, and 8 months. John's SPSS data file contains eight variables, including novel and familiar visual attention scores in seconds for each of the time periods. There are two within-subjects factors: time with four levels (3, 4, 6, and 8 months) and stimulus type with two levels (novel and familiar).

1. Conduct a two-way within-subjects ANOVA to evaluate these data. From the output, identify the following:
   a. *p* value associated with the Wilks's lambda for the time effect
   b. *F* value associated with the Wilks's lambda for the interaction effect
   c. Effect size associated with the Wilks's lambda for the stimulus effect
   d. Mean visual attention score on the novel stimuli for 6-month-old infants
2. What should John conclude about the interaction between time and stimulus type? Report the appropriate information from the output to justify your conclusion.
3. What follow-up tests should be conducted?
4. Write a Results section reporting your conclusions.
5. Create a profile plot to show the interaction between stimuli types and time.

The data for Exercises 6 through 8 are in the data file named *Lesson 30 Exercise File 2* on the Web at http://www.prenhall.com/greensalkind. The data are from the following research problem.

Peter, a child psychologist, is interested in the relationship between physical characteristics and popularity. Specifically, he believes that girls are more likely to find better-dressed boys and girls more likeable, although to a lesser extent if the better-dressed children are males. He conducts the following study using a group of 25 girls in the first grade. He obtains 100 pictures of first-graders whom the children in this study do not know. Twenty-five of the pictures are of well-dressed boys, 25 are of well-dressed girls, 25 are of poorly dressed boys, and 25 are of poorly dressed girls. Before having each girl come into the testing room, the experimenter shuffles the 100 cards 15 times. The child is told that she will see a series of pictures. If the child likes the person in the photographs, she is to cheer. If she dislikes the child, she is to boo. A score of 1 is assigned to a cheer and a score of −1 is assigned to a boo. Peter computes an overall scale for each child by summing the scores for the 25 cards in each of the four conditions (e.g., well-dressed boys).

6. Conduct a two-way within-subjects ANOVA to evaluate these data. One factor represents gender of the child in the picture, and the second factor represents his or her dress. What should Peter conclude about the interaction between gender and dress type? Report the appropriate information from the output to justify your conclusion.
7. Should Peter conduct any follow-up tests? Why or why not?
8. Write a Results section for these data.

PART II

UNIT 8 Correlation, Regression, and Discriminant Analysis Procedures

Unit 8 includes lessons on the Pearson product-moment correlation coefficient, partial correlation coefficient, bivariate linear regression, multiple linear regression, and discriminant analysis. We describe how to use these statistical techniques to analyze a wide range of research problems. In some applications, the analyses may appear very simple, such as computing a Pearson product-moment correlation coefficient to assess the linear relationship between two quantitative variables. In other applications, the analyses may seem complex, such as computing partial and multiple correlations to evaluate the relative importance of various predictors in a multiple regression equation. For all techniques, it is important to have an in-depth understanding of them. If you are unfamiliar with these analytic methods, please read one or more of the following sources on these methods: Cohen, Cohen, West, and Aiken (2003), Darlington (1990), and Pehazur (1997).

The lessons in Unit 8 address the following topics:

- Lesson 31 deals with bivariate correlational analysis. We will restrict our discussion to linear relationships between two quantitative variables.
- Lesson 32 describes partial correlational analysis. We describe how to assess the relationship between two quantitative variables, statistically controlling for or partialling out differences on one or more potential confounding variables.
- Lesson 33 presents bivariate linear regression. We will focus our discussion on applications with a quantitative predictor variable and a quantitative criterion variable.
- Lesson 34 describes multiple linear regression. We'll talk about applications with two or more quantitative predictors and a quantitative criterion variable.
- Lesson 35 discusses discriminant analysis and includes applications for a qualitative criterion variable and one or more quantitative predictor variables.

LESSON

31

Pearson Product-Moment Correlation Coefficient

The Pearson product-moment correlation coefficient (r) assesses the degree that quantitative variables are linearly related in a sample. Each individual or case must have scores on two quantitative variables. The significance test for r evaluates whether there is a linear relationship between the two variables in the population. In this lesson, we will work on understanding Pearson correlations between quantitative variables. We will not discuss correlations in which one or both variables are categorical or ordinal.

Applications of the Pearson Correlation Coefficient

We will illustrate three types of research studies in which the Pearson correlations are computed.

- Studies with a correlation between two variables
- Studies with correlations among three or more variables
- Studies with correlations within and between sets of variables

Study with a Correlation between Two Variables

Annette is interested in the relationship between leg strength and running agility for male college students. She obtains 40 male students from undergraduate physical education classes. Each student must complete a series of leg strength exercises on a weight machine. Annette computes a total leg strength index that takes into account a student's performance across all of the exercises. In addition, each student must run 200 meters along a straight line and 200 meters along a curving, twisting line. Running clumsiness (the opposite of running agility) is the number of seconds to complete the curving, twisting course minus the number of seconds to complete the straight course. Annette's SPSS data file for this analysis includes scores on both the leg strength and running clumsiness variables for the 40 cases.

Study with Correlations among Three or More Variables

John is interested in whether people who have a positive view of themselves in one aspect of their lives also tend to have a positive view of themselves in other aspects of their lives. To address this question, he has 80 men complete a self-concept inventory that contains five scales. Four scales involve questions about how competent respondents feel in the areas of intimate relationships, relationships with friends, commonsense reasoning and everyday knowledge, and academic reasoning and scholarly knowledge. The fifth scale includes items about how competent a person feels in general. John's SPSS data file includes the five self-concept variables for the 80 cases. John is interested in determining the correlations between all possible pairs of variables, that is, 10 correlations.

Study with Correlations within and between Sets of Variables

Cindy, director of personnel at an insurance company, wants to know whether a personality measure predicts on-the-job performance. She has the records of 50 life insurance sales people who took a personality measure when they were hired. The personality measure has three scales: extroversion, conscientiousness, and openness. Their on-the-job performance was assessed using two measures: amount of insurance sold in first year and ratings by supervisors. Cindy's SPSS data file has two sets of variables for the 50 cases: one set containing the three personality scales and the second set having the two on-the-job performance measures. Cindy thinks of the personality scales as predictor variables and the on-the-job performance measures as criteria. Although Cindy would undoubtedly compute the correlations within sets, she is primarily interested in the six correlations of the three personality measures with the two criteria.

Understanding the Pearson Correlation Coefficient

We will first consider the two assumptions underlying the significance test for the Pearson correlation. We will then examine the meaning of the Pearson correlation as an effect size statistic.

Assumptions Underlying the Significance Test

There are two assumptions underlying the significance test associated with a Pearson correlation coefficient between two variables.

Assumption 1: The Variables Are Bivariately Normally Distributed

If the variables are bivariately normally distributed, each variable is normally distributed ignoring the other variable and each variable is normally distributed at all levels of the other variable. If the bivariate normality assumption is met, the only type of statistical relationship that can exist between two variables is a linear relationship. However, if the assumption is violated, a nonlinear relationship may exist. It is important to determine if a nonlinear relationship exists between two variables before describing the results with the Pearson correlation coefficient. Nonlinearity can be assessed visually by examining a scatterplot of the data points as we illustrate later in this lesson.

Assumption 2: The Cases Represent a Random Sample from the Population and the Scores on Variables for One Case Are Independent of Scores on These Variables for Other Cases

The significance test for a Pearson correlation coefficient is not robust to violations of the independence assumption. If this assumption is violated, the correlation significance test should not be computed.

An Effect Size Statistic: A Pearson Correlation Coefficient

SPSS computes the Pearson correlation coefficient, an index of effect size. The index ranges in value from -1 to $+1$. This coefficient indicates the degree that low or high scores on one variable tend to go with low or high scores on another variable. A score on a variable is a low (or high) score to the extent that it falls below (or above) the mean score on that variable.

To understand how the correlation coefficient is interpreted, let's use our example involving leg strength and running clumsiness. A correlation of $+1$ indicates that as scores on strength increase across cases, the scores on clumsiness increase precisely at a constant rate. If r is positive, low scores on strength tend to be associated with low scores on clumsiness, and high scores on strength tend to be associated with high scores on clumsiness. If r is zero, low scores on strength tend to be associated equally with low and high scores on clumsiness, and high scores on strength tend to be associated equally with low and high scores on clumsiness. In other words, as the scores on strength increase across cases, the scores on clumsiness tend neither to increase nor to decrease. If r is negative, low scores on strength tend to be associated with high scores on clumsiness, and high scores on strength tend to be associated with low scores on clumsiness. A correlation of -1 indicates that as scores on strength increase across cases, the scores on clumsiness decrease precisely at a constant rate.

As with all effect size indices, there is no good answer to the question, "What value indicates a strong relationship between two variables?" What is large or small depends on the discipline within which the research question is being asked. However, for the behavioral sciences, correlation coefficients of .10, .30, and .50, irrespective of sign, are, by convention, interpreted as small, medium, and large coefficients, respectively.

If one variable is thought of as the predictor and another variable as the criterion, we can square the correlation coefficient to interpret the strength of the relationship. The square of the correlation gives the proportion of criterion variance that is accounted for by its linear relationship with the predictor. For our example, if strength is the predictor and clumsiness is the criterion and the correlation between these variables is .40, we would conclude that 16% of the variance ($.40^2$) of the clumsiness variable is accounted for by its linear relationship with strength.

The Data Set

The data set used to illustrate the Pearson correlation coefficient is named *Lesson 31 Data File 1* on the Web at http://www.prenhall.com/greensalkind. It has data coming from the self-concept example we presented earlier in this lesson. The variables in the data set are in Table 35.

Table 35
Variables in Lesson 31 Data File 1

Variables	Definition
intimate	High scores on this variable indicate that respondents are self-confident in intimate relationships.
friend	High scores on this variable indicate that respondents are self-confident in relationships among friends.
common	High scores on this variable indicate that respondents are self-confident in their knowledge of everyday events and their use of commonsense reasoning.
academic	High scores on this variable indicate that respondents are self-confident in their scholarly knowledge and their ability to reason in a rigorous, formal manner.
general	High scores on this variable indicate that respondents are self-confident in the way they conduct their lives. This variable is not a sum of the other four variables, but is based on items specific to this variable. The items ask how competent a person feels in general, and they do not specify a particular life domain.

The Research Question

The research question concerns the relationship between the five self-concept variables. Do men who feel confident in one life domain tend to feel confident in other domains, and conversely, do men who feel insecure in one life domain also tend to feel insecure in other domains?

Conducting Pearson Correlation Coefficients

We first conduct Pearson correlation coefficients among variables within a set and then between variables from different sets.

Conducting Pearson Correlation Coefficients among Variables within a Set

To compute Pearson correlation coefficients among the five self-concept variables, follow these steps:

1. Click **Analyze**, click **Correlate**, then click **Bivariate**. You will see the Bivariate Correlations dialog box as shown in Figure 187.

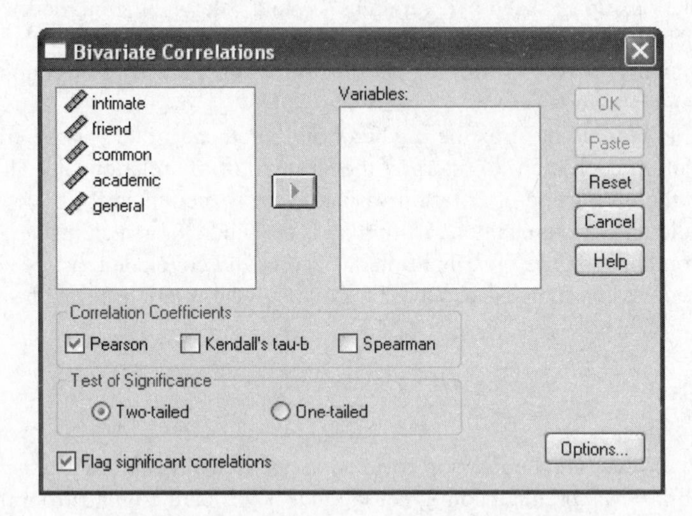

***Figure 187.* The Bivariate Correlations dialog box.**

2. Holding down the Ctrl key, click **intimate**, **friend**, **common**, **academic**, and **general**, and click ▶ to move them to the Variables box.
3. Make sure **Pearson** is selected in the Correlation Coefficients area.
4. Make sure the **Two-tailed** option is selected in the Test of Significance box (unless you have some a priori reason to select one-tailed).
5. Make sure **Flag significant correlations** is selected.
6. Click **Options**. You'll see the Bivariate Correlations: Options dialog box as shown in Figure 188.

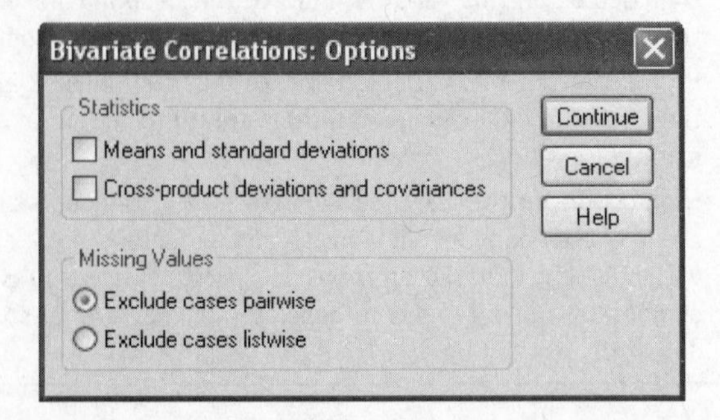

***Figure 188.* The Bivariate Correlations: Options dialog box.**

7. Click **Means and standard deviations** in the Statistics box.
8. Click **Continue**.
9. Click **OK**.

Correlations

		Intimate Relationships	Friends	Common Sense	Academic Intelligence	General
Intimate Relationships	Pearson Correlation	1	.552**	.351**	.218	.393**
	Sig. (2-tailed)		.000	.001	.052	.000
	N	80	80	80	80	80
Friends	Pearson Correlation	.552**	1	.462**	.244*	.546**
	Sig. (2-tailed)	.000		.000	.029	.000
	N	80	80	80	80	80
Common Sense	Pearson Correlation	.351**	.462**	1	.400**	.525**
	Sig. (2-tailed)	.001	.000		.000	.000
	N	80	80	80	80	80
Academic Intelligence	Pearson Correlation	.218	.244*	.400**	1	.261*
	Sig. (2-tailed)	.052	.029	.000		.019
	N	80	80	80	80	80
General	Pearson Correlation	.393**	.546**	.525**	.261*	1
	Sig. (2-tailed)	.000	.000	.000	.019	
	N	80	80	80	80	80

**. Correlation is significant at the 0.01 level (2-tailed).

*. Correlation is significant at the 0.05 level (2-tailed).

***Figure 189.* The bivariate correlations among the five self-concept measures.**

Selected SPSS Output for Pearson Correlations

The results of the correlational analyses are shown in Figure 189. The Correlations table in Figure 189 presents the correlations, asterisks (*) indicating whether a particular correlation is significant at the .05 level (*) or the .01 level (**), p values associated with the significance tests for these correlations, and sample size (N). Note that the information in the upper-right triangle of the matrix is redundant with the information in the lower-left triangle of the matrix and can be ignored.

If several correlations are computed, you may wish to consider a corrected significance level to minimize the chances of making a Type I error. One possible method is the Bonferroni approach, which requires dividing .05 by the number of computed correlations. A correlation coefficient would not be significant unless its p value is less than the corrected significance level. In this example, a p value would have to be less than .05 divided by 10 or .005 to be declared significant.

SPSS presents the correlations in tabular form. However, correlations are often presented within the text of a manuscript. For example, "The correlation between the friendships and general self-concept scales was significant, $r(78) = .55, p < .001$." The number in parentheses represents the degrees of freedom associated with the significance test, which is equal to the number of cases minus 2 (or $N - 2$). As shown on the output, the number of cases for this correlation is 80, and therefore the degrees of freedom are $80 - 2 = 78$.

Conducting Pearson Correlation Coefficients between Variables from Different Sets

If you are interested in computing correlations between variables from two different sets but not within each set, the method changes slightly. Let's say that we were interested in the correlations of the four self-concept scales with the general self-concept scale. To conduct this analysis, perform the following steps:

1. Follow the same instructions as those specified earlier for computing a Pearson correlation coefficient, but at Step 9, click **Paste** rather than OK. The syntax window shows the syntax for the bivariate correlation program. One of the syntax lines looks similar to this:

   /VARIABLES=academic common friend intimate general

2. Depending on how you selected variables in the Bivariate Correlation dialog box, the variable **general** may or may not be listed last. If it is not last, retype the statement so that **general** is last, and put the word **with** before the word **general**. Your syntax should look like this:

   /VARIABLES=academic common friend intimate with general

3. Highlight the appropriate syntax, click **Run**, and then click **Selection**. The output will give the correlations of the first four self-concept variables with general self-concept.

Using SPSS Graphs to Display the Results

Scatterplots are rarely included in results sections of manuscripts, but they should be included more often because they visually represent the relationship between variables. While a correlation coefficient tries to summarize the relationship between two variables with a single value, a scatterplot gives a rich descriptive picture of this relationship. In addition, the scatterplot can show whether a few extreme scores (outliers) overly influence the value of the correlation coefficient or whether nonlinear relationships exist between variables.

To create scatterplots among multiple variables, follow these steps:

1. Click **Graphs**, click **Legacy Dialogs**, and then click **Scatter/Dot**. You'll see the Scatter/Dot dialog box as shown in Figure 190.

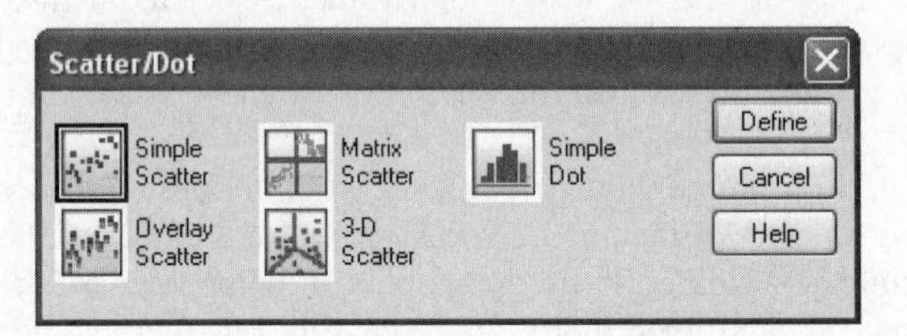

Figure 190. **The Scatterplot dialog box.**

2. Click **Matrix Scatter**, then click **Define**. You'll see the Scatterplot Matrix dialog box as shown in Figure 191.
3. Holding down the Ctrl key, click **intimate**, **friends**, **common**, **academic**, and **general**.
4. Click ▶ to move them to the Matrix Variables box.
5. Click **OK**. The edited graph is shown in Figure 192.

An APA Results Section

Correlation coefficients were computed among the five self-concept scales. Using the Bonferroni approach to control for Type I error across the 10 correlations, a p value of less than .005 (.05/10 = .005) was required for significance. The results of the correlational analyses presented in Table 36 show that 7 out of the 10 correlations were statistically significant and were greater than or equal to .35. The correlations of scholarly knowledge self-concept with the other self-concept measures tended to be lower and not significant. In general, the results suggest that if men say that they are self-confident in one area, they tend to state that they are self-confident in other areas except for scholarly knowledge.

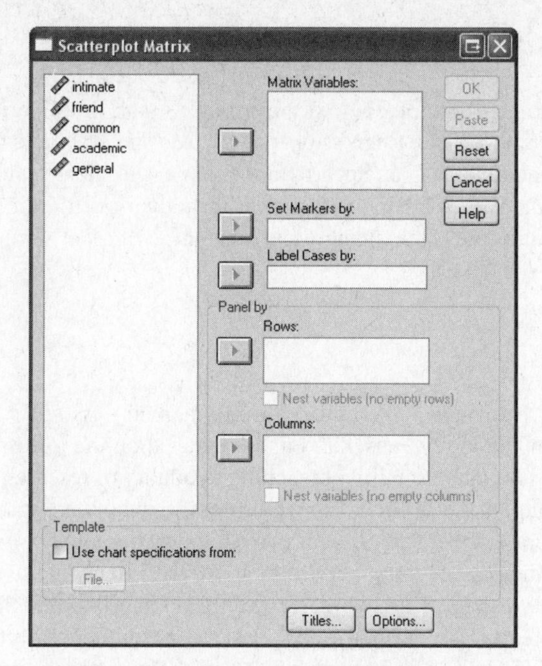

Figure 191. **The Scatterplot Matrix dialog box.**

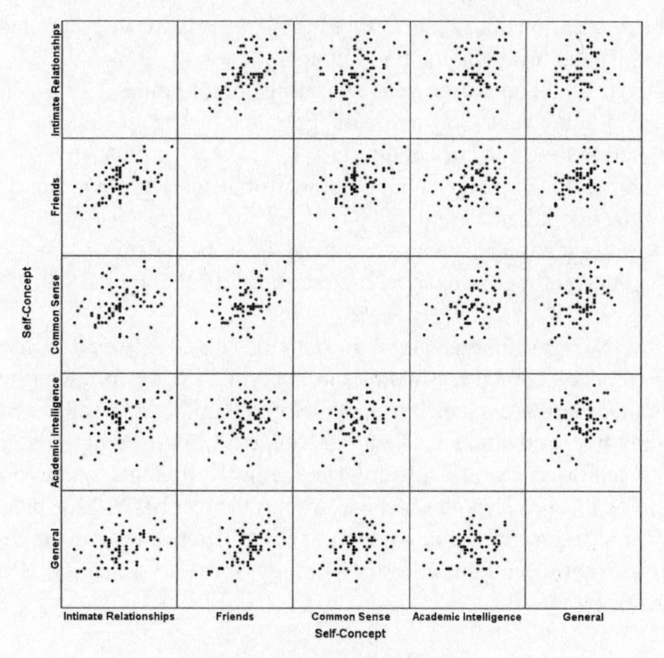

Figure 192. **Scatterplot matrix.**

Table 36

Correlations among the Five Self-Concept Scales (N = *80*)

	Scholarly knowledge	Everyday knowledge	Friendships	Intimate relationships
Everyday knowledge	.40*			
Friendships	.24	.46*		
Intimate relationships	.22	.35*	.55*	
General	.26	.52*	.55*	.39*

* $p < .005$

Alternative Analyses

We can compute Pearson correlations by other procedures than the Bivariate Correlation procedure, such as the Bivariate Linear Regression procedure. The Bivariate Correlation procedure can also compute a Kendall's tau-b or Spearman if the measurement scales underlying the variables are ordinal (i.e., the values for the variable indicate their position in relation to each other, but the intervals between scores lack quantitative meaning).

Exercises

The data for Exercises 1 through 4 are in the data file named *Lesson 31 Exercise File 1* on the Web at http://www.prenhall.com/greensalkind and are based on the following research problem.

Betsy is interested in relating quality of teaching to quality of research by college professors. She has access to a sample of 50 social science professors who were teaching at the same university for a 10-year period. Over this 10-year period, the professors were evaluated on a 5-point scale on quality as instructors and on quality of their courses. Betsy has averaged these ratings to obtain an overall quality rating as an instructor (rating_1) and the overall quality of the course (rating_2) for each professor. In addition, Betsy also has the number of articles that each professor published during this time period (num_pubs) and the number of times these articles were cited by other authors (cites).

1. Conduct a correlational analysis to investigate the relationships among these variables. Identify the following on the output:
   a. p value for the correlation between rating_1 and rating_2
   b. Correlation between cites and num_pubs
   c. Correlation between cites and rating_1
2. What is the relationship between the number of articles published and the overall quality of the instructor?
3. Write a Results section based on your analysis of these data.
4. Create a scatterplot matrix to show the relationships among the four variables.

The data for Exercises 5 through 8 are in the data file named *Lesson 31 Exercise File 2* on the Web at http://www.prenhall.com/greensalkind and are based on the following research problem.

Fred believes that there are excellent students, good students, mediocre students, bad students, and extremely bad students. Excellent students tend to do exceptionally well in all subjects; good students tend to do well in all subjects; mediocre students tend to do mediocre on all subjects; and so on. To test this hypothesis, he examines the records of 120 students who have recently graduated from high school. For each student, he determines their high school GPAs in five types of courses: math (mathgpa), English (enggpa), history (histgpa), science (sciengpa), and social sciences (socgpa).

5. Conduct a correlational analysis to investigate the relationships of students' GPA in math and science with students' GPA in social sciences and humanities (English and history).
6. What should Fred conclude from the correlations between the two sets of variables?
7. Create two new variables: (1) an average GPA in math and science and (2) an average GPA in social science and humanities. Conduct a correlational analysis on the two averaged scores. What is the resulting correlation?
8. What conclusions would you draw from the correlation between the two averaged scores? Would they be different from the conclusion you would draw from Exercise 6?

LESSON

32 Partial Correlations

The partial correlation (r_p), an effect size index, indicates the degree that two variables are linearly related in a sample, partialling out the effects of one or more control variables. The partial correlation attempts to estimate the correlation between two variables given all cases had exactly the same scores on the control variables, that is, holding constant the values of the control variables. The significance test for a partial correlation evaluates whether in the population the partial correlation is equal to zero.

It makes little sense to interpret a partial correlation between two variables controlling for a third variable without considering the zero-order or bivariate correlation between the two variables. In other words, to interpret properly a partial correlation between two variables, we must also know the magnitude of the bivariate correlation between these two variables. Because there is a strong link between partial and bivariate correlations, it may be useful to reread Lesson 31 on bivariate correlations before proceeding with this lesson.

The SPSS data file associated with a partial correlation coefficient contains scores on at least three variables for each individual or case. For our purposes in this lesson, all the variables are quantitative.

Applications of Partial Correlations

Partial correlations can be used in several ways, such as the following:

- Partial correlation between two variables
- Partial correlations among multiple variables within a set
- Partial correlations between sets of variables

We illustrate these three applications by using the studies described in Lesson 31, except we will add a control variable to each study.

Partial Correlation between Two Variables

Annette is conducting research in physical education and believes that there is a relationship between leg strength and running clumsiness for male college students. She believes that this relationship is a function of the amount of physical activity in which students engage. In other words, those students who tend to exercise regularly also tend to have greater leg strength and to demonstrate less running clumsiness. To evaluate her hypothesis, she obtains scores for leg strength and running clumsiness on 40 male students from undergraduate physical education classes. In addition, she records the number of hours that each student is physically active per week. Annette is interested in the correlation between leg strength and running clumsiness partialling out the effects of physical activity, the control variable. Annette's SPSS data file includes leg strength, running clumsiness, and physical activity variables for the 40 cases.

Partial Correlations among Multiple Variables within a Set

John thinks that people who have a positive view of themselves in one specific area of their life tend to have a positive view of themselves in other specific areas of their life. He believes further that the tendency of people to evaluate themselves consistently across specific life domains is a function of their general self-concept. To address these issues, he has 80 male adults complete a self-concept inventory that contains the five scales described in Lesson 31. John is interested in the correlations among the first four scales that assess self-concept in specific life domains, partialling out the effects of the fifth scale, general self-concept. John's SPSS data file includes the five self-concept variables for the 80 cases.

Partial Correlations between Sets of Variables

Cindy, the director of personnel at an insurance company, theorizes that salespeople with particular types of personalities are more likely to develop a positive perspective toward the product that they sell, and this positive perspective is likely to spill over to their on-the-job performance. To address this issue, Cindy obtains scores on six variables from 50 life insurance salespeople. Three personality measures—extroversion, conscientiousness, and openness—are collected at the time of hiring. An attitudinal measure that assesses positive perspective toward the company's life insurance is obtained in the middle of their first year. Finally, two performance measures, amount of insurance sold in the first year and ratings by supervisors, are obtained at the end of their first year of selling insurance. Cindy is interested in the correlations between the personality variables and the on-the-job performance variables, partialling out the effects of perspective toward the company's life insurance, the control variable. Cindy's SPSS data file has scores on the three personality scales, the attitudinal measure, and the two on-the-job measures.

Understanding Partial Correlations

Before discussing the assumptions underlying the significance test for a partial correlation coefficient and the interpretation of a partial correlation coefficient, we discuss the relationship between a partial correlation coefficient and two research hypotheses.

Research Hypotheses and Partial Correlations

The partial correlation is calculated to evaluate why two variables are correlated. Two possible explanations are the common cause and the mediator variable hypotheses.

Common Cause Hypothesis

The common cause hypothesis says that variables A and B are correlated because they share the same causal variable or variables. The common cause hypothesis is illustrated in Figure 193.

If this hypothesis is correct, then the correlation between A and B should be nonzero in value ($r \neq 0$), but the correlation between A and B partialling out the effects of the common causal variable or variables should be equal to zero ($r_p = 0$).

In the first two applications described earlier, Annette and John are interested in evaluating the common cause hypothesis. Annette believes that the reason leg strength and running clumsiness are correlated is that they share a common cause, physical activity. If she is correct, the correlation between leg strength and running clumsiness should be negative, but the correlation

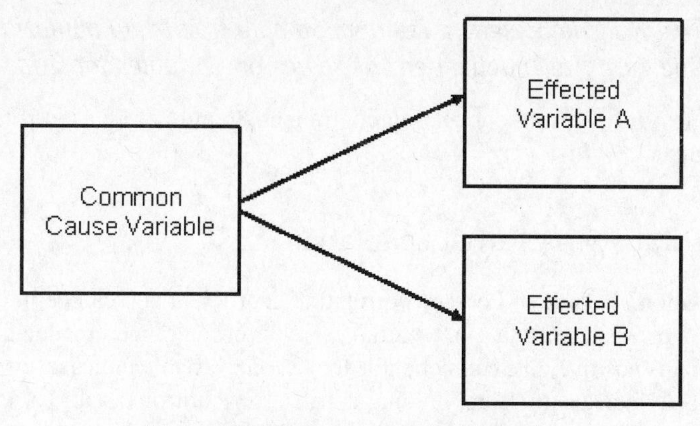

Figure 193. **Common cause hypothesis.**

between these variables partialling out the effects of physical activity should approach zero. In the second application, John believes that self-concept in specific life domains are correlated because they share a common cause, general self-concept. Accordingly, the self-concept measures across specific domains should be positively correlated, but their partial correlations, controlling for general self-concept, should be equal to zero.

Mediator Variable Hypothesis

The mediator variable hypothesis says that variables A and B are correlated because A causes B through one or more mediator variables. Figure 194 illustrates the mediator variable hypothesis.

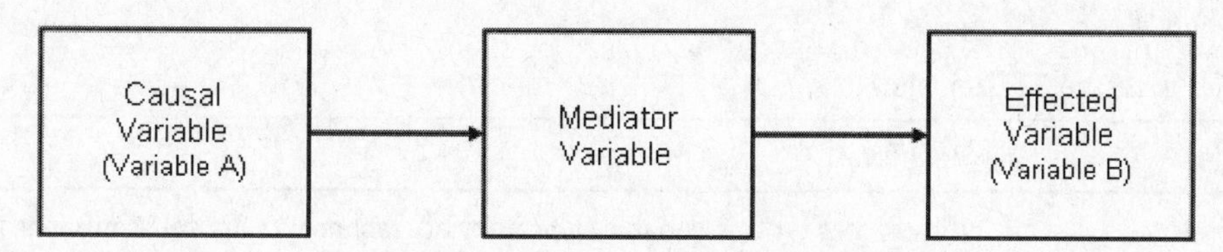

Figure 194. **Mediator variable hypothesis.**

If this hypothesis is correct, then the correlation between A and B should be nonzero in value ($r \neq 0$), but the correlation between A and B partialling out the effects of the mediator variable or variables should be equal to zero ($r_p = 0$).

In the third application, Cindy appears to be interested in evaluating the mediator variable hypothesis. She believes that the reason the personality and the on-the-job performance variables are correlated is that a salesperson's personality affects his or her perspective toward the company's insurance and that attitude subsequently affects how effective the salesperson is. If Cindy is correct, the correlation between the personality and the on-the-job performance variables should be positive, but the correlation between these variables partialling out the effects of attitude toward the company's insurance should approach zero.

Assumptions Underlying the Significance Test for a Partial Correlation Coefficient

Assumption 1: The Variables Are Multivariately Normally Distributed

If the variables are multivariately normally distributed, each variable is normally distributed ignoring the other variables, and each variable is normally distributed at every combination of values of the other variables. If the multivariate normality assumption is met, the only type of statistical relationship that can exist between variables is linear. If this assumption is violated, however, a nonlinear relationship may be present. It is important to determine if a nonlinear relationship exists between variables before describing the results with the partial correlation coefficient. Nonlinearity between any two variables can be evaluated by visually inspecting a scatterplot, as we will show you later in this lesson.

Assumption 2: The Cases Represent a Random Sample from the Population, and Scores for One Case Are Independent of Scores on Variables for Other Cases

The significance test for a partial correlation coefficient should not be used if the independence assumption is violated.

An Effect Size Statistic: A Partial Correlation

The partial correlation is a type of Pearson correlation coefficient and can be interpreted in a similar manner. Like a bivariate correlation, a partial correlation can range in value from -1 to $+1$. If a partial correlation is positive, one can conclude that as one variable increases in value, the second variable also tends to increase in value, holding constant the control variable or variables. If a partial correlation is zero, one can conclude that as one variable increases in value, the second variable tends neither to increase nor decrease in value, holding constant the control variable or variables. Finally, if a partial correlation is negative, one can conclude that as one variable increases in value, the second variable tends to decrease in value, holding constant the control variable or variables.

The Data Set

The data set used to illustrate the partial correlation coefficient is named *Lesson 32 Data File 1* on the Web at http://www.prenhall.com/greensalkind and represents data coming from the self-concept example discussed earlier in this lesson. Table 37 shows the variables in the data file.

Table 37
Variables in Lesson 32 Data File 1

Variables	Definition
intimate	High scores on this variable indicate that respondents are self-confident in intimate relationships.
friend	High scores on this variable indicate that respondents are self-confident in relationships among friends.
common	High scores on this variable indicate that respondents are self-confident in their knowledge of everyday events and their use of commonsense reasoning.
academic	High scores on this variable indicate that respondents are self-confident in their scholarly knowledge and their ability to reason in a rigorous, formal manner.
general	High scores on this variable indicate that respondents are self-confident in the way they conduct their lives. This variable is not a sum of the other four variables, but is based on items specific to this variable. The items ask how competent a person feels in general, and they do not specify a particular life domain.

The Research Question

The research questions for partial correlations address sequentially the magnitude of the bivariate correlations among variables and the magnitude of the correlations between variables partialling out the effects of a third (control) variable. Here are the research questions for our example.

1. Bivariate correlations among variables: Do men who feel confident in one specific life domain tend to feel confident in other specific domains, and do men who feel insecure in one life domain also tend to feel insecure in other domains?
2. Correlations among variables partialling out the effects of a control variable: If men have the same level of general self-confidence, do they tend to feel more

confident in one specific life domain when they feel more confident in another domain, and do they tend to feel insecure in one specific life domain when they feel insecure in another domain?

Conducting Partial Correlations

To evaluate fully the results of the questions, we must compute both bivariate and partial correlations. Both sets of statistics can be computed with the Partial Correlation procedure. To compute these bivariate and partial correlations, follow these steps:

1. Click **Analyze**, click **Correlate**, and then click **Partial**. You will see the Partial Correlations dialog box as shown in Figure 195.

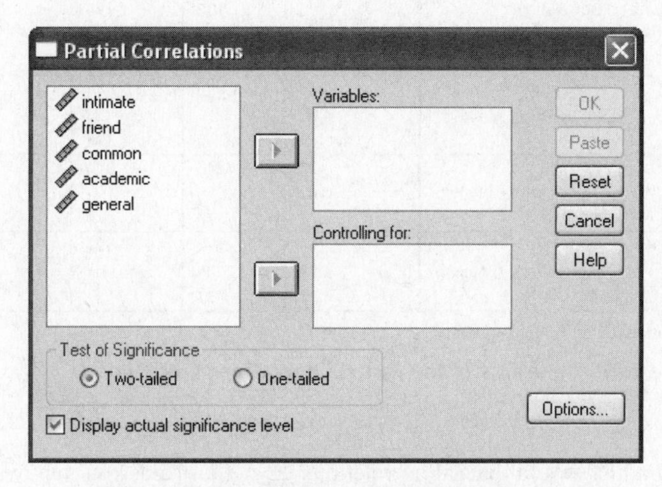

***Figure 195.* The Partial Correlations dialog box.**

2. Holding down on the Ctrl key, click **intimate**, **friend**, **common**, and **academic**.
3. Click ▶ to move them to the Variables box.
4. Click **general**, then click ▶ to move it to the Controlling for box.
5. Make sure the **Two-tailed** option is selected in the Test of Significance box (unless you have some a priori reason to select one-tailed).
6. Make sure **Display actual significance level** is selected.
7. Click **Options**, then click **Means and standard deviations** and **Zero-order correlations** in the Statistics box.
8. Click **Continue**.
9. Click **OK**.

Selected SPSS Output for Partial Correlations

The partial correlations from the output are shown in Figure 196 (on page 270). The output from the Partial Correlations procedure gives the means and standard deviations, the bivariate correlations among all the self-concept variables, and the partial correlations.

Each cell in the partial correlation table presents the partial correlation coefficient, the degrees of freedom for the significance test (in parentheses), and the *p* value. If the significance level is set at .05 for each partial correlation coefficient and a *p* value is less than .05, the null hypothesis that the population partial correlation coefficient is equal to zero is rejected, and the partial correlation coefficient is significant. The values in the lower-left triangle of the matrix are the same as the values in the upper-right triangle.

Correlations

Control Variables			Intimate Relationships	Friends	Common Sense	Academic Intelligence	General
-none-[a]	Intimate Relationships	Correlation	1.000	.552	.351	.218	.393
		Significance (2-tailed)	.	.000	.001	.052	.000
		df	0	78	78	78	78
	Friends	Correlation	.552	1.000	.462	.244	.546
		Significance (2-tailed)	.000	.	.000	.029	.000
		df	78	0	78	78	78
	Common Sense	Correlation	.351	.462	1.000	.400	.525
		Significance (2-tailed)	.001	.000	.	.000	.000
		df	78	78	0	78	78
	Academic Intelligence	Correlation	.218	.244	.400	1.000	.261
		Significance (2-tailed)	.052	.029	.000	.	.019
		df	78	78	78	0	78
	General	Correlation	.393	.546	.525	.261	1.000
		Significance (2-tailed)	.000	.000	.000	.019	.
		df	78	78	78	78	0
General	Intimate Relationships	Correlation	1.000	.438	.186	.130	
		Significance (2-tailed)	.	.000	.102	.253	
		df	0	77	77	77	
	Friends	Correlation	.438	1.000	.246	.126	
		Significance (2-tailed)	.000	.	.029	.268	
		df	77	0	77	77	
	Common Sense	Correlation	.186	.246	1.000	.321	
		Significance (2-tailed)	.102	.029	.	.004	
		df	77	77	0	77	
	Academic Intelligence	Correlation	.130	.126	.321	1.000	
		Significance (2-tailed)	.253	.268	.004	.	
		df	77	77	77	0	

a. Cells contain zero-order (Pearson) correlations.

Figure 196. **The results of the Partial Correlations procedure.**

When you conduct several partial correlations, as we just did, you may want to consider a corrected significance level to minimize the chances of incorrectly rejecting the null hypotheses (Type I errors) across the multiple tests. One method that might be used is the Bonferroni approach, which requires dividing .05 by the number of computed partial correlations. A partial correlation coefficient is not significant unless its p value is less than the corrected significance level. In our example, a p value must be less than .05 divided by 6, or .008, to be declared significant. Using this criterion, two of the six partial correlations are significant.

Using SPSS Graphs to Display the Results

The simplest partial correlation involves three variables including the control variable. A graph associated with such a partial correlation should display the results for all three variables. SPSS offers several possibilities. Let's look at steps for creating two types of scatterplots—a three-dimensional scatterplot and a simple scatterplot with markers.

Creating a 3-D Scatterplot

You could create a three-dimensional scatterplot for three of the self-concept scales, but, as with many three-dimensional graphs, it might be difficult to interpret. One solution is to dichotomize scores on the control variable and use the dichotomized score in the three-dimensional scatterplot.

We will illustrate this strategy by creating a three-dimensional scatterplot of the relationship between intimate relationship and friendship self-concepts for two levels of general self-concept, high and low. To create this graph, you will need to create a new variable (rgeneral) such that rgeneral = 0 if values on the general scale fall at or below its median score of 54 and rgeneral = 1 if values on the general scale fall above its median. Lesson 19 explains how to create a qualitative variable from a quantitative variable.

To produce the three-dimensional scatterplot, follow these steps:

1. Create the variable **rgeneral** by using the steps described in Lesson 19 for creating a qualitative variable from a quantitative variable.
2. Click **Graphs**, click **Legacy Dialogs**, and then click **Scatter/Dot**.
3. Click **3-D Scatter**, then click **Define**. You'll see the 3-D Scatterplot dialog box as shown in Figure 197.

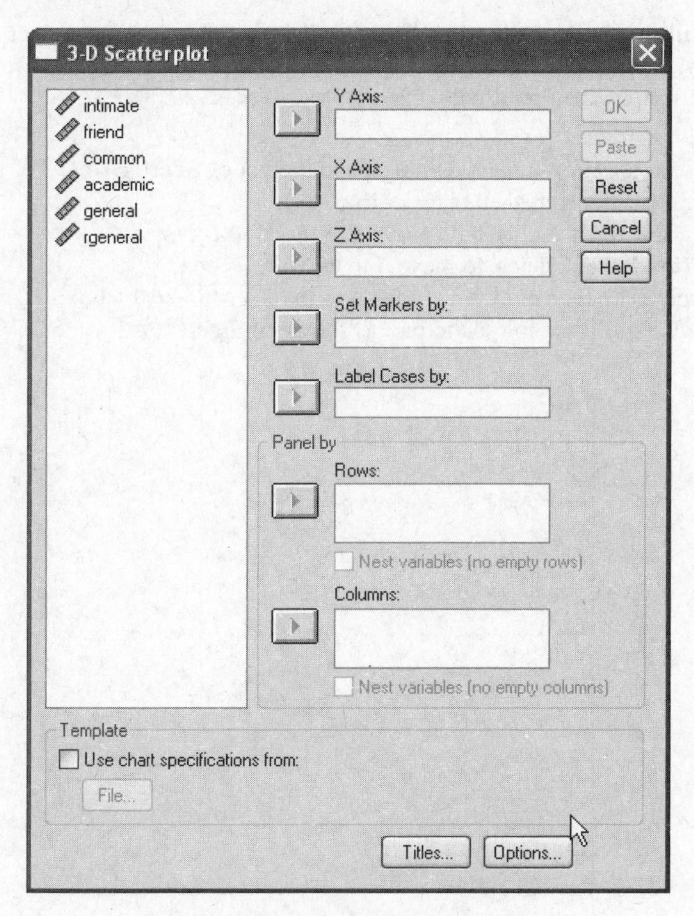

Figure 197. **The 3-D Scatterplot dialog box.**

4. Click **intimate**, then click ▶ to move it to the *Y*-Axis box.
5. Click **friend**, then click ▶ to move it to the *X*-Axis box.
6. Click **rgeneral**, then click ▶ to move it to the *Z*-Axis box.
7. Click **OK**. The edited graph is shown in Figure 198.

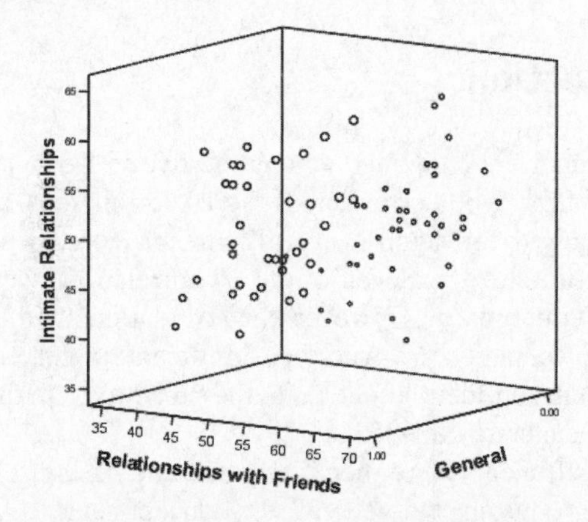

Figure 198. **A three-dimensional scatterplot.**

A moderate linear relationship appears to exist between the Friendship and the Intimate Relationships scales for both low and high scorers on the General scale. This interpretation is consistent with the partial correlation of .44 between Friendships and Intimate Relationships scales holding constant the General scale.

Creating a Simple Scatterplot with Markers

Another way to illustrate the relationship between these variables is to construct a simple bivariate scatterplot with separate markers for low and high scorers on the general self-concept variable. To create a simple scatterplot with markers, follow these steps:

1. Click **Graphs**, click **Legacy Dialogs**, and then click **Scatter/Dot**.
2. Click **Simple Scatter**, then click **Define**.
3. Click **intimate**, then click ▶ to move it to the *Y*-Axis box.
4. Click **friend**, then click ▶ to move it to the *X*-Axis box.
5. Click **rgeneral**, then click ▶ to move it to the Set Markers by box.
6. Click **OK**. You'll see the scatterplot as shown in Figure 199.

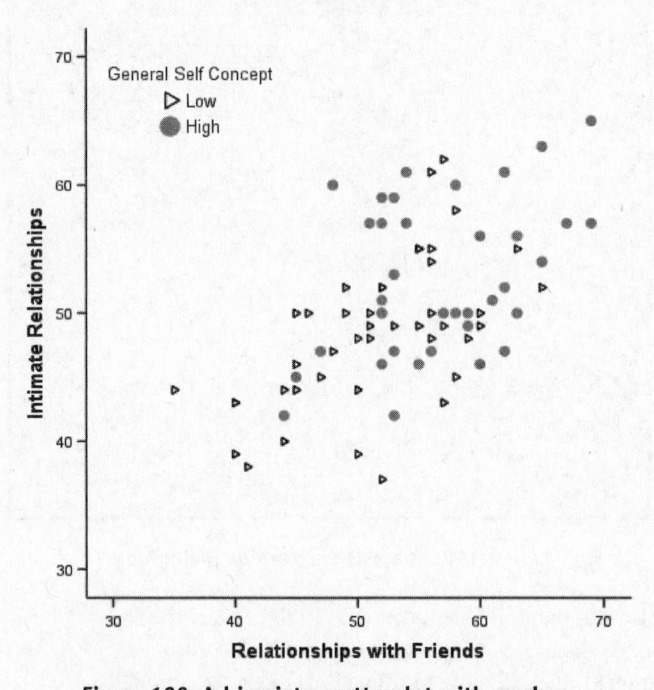

Figure 199. **A bivariate scatterplot with markers.**

An APA Results Section

Correlation coefficients were computed among the five self-concept scales. Using the Bonferroni approach to control for Type I error across the 10 correlations, a p value of less than .005 (.05/10 = .005) was required for significance. The results of the correlational analyses are presented in the top half of Table 38. Seven of the 10 correlations were statistically significant and were greater than or equal to .35. However, the correlations with self-concept in academic knowledge tended to be lower, and many were nonsignificant. In general, the results suggest that if men say that they are self-confident in one area, they tend to state that they are self-confident in other areas except for academic knowledge.

Partial correlation coefficients were then computed among the self-concept scales for specific life domains holding constant the general self-concept scale. A p value of less than .008

Table 38

Bivariate and Partial Correlations among the Self-Concept Scales (N = 80)

	Scholarly knowledge	Everyday knowledge	Friendships	Intimate relationships
		Bivariate correlations		
Everyday knowledge	.40*			
Friendships	.24	.46*		
Intimate relationships	.22	.35*	.55*	
General	.26	.52*	.55*	.39*
	Partial correlations controlling for general self-concept			
Everyday knowledge	.32*			
Friendships	.13	.25		
Intimate relationships	.13	.19	.44*	

* p < .005 for bivariate correlations and p < .008 for partial correlations

(.05/6 = .008) was required for significance while using the Bonferroni approach to control for Type I error across the six partial correlations. The partial correlations are reported in the second half of Table 38. Two of the six partial correlations were significant and moderately large in magnitude. One of the significant partial correlations assessed the correlation between the two knowledge self-concept measures, while the other evaluated the correlation between the two relationship-oriented self-concept measures. If general self-concept is the sole determinant of self-concept in specific areas, all the partial correlations would be equal to zero. The results do not support this hypothesis.

Alternative Analyses

Partial correlations may also be computed using the Multiple Regression procedure (Lesson 34).

Exercises

Exercises 1 through 4 are based on data in the data file named *Lesson 32 Exercise File 1* on the Web at http://www.prenhall.com/greensalkind. The data are from the following research study.

Betsy does not believe that quality of research by college professors leads to quality of teaching. She believes that any relationship between these two variables is due to work ethic. In other words, professors who are dedicated to their work will be better researchers and better teachers. Betsy has access to a sample of 50 social science professors who were teaching at the same university for a 10-year period. Over this 10-year period, the professors were evaluated on a 5-point scale on their quality as instructors and the quality of their courses. She has averaged these ratings to obtain an overall quality rating as an instructor (rating_1) and the overall quality of the course (rating_2) for each professor. Betsy also has the number of articles that each published during this time period (num_pubs) and the number of times these articles were cited by other authors (cites). Finally, Betsy has scores that reflect professors' work ethic (work_eth). These scores range from 1 to 50, with 50 indicating a very strong work ethic.

1. Evaluate Betsy's hypothesis by computings correlations between the two teaching variables and the two research variables controlling for the work ethic variable. From the output, identify the following:
   a. Bivariate correlation between the number of publications and number of citations

   b. p value associated with the bivariate relationship between number of publications and quality rating as an instructor
   c. Correlation between quality of the course and number of citations partialling out the effect of work ethic
2. What effect does partialling out the effects of work ethic have on the relationship between quality rating as an instructor and number of publications?
3. Write a Results section based on your analyses.
4. Create a 3-D scatterplot to display the relationship between quality rating as an instructor and number of publications.

Exercises 5 through 8 are based on data in the data file named *Lesson 32 Exercise File 2* on the Web at http://www.prenhall.com/greensalkind. The data are from the following research study.

A local newspaper reports that the number of violent crimes committed in a region is strongly related to the amount of beer drunk in the region. Susan, a sociologist interested in crime prevention, elects to investigate this relationship. She decides that the relationship between beer drinking and violent crimes is spurious and that a third variable, air temperature, could explain this relationship. She collects data on the amount of beer purchased (in hundreds of gallons), the number of violent crimes committed, and the average daily high temperatures for the month of July in 30 U.S. cities.

5. What is the bivariate correlation between amount of beer purchased and violent crimes?
6. What is the partial correlation between amount of beer purchased and violent crimes, holding temperature constant?
7. What should Susan conclude about the relationship between amount of beer purchased and the number of violent crimes committed?
8. Write a Results section based on your analyses. Be sure to include a graphical display of your data.

LESSON 33

Bivariate Linear Regression

For a bivariate linear regression problem, data are collected on an independent or predictor variable (X) and a dependent or criterion variable (Y) for each individual. Bivariate linear regression computes an equation that relates predicted Y scores ($\hat{Y}$) to X scores. The regression equation includes a slope weight for the independent variable, B_{slope}, and an additive constant, $B_{constant}$:

$$\hat{Y} = B_{slope}X + B_{constant}$$

Indices are computed to assess how accurately the Y scores are predicted by the linear equation.

This lesson focuses on applications in which both the predictor and criterion are quantitative variables. However, bivariate regression analysis may be used in other applications. For example, a predictor could have two levels like gender and be scored 0 for male and 1 for female. A criterion may also have two levels like pass–fail performance, scored 0 for fail and 1 for pass.

Linear regression can be used to analyze data from experimental or nonexperimental designs. If the data are collected using experimental methods (e.g., a tightly controlled study in which participants have been randomly assigned to different treatment groups), the X and Y variables may be referred to appropriately as the independent and the dependent variables, respectively. SPSS uses these terms. However, if the data are collected using nonexperimental methods (e.g., a study in which subjects are measured on a variety of variables), the X and the Y variables are more appropriately referred to as the predictor and the criterion, respectively.

Applications of Bivariate Linear Regression

We will illustrate two types of studies that can be analyzed using bivariate linear regression.

- Nonexperimental studies
- Experimental studies

Nonexperimental Study

Mary conducts a nonexperimental study to evaluate what she refers to as the strength–injury hypothesis. It states that overall body strength in elderly women determines the number and severity of accidents that cause bodily injury. If the results of her prediction study support her strength–injury hypothesis, she plans to conduct an experimental study to assess whether weight training reduces injuries in elderly women. In the prediction study, Mary collects data from 100 women who range in age from 60 to 75 years old at the time the study begins. The women initially undergo a number of measures that assess lower- and upper-body strength, and these measures are summarized using an overall index of body strength. Over the next five years, the women record each time they have an accident that results in a bodily injury and describe fully the extent of the injury. On the basis of these data, Mary calculates an overall injury index for each woman. Mary is interested in conducting a regression analysis with the overall index of body strength as the predictor (independent) variable and the overall injury index as the criterion (dependent) variable. Mary's SPSS data file has 100 cases and scores on six variables, five individual strength measures (quads, gluts, abdoms, grip, and arms) that are to be combined to yield an overall index of body strength and the criterion variable, the overall injury index.

Experimental Study

Jack is interested in the effect of coffee drinking on cigarette smoking. He gains the cooperation of 40 men and women who smoke 20 or more cigarettes and drink 2 or more cups of coffee a day. He randomly assigns 10 participants to each of the four treatment conditions: 0 cups of coffee, 1 cup of coffee, 2 cups of coffee, and 3 cups of coffee. Each participant is asked to sit in a small room and watch TV for 45 minutes. Those in the coffee drinking conditions are asked to drink their coffee at a uniform pace over the first 35 minutes of the 45-minute period. All subjects are told they may not smoke their own cigarettes, but they may smoke the cigarettes that are on the desk in the room. On the basis of a videotape of each session and the cigarette butts left in the ashtray on the desk, Jack estimates the number of millimeters of cigarettes smoked by each research subject. Jack is interested in conducting a regression analysis with the amount of coffee drunk as the independent variable and the number of millimeters of cigarettes smoked as the dependent variable. Jack's SPSS data file has scores on these two variables for the 40 cases.

Understanding Bivariate Linear Regression

A significance test can be conducted to evaluate whether X is useful in predicting Y. This test can be conceptualized as evaluating either of the following null hypotheses: the population slope weight is equal to zero or the population correlation coefficient is equal to zero.

The significance test can be derived under two alternative sets of assumptions, assumptions for a fixed-effects model and those for a random-effects model. The fixed-effects model is probably more appropriate for experimental studies, while the random-effects model seems more appropriate for nonexperimental studies. If the fixed-effects assumptions hold, linear or nonlinear relationships can exist between the predictor and criterion. On the other hand, if the random-effects assumptions hold, the only type of statistical relationship that can exist between two variables is a linear one.

Regardless of choice of assumptions, it is important to examine a bivariate scatterplot of the predictor and the criterion variables prior to conducting a regression analysis to assess if a nonlinear relationship exists between X and Y and to detect outliers. If the relationship appears to be nonlinear based on the scatterplot, you should not conduct a simple bivariate regression analysis but should evaluate the inclusion of higher-order terms (variables that are squared, cubed, and so on) in your regression equation. (See a regression textbook, such as Pedhazur, 1997.) Outliers should be checked to ensure that they were not incorrectly entered in the data set and, if correctly entered, to determine their effect on the results of the regression analysis.

Fixed-Effects Model Assumptions for Bivariate Linear Regression

As discussed earlier, there are two potential sets of assumptions to be considered—those for a fixed-effects model and those for a random-effects model. The following are assumptions for a fixed-effects model.

Assumption 1: The Dependent Variable Is Normally Distributed in the Population for Each Level of the Independent Variable

In many applications with a moderate or larger sample size, the test of the slope may yield reasonably accurate *p* values even when the normality assumption is violated. To the extent that population distributions are not normal and sample sizes are small, the *p* values may be invalid. In addition, the power of this test may be reduced if the population distributions are nonnormal.

Assumption 2: The Population Variances of the Dependent Variable Are the Same for All Levels of the Independent Variable

To the extent that this assumption is violated and the sample sizes differ among the levels of the independent variable, the resulting *p* value for the overall *F* test is not trustworthy.

Assumption 3: The Cases Represent a Random Sample from the Population, and the Scores Are Independent of Each Other from One Individual to the Next

The significance test for regression analysis will yield inaccurate *p* values if the independence assumption is violated.

Random-Effects Model Assumptions for Bivariate Linear Regression

Assumption 1: The X *and* Y *Variables Are Bivariately Normally Distributed in the Population*

If the variables are bivariately normally distributed, each variable is normally distributed ignoring the other variable and each variable is normally distributed at every level of the other variable. The significance test for bivariate regression yields, in most cases, relatively valid results in terms of Type I errors when the sample is moderate to large in size. If *X* and *Y* are bivariately normally distributed, the only type of relationship that exists between these variables is linear.

Assumption 2: The Cases Represent a Random Sample from the Population, and the Scores on Each Variable Are Independent of Other Scores on the Same Variable

The significance test for regression analysis will yield inaccurate *p* values if the independence assumption is violated.

Effect Size Statistics for Bivariate Linear Regression

This lesson focuses on using linear regression to evaluate how well a single independent variable predicts a dependent variable. However, linear regression is a more general procedure that assesses how well one or more independent variables predict a dependent variable. Consequently, SPSS reports strength-of-relationship statistics that are useful for regression analyses with multiple predictors. Four correlational indices are presented in the output for the Linear Regression procedure: the Pearson product-moment correlation coefficient (r), the multiple correlation coefficient (R), its squared value (R^2), and the adjusted R^2. However, there is considerable redundancy among these statistics for the single-predictor case: $R = |r|$, $R^2 = r^2$, and the adjusted R^2 is approximately equal to R^2. Accordingly, the only correlational indices we need to report in our manuscript for a bivariate regression model are r and r^2.

The Pearson product-moment correlation coefficient ranges in value from -1 to $+1$. A positive value suggests that as the independent variable *X* increases, the dependent variable *Y* increases. A zero value indicates that as *X* increases, *Y* neither increases nor decreases. A negative value indicates that as *X* increases, *Y* decreases. Values closer to -1 or $+1$ indicate stronger linear relationships. By convention, correlation coefficients of .10, .30, and .50, irrespective of sign, are interpreted as small, medium, and large coefficients, respectively. However, the interpretation of strength of relationship should depend on the research context.

By squaring r, we obtain an index that directly tells us how well we can predict Y from X. r^2 indicates the proportion of Y variance that is accounted for by its linear relationship with X. Alternatively, r^2 can be conceptualized as the proportion reduction in error that we achieve by including X in the regression equation in comparison with not including X in the regression equation.

Other strength-of-relationship indices may be reported for bivariate regression problems. For example, SPSS gives Std. Error of the Estimate on the output. The standard error of estimate is an index indicating how large the typical error is in predicting Y from X. It is a useful index over and above correlational indices because it indicates how badly we predict the dependent variable scores in the metric of these scores. In comparison, correlational statistics are unitless indices and, therefore, are abstract and difficult to interpret.

The Data Set

The data set used to illustrate bivariate linear regression is named *Lesson 33 Data File 1* on the Web at http://www.prenhall.com/greensalkind. It presents data from our strength–injury example. The variables in the data set are in Table 39.

Table 39
Variables in Lesson 33 Data File 1

Variable	Definition
quads	A measure of strength primarily associated with the quadriceps
gluts	A measure of strength of the muscles in the upper part of the back of the leg and the buttocks
abdoms	A measure of strength of the muscles of the abdomen and the lower back
arms	A measure of strength of the muscles of the arms and the shoulders
grip	An assessment of the hand-grip strength
injury	Overall injury index based on the records kept by the participants

The Research Question

The research question addresses the relationship between two variables. What is the linear equation that predicts the extent of physical injury from body strength for elderly women, and how accurately does this equation predict the extent of physical injuries?

Conducting a Bivariate Linear Regression Analysis

The data set includes the dependent variable for the regression analysis (injury) but does not include the independent variable, an overall index of body strength. This index can be calculated by creating z scores for the five strength measures and adding them together. See Lesson 19 for a more detailed explanation for creating an overall variable from standardized variables. To conduct a bivariate linear regression analysis, follow these steps:

1. Create the total strength variable named **ztotstr** by z-scoring the five individual strength measures and summing them.
2. Click **Analyze**, click **Regression**, then click **Linear**. You will see the Linear Regression dialog box shown in Figure 200.
3. Click **injury**, then click ▶ to move it to the Dependent box.
4. Click **ztotstr**, then click ▶ to have it appear in the Independent(s) box.
5. Click **Statistics**. You will see the Linear Regression: Statistics dialog box shown in Figure 201.

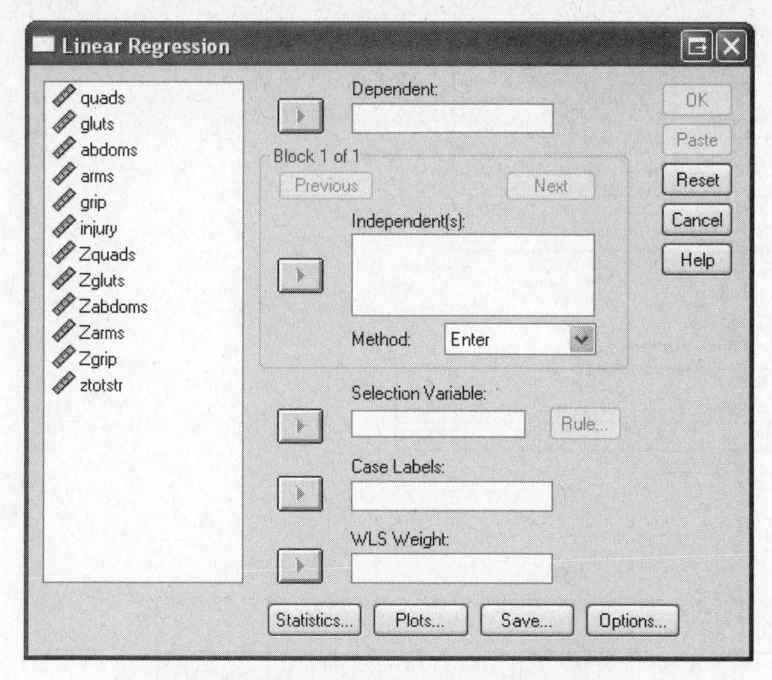

***Figure 200.* The Linear Regression dialog box.**

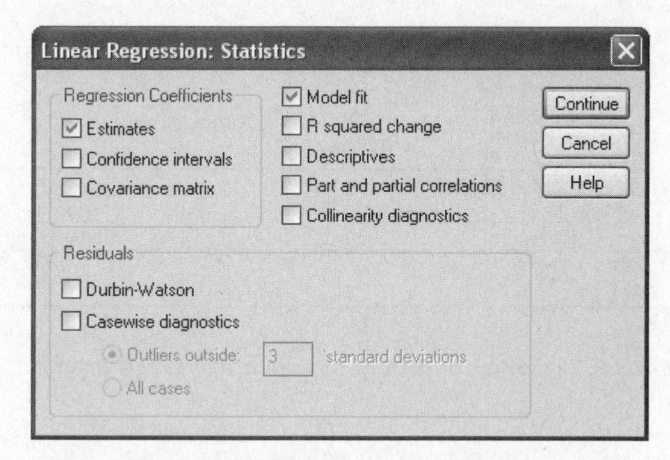

***Figure 201.* The Linear Regression: Statistics dialog box.**

6. Click **Confidence intervals** and **Descriptives**. Make sure that **Estimates** and **Model Fit** are also selected.
7. Click **Continue**.
8. Click **OK**.

Selected SPSS Output for Bivariate Linear Regression

The results of the bivariate linear regression analysis are shown in Figure 202 (on page 280). The *B*s, as labeled on the output in the Unstandardized Coefficients box, are the additive constant (145.80) and the slope weight (–4.89) of the regression equation used to predict the dependent variable from the independent variable. Accordingly, the regression or prediction equation is as follows:

$$\textit{Predicted Overall Injury} = -4.89\ \textit{Overall Strength} + 145.80$$

The slope weight indicates that greater overall strength predicts lower scores on the overall injury index. It should be noted that the 95% confidence interval for the slope is fairly wide (–7.74 to –2.04), but is negative throughout the range of the interval.

Correlations

		injury	ztotstr
Pearson Correlation	injury	1.000	-.325
	ztotstr	-.325	1.000
Sig. (1-tailed)	injury	.	.000
	ztotstr	.000	.
N	injury	100	100
	ztotstr	100	100

Model Summary

Model	R	R Square	Adjusted R Square	Std. Error of the Estimate
1	.325[a]	.106	.097	49.610

a. Predictors: (Constant), ztotstr

ANOVA[b]

Model		Sum of Squares	df	Mean Square	F	Sig.
1	Regression	28520.191	1	28520.191	11.588	.001[a]
	Residual	241193.8	98	2461.161		
	Total	269714.0	99			

a. Predictors: (Constant), ztotstr

b. Dependent Variable: injury

Coefficients[a]

Model		Unstandardized Coefficients		Standardized Coefficients	t	Sig.	95% Confidence Interval for B	
		B	Std. Error	Beta			Lower Bound	Upper Bound
1	(Constant)	145.800	4.961		29.389	.000	135.955	155.645
	ztotstr	-4.891	1.437	-.325	-3.404	.001	-7.743	-2.040

a. Dependent Variable: injury

***Figure 202.* The results of the bivariate linear regression analysis.**

A standardized regression equation can be computed if the independent and dependent variables are transformed to z scores with a mean of 0 and a standard deviation of 1:

$$Predicted\ Z_{Injury} = -.32\ Z_{Overall\ Strength}$$

For bivariate regression analysis based on standardized scores, the slope (in the standardized coefficients box) is always equal to the correlation coefficient and the additive constant must be equal to zero. The standardized slope is labeled beta on the SPSS output. Based on the magnitude of the correlation coefficient, we can conclude that overall strength is moderately related to injury level in this elderly sample. Eleven percent ($r^2 = .106$) of the variance of the injury index is associated with overall strength.

The hypothesis test of interest evaluates whether the independent variable predicts the dependent variable in the population. More specifically, it assesses whether the population correlation coefficient is equal to zero or, alternatively, whether the population slope is equal to zero. This significance test appears in two places for a bivariate regression analysis: the F test reported as part of the ANOVA table and the t test associated with the independent variable in the Coefficients table. They yield the same p value because they are identical tests: $F(1, 98) = 11.59, p < .01$ and $t(98) = -3.40, p < .01$. In addition, the fact that the 95% confidence interval for the slope does not contain the value of zero indicates that the hypothesis should be rejected at the .05 level.

Using SPSS Graphs to Display the Results

A variety of graphs have been suggested for interpreting linear regression results. Below, we discuss two types of graphs, the bivariate scatterplot and the plot of predicted and residual values.

Creating a Bivariate Scatterplot

The results of the bivariate regression analysis can be summarized using a bivariate scatterplot. Conduct the following steps to create a simple bivariate scatterplot:

1. Click **Graphs**, click **Legacy Dialogs**, and then click **Scatter/Dot**.
2. Click **Simple Scatter**, and then click **Define**.
3. Click **injury** and click ▶ to move it to the *Y*-axis box.
4. Click **ztotstr** and click ▶ to move it to the *X*-axis box.
5. Click **OK**.

Once you have created a scatterplot showing the relationship between the number of injuries and the total strength index, you can add a regression line by conducting the following steps:

1. Double-click on the chart to select it for editing, and maximize the chart editor.
2. Click on any of the data points in the scatterplot to highlight data points.
3. Click **Elements** from the main menu, and click on **Fit Line at Total**.
4. Click **Close** if a Properties dialog box is open. You will see the scatterplot (with some additional edits) in Figure 203.

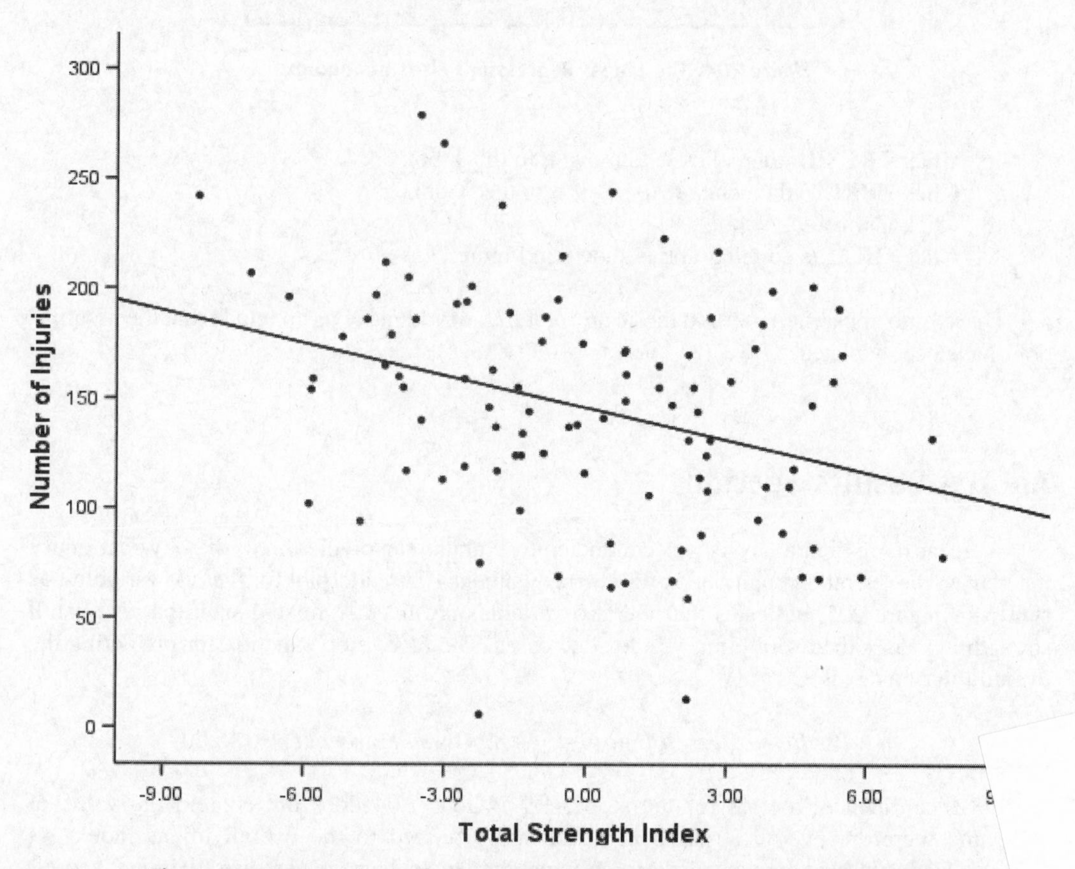

Figure 203. **Scatterplot between number of injuries and total body strength.**

An examination of the plot allows us to assess how accurately the regression equation pr dicts the dependent variable scores. In this case, the equation offers some predictability, b many points fall far off the line, indicating poor prediction for those points.

Creating a Plot of Predicted and Residual Values

The plot of predicted and residual values may form a pattern that indicates that an assumption has been violated. For example, a U-shaped plot would suggest that the two variables are nonlinearly related. Or, if the residuals are tightly clustered for some values of the predicted scores and widely varying for other values, one might suspect that the homogeneity-of-variance assumption has been violated. A plot of predicted and residual values can be created by following these steps:

1. Click **Analyze**, click **Regression**, and then click **Linear**.
2. The appropriate options should already be selected. If not, conduct steps 3 through 7 as described in the section labeled Conducting a Bivariate Linear Regression Analysis.
3. Click **Plots** in the Linear Regression dialog box. You will see the Linear Regression: Plots dialog box shown in Figure 204.

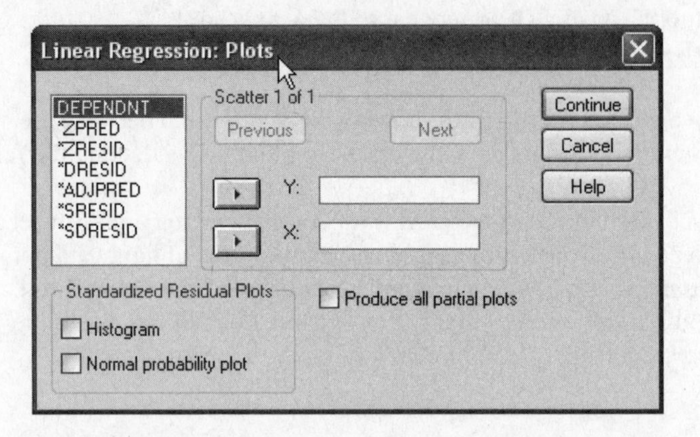

Figure 204. **The Linear Regression: Plots dialog box.**

4. Click **ZRESID**, then click ▶ to move it to the *Y* box.
5. Click **ZPRED**, then click ▶ to move it to the *X* box.
6. Click **Continue**.
7. Click **OK**. The edited graph is shown in Figure 205.

There is no apparent pattern to the scatterplot that would make us conclude that the assumptions have been violated.

An APA Results Section

A linear regression analysis was conducted to evaluate the prediction of the physical injury index from the overall strength index for elderly women. The scatterplot for the two variables, as shown in Figure 205, indicates that the two variables are linearly related such that as overall strength increases the overall injury index decreases. The regression equation for predicting the overall injury index is

$$\textit{Predicted Overall Injury} = -4.89\ \textit{Overall Strenght} + 145.80$$

The 95% confidence interval for the slope, −7.74 to −2.04 does not contain the value of zero, and therefore overall strength is significantly related to the overall injury index. As hypothesized, elderly women who are stronger tended to have lower overall injury scores. Accuracy in predicting the overall injury index was moderate. The correlation between the strength index and the injury index was −.32. Approximately 11% of the variance of the injury index was accounted for by its linear relationship with the strength index.

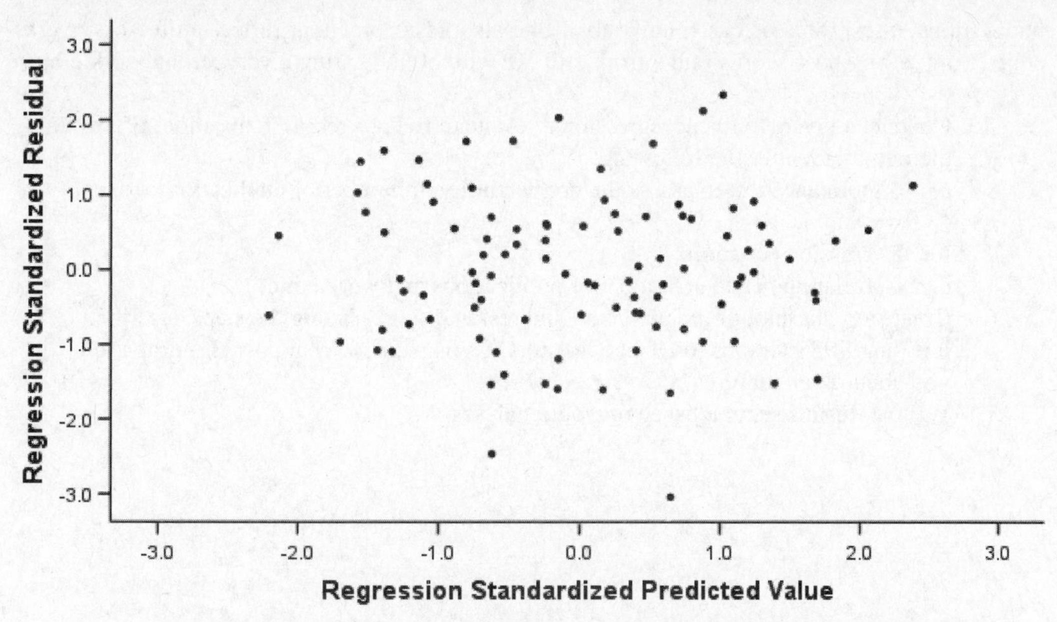

Figure 205. **Scatterplot depicting the relationship between standardized predicted and residual injury scores.**

Exercises

The data for Exercises 1 through 4 are in the data file named *Lesson 33 Exercise File 1* on the Web at http://www.prenhall.com/greensalkind. The data are based on the following research problem.

Peter was interested in determining if children who hit a bobo doll more frequently would display more or less aggressive behavior on the playground. He was given permission to observe 10 boys in a nursery school classroom. Each boy was encouraged to hit a bobo doll for 5 minutes. The number of times each boy struck the bobo doll was recorded (bobo). Next, Peter observed the boys on the playground for an hour and recorded the number of times each boy struck a classmate (peer).

1. Conduct a linear regression to predict the number of times a boy would strike a classmate from the number of times the boy hit a bobo doll. From the output, identify the following:
   a. Slope associated with the predictor
   b. Additive constant for the regression equation
   c. Mean number of times they struck a classmate
   d. Correlation between the number of times they hit the bobo doll and the number of times they struck a classmate
   e. Standard error of estimate
2. What is the relationship between the multiple *R* and the bivariate correlation between the predictor and the criterion?
3. Create a scatterplot of the relationship between the two variables. Plot the regression line on the graph. What can you tell from this graph about the predictability of the dependent variable?
4. Write a Results section based on your analyses.

The data for Exercises 5 through 7 are found in the data file named *Lesson 33 Exercise File 2* on the Web at http://www.prenhall.com/greensalkind. The data are based on the following research problem.

Betsy is interested in determining whether the number of publications by a professor can be predicted from work ethic. She has access to a sample of 50 social science professors who were teaching at the same university for a 10-year period. Betsy has collected data on the number of publications each professor has (num_pubs). She also has scores that reflect professors' work ethic (work_eth). These scores range from 1 to 50, with 50 indicating a very strong work ethic.

5. Conduct a bivariate linear regression to evaluate Betsy's research question. From the output, identify the following:
   a. Significance test to assess the predictability of number of publications from work ethic
   b. Regression equation
   c. Correlation between number of publications and work ethic
6. Create a scatterplot of the predicted and residual scores, using the steps described in Using SPSS Graphs to Display Results (on page 281). What does this graph tell you about your analyses?
7. Write a Results section based on your analyses.

LESSON

34

Multiple Linear Regression

In Lesson 33, you learned about conducting a bivariate linear regression where one variable can be used to predict another. With multiple regression analysis, each individual or case has scores on multiple independent variables (e.g., X_1, X_2, and X_3 if there are three independent variables) and on a dependent variable (Y). A predicted dependent variable ($\hat{Y}$) is formed that is a linear combination of the multiple independent variables. With three predictors, the linear combination or regression equation is

$$\hat{Y} = B_1X_1 + B_2X_2 + B_3X_3 + B_0$$

Here B_1 through B_3 are slope weights for the three independent variables X_1 through X_3 and B_0 is an additive constant. The values for B_0 through B_3 are calculated so that the actual dependent variable scores (Y) and the predicted dependent variable scores ($\hat{Y}$) are as similar as possible for the sample data.

The multiple correlation (R) is a strength-of-relationship index that indicates the degree that the predicted scores are correlated with the Y scores (observed scores) for a sample. The significance test for R evaluates whether the population multiple correlation coefficient is equal to zero (i.e., $\hat{Y}$ and Y are uncorrelated in the population).

Multiple regression/correlation analyses are extensions of bivariate regression/correlation analyses and are related to partial correlation analysis. In many applications, when researchers present multiple regression results, they also introduce bivariate and partial correlation results. Consequently, it is important to have read Lessons 31 through 33 on bivariate correlation coefficients, partial correlations, and bivariate linear regression before reading Lesson 34.

Multiple regression is used to analyze data from studies with experimental or nonexperimental designs. If data are collected using experimental methods (e.g., a tightly controlled study in which subjects have been randomly assigned to different treatment groups), the variables in a regression analysis may be called independent and dependent variables. SPSS uses these terms. However, if data are collected using nonexperimental methods (e.g., a study in which subjects are measured on a variety of variables), the variables in the regression analysis may be called the predictors and the criterion rather than the independent variables and the dependent variable, respectively.

Our presentation of multiple regression analysis focuses on applications in which both the independent and dependent variables are quantitative. However, the regression procedure may also be used if one or more of the independent variables or the dependent variable has only two values or levels. For example, a variable could be gender or pass–fail performance on a task. Also, a multiple regression analysis could be conducted with qualitative independent variables that have multiple categories by using coded variables (see Pedhazur, 1997).

Applications of Multiple Regression

We can differentiate multiple regression analyses based on whether data are from studies with experimental or nonexperimental designs. This distinction is useful because the design of the study suggests which set of assumptions is most applicable. We'll talk more about those issues in a moment.

We'll present three applications that require different analytical strategy. We differentiate the applications based on whether the predictors can be divided into sets and, if so, whether the sets can be ordered.

- One set of predictors
- Unordered sets of predictors
- Ordered sets of predictors

In studies with one set of predictors, the research questions do not differentiate among predictors and, therefore, the predictors are treated as a single set. In studies with unordered sets of predictors, the research question divides predictors into two or more sets, but there is no particular ordering of these sets. The multiple regression analysis for it examines the validity of each set of predictors, the incremental validity of each set of predictors over other sets of predictors, and the validity of all sets in combination. In studies with ordered sets of predictors, the research question indicates not only sets of predictors, but also suggests an order to the sets. The multiple regression analysis for the third application examines the validity of the first set of predictors, the incremental validity of the second set of predictors over and above the first set, and so on. In addition, the analysis yields results that assess the validity of all predictors.

One Set of Predictors

Mary conducts a study to evaluate her strength–injury hypothesis. Her hypothesis states that overall body strength in elderly women determines the number and severity of accidents these women have. In her field study to evaluate this hypothesis, she collects data on a group of 100 elderly women. The women initially undergo a number of measures that assess lower- and upper-body strength. Over the next five years, the women record each time they have an accident that results in a bodily injury and a full description of the injury. On the basis of these data, Mary calculates for these women scores on an overall injury index. In Lesson 33, Mary conducted analyses that related an overall index of body strength (the independent variable) to the overall injury index (the dependent variable).

Laurie, Mary's research colleague, feels that Mary should not use the overall strength index as the predictor variable but that she should use the individual strength measures as separate predictors. Laurie argues that some of the strength variables are bound to be better predictors than others; however, she makes no a priori hypotheses about what predictors will be better predictors. Based on Laurie's suggestion, Mary decides to conduct a multiple regression analysis with five predictor variables, assessing strength in quadriceps, gluteus/hamstrings, abdominals/lower back, arms/shoulders, and grip strength. The dependent variable remains the same, the overall injury index. Mary's SPSS data file has 100 cases and six variables containing scores on the five strength variables and the overall injury index.

Unordered Sets of Predictors

Once Mary begins to consider the usefulness of the individual strength measures rather than the overall strength index, she formulates a corollary to her main hypothesis: lower-body strength in comparison with upper-body strength is a better determinant of the number and severity of accidents that older women have that cause bodily injury. Therefore, Mary decides to divide her predictors into two sets: lower-body and upper-body strength predictors. The lower-body predictors include the quadriceps, the gluteus/hamstrings, and the abdominals/lower back. The upper-body predictors include the arms/shoulders and grip strength. Her SPSS data file contains the same six variables described in the first example. The regression analyses evaluate the contribution of upper body and lower body strength sets of predictors, each by itself and over and above the other set.

Ordered Sets of Predictors

Laurie reminds Mary that they collected not only strength measures at the beginning of the study but also age and measures of general medical history. Laurie believes that age and general medical history should be included in the regression analyses. She thinks that women who are having medical problems, partially as a function of their age, may be weaker. They also may be less agile, more likely to have serious accidents, and less likely to recover quickly from these accidents. If this disease-oriented hypothesis is true, then it is not the lack of strength that is causing the serious accidents. Instead, the medical difficulties generate both the lack of strength and the injuries from the accidents. In addition, strength training with women who have medical problems may not be an appropriate or effective intervention strategy. To investigate this alternative, disease-oriented hypothesis, Laurie suggests that Mary determine first if age and medical difficulties predict the overall injury index. If it does, then Mary needs to assess whether the strength indices predict the overall injury index over and above age and medical difficulties. Mary's SPSS data file contains not only the five strength variables and the overall injury index, but also variables containing age and scores from the medical history measure for all 100 women.

Understanding Multiple Regression

Although linear regression analysis can be conducted with a single independent variable, it becomes a more complex and delicate tool with multiple independent variables. How a multiple regression analysis is carried out depends on whether the independent variables can be divided into sets. For example, if we are interested in predicting on-the-job performance, we may be unable to differentiate our independent variables into different sets of variables. On the other hand, if we are interested in predicting adolescent behavior, we may be able to differentiate the independent variables into different sets of variables, one set of family variables and a second set of peer-group variables. By dividing the predictors into sets, we can attempt to answer questions like "Are family or peer-group variables more useful in predicting adolescent behavior?" or "Do the family variables help in the prediction of adolescent behavior over and above the peer-group variables and vice versa?"

The specific analytical steps taken in conducting multiple regression analyses also depend on whether the sets are unordered or ordered. If the sets are unordered, we might be interested in many types of relationships. With our adolescent behavior example, we could handle the family and peer-group variable sets as unordered. Under these conditions, we might conduct analyses to assess how well we predict adolescent behavior (1) from family variables, (2) from peer-group variables, (3) from family variables over and above peer-group variables, (4) from peer-group variables over and above family variables, and (5) from family and peer-group variables.

If the sets are ordered, we could focus our interests on a more limited number of relationships. For example, we may have a second data set in which we are interested in predicting late adolescent behavior from two sets of variables—childhood variables and early adolescent variables. Because of the longitudinal nature of the data, we could restrict our analyses to assess how well we predict late adolescent behavior (1) from childhood variables, (2) from early adolescent variables over and above childhood variables, and (3) from the combination of childhood and early adolescent variables.

Assumptions Underlying the Significance Test for the Multiple Correlation Coefficient

The significance test for a multiple correlation is based on two alternative sets of assumptions: assumptions for the fixed-effects model and assumptions for the random-effects model. The fixed-effects model is probably more appropriate for experimental studies, while the random-effects model seems more appropriate for nonexperimental studies.

Fixed-Effects Model Assumption 1: The Dependent Variable Is Normally Distributed in the Population for Each Combination of Levels of the Independent Variables

In many applications with a moderate or larger sample size, the test of a multiple correlation coefficient may yield reasonably accurate p values even when the normality assumption is violated. To the extent that population distributions are not normal and sample sizes are small, the p values may be invalid. In addition, the power of this test may be reduced if the population distributions are nonnormal.

Fixed-Effects Model Assumption 2: The Population Variances of the Dependent Variable Are the Same for All Combinations of Levels of the Independent Variables

To the extent that this assumption is violated and the sample sizes differ among the levels of the independent variable, the resulting p value for the overall F test is not trustworthy.

Fixed-Effects Model Assumption 3: The Cases Represent a Random Sample from the Population, and the Scores Are Independent of Each Other from One Individual to the Next

The F test for regression analyses yields inaccurate p values if the independence assumption is violated.

Random-Effects Model Assumption 1: The Variables Are Multivariately Normally Distributed in the Population

If the variables are multivariately normally distributed, each variable is normally distributed ignoring the other variables and each variable is normally distributed at every combination of values of the other variables. If the multivariate normality assumption is met, the only type of statistical relationship that can exist between variables is a linear one.

Random-Effects Model Assumption 2: The Cases Represent a Random Sample from the Population, and the Scores on Variables Are Independent of Other Scores on the Same Variables

The F test for regression analyses yields inaccurate p values if the independence assumption is violated.

It is important to evaluate whether nonlinear relationships exist between the predictors and the criterion, regardless of choice of models. With the fixed-effects model, either linear or nonlinear relationships may exist between the predictors and the criterion. With the random-effects model, nonlinear relationships may be present if the assumption of multivariate normality is violated. Accordingly, for all data sets, we should assess whether nonlinear relationships exist. At an absolute minimum, scatterplots between each predictor and the criterion should be scrutinized for nonlinearity.

Effect Size Statistics for Multiple Regression

We limit our discussion to two types of effect sizes for multiple regression analysis: multiple correlation indices to assess the overall effect of the predictors on the dependent variable and part and partial correlations to assess the relative effects of individual predictors. However, other indices are useful to understand how well we can predict with a multiple regression equation. For example, the standard error of estimate indicates how badly we predict the criterion variable in the units of this variable. (See Lesson 33.)

Multiple Correlation Indices

SPSS computes a multiple correlation (R), a squared multiple correlation (R^2), and an adjusted squared multiple correlation (R^2_{adj}). All three indices assess how well the linear combination of predictor variables in the regression analysis predicts the criterion variable. In addition, SPSS calculates changes in R^2 if there are multiple sets of predictors (blocks).

The multiple correlation is a Pearson product-moment correlation coefficient between the predicted criterion scores ($\hat{Y}$) and the actual criterion scores (Y). The multiple correlation could be symbolized as $r_{Y\hat{Y}}$. Because $\hat{Y}$ is based on results of multiple regression analysis, $r_{Y\hat{Y}}$ has some special properties and is denoted as R.

R ranges in value from 0 to 1. A value of 0 means there is no linear relationship between the predicted scores and the criterion scores. A value of 1 implies that the linear combination of the predictor variables perfectly predicts the criterion variable; that is, $\hat{Y}$ is equal to Y for all individuals in the sample. Values between 0 and 1 indicate a less-than-perfect linear relationship between the predicted and criterion scores, but that one or more of the Xs are useful to some extent in predicting Y.

To interpret the values of R between 0 and 1, R may be squared and multiplied by 100. The resulting statistic may now be interpreted as the percent of criterion variance accounted for by the linear combination of the predictors. For example, if an R is .5, R^2 is .25, and we can make the following interpretation: The R^2 of .25 indicates that 25% of the criterion variance can be accounted for by its linear relationship with the predictor variables.

The sample multiple correlation and the squared multiple correlation are biased estimates of their corresponding population values. The sample R^2 typically overestimates the population R^2 and needs to be adjusted downward. The adjusted R^2 reported by SPSS makes the adjustment by assuming a fixed-effects model that uses the equation

$$\text{Adjusted } R^2 = 1 - (1 - R^2)\frac{N - 1}{N - k - 1}$$

where N = sample size and k = number of predictors. Based on this equation, it is apparent that the sample R^2 shows greater bias when the sample size is small and the number of predictors is large. When comparing regression equations with different numbers of predictors and a small sample size, it is particularly important to report the adjusted R^2 as well as the R^2 because of the effect of the number of predictors on R^2.

A change in R^2 is the difference between an R^2 for one set of predictors and an R^2 for a subset of these predictors. An example of this change would be an R^2 based on a combination of two sets of predictors (e.g., lower-body and upper-body strength variables) minus an R^2 based on only one of the two sets of predictors (e.g., upper-body strength variables). The value of the change in R^2 ranges from 0 to 1 and is interpreted as an increment in the percent of criterion variance accounted for by including two sets of predictors in the regression equation versus only a single set of predictors. For example, if an R^2 change is .25, you can interpret this finding to mean that an additional 25% of the variance of the criterion (e.g., overall injury index) is contributed by having both sets of variables in the regression equation predictors (e.g., lower-body and upper-body strength variables) versus having only a single set of predictor variables in the equation (e.g., upper-body strength variables). Sometimes more abbreviated, less accurate language is used to describe this R^2 change: "An additional 25% of the criterion (e.g., overall injury index) variance is contributed by one set of predictors (e.g., lower-body strength variables) over and above a second set of predictors (e.g., upper-body strength variables)."

The Relative Importance of Predictors: Using Part and Partial Correlations

On the basis of correlation-regression analyses, we may attempt to reach conclusions about the relative importance of variables in predicting a criterion. These judgments may be made based on the size of the bivariate correlation between each predictor and the criterion, the standardized regression weight for each predictor within a regression equation, the part or partial correlations between each predictor and the criterion (partialling out the effect of all other predictors in the regression equation), or any combinations of these indices. None of these indices, alone or together, however, yields an unequivocal answer to the question of relative importance.

Now we will discuss the interpretation of part and partial correlations, two indices used in reaching a decision about the importance of any one predictor. We'll discuss the interpretation of standardized regression weights in the section labeled Selected SPSS Output for One Set of Predictors.

As explained in Lesson 32, a partial correlation is the correlation between two variables while partialling out the effects of one or more control variables. Within the regression procedure, a partial correlation is computed between a predictor and criterion variable partialling out the effects of all other predictor variables. In contrast, the part correlation gives the correlation between a predictor and a criterion variable, partialling out the effects of all other predictors in the regression equation from the predictor but not the criterion.

A part correlation or a partial correlation between a predictor and a criterion variable may range in value from -1 to $+1$. A positive sign indicates that the predictor and criterion are directly related after partialling out the effects of the other predictors, while a negative sign indicates that they are inversely related after the partialling process.

The same decision about the relative importance of predictors is made using either partial or part correlation. The relative magnitudes of the partial correlations between each predictor and the criterion (partialling out the effects of the other predictors) are identical to the relative magnitudes of the part correlations.

The Data Set

The data we will use to illustrate multiple regression comes from the strength and injury study described earlier in this lesson. They are in the data file named *Lesson 34 Data File 1* on the Web at http://www.prenhall.com/greensalkind. Table 40 shows the variables in the data set.

Table 40
Variables in Lesson 34 Data File 1

Variable	Definition
quads	A measure of strength primarily associated with the quadriceps
gluts	A measure of strength of the muscles in the upper part of the back of the leg and the buttocks
abdoms	A measure of strength of the muscles of the abdomen and the lower back
arms	A measure of strength of the muscles of the arms and the shoulders
grip	An assessment of the hand-grip strength
injury	Overall injury index based on the records kept by the participants
age	The age of the woman when the study began
medindex	A general medical history index based on information obtained from an interview about the health of the woman at the beginning of the study

The Research Question

Research questions are asked for the three different applications we described earlier.

1. One set of predictors: How accurately can a physical injury index be predicted from a linear combination of strength measures for elderly women?
2. Two unordered sets of predictors: How well do the lower-body strength measures predict the total injury index for elderly women? How well do the upper-body strength measures predict the total injury index for elderly women? For elderly

women, how well do the lower-body strength measures predict the total injury index over and above the upper-body strength measures? For elderly women, how well do the upper-body strength measures predict the total injury index over and above the lower-body strength measures?
3. Two ordered sets of predictors: How well do previous medical difficulties and age predict total injuries for elderly women? For elderly women, how well do the strength measures predict total injuries controlling for previous medical difficulties and age?

Conducting a Multiple Regression

We will demonstrate how to conduct analyses for a single set of predictors, two unordered sets of predictors, and two ordered sets of predictors.

Conducting Multiple Regression with One Set of Predictors

To conduct the multiple regression analysis with one set of predictors, follow these steps:

1. Click **Analyze**, click **Regression**, and click **Linear**.
2. Click **injury**, and click ▶ to move it to the Dependent box.
3. Holding down the Ctrl key, click **quads**, **gluts**, **abdoms**, **arms**, and **grip**, and click ▶ to have them appear in the Independent(s) box.
4. Click **Statistics**. The Linear Regression: Statistics dialog box is shown in Figure 206. The default options of **Estimates** and **Model Fit** should already be selected.

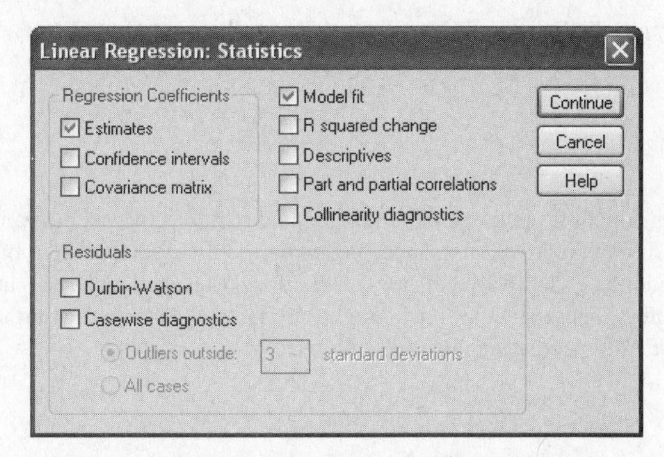

***Figure 206.* The Linear Regression: Statistics dialog box.**

5. Click **Confidence intervals**, **Descriptives**, and **Part and partial correlations**.
6. Click **Continue**.
7. Click **OK**.

Selected SPSS Output for One Set of Predictors

The results of the multiple regression analysis are shown in Figure 207 (on page 292). The descriptive statistics and the bivariate correlations are not presented. The regression equation with all five strength predictors was significantly related to the injury index, $R^2 = .18$, adjusted $R^2 = .14$, $F(5, 94) = 4.18$, $p = .002$. The *B*s, as labeled on the output, are the weights associated with the regression equation. According to these *B* weights, the regression equation is as follows:

$$Predicted\ Injury = .63\ Quads - 3.24\ Gluts - .56\ Abdoms - 1.13\ Arms + .79\ Grip + 260.39$$

Model Summary

Model	R	R Square	Adjusted R Square	Std. Error of the Estimate	Change Statistics				
					R Square Change	F Change	df1	df2	Sig. F Change
1	.426[a]	.182	.138	48.450	.182	4.180	5	94	.002

a. Predictors: (Constant), grip, quads, gluts, arms, abdoms

ANOVA[b]

Model		Sum of Squares	df	Mean Square	F	Sig.
1	Regression	49058.061	5	9811.612	4.180	.002[a]
	Residual	220655.9	94	2347.404		
	Total	269714.0	99			

a. Predictors: (Constant), grip, quads, gluts, arms, abdoms

b. Dependent Variable: injury

Coefficients[a]

Model		Unstandardized Coefficients		Standardized Coefficients	t	Sig.	Correlations		
		B	Std. Error	Beta			Zero-order	Partial	Part
1	(Constant)	260.393	30.170		8.631	.000			
	quads	.628	.645	.116	.973	.333	-.162	.100	.091
	gluts	-3.245	1.038	-.360	-3.125	.002	-.393	-.307	-.292
	abdoms	-.563	.674	-.097	-.836	.406	-.232	-.086	-.078
	arms	-1.130	.702	-.185	-1.609	.111	-.243	-.164	-.150
	grip	.794	1.083	.079	.733	.465	-.099	.075	.068

a. Dependent Variable: injury

***Figure 207*. The results of the multiple linear regression analysis for a single set of predictors.**

Although this equation yields predicted dependent variable scores, the weights are not useful for understanding the relative importance of the predictors. Weights are more interpretable if the independent and dependent variables are standardized to have a mean of 0 and a standard deviation of 1 (*z* scores). These standardized weights are labeled Beta on the output. The prediction equation for the standardized variables is as follows:

$$Z_{Predicted\ Injury} = .12\ Z_{Quads} - .36\ Z_{Gluts} - .10\ Z_{Abdoms} - .18\ Z_{Arms} + .08\ Z_{Grip}$$

It should be noted that the next set of statistics—the *t* and the Sig—are *t* values and *p* values to evaluate the significance of the B weights. These *t* and *p* values can also be used to assess the significance of the beta weights, the part correlations, and the partial correlations. The other three indices—the bivariate correlations (labeled Zero-order), the part correlations (labeled Part), the partial correlations (labeled Partial)—are presented in the Coefficients box next to the Sig. and the *t*-value.

Conducting Multiple Regression with Two Unordered Sets of Predictors

We must determine how well the criterion is predicted by each set of predictors, which in this example are the lower-body and upper-body measures of strength. In addition, we want to determine how well each set of variables predicts the criterion over and above the other set. We will do this by examining the change in R^2.

To evaluate how well the criterion is predicted by the lower-body strength measures (called Set 1), and by the upper-body strength measures (Set 2) over and above the lower-body strength measures, follow the steps on page 293.

1. Click **Analyze**, click **Regression**, and click **Linear**.
2. Click **Reset** to clear the dialog box.
3. Holding down the Ctrl key, click **quads**, **gluts**, and **abdoms**, and click ▶ to have them appear in the Independent(s) box. This is the first set of predictors.
4. Click **Next**.
5. Holding down the Ctrl key, click **arms** and **grips**, and click ▶ to have them appear in the Independent(s) box. This is the second set of predictors.
6. Click **injury**, and click ▶ to move it to the Dependent box.
7. Click **Statistics**. The default options of **Estimates** and **Model Fit** should already be selected.
8. Click **Confidence intervals**, **R Squared Change**, **Descriptives**, and **Part and partial correlations**.
9. Click **Continue**.
10. Click **OK**.

Because the sets are unordered, you will also want to evaluate how well the criterion is predicted by the upper-body strength measures (Set 2) and how well the criterion is predicted by the lower-body strength measures (Set 1) over and above the upper-body strength measures. To conduct these analyses, follow the same steps as before, except in Steps 3 and 5, reverse the sets of predictors.

Selected SPSS Output for Two Unordered Sets of Predictors

The results of the first run are shown in Figure 208 (on page 294). The analysis for the first run evaluates how well the criterion is predicted by the lower-body strength measures (Set 1) and how well the criterion is predicted by the upper-body strength measures (Set 2) over and above the lower-body strength measures. The relationship between the lower-body measures and the injury index was significant, $R^2 = .16$, adjusted $R^2 = .13$, $F(3, 96) = 6.07$, $p < .01$. The upper-body strength measures did not predict significantly over and above the lower-body measures, R^2 change $= .02$, $F(2, 94) = 1.29$, $p = .28$. The numerator degrees of freedom is the number of predictors in a set, while the denominator degrees of freedom is the degrees of freedom associated with the residual source.

The results of the second run are not shown. If you conducted these analyses, you discovered that the regression equation with the upper-body measures as predictors was not significant, $R^2 = .06$, adjusted $R^2 = .04$, $F(2, 97) = 3.06$, $p = .051$. The lower-body strength measures predicted significantly over and above the upper-body measures, R^2 change $= .12$, $F(3, 94) = 4.69$, $p < .01$.

Conducting Multiple Regression with Two Ordered Sets of Predictors

To answer the research questions posed in the third application, we must evaluate whether the total number of injuries is predicted by previous medical difficulties and age as well as whether the strength measures predict overall injuries after controlling for the effects of previous medical difficulties and age. To conduct the analyses, follow these steps:

1. Click **Analyze**, click **Regression**, and click **Linear**.
2. Click **Reset** to clear the dialog box.
3. Holding down the Ctrl key, click **medindex** and **age**, and click ▶ to have them appear in the Independent(s) box. This is the first set of predictors.
4. Click **Next**.
5. Holding down the Ctrl key, click **quads**, **gluts**, **abdoms**, **arms**, and **grips**, and click ▶ to have them appear in the Independent(s) box. This is the second set of predictors.
6. Click **injury**, and click ▶ to move it to the Dependent box.
7. Click **Statistics**. The default options of **Estimates** and **Model Fit** should already be selected.
8. Click **Confidence intervals**, **R Squared Change**, **Descriptives**, and **Part and partial correlations**.
9. Click **Continue**.
10. Click **OK**.

Model Summary

Model	R	R Square	Adjusted R Square	Std. Error of the Estimate	Change Statistics				
					R Square Change	F Change	df1	df2	Sig. F Change
1	.399[a]	.159	.133	48.598	.159	6.066	3	96	.001
2	.426[b]	.182	.138	48.450	.023	1.294	2	94	.279

a. Predictors: (Constant), abdoms, gluts, quads

b. Predictors: (Constant), abdoms, gluts, quads, grip, arms

ANOVA[c]

Model		Sum of Squares	df	Mean Square	F	Sig.
1	Regression	42981.028	3	14327.009	6.066	.001[a]
	Residual	226733.0	96	2361.802		
	Total	269714.0	99			
2	Regression	49058.061	5	9811.612	4.180	.002[b]
	Residual	220655.9	94	2347.404		
	Total	269714.0	99			

a. Predictors: (Constant), abdoms, gluts, quads

b. Predictors: (Constant), abdoms, gluts, quads, grip, arms

c. Dependent Variable: injury

Coefficients[a]

Model		Unstandardized Coefficients		Standardized Coefficients	t	Sig.	95% Confidence Interval for B		Correlations		
		B	Std. Error	Beta			Lower Bound	Upper Bound	Zero-order	Partial	Part
1	(Constant)	250.546	29.615		8.460	.000	191.761	309.331			
	quads	.355	.624	.066	.570	.570	-.883	1.593	-.162	.058	.053
	gluts	-3.489	1.016	-.387	-3.433	.001	-5.506	-1.472	-.393	-.331	-.321
	abdoms	-.454	.672	-.078	-.676	.501	-1.788	.879	-.232	-.069	-.063
2	(Constant)	260.393	30.170		8.631	.000	200.491	320.295			
	quads	.628	.645	.116	.973	.333	-.654	1.909	-.162	.100	.091
	gluts	-3.245	1.038	-.360	-3.125	.002	-5.306	-1.183	-.393	-.307	-.292
	abdoms	-.563	.674	-.097	-.836	.406	-1.902	.775	-.232	-.086	-.078
	arms	-1.130	.702	-.185	-1.609	.111	-2.524	.264	-.243	-.164	-.150
	grip	.794	1.083	.079	.733	.465	-1.356	2.944	-.099	.075	.068

a. Dependent Variable: injury

Figure 208. **Selected results of the multiple linear regression analysis for two unordered sets of predictors.**

Selected SPSS Output for Two Ordered Sets of Predictors

The results of the analysis are shown in Figure 209. The results present an evaluation of how well the number of injuries is predicted by previous medical difficulties and age (Set 1), and how well the set of strength measures (Set 2) predicts injuries over and above medical difficulties and age. The first set of predictors, previous medical difficulties and age, accounted for a significant amount of the injury variability, $R^2 = .16$, $F(2, 97) = 9.35$, $p < .01$. The five strength measures accounted for a significant proportion of the injury variance after controlling for the effects of medical history and age, R^2 change $= .15$, $F(5, 92) = 3.97$, $p < .01$.

The research question for this example was "How well do the strength measures predict total injuries, controlling for previous medical difficulties and age?" This question indicates that the sets of predictors should be ordered such that medical difficulties and age are entered first in the equation and then the strength predictors are added to these initial two predictors. The research question is addressed by the R^2 change for the two equations. Because these

Model Summary

Model	R	R Square	Adjusted R Square	Std. Error of the Estimate	Change Statistics				
					R Square Change	F Change	df1	df2	Sig. F Change
1	.402[a]	.162	.144	48.282	.162	9.349	2	97	.000
2	.557[b]	.310	.258	44.967	.149	3.966	5	92	.003

a. Predictors: (Constant), medindex, age

b. Predictors: (Constant), medindex, age, grip, quads, abdoms, arms, gluts

ANOVA[c]

Model		Sum of Squares	df	Mean Square	F	Sig.
1	Regression	43589.135	2	21794.568	9.349	.000[a]
	Residual	226124.9	97	2331.184		
	Total	269714.0	99			
2	Regression	83683.975	7	11954.854	5.912	.000[b]
	Residual	186030.0	92	2022.065		
	Total	269714.0	99			

a. Predictors: (Constant), medindex, age

b. Predictors: (Constant), medindex, age, grip, quads, abdoms, arms, gluts

c. Dependent Variable: injury

Coefficients[a]

Model		Unstandardized Coefficients		Standardized Coefficients	t	Sig.	Correlations		
		B	Std. Error	Beta			Zero-order	Partial	Part
1	(Constant)	-244.230	104.141		-2.345	.021			
	age	3.628	1.541	.225	2.353	.021	.290	.232	.219
	medindex	284.142	95.017	.286	2.990	.004	.337	.291	.278
2	(Constant)	-141.553	101.652		-1.393	.167			
	age	5.331	1.501	.330	3.553	.001	.290	.347	.308
	medindex	89.909	99.132	.090	.907	.367	.337	.094	.079
	quads	.540	.600	.100	.900	.371	-.162	.093	.078
	gluts	-3.386	1.042	-.375	-3.250	.002	-.393	-.321	-.281
	abdoms	-.682	.627	-.117	-1.087	.280	-.232	-.113	-.094
	arms	-.842	.657	-.138	-1.281	.203	-.243	-.132	-.111
	grip	.805	1.005	.081	.801	.425	-.099	.083	.069

a. Dependent Variable: injury

Figure 209. **The results of the multiple linear regression analysis for two ordered sets of predictors.**

analyses answer the research question, it is unnecessary to reverse the order of the two sets of variables and to reanalyze the data.

Using SPSS Graphs to Display the Results

Scatterplots can be used to display the relationships between variables. To look at the relationship between two variables, a bivariate scatterplot (as demonstrated in Lesson 33) can be created for each pair of variables. The matrix scatterplot described in Lesson 31 can be used to look at the relationships between all pairs of variables. To look at relationships among three variables, the 3-D scatterplots described in Lesson 32 can be created.

Other types of graphs are used to evaluate the assumptions of multiple linear regression. They are quite complex and beyond the scope of this book.

Three APA Results Sections

Results sections were written for each of the three applications.

Results for One Set of Predictors

A multiple regression analysis was conducted to evaluate how well the strength measures predicted physical injury level. The predictors were the five strength indices, while the criterion variable was the overall injury index. The linear combination of strength measures was significantly related to the injury index, $F(5, 94) = 4.18$, $p < .01$. The sample multiple correlation coefficient was .43, indicating that approximately 18% of the variance of the injury index in the sample can be accounted for by the linear combination of strength measures.

In Table 41, we present indices to indicate the relative strength of the individual predictors. All the bivariate correlations between the strength measures and the injury index were negative, as expected, and three of the five indices were statistically significant ($p < .05$). Only the partial correlation between the strength measure for the gluteus/hamstring muscles and the injury index was significant. On the basis of these correlational analyses, it is tempting to conclude that the only useful predictor is the strength measure for the gluteus/hamstring muscles. It alone accounted for 15% ($-.392 = .15$) of the variance of the injury index, while the other variables contributed only an additional 3% (18% − 15% = 3%). However, judgments about the relative importance of these predictors are difficult because they are correlated. The correlations among the strength measures, except grip, ranged from .37 to .52.

Table 41

The Bivariate and Partial Correlations of the Predictors with Injury Index

Predictors	Correlation between each predictor and the injury index	Correlation between each predictor and the injury index controlling for all other predictors
Quadriceps	−.16	.10
Gluteus/Hamstrings	−.39**	−.31**
Abdominals/Lower back	−.23*	−.09
Arms/Shoulder	−.24*	−.16
Grip	−.10	.08

* $p < .05$, ** $p < .01$

Results for Two Unordered Sets of Predictors

Two multiple regression analyses were conducted to predict the overall injury index. One analysis included the three lower-body strength measures as predictors (quadriceps, gluteus/hamstrings, and the abdominal/lower back), while the second analysis included the two upper-body strength measures (arms/shoulders and grip strength). The regression equation with the lower-body measures was significant, $R^2 = .16$, adjusted $R^2 = .13$, $F(3, 96) = 6.07$, $p < .01$. However, the regression equation with the upper-body measures was not significant, $R^2 = .06$, adjusted $R^2 = .04$, $F(2, 97) = 3.06$, $p = .051$. Based on these results, the lower-body measures appear to be better predictors of the injury index.

Next, a multiple regression analysis was conducted with all five strength measures as predictors. The linear combination of the five strength measures was significantly related to the injury index, $R^2 = .18$, adjusted $R^2 = .14$, $F(5, 94) = 4.18$, $p < .01$. The lower-body strength measures predicted significantly over and above the upper-body measures, R^2 change $= .12$, $F(3, 94) = 4.69$, $p < .01$, but the upper-body strength measures did not predict significantly over and above the lower-body measures, R^2 change $= .02$, $F(2, 94) = 1.29$, $p = .28$. Based on these results, the upper-body measures appear to offer little additional predictive power beyond that contributed by a knowledge of lower-body measures.

Of the lower-body strength measures, the strength measure for the gluteus/hamstring muscles was most strongly related to the injury index. Supporting this conclusion is the strength of the bivariate correlation between the gluteus/hamstring measure and the injury index, which was

$-.39$, $p < .01$, as well as the comparable correlation partialling out the effects of the other two lower-body measures, which was $-.33$, $p < .01$.

Results for Two Ordered Sets of Predictors

A multiple regression analysis was conducted to predict the overall injury index from previous medical difficulties and age. The results of this analysis indicated that medical difficulties and age accounted for a significant amount of the injury variability, $R^2 = .16$, $F(2, 97) = 9.35$, $p < .01$, indicating that older women who had more medical problems tended to have higher scores on the overall injury index.

A second analysis was conducted to evaluate whether the strength measures predicted injury over and above previous medical difficulties and age. The five strength measures accounted for a significant proportion of the injury variance after controlling for the effects of medical history and age, R^2 change $= .15$, $F(5, 92) = 3.97$, $p < .01$. These results suggest that women who have similar medical histories and are the same age are less likely to have injuries if they are stronger.

Tips for Writing an APA Results Section for Multiple Regression

Here are some guidelines for writing a Results section for multiple regression analyses.

1. It may be appropriate to include an initial description of the analytic strategy used to answer the research questions. This general description is necessary to the degree that the analyses include a number of multiple regression analyses and are unconventional.
2. Report descriptive statistics (e.g., means, standard deviations, and bivariate correlations). These statistics may be reported prior to the presentation of the multiple regression analyses or in conjunction with them.
   - If you are reporting only a few correlations, report them in the text (e.g., $r(98) = .39$, $p < .01$).
   - For regression analyses, there are often many bivariate correlations. These statistics can be summarized in tabular form. Typically, you would present the lower triangle of a correlation matrix. Means and standard deviations for the variables also can be reported in the table.
3. When presenting the results of a particular regression analysis, describe the variables. For example: "A multiple regression analysis was conducted to predict the overall injury index from previous medical difficulties and age."
   - Describe the independent or predictor variables. If the variables can be divided into conceptually distinct sets, describe the predictors in each set. In addition, indicate whether the sets are ordered or nonordered.
   - Describe what the criterion variable is.
4. Report the overall strength of the relationship between the predictors and the criterion as well as the results of the overall significance test. If several analyses have been conducted, report the results for each one separately.
   - Besides the R^2, consider reporting the adjusted R^2.
   - Report the standard error of estimate if the dependent variable has a meaningful metric.
5. If multiple sets of predictors are evaluated, report the changes in R^2 and the significance tests associated with those changes in R^2.
6. Report the contributions of the individual predictors.
   - Consider relevant statistics to evaluate the relative importance of each predictor. The bivariate correlations, the partial correlations, and the standardized regression coefficients might be presented.
   - Report whether the individual variables make a significant contribution to the prediction equation (e.g., $t(98) = -3.13$, $p < .01$).

7. Describe the specific research conclusions that should be drawn from the regression analyses. For example: "The multiple regression results suggest that women who have similar medical histories and are the same age are less likely to have injuries if they are stronger."
8. Consider presenting relevant scatterplots.

Exercises

Exercises 1 through 5 are based on the data set named *Lesson 34 Exercise File 1* on the Web at http://www.prenhall.com/greensalkind. The data are from the following research study.

Jenna is interested in understanding who does well and who does poorly in statistics courses. Jenna collects data from the 100 students in her statistics course. Besides their performance on her exams in the course, she obtains students' scores on math and English aptitude tests that students took in their senior year of high school and their high school grade point averages in math, English, and all other courses. The variables in the data set are shown in Table 42.

Table 42
Variables in Lesson 34 Exercise File 1

Variable	Definition
mathtest	Score on a math aptitude test taken senior year of high school
engtest	Score on an English aptitude test taken senior year of high school
eng_gpa	High school GPA in English courses
math_gpa	High school GPA in math courses
othr_gpa	High school GPA in courses other than English and math
statexam	Average percentage correct on exams in a college statistics course

Jenna wants to address the following research questions: (a) How well do high school test scores and grade point averages predict test performance in a statistics course? (b) Is it necessary to have both high school test scores and grade point averages as predictors of exams scores in statistics? Conduct the appropriate regression analysis to answer these questions.

1. Should the predictor variables be divided into sets and, if so, what variables should be included in each set? What is the criterion variable? If you divided the predictors into sets, do you think that they should be ordered?
2. What is the regression equation for all predictors?
3. What is the contribution of high school test scores over and above high school grade point averages? Report the appropriate statistic from the output.
4. What is the contribution of high school grade point averages over and above high school test scores? Report the appropriate statistic from the output.
5. Write a Results section based on your analyses that addresses the two research questions.

Exercises 6 through 10 are based on the data set named *Lesson 34 Exercise File 2* on the Web at http://www.prenhall.com/greensalkind. The data are from the following research study.

Sally is interested in predicting the number of relapses that occur after participation in an intensive inpatient substance abuse program. She has data from 120 women who are at least five years beyond their initial admission to the program. Upon entry into the treatment program, the women were evaluated with respect to their psychological state and, in particular, on depression, propensity for substance abuse, and daily life stress. In addition, they completed two measures of social support—amount of social support available to them (number of close relationships) and perceptions of the quality of their social support. She also has records that indicate the number of relapses that each woman has had in the first five years after entry into the treatment program. The SPSS file contains six variables: three psychological state variables (named depress for depression, subst for substance abuse, and stress for daily life stress), two social support predictors (named amt_ss for amount of social support and per_ss for perception of social support), and the number of relapses (named relapses).

6. Conduct a regression analysis to evaluate the relative importance of the psychological state predictors and the social support predictors in predicting the number of substance abuse relapses.
7. What is the contribution of the set of social support predictors? Report the appropriate statistics.
8. What is the contribution of the psychological state predictors? Report the appropriate statistics.
9. Which variables in each set are the best predictors of the number of relapses? How did you arrive at your answer?
10. Write a Results section reporting the analyses you conducted.

LESSON

35 Discriminant Analysis

Discriminant analysis can be used to classify individuals into groups on the basis of one or more measures or to distinguish groups based on linear combinations of measures, possibly after obtaining a significant *F* test with a MANOVA. (See Lesson 28 for a discussion of MANOVA.) For discriminant analysis, each case must have a score or scores on one or more quantitative variables and a value on a classification variable that indicates group membership.

The quantitative variables for discriminant analysis are frequently called independent variables or predictors, and the group membership variable is referred to as a dependent variable or a criterion variable. This terminology may be confusing when discriminant analysis is used as a follow-up procedure to a significant MANOVA. In contrast to discriminant analysis, the quantitative variables for a MANOVA are typically referred to as dependent or criterion variables, while the classification variable is called an independent variable or a factor.

Applications of Discriminant Analysis

We'll discuss only two types of applications of discriminant analysis:

- Prediction of group membership on the basis of quantitative predictor variables
- Discriminant analysis as a follow-up procedure after conducting a one-way MANOVA

Predicting Group Membership

Mike has applicants for engineering positions in a large company complete three measures: a job history, a job test that assesses knowledge in engineering, and a personality measure that assesses friendliness. He also asks the applicants for college transcripts. Four scores are derived from these measures: a quality rating of job history, a test score of engineering knowledge, a friendliness score, and a grade point average. Mike is interested in determining whether these four scores are useful in predicting job performance. One-hundred-and-twenty-four applicants are hired and work for the company for at least one year. At the end of the year, a panel evaluates the employees and classifies them into one of three categories: a poor performer (= 1), a good individual achiever (= 2), or a good team player (= 3). Mike's SPSS data file contains 124 cases and five variables, the four predictor variables and the grouping variable distinguishing among the three job-performance groups.

Discriminant Analysis as a Follow-Up Procedure to MANOVA

Diane wants to determine which of several therapeutic approaches is most effective in remediating articulation disorders. She randomly assigns 20 children to each of the following intervention programs: (1) training program using videotapes, (2) individual therapy, and (3) parent training procedures. At the end of the semester, she tests the children by using three different articulation measures. Diane conducts a one-way MANOVA. The factor reflects the three types of therapy, and the dependent variables are the three articulation measures. She conducts a discriminant analysis as a follow-up procedure to the significant MANOVA. For the discriminant analysis, the grouping variable is type of therapy and the independent variables are the three articulation measures. Diane's SPSS data file contains four variables: one variable indicating type of intervention program and the three articulation measures.

Understanding Discriminant Analysis

With discriminant analysis, one or more linear combinations of quantitative predictors are created that are called discriminant functions. The number of possible discriminant functions for an analysis with N_g groups and p quantitative variables is either $(N_g - 1)$ or p, whichever is smaller. For example, our first application has three groups and four quantitative variables. Consequently, the number of functions is 2 because 2 is the smaller of the two values, $(N_g - 1) = (3 - 1) = 2$ and $p = 4$.

The first discriminant function is extracted such that it maximizes the differences on this function among groups. A second discriminant function may then be extracted that maximizes the differences on this function among groups but with the added constraint that it is uncorrelated with the first discriminant function. Additional discriminant functions may be extracted that maximize the differences among groups but always with the constraint that they are uncorrelated with all previously extracted functions.

Eigenvalues associated with discriminant functions indicate how well the functions differentiate the groups; the larger the eigenvalue, the better the groups are discriminated. An eigenvalue for a discriminant function is the ratio of the between-groups sums of squares to the within-group sums of squares for an ANOVA that has the discriminant function as the dependent variable and groups as levels of a factor. Because eigenvalues reflect how well the functions discriminate the groups, the largest eigenvalue is associated with the first discriminant function, the second largest eigenvalue is associated with the second discriminant function, and so on.

A frequently used option within the SPSS discriminant analysis procedure is the classification of cases into groups. With classification, the predictors are linearly combined together to predict group membership, as defined by the grouping variable. These linear combinations of predictor variables are called Fisher's linear discriminant functions (or classification functions) by SPSS, and their coefficients are referred to as Fisher's function coefficients. Accuracy in classification is appraised by computing the percent of cases correctly classified into groups based on the classification functions. An alternative statistic, kappa, also evaluates the percent of cases correctly classified except that it corrects for chance agreements.

Assumptions Underlying Discriminant Analysis

There are three assumptions associated with the significance tests for the discriminant analysis.

Assumption 1: The Quantitative Variables Are Multivariately Normally Distributed for Each of the Populations, with the Different Populations Being Defined by the Levels of the Grouping Variable

If the dependent variables are multivariately normally distributed, each variable is normally distributed ignoring the other variables, and each variable is normally distributed at every combination of values of the other variables. It is difficult to imagine that this assumption could be met. To the extent that population distributions are not multivariate normal and sample sizes are small, the p values may be invalid. In addition, the power of this test may be reduced considerably if the population distributions are nonnormal and, more specifically, thick-tailed or heavily skewed.

Assumption 2: The Population Variances and Covariances among the Dependent Variables Are the Same across All Levels of the Factor

To the extent that the sample sizes are disparate and the variances and covariances are unequal, the p values yield invalid results. SPSS allows us to test the assumption of homogeneity of the variance–covariance matrices with Box's M statistic. The F test from Box's M statistic should be interpreted cautiously in that a significant result may be due to violation of the multivariate normality assumption associated with Box's test and a nonsignificant result may be due to small sample size and a lack of power.

Assumption 3: The Participants Are Randomly Sampled, and the Score on a Variable for any One Participant Is Independent from the Scores on this Variable for All Other Participants

The significance test for the discriminant analyses should not be trusted if the independence assumption is violated.

Effect Size Statistics for Discriminant Analysis

SPSS prints out a number of indices to help assess the effect size for each discriminant function. An eigenvalue (λ) is listed for each discriminant function. An eigenvalue is the ratio of the between-groups sums of squares to the within-group sums of squares for an ANOVA that has groups as levels of a factor and the discriminant function as the dependent variable. Accordingly, an eigenvalue can assume any value that is greater than or equal to zero.

A difficulty with interpreting an eigenvalue is that it has no upper limit. Most researchers would prefer to have an index like R^2 that ranges in value from 0 to 1. A more interpretable index than an eigenvalue may be a ratio of the between-groups sums of square to the total sums of squares. It is a function of the eigenvalue: $\lambda/(1 + \lambda)$. The square root of this index is labeled on the SPSS output as the canonical corr. (canonical correlation).

SPSS also reports for each discriminant function a percent of variance. It is computed by dividing the eigenvalue for the discriminant function by the sum of the eigenvalues, and multiplying the resulting value by 100. This index tells us how strong the prediction is for a discriminant function relative to all discriminant functions.

The Data Set

The data set used to illustrate discriminant analysis is based on our example about predicting job performance of engineers and is in *Lesson 35 Data File 1* on the Web at http://www.prenhall.com/greensalkind. The variables in the data set are presented in Table 43.

Table 43

Variables in Lesson 35 Data File 1

Variable	Definition
friendly	Friendliness scale
gpa	College GPA
job_cat	Employees classified into one of three categories based on their job performance: 1 = Poor performer 2 = Individual achiever 3 = Team player
job_hist	Rating of job history on a scale from 1 = Unimpressive to 10 = Extremely impressive
job_test	Percent correct on a job test

The Research Question

The research question can be asked for group differences or classification into appropriate groups.

1. Group differences: Are there differences among the three job-performance groups in the population on linear combinations of the four predictor variables?
2. Classification into groups: Can the individuals in the three job-performance groups be correctly classified into these three categories based on their scores on the four predictor variables?

Conducting a Discriminant Analysis

To conduct a discriminant analysis, follow these steps:

1. Click **Analyze**, click **Classify**, and click **Discriminant**. You will see the Discriminant Analysis dialog box in Figure 210.

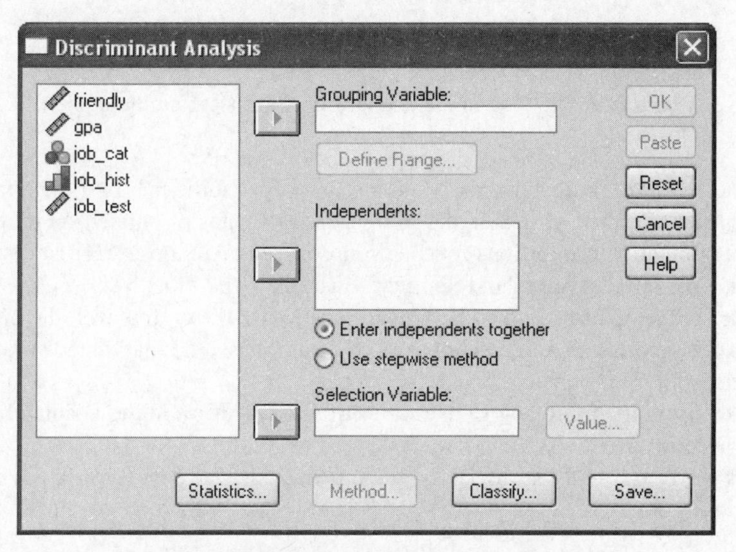

Figure 210. **The Discriminant Analysis dialog box.**

2. Click **job_cat**, and click ▶ to move it to the Grouping Variable text box.
3. Click **Define Range**. In the Define Range dialog box, type in **1** for Minimum and **3** for Maximum.
4. Click **Continue**.
5. Holding down the Ctrl key, click **friendly**, **gpa**, **job_hist**, and **job_test**, and click ▶ to move them to the Independents box.
6. Click **Statistics**. You will see the Discriminant Analysis: Statistics dialog box in Figure 211.

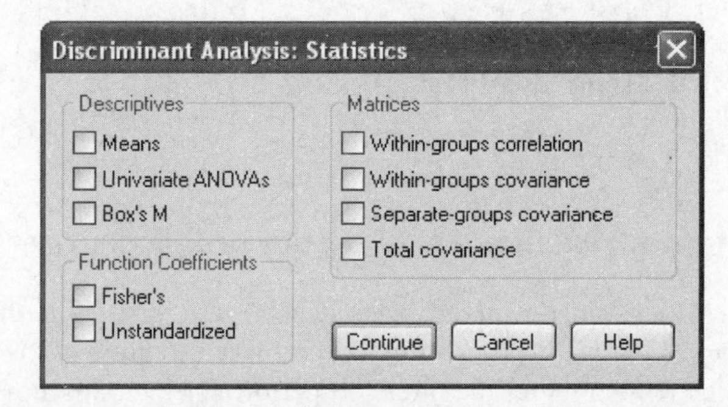

Figure 211. **The Discriminant Analysis: Statistics dialog box.**

7. Click **Means**, **Univariate ANOVAs**, and **Box's M** in the Descriptives box.
8. Click **Fisher's** and **Unstandardized** in the Function Coefficients box.
9. Click all four options in the Matrices box.
10. Click **Continue**.
11. Click **Classify**. You will see the Discriminant Analysis: Classification dialog box in Figure 212.

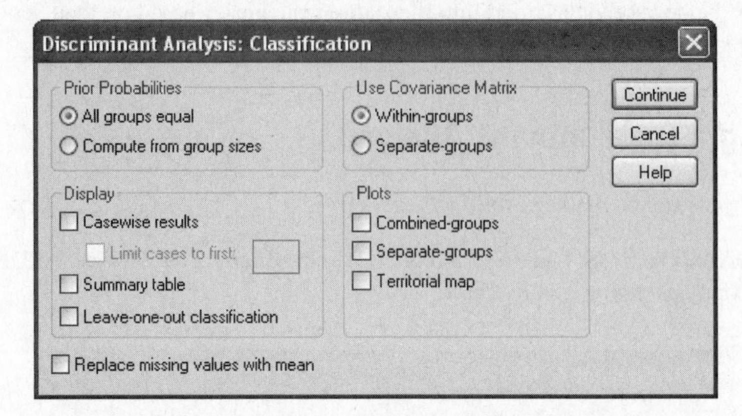

Figure 212. **The Discriminant Analysis: Classification dialog box.**

12. Click **Compute from group sizes** in the Prior Probabilities box if the relative sample sizes for the groups in the sample are estimates of population proportions. If the population proportions for the groups are equal, choose **All groups equal**.
13. Click **Combined-groups** and **Separate-groups** in the Plots box.
14. Click **Within-groups** in the Use Covariance Matrix box. However, the option **Separate-groups** should be chosen if you cannot assume that the covariance matrices for all groups are equal.
15. Click **Summary table** and **Leave-one-out classification** in the Display box.
16. Click **Continue**.
17. Click **Save**. You will see the Discriminant Analysis: Save New Variables dialog box in Figure 213.

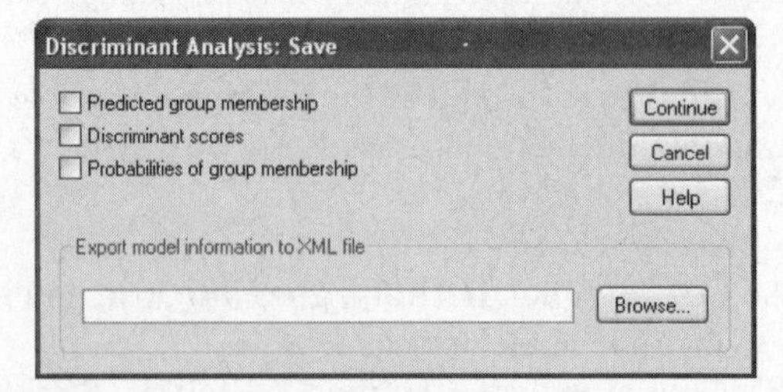

Figure 213. **The Discriminant Analysis: Save New Variables dialog box.**

18. Click **Predicted Group Membership**.
19. Click **Continue**.
20. Click **OK**.

Selected SPSS Output for Preliminary Statistics of Discriminant Analysis

As shown in Figure 214, a number of preliminary statistics are reported in the output: the means and standard deviations of the predictors within job-performance groups, ANOVAs assessing differences among the four predictors for the three job-performance groups, covariance matrices for the three job-performance groups, and a test of equality of the within-group covariance matrices.

These statistics indicate significant differences in means on the predictors among the three job-performance groups (p values range from .00 to .02). There were not significant differences in the covariance matrices among the three groups (p value of .19 for the Box's M test). However, some of the covariances do appear to differ across groups. For example, the covariance between friendly and job_test varies from −4.23 to 19.04.

Group Statistics

				Valid N (listwise)	
Job Category		Mean	Std. Deviation	Unweighted	Weighted
Poor Performer	Friendliness Scale	44.2963	9.87933	27	27.000
	College GPA	2.9517	.64832	27	27.000
	Impressiveness of Job History	5.5556	1.78311	27	27.000
	Job Test	68.2593	9.02387	27	27.000
Individual Achiever	Friendliness Scale	51.8372	12.15756	43	43.000
	College GPA	3.1318	.55756	43	43.000
	Impressiveness of Job History	6.4419	1.63740	43	43.000
	Job Test	79.8605	9.97039	43	43.000
Team Player	Friendliness Scale	55.1111	8.64317	54	54.000
	College GPA	2.7612	.47821	54	54.000
	Impressiveness of Job History	5.5741	1.65520	54	54.000
	Job Test	69.2037	9.23770	54	54.000
Total	Friendliness Scale	51.6210	10.97417	124	124.000
	College GPA	2.9312	.56603	124	124.000
	Impressiveness of Job History	5.8710	1.71539	124	124.000
	Job Test	72.6935	10.74931	124	124.000

Tests of Equality of Group Means

	Wilks' Lambda	F	df1	df2	Sig.
Friendliness Scale	.858	10.040	2	121	.000
College GPA	.916	5.533	2	121	.005
Impressiveness of Job History	.941	3.813	2	121	.025
Job Test	.761	19.004	2	121	.000

Covariance Matrices[a]

Job Category		Friendliness Scale	College GPA	Impressiveness of Job History	Job Test
Poor Performer	Friendliness Scale	97.601	2.221	-1.479	19.036
	College GPA	2.221	.420	.384	.584
	Impressiveness of Job History	-1.479	.384	3.179	-3.496
	Job Test	19.036	.584	-3.496	81.430
Individual Achiever	Friendliness Scale	147.806	1.822	3.978	7.905
	College GPA	1.822	.311	.312	3.000
	Impressiveness of Job History	3.978	.312	2.681	6.849
	Job Test	7.905	3.000	6.849	99.409
Team Player	Friendliness Scale	74.704	.052	2.671	-4.231
	College GPA	.052	.229	.242	1.429
	Impressiveness of Job History	2.671	.242	2.740	1.843
	Job Test	-4.231	1.429	1.843	85.335
Total	Friendliness Scale	120.432	.836	2.292	7.224
	College GPA	.836	.320	.353	2.507
	Impressiveness of Job History	2.292	.353	2.943	4.586
	Job Test	7.224	2.507	4.586	115.548

a. The total covariance matrix has 123 degrees of freedom.

Test Results

Box's M		26.716
F	Approx.	1.265
	df1	20
	df2	28453.612
	Sig.	.190

Tests null hypothesis of equal population covariance matrices.

***Figure 214.* Preliminary statistics of the discriminant analysis.**

Selected SPSS Output for Significance Tests and Strength-of-Relationship Statistics

The output for significance tests and strength-of-relationship statistics for the discriminant analysis is shown in Figure 215. In the box labeled Wilks's Lambda, a series of chi-square significance tests are reported. These tests assess whether there are significant differences among groups across the predictor variables, after removing the effects of any previous discriminant functions. In our example in the first row, SPSS reports the overall Wilks's lambda, $\Lambda = .61$, $\chi^2(8, N = 124) = 59.53$, $p < .01$. This test is significant at the .05 level and indicates that there are differences among groups across the four predictor variables in the population. In the second row, SPSS reports $\Lambda = .82$, $\chi^2(3, N = 124) = 24.24$, $p < .01$. This test is significant at the .05 level and indicates that there is a significant difference among groups across all predictor variables in the population, after removing the effects associated with the first discriminant function.

Eigenvalues

Function	Eigenvalue	% of Variance	Cumulative %	Canonical Correlation
1	.343[a]	60.4	60.4	.506
2	.225[a]	39.6	100.0	.429

a. First 2 canonical discriminant functions were used in the analysis.

Wilks's Lambda

Test of Function(s)	Wilks' Lambda	Chi-square	df	Sig.
1 through 2	.608	59.527	8	.000
2	.816	24.243	3	.000

Figure 215. **Significance tests and strength-of-relationship statistics for discriminant analysis.**

Significance tests can help determine how many discriminant functions should be interpreted. For example, if the overall Wilks's lambda is significant but none of the remaining functions is significant, only the first discriminant function is interpreted. If the first two Wilks's lambdas are significant but none of the remaining ones are significant, then only the first two discriminant functions are interpreted. For our example, because the overall lambda and the lambda after removing the first discriminant function are significant, both discriminant functions could be interpreted according to the chi-square tests.

A series of statistics associated with each discriminant function are reported in the box labeled Eigenvalues. The first discriminant function has an eigenvalue of .34 and a canonical correlation of .51. By squaring the canonical correlation for the first discriminant function ($.51^2 = .26$), you obtain the eta square that would result from conducting a one-way ANOVA on the first discriminant function. Accordingly, 26% of the variability of the scores for the first discriminant function is accounted for by differences among the three job-performance groups. In comparison, the second discriminant function has an eigenvalue of .22 and a canonical correlation of .43. Therefore, 18% ($.43^2 = .18$) of the variability of the scores for the second discriminant function is accounted for by the job-performance factor.

Selected SPSS Output of Coefficients for the Discriminant Functions

The coefficients for the discriminant functions are shown in Figure 216. Each discriminant function is named by determining which variables are most strongly related to it. Strength of relationship is assessed by the magnitudes of the standardized coefficients for the predictor variables in the function and the correlation coefficients between the predictor variables and the function within a group (coefficients in the structure matrix). For the first discriminant

Standardized Canonical Discriminant Function Coefficients

	Function	
	1	2
Friendliness Scale	-.087	.955
College GPA	.088	-.563
Impressiveness of Job History	.275	.040
Job Test	.884	.245

Structure Matrix

	Function	
	1	2
Job Test	.953*	.105
College GPA	.467*	-.272
Impressiveness of Job History	.428*	.020
Friendliness Scale	.012	.859*

Pooled within-groups correlations between discriminating variables and standardized canonical discriminant functions
Variables ordered by absolute size of correlation within function.

*. Largest absolute correlation between each variable and any discriminant function

***Figure 216.* Standardized coefficients and the pooled within-groups correlations for discriminant analysis.**

function in our application, job_test has a relatively large positive coefficient (standardized function and structure coefficients), while gpa and job_hist have weaker coefficients. For the second discriminant function, the largest positive coefficient is for friendly, while a negative relationship exists for gpa. On the basis of these standardized function and structure coefficients, we will name the first and second discriminant functions engineering knowledge and gregariousness, respectively.

Selected SPSS Output for Group Centroids

The output for the group centroids is shown in Figure 217. The values labeled group centroids are the mean values on the discriminant functions for the three groups. In our illustration, based on our interpretation of the discriminant function, the individual achiever group (Group 2) had the highest mean score on the engineering knowledge dimension (Function 1). On the other hand, the team player group (Group 3) had the highest mean score on the gregariousness dimension (Function 2). The pattern of the means for the discriminant functions appears consistent with our interpretation of the two functions.

Functions at Group Centroids

	Function	
Job Category	1	2
Poor Performer	-.401	-.826
Individual Achiever	.794	.013
Team Player	-.432	.403

Unstandardized canonical discriminant functions evaluated at group means

***Figure 217.* Group centroids for discriminant functions.**

Selected SPSS Output for Group Classification

The output for group classification is shown in Figure 218. The classification results allow us to determine how well we can predict group membership by using a classification function. The top part of the table (labeled Original) indicates how well the classification function predicts in the sample. Correctly classified cases appear on the diagonal of the classification table. For example, of the 27 cases in the poor performer group, 12 (44%) were predicted correctly. In the individual achiever group, 27 of 43 cases (63%) were classified correctly. In the team player group, 43 of 54 (80%) were classified correctly. Of the total sample of 124 cases, the overall number of cases classified correctly was 82 or 66% of the sample.

Classification Results[b,c]

			Predicted Group Membership			
		Job Category	Poor Performer	Individual Achiever	Team Player	Total
Original	Count	Poor Performer	12	5	10	27
		Individual Achiever	4	27	12	43
		Team Player	4	7	43	54
	%	Poor Performer	44.4	18.5	37.0	100.0
		Individual Achiever	9.3	62.8	27.9	100.0
		Team Player	7.4	13.0	79.6	100.0
Cross-validated[a]	Count	Poor Performer	10	5	12	27
		Individual Achiever	5	25	13	43
		Team Player	5	7	42	54
	%	Poor Performer	37.0	18.5	44.4	100.0
		Individual Achiever	11.6	58.1	30.2	100.0
		Team Player	9.3	13.0	77.8	100.0

a. Cross validation is done only for those cases in the analysis. In cross validation, each case is classified by the functions derived from all cases other than that case.

b. 66.1% of original grouped cases correctly classified.

c. 62.1% of cross-validated grouped cases correctly classified.

Figure 218. **Group classification for discriminant analysis.**

The bottom part of the table (labeled Cross-validated) is generated by choosing the leave-one-out option within the classification dialog box. With the leave-one-out option, classification functions are derived on the basis of all cases except one, and then the left-out case is classified. This is repeated N times, until all cases have been left out once and classified based on classification functions for the $N - 1$ cases. The results for how well the classification functions predicted the N left out cases are reported in the cross-validated table. These results can be used to estimate how well the classification functions derived on all N cases should predict with a new sample. As shown in the cross-validated table, 10 of the poor performers, 25 of the individual achievers, and 42 of the team players were correctly classified. Overall, 62% of the cases were correctly classified.

Computing Kappa to Assess Classification Accuracy

The overall percent of cases correctly classified is 66% in the sample. This value is affected by chance agreement. Kappa, an index that corrects for chance agreements, could be reported in your Results section along with the proportion of individuals who are correctly classified.

We will compute kappa to assess the accuracy in prediction of group membership. To compute kappa, we use the variable, dis_1, which was created by choosing Predicted Group Membership in the Discriminant Analysis: Save New Variables dialog box. To calculate kappa, conduct the following steps.

1. Click **Analyze**, click **Descriptive Statistics**, and click **Crosstabs**.
2. Click **job_cat**, and then click ▶ to move it to the box labeled Row(s).

3. Click **dis_1**, and then click ▶ to move it to the box labeled Column(s).
4. Click **Statistics**.
5. Click **Kappa**.
6. Click **Continue**.
7. Click **Cells**.
8. Click **Expected** in the Counts box. Be sure Observed is selected.
9. Click **Row** in the Percentages box.
10. Click **Continue**.
11. Click **OK**.

Selected SPSS Output for Kappa

The output presenting the kappa coefficient is shown in Figure 219. As shown in the figure, kappa is .46. A kappa of .46 indicates moderate accuracy in prediction. Kappa ranges in value from −1 to +1. A value of 1 for kappa indicates perfect prediction, while a value of 0 indicates chance-level prediction. Values that are less than 0 indicate poorer than chance-level prediction, and those that are greater than 0 indicate better than chance-level prediction.

Symmetric Measures

		Value	Asymp. Std. Error[a]	Approx. T[b]	Approx. Sig.
Measure of Agreement	Kappa	.460	.066	7.074	.000
N of Valid Cases		124			

a. Not assuming the null hypothesis.

b. Using the asymptotic standard error assuming the null hypothesis.

Figure 219. **Results of kappa analysis.**

Using SPSS Graphs to Display the Results

We requested a combined groups plot to illustrate the classification of the three groups. Figure 220 (on page 310) shows the edited combined groups plot for our example.

An APA Results Section

A discriminant analysis was conducted to determine whether four predictors—friendliness scale scores, college grade point average, ratings of job history, and job test scores—could predict on-the-job performance. The overall Wilks's lambda was significant, $\Lambda = .61$, $\chi^2(8, N = 124) = 59.53$, $p < .01$, indicating that overall the predictors differentiated among the three on-the-job performance groups. In addition, the residual Wilks's lambda was significant, $\Lambda = .82$, $\chi^2(3, N = 124) = 24.24$, $p < .01$. This test indicated that the predictors differentiated significantly among the three on-the-job performance groups after partialling out the effects of the first discriminant function. Because these tests were significant, we chose to interpret both discriminant functions.

In Table 44 (on page 310), we present the within-groups correlations between the predictors and the discriminant functions as well as the standardized weights. Based on these coefficients, the job test scores demonstrate the strongest relationship with the first discriminant function, while college grade point average and job-history ratings show a weaker relationship. On the other hand, the friendliness scale shows the strongest relationship with the second discriminant function, while college GPA demonstrates a negative relationship with this function. On the basis

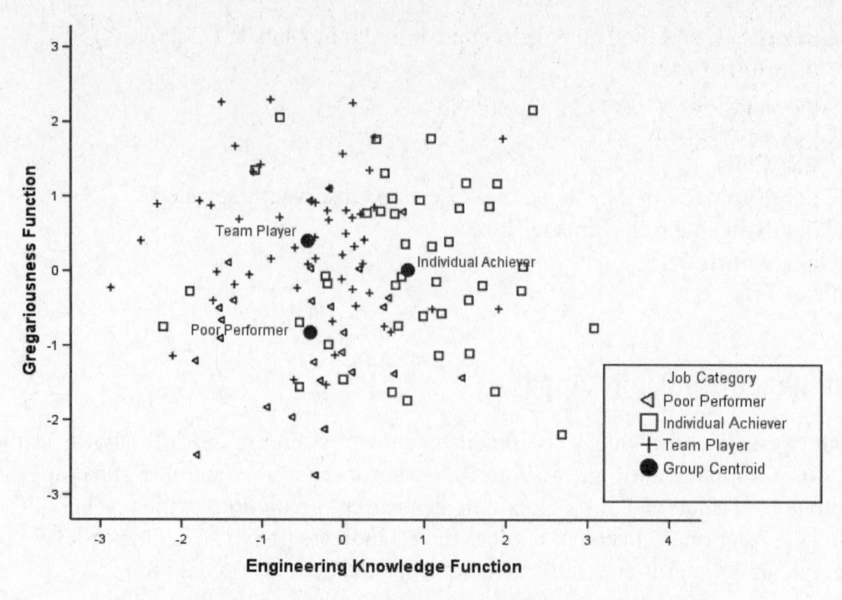

Figure 220. **Separation of groups on discriminant functions.**

of the results presented in Table 44, we labeled the first and second discriminant functions engineering knowledge and gregariousness, respectively.

Table 44

Standardized Coefficients and Correlations of Predictor Variables with the Two Discriminant Functions

	Correlation coefficients with discriminant functions		Standardized coefficients for discriminant functions	
Predictors	Function 1	Function 2	Function 1	Function 2
Friendliness scale	.01	.86	−.09	.96
College GPA	.47	−.27	.09	−.56
Rating of job history	.43	.02	.28	.04
Job test	.95	.10	.88	.24

The means on the discriminant functions are consistent with this interpretation. The individual achievers ($M = .79$) had the highest mean on the engineering knowledge dimension (the first discriminant function), while the poor performers ($M = -.40$) and the team players ($M = -.43$) had lower means. On the other hand, the team players ($M = .40$) had the highest mean on the gregariousness dimension, the individual performers ($M = .01$) the next highest mean, and the poor performers ($M = -.83$) the lowest mean scores.

When we tried to predict on-the-job group membership, we were able to classify correctly 66% of the individuals in our sample. In order to take into account chance agreement, we computed a kappa coefficient and obtained a value of .46, a moderate value. Finally, to assess how well the classification procedure would predict in a new sample, we estimated the percent of employees accurately classified by using the leave-one-out technique and correctly classified 62% of the cases.

Alternative Analyses

Logistic regression can be used as an alternative to discriminant analysis to classify individuals into groups on the basis of multiple measures. Logistic regression is available in SPSS under Analyze—Regression.

Exercises

Exercises 1 through 5 are based on data found in the data file named *Lesson 35 Exercise File 1* on the Web at http://www.prenhall.com/greensalkind. The data are from the following study.

Dave wants to classify college professors into two groups—research scientists and teaching moguls. He obtains scores on four variables from 50 college professors: number of publications in last two years, amount of grant funding generated in last five years, mean teaching evaluation score for the last three semesters, and number of student committees served on during last five years. Dave then has three university administrators evaluate all 50 professors as one of two types, research scientists and teaching moguls. The SPSS data file has 50 cases and five variables. The five variables are the four predictors and a grouping variable distinguishing research scientists and teaching moguls based on the judgments of the university administrators.

1. Conduct a discriminant analysis to distinguish professors who are research scientists from professors who are teaching moguls. From the output, identify the following:
   a. Research scientist group mean for number of publications
   b. Univariate ANOVA F value for grant money
   c. χ^2 associated with the discriminant function
2. What percent of research scientists are correctly classified? What is the overall percent of professors correctly classified as teaching moguls and research scientists?
3. Which predictor variables contribute the most to discriminating between research scientists and teaching moguls?
4. Create a combined groups plot to show the classification of the two groups.
5. Write a Results section based on your analyses.

Exercises 6 through 8 are based on data found in the data file named *Lesson 35 Exercise File 2* on the Web at http://www.prenhall.com/greensalkind. The data are from the following study.

Jacki is interested in differentiating successful women basketball coaches into three categories: successful coaches as character builders, successful winning coaches, and successful winning coaches and character builders. Based on interview data, she is able to differentiate reliably 100 successful women coaches into these three categories. She collects a series of measures for each coach: average number of points per game over last five years, number of ranked players recruited from high schools for the last five years, a measure of problem-focused coping, a rating of likability by the team, a measure of social support offered by the coach, a measure of coach persistence, and a measure reflecting the continuum of authoritative–authoritarian discipline styles. The SPSS file contains 100 cases, seven predictors, and the grouping variable.

6. Conduct a discriminant analysis. Determine whether both discriminant functions should be interpreted.
7. Interpret the discriminant function or functions.
8. Write a Results section based on your analysis.

PART II

UNIT 9 Scaling Procedures

Unit 9 discusses methods involved in evaluating the psychometric quality of measures. These techniques focus on relationships among scales, items, and other evaluation instruments to assess their psychometric properties.

- Lesson 36 introduces exploratory factor analysis. Factor analysis is an exploratory multivariate technique used to assess the dimensionality of a set of variables. For example, a common application of factor analysis would be to assess whether a number of scales have a single dimension underlying them. The Factor Analysis program in SPSS is relatively flexible and allows for a variety of factor extraction and rotation options.
- Lesson 37 discusses internal consistency estimates of reliability. Internal consistency coefficients obtained through the Reliability Analysis program are estimates of a scale's reliability. Reliability is computed for these estimates based on the relationships among items on a measure. In this lesson, we focus on split-half and coefficient alpha estimates.
- Lesson 38 addresses item analysis procedures. Item analysis procedures evaluate whether items should be maintained on a scale or whether revisions to a scale are necessary. Revisions to a unidimensional scale are based on the relationships between the items and their total scale (e.g., corrected item total correlations) as well as the researcher's understanding of what he or she is trying to measure. Revisions to multidimensional scales are based on similar information. In addition, for multidimensional scales, item scores assessing a particular dimension are correlated with total scale scores assessing different dimensions to evaluate discriminant validity.

LESSON

Factor Analysis 36

Factor analysis is a technique used to identify factors that statistically explain the variation and covariation among measures. Generally, the number of factors is considerably smaller than the number of measures and, consequently, the factors succinctly represent a set of measures. From this perspective, factor analysis can be viewed as a data-reduction technique since it reduces a large number of overlapping measured variables to a much smaller set of factors. If a study is well designed so that different sets of measures reflect different dimensions of a broader conceptual system, factor analyses can yield factors that represent these dimensions. More specifically, the factors can correspond to constructs (i.e., unobservable latent variables) of a theory that helps us understand behavior. Examples of constructs that might emerge as factors from factor analyses include altruism, test anxiety, mechanical aptitude, attention span, and academic self-esteem.

Ideally, variables to be factor analyzed should be quantitative, have a wide range of scores, and be unimodally, symmetrically distributed. However, researchers, for better or for worse, do sometimes analyze variables with a very limited range of scores (e.g., 3-point Likert scales and right–wrong items). You should consult textbooks on factor analysis to understand the difficulties with analyzing such variables.

Factor analysis can be conducted directly on the correlations among the variables. In this lesson, we will learn how to create a data file containing a correlation matrix for factor analysis.

Applications of Factor Analysis

We will examine three applications of factor analysis. Each application illustrates a different objective for factor analysis.

- Defining indicators of constructs
- Defining dimensions for an existing measure
- Selecting items or scales to be included on a measure

Defining Indicators of Constructs

Dave believes that our verbal and nonverbal reactions to art are a function of two underlying dimensions, emotional reactions and intellectual reactions. Dave conducts a study to examine nonverbal behavior displayed when viewing art. He has 300 students view 10 different pieces of art while being videotaped. Two judges evaluate the tapes by rating the degree to which students engaged in 30 specific behaviors across the 10 art pieces. For example, judges rate the degree to which students sit and stare at an art piece, the degree that students inspect specific portions of an art piece, the degree that they read about an art piece, the degree that they change their facial expressions when they first view an art piece, and how frequently they return to an art piece after viewing it. Dave selected 15 of the behaviors because he thought they represented intellectual reactions, while the other 15 behaviors represented emotional reactions. For each student, 30 scale scores are obtained by totaling across the 10 art pieces and two judges for each type of behavior. Dave is interested in determining if these 30 variables are indicators of the two constructs, emotional and intellectual reactions. Dave's SPSS data file includes the scores on these 30 variables for the 300 students.

Defining Dimensions for an Existing Measure

George had developed a 10-item questionnaire concerning how teachers cope with difficult students. When completing the questionnaire, teachers indicate the degree to which they use 10 different approaches to deal with problems between themselves and their most recalcitrant students. Teachers respond to the items on a 4-point scale, with 1 = Did not try this approach to 4 = Used this approach extensively. A single score for this measure is computed by simply summing the 10 items. By recommending a single scale score, George implicitly assumed that a single dimension underlies the coping measure. Kandace decides to conduct a study to gain some insights into the dimensionality of George's coping measure. She administers the coping measure to 91 teachers. Kandace's SPSS data file has the scores on these 10 variables for the 91 teachers.

Selecting Items or Scales to Be Included in a Measure

Kandace decides to revise the coping measure on the basis of her factor analysis. Based on her factor analysis, she may decide to delete some items and add others. Kandace's SPSS data file is the same as in the previous application and includes the scores on the 10 items for the 91 teachers.

Understanding Factor Analysis

The results of factor analyses are controlled through our choice of measures and research participants. Ideally, four or more measures should be chosen to represent each construct of interest. Accordingly, if the focus is on identifying measures to assess two constructs (as in Dave's art study), at least eight measures should be included in the factor analysis, four measures for each of the two constructs that might emerge as factors from the analysis. Because the dimensionality of the measures may vary as a function of the participants sampled in the study, it is important to consider not only the measures but also the respondents when designing factor analytic studies.

A common use of factor analysis is to define dimensions underlying existing measurement instruments (as in Kandace's initial problem). In this case, we have not chosen the variables to be analyzed, but rather are stuck with analyzing a predetermined set of items or scales. The subsequent results of the factor analysis may be difficult to interpret for several reasons. The items or scales may not have been created to reflect a construct or constructs, and therefore they may be poor indicators of the construct or constructs. There may be too few items or scales to represent each underlying dimension. Finally, the items or scales may be complexly determined in that they may be a function of multiple factors.

Factor analysis also may be used to determine what items or scales should be included on and excluded from a measure (as in Kandace's second use). The results of the factor analysis should not be used alone in making these decisions, but in conjunction with what is known about the construct or constructs that the items or scales assess. (See Lesson 38 for a discussion of why item analysis decisions should not be made based on empirical results alone.)

Factor analysis requires two stages, factor extraction and factor rotation. The primary objective of the first stage is to make an initial decision about the number of factors underlying a set of measured variables. The goal of the second stage is twofold: (1) to statistically manipulate (i.e., to rotate factors) the results to make the factors more interpretable and (2) to make final decisions about the number of underlying factors.

The first stage involves extracting factors from a correlation matrix to make initial decisions about the number of factors underlying a set of measures. Principal components analysis (a type of factor analysis) is used to make these decisions. The first extracted factor of a principal components analysis (unrotated solution) accounts for the largest amount of the variability among the measured variables, the second factor the next most variability, and so on. The variability of a factor is called an eigenvalue. In Stage 1, two statistical criteria are used to determine the number of factors to extract: (1) the absolute magnitude of the eigenvalues of factors (e.g., eigenvalue-greater-than-one criterion) and (2) the relative magnitudes of the eigenvalues (e.g., scree test). In addition to these two statistical criteria, one should make initial decisions about the number of factors based on a priori conceptual beliefs about the number of underlying dimensions. Because the decision about the number of factors is not a final one, we do not have to choose a single value for the number of factors (e.g., six factors), but instead may choose a range of values (e.g., four to six factors).

In Stage 2, we rotate the factors. Unrotated factors are typically not very interpretable. Factors are rotated to make them more meaningful. The rotated factors may be uncorrelated (orthogonal) or correlated (oblique). The most popular rotational method, VARIMAX, yields orthogonal factors. Oblique rotations are less frequently applied, perhaps because their results are more difficult to summarize. A factor is interpreted or named by examining the largest values linking the factor to the measured variables in the rotated factor matrix.

In Stage 2, we rotate factor solutions with different numbers of factors if the decision in Stage 1 was to consider a range of values for the number of factors. The final decision about the number of factors to choose is the number of factors for the rotated solution that is most interpretable.

Across Stages 1 and 2, four criteria determine the number of factors to include in a factor analysis: (1) a priori conceptual beliefs about the number of factors based on past research or theory, (2) the absolute values of the eigenvalues computed in Stage 1, (3) the relative values of the eigenvalues computed in Stage 1, and (4) the relative interpretability of rotated solutions computed in Stage 2.

Assumptions Underlying Factor Analysis

A single assumption underlies factor analysis extraction in general, while an additional assumption is required for maximum likelihood methods.

Assumption 1: The Measured Variables Are Linearly Related to the Factors Plus Errors

This assumption is likely to be violated if items are factor-analyzed, particularly if the items have very limited response scales (e.g., a two-point response scale like those associated with right-wrong or true–false items), and the item distributions vary in skewness. Violation of this assumption may lead to the identification of spurious factors. If the assumption is met, the relationships among the measured variables are also linear. To determine whether the variables are linearly related, you can evaluate scatterplots for pairs of items. (See Lesson 31.)

Assumption 2: The χ^2 Test for the Maximum Likelihood Solution Assumes That the Measured Variables Are Multivariately Normally Distributed

This assumption is problematic when the factor analysis is conducted on items.

The Data Set

The data set for illustrating factor analysis is named *Lesson 36 Data File 1* on the Web at http://www.prenhall.com/greensalkind. The data set contains the results of a survey of 91 teachers who were asked to indicate the degree to which they used 10 different approaches to deal with problems between themselves and their most difficult students. They made their responses to each of these items on a 4-point scale, with 1 = Did not try this approach to 4 = Used this approach extensively. The 10 items on the coping questionnaire are shown in Table 45.

Table 45
Variables in Lesson 36 Data File 1

Variables	Definition
Item_a	I discussed my frustrations and feelings with person(s) at school.
Item_b	I tried to develop a step-by-step plan of action to remedy the problems.
Item_c	I expressed my emotions to my family and close friends.
Item_d	I read, attended workshops, or sought some other educational approach to correct them.
Item_e	I tried to be emotionally honest with myself about the problems.
Item_f	I sought advice from others on how I should solve the problems.
Item_g	I explored the emotions caused by the problems.
Item_h	I took direct action to try to correct the problems.
Item_i	I told someone I could trust about how I felt about the problems.
Item_j	I put aside other activities so that I could work to solve the problems.

The Research Question

Research questions in factor analysis are framed to reflect the purpose underlying the application. The purpose of our factor analysis is to define dimensions for an existing measure. Is there a single dimension or are multiple dimensions underlying the 10 coping items?

Conducting Factor Analysis

Factor analysis is conducted in two stages: factor extraction and factor rotation.

Conducting Factor Extraction

As part of the first decision to determine the number of extracted factors, we want to obtain the eigenvalues based on the principal components solution to assess their absolute and relative magnitudes.

To conduct the initial analysis, follow these steps:

1. Click **Analyze**, click **Data Reduction**, and then click **Factor**. You will see the Factor Analysis dialog box shown in Figure 221.

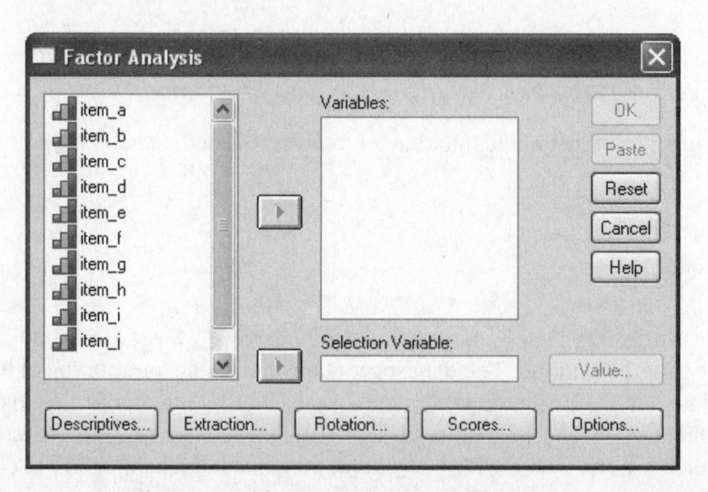

***Figure 221.* The Factor Analysis dialog box.**

2. Holding down the Ctrl key, click the 10 coping variables (items a through j). Then click ▶ to move them to the Variables box in the Factor Analysis dialog box.
3. Click **Extraction**. You will see the Factor Analysis: Extraction dialog box shown in Figure 222.

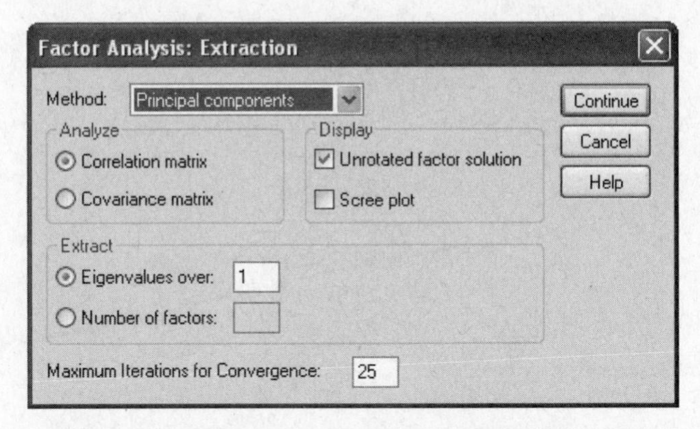

Figure 222. **The Factor Analysis: Extraction dialog box.**

4. Click **Scree Plot**.
5. Click **Continue**.
6. Click **OK**.

Selected SPSS Output for the Initial Factor Extraction

The output showing the initial statistics and the scree plot from the principal components analysis is shown in Figure 223 and Figure 224 (on page 318).

Total Variance Explained

Component	Initial Eigenvalues			Extraction Sums of Squared Loadings		
	Total	% of Variance	Cumulative %	Total	% of Variance	Cumulative %
1	3.046	30.465	30.465	3.046	30.465	30.465
2	1.801	18.011	48.476	1.801	18.011	48.476
3	1.009	10.091	58.566	1.009	10.091	58.566
4	.934	9.336	67.902			
5	.840	8.404	76.307			
6	.711	7.107	83.414			
7	.574	5.737	89.151			
8	.440	4.396	93.547			
9	.337	3.368	96.915			
10	.308	3.085	100.000			

Extraction Method: Principal Component Analysis.

Figure 223. **Initial statistics from factor extraction procedure.**

In Figure 223, the eigenvalues are listed for components 1 through 10. These are important quantities. The total amount of variance of the variables in an analysis is equal to the number of variables (in our example, 10). The extracted factors (or components because principal components was used as the extraction method) account for the variance among these variables. An eigenvalue is the amount of variance of the variables accounted for by a factor. An eigenvalue for a factor should be greater than or equal to zero and cannot exceed the total variance (in our

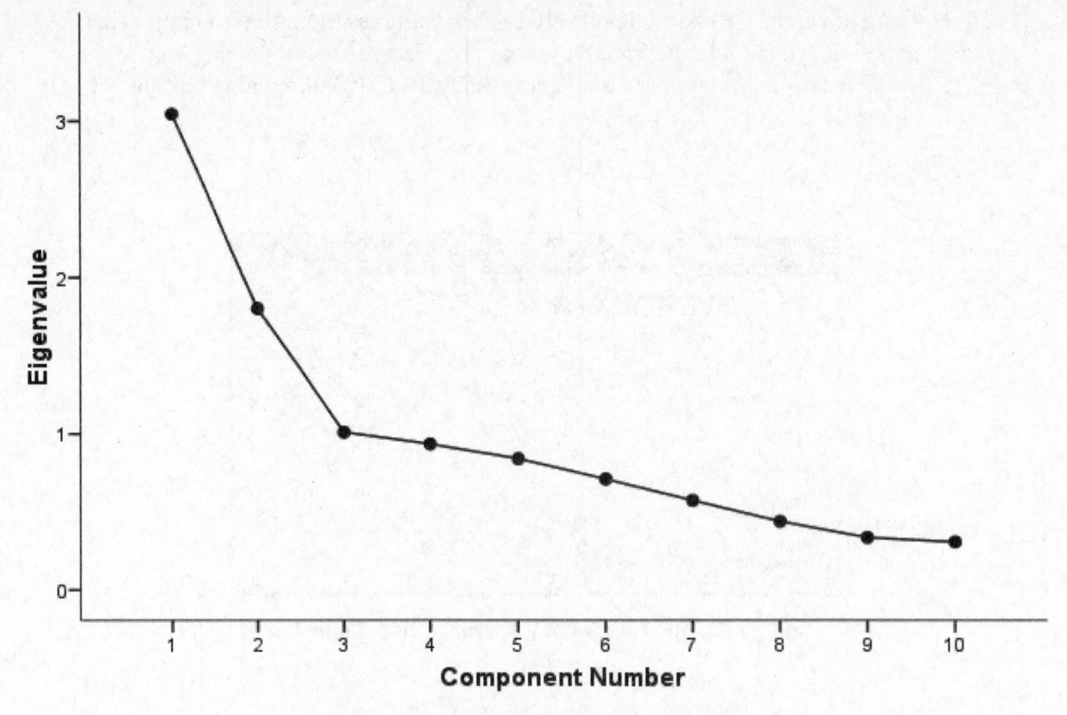

Figure 224. **Scree plot of the eigenvalues.**

example, 10). The percent of variance of the variables accounted for by the factor, as shown in the output, is equal to the eigenvalue divided by the total amount of variance of the variables times 100. For example, the eigenvalue associated with the first factor is 3.046 and the percent of total variance accounted for by the first factor is (3.046 / 10) 100 = 30.46 (as shown in the first row in Figure 223).

Eigenvalues are helpful in deciding how many factors should be used in the analysis. Many criteria have been proposed in the literature for deciding how many factors to extract based on the magnitudes of the eigenvalues. One criterion is to retain all factors that have eigenvalues greater than 1. This criterion is the default option in SPSS. It may not always yield accurate results. Another criterion is to examine the plot of the eigenvalues, also known as the scree test, and to retain all factors with eigenvalues in the sharp descent part of the plot before the eigenvalues start to level off. This criterion yields accurate results more frequently than the eigenvalue-greater-than-1 criterion. Based on the scree plot in Figure 224, we concluded that two factors should be rotated.

Conducting Factor Rotation

Our next stage is to rotate a two-factor solution. To conduct a factor analysis with two rotated factors, follow these steps:

1. Click **Analyze**, click **Data Reduction**, and then click **Factor**. The 10 coping items should already be visible in the Variables box.
2. Click **Extraction**.
3. Click next to **Number of factors**. Type **2** in the box next to the number of factors that you wish to extract and rotate. We chose two factors based on the scree plot.
4. Choose **Maximum likelihood** in the Method drop-down menu. Other extraction methods may be reasonable alternatives.
5. Click next to **Scree plot** so that it contains no check.
6. Click **Continue**.

7. Click **Rotation**. You will see the Factor Analysis: Rotation dialog box shown in Figure 225.

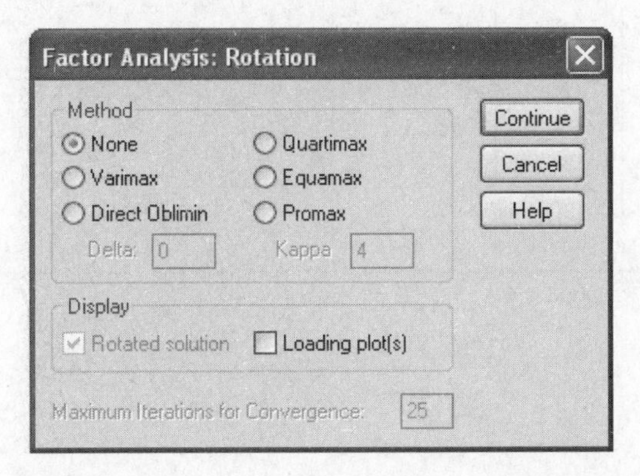

Figure 225. **The Factor Analysis: Rotation dialog box.**

8. Click **Varimax** in the Method box to choose an orthogonal rotation of the factors. Click **Direct oblimin** if you would prefer an oblique rotation.
9. Click **Continue**.
10. Click **Descriptives**. You will see the Factor Analysis: Descriptives dialog box shown in Figure 226.

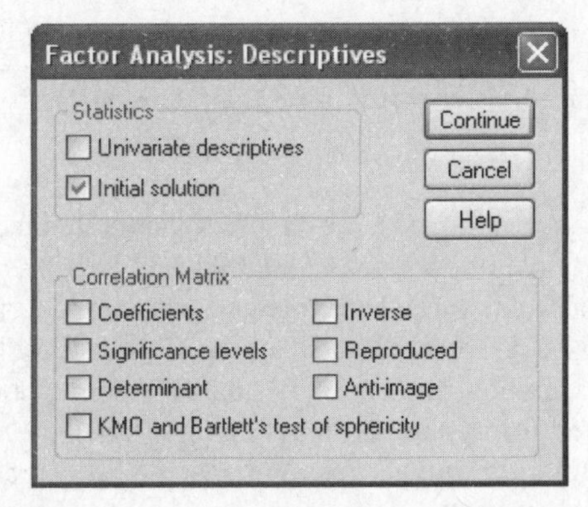

Figure 226. **The Factor Analysis: Descriptives dialog box.**

11. Click **Univariate descriptives** in the Statistics box.
12. Click **Continue**.
13. Click **OK**.

Selected SPSS Output for Rotated Factors

The rotated factor matrix is shown in Figure 227 (on page 320). This matrix shows factor loadings, which are the correlations between each of the variables and the factors for a Varimax rotation. The factors are interpreted by naming them based on the size of the loadings. In our example, items a, c, f, and i are associated the most with the first factor, and items b, d, e, h, and j are associated the most with the second factor. On the basis of looking at the content of these two sets of items, we named the two factors other-directed coping and self-directed coping.

Total Variance Explained

Factor	Initial Eigenvalues			Extraction Sums of Squared Loadings			Rotation Sums of Squared Loadings		
	Total	% of Variance	Cumulative %	Total	% of Variance	Cumulative %	Total	% of Variance	Cumulative %
1	3.046	30.465	30.465	2.495	24.954	24.954	2.262	22.616	22.616
2	1.801	18.011	48.476	1.173	11.734	36.688	1.407	14.072	36.688
3	1.009	10.091	58.566						
4	.934	9.336	67.902						
5	.840	8.404	76.307						
6	.711	7.107	83.414						
7	.574	5.737	89.151						
8	.440	4.396	93.547						
9	.337	3.368	96.915						
10	.308	3.085	100.000						

Extraction Method: Maximum Likelihood.

Rotated Factor Matrix[a]

	Factor	
	1	2
item_a	.773	.147
item_b	.218	.708
item_c	.535	.042
item_d	.166	.354
item_e	.155	.437
item_f	.747	.059
item_g	.341	.341
item_h	-.171	.475
item_i	.759	.056
item_j	-.029	.467

Extraction Method: Maximum Likelihood.
Rotation Method: Varimax with Kaiser Normalization.
a. Rotation converged in 3 iterations.

Figure 227. **The rotated factor matrix from the factor analysis.**

The proportion of variance accounted for by each of the rotated factors is frequently reported in articles to indicate the relative importance of each factor. SPSS reports these statistics in the right side of the table labeled Total Variance Explained. As reported, the first and second factors accounted for 22.62% and 14.07% of the variance of the 10 variables. In total, the two factors accounted for 36.69% of the variable variance. Note that this percentage must be identical to the percent of variance accounted for by the unrotated factors (labeled Extraction Sums of Squared Loadings). Also note that this percentage does not match the percentage based on the initial eigenvalues. It does not match because the initial extraction is based on principal components and not maximum likelihood extraction methods.

If an oblique rotation such as Direct Oblimin is chosen in Step 8 rather than Varimax, be sure to refer to a factor analysis book to see how to interpret the results.

Factor Analysis of a Correlation Matrix

In some instances, you may not have the raw data available to conduct a factor analysis, but you may have the correlation matrix among variables. Under these circumstances, the correlation matrix can be read into SPSS with the Matrix Data command in the Syntax Editor and then factor analyzed.

Creating Correlation Matrices as Data

To illustrate how a correlation is read by SPSS, let's show how we would use SPSS syntax to read in the correlation matrix for our coping example.

1. Click **File**, click **New**, and then click **SPSS Syntax**.
2. Type in the syntax statements. The values in the correlation matrix embedded in the syntax statements have been rounded to three decimal places.

```
MATRIX DATA VARIABLES=ROWTYPE_ item_a item_b item_c item_d
    item_e item_f item_g item_h item_i item_j.
BEGIN DATA
N 91 91 91 91 91 91 91 91 91 91
CORR  1.000
CORR   .284  1.000
CORR   .384   .144  1.000
CORR   .161   .257   .008  1.000
CORR   .161   .307   .060   .173  1.000
CORR   .631   .251   .327   .223   .053  1.000
CORR   .292   .284   .256   .313   .384   .209  1.000
CORR  –.128   .356  –.021   .138   .156  –.085   .001  1.000
CORR   .566   .173   .526   .107   .232   .555   .300  –.052  1.000
CORR   .138   .332   .033   .159   .210  –.073   .097   .202  –.031  1.000
END DATA.
```

The variable ROWTYPE_ is required to tell SPSS the meaning of the information in each row of the data (the lines between BEGIN DATA and END DATA). The *N* in the row immediately after Begin Data indicates that what follows in this row are the sample sizes, while CORR at the beginning of the next 10 rows indicates that correlations are in these rows.

3. Highlight all syntax commands.
4. Click **Run** and then click **Selection**. The Data Editor contains your correlation matrix so now you can factor analyze this matrix.

Warning! Because the data in the Data Editor are different from standard data, SPSS must be told that it is analyzing a correlation matrix. You must follow the steps under Conducting Factor Analysis with a Correlation Matrix each time you conduct a factor analysis. For example, you may run two separate analyses on the same data, one to make initial factor decisions and one to rotate the extracted factors. For both analyses, you must follow the steps to tell SPSS that it is analyzing a correlation matrix.

Conducting Factor Analysis of a Correlation Matrix

To conduct a factor analysis on a correlation matrix, follow these steps:

1. Conduct all steps up to clicking OK in the Factor Analysis dialog box. For example, for factor extraction, follow Steps 1 through 5 presented earlier in this lesson.
2. Instead of **OK**, click **Paste**. The first four lines of the Syntax Editor should look more or less like this:

```
FACTOR
/VARIABLES item_a item_b item_c item_d item_e item_f item_g item_h item_i
    item_j
/MISSING LISTWISE
/ANALYSIS item_a item_b item_c item_d item_e item_f item_g item_h item_i
    item_j
```

3. Delete the /VARIABLES and the /MISSING statements and substitute for these two statements a single statement, /MATRIX IN(COR=*). The intial lines should now read as follows:

```
FACTOR
/MATRIX IN(COR=*)
/ANALYSIS item_a item_b item_c item_d item_e item_f item_g item_h item_i
    item_j
```

4. You should not have to make any changes to any later statements in the syntax window.
5. Highlight the syntax commands for the factor analysis.
6. Click **Run** and click **Selection**.

An APA Results Section

The dimensionality of the 10 items from the teacher coping measure was analyzed using maximum likelihood factor analysis. Three criteria were used to determine the number of factors to rotate: the a priori hypothesis that the measure was unidimensional, the scree test, and the interpretability of the factor solution. The scree plot indicated that our initial hypothesis of unidimensionality was incorrect. Based on the plot, two factors were rotated using a Varimax rotation procedure. The rotated solution, as shown in Table 46, yielded two interpretable factors, other-directed and self-directed coping. The other-directed coping factor accounted for 22.6% of the item variance, and the self-directed coping factor accounted for 14.1% of the item variance. Only one item loaded on both factors.

Table 46

Correlations between the Coping Items and the Coping Factors

Items	Factors	
	Self-Directed	Other-Directed
Self-Directed (SD) items		
I tried to develop a step-by-step plan of action to remedy the problems.	**.71**	.22
I read, attended workshops, or sought some other educational approach to correct them.	**.35**	.17
I tried to be emotionally honest with myself about the problems.	**.44**	.16
I took direct action to try to correct the problems.	**.48**	−.17
I put aside other activities so that I could work to solve the problems.	**.47**	−.03
Other-Directed (OD) items		
I discussed my frustrations and feelings with person(s) at school.	.15	**.77**
I expressed my emotions to my family and close friends.	.04	**.54**
I sought advice from others on how I should solve the problems.	.06	**.75**
I told someone I could trust about how I felt about the problems.	.06	**.76**
Complexly Determined Item		
I explored the emotions caused by the problems.	.34	.34

Alternative Analyses

Multidimensional scaling can be used as an alternative to factor analysis to assess the dimensionality among measures. Multidimensional scaling is available in SPSS under the Statistics—Scale—Multidimensional Scaling option, but the topic is beyond the scope of this book. Item analysis procedures can be used when making decisions about what items to delete from an existing measure. See Lesson 38 for a discussion of item analysis.

Exercises

The data for Exercises 1 through 3 are in the data file named *Lesson 36 Exercise File 1* on the Web at http://www.prenhall.com/greensalkind. The data are based on the following research problem.

Terrill is interested in assessing how much women value their careers. He develops a 12-item scale, the Saxon Career Values Scale (SCVS). He has 100 college women take the SCVS. All of the items reflect the value women place on having a career versus having a family. Students are asked to respond to each on a 4-point scale, with 0 indicating "disagree" and 3 indicating "agree." The items are shown in Table 47.

Table 47
Items on the Saxon Career Values Scale

Variables	Definition
q01	I consider marriage and having a family to be more important than a career.
q02	To me, marriage and family are as important as having a career.
q03	I prefer to pursue my career without the distractions of marriage, children, or a household.
q04	I would rather have a career than a family.
q05	I often think about what type of job I'll have 10 years from now.
q06	I could be happy without a career.
q07	I would feel unfulfilled without a career.
q08	I don't need to have a career to be fulfilled.
q09	I would leave my career to raise my children.
q10	Having a career would interfere with my family responsibilities.
q11	Planning for and succeeding in a career is one of my primary goals.
q12	I consider myself to be very career-minded.

1. Conduct a factor analysis. How many factors underlie the SCVS based on the scree plot?
2. How many factors underline the SCVS based on the eigenvalue-greater-than-1 criterion?
3. Write a Results section reporting your analyses.

The data for Exercises 4 through 6 are the correlations in the correlation matrix. The data are based on the following research problem.

Jessica is interested in assessing humor. She develops a 10-item measure in which some items represent humor demeaning to others (Don Rickles items), while other items reflect self-deprecating humor (Woody Allen items). She administers her measure to 100 college students. Students are asked to respond to each on a 5-point scale with 1 indicating "disagree" and 5 indicating "agree." The variables in the data set are shown in Table 48 (on page 324).

Table 48

Items from a Humor Measure

Variables	Definition
Item01	I like to make fun of others. (Don Rickles)
Item02	I make people laugh by making fun of myself. (Woody Allen)
Item03	People find me funny when I make jokes about others. (Don Rickles)
Item04	I talk about my problems to make people laugh. (Woody Allen)
Item05	I frequently make others the target of my jokes. (Don Rickles)
Item06	People find me funny when I tell them about my failings. (Woody Allen)
Item07	I love to get people to laugh by using sarcasm. (Don Rickles)
Item08	I am funniest when I talk about my own weaknesses. (Woody Allen)
Item09	I make people laugh by exposing other people's stupidities. (Don Rickles)
Item10	I am funny when I tell others about the dumb things I have done. (Woody Allen)

4. Using the syntax window ($N = 100$), create an SPSS data file containing the following correlations among items:

	Item01	Item02	Item03	Item04	Item05	Item06	Item07	Item08	Item09	Item10
Item01	1.00									
Item02	−.22	1.00								
Item03	.27	−.10	1.00							
Item04	−.12	.58	.12	1.00						
Item05	.22	−.02	.59	.16	1.00					
Item06	−.09	.29	.17	.26	.07	1.00				
Item07	.23	−.16	.42	−.06	.30	.06	1.00			
Item08	−.02	.24	.04	.43	.04	.38	−.15	1.00		
Item09	.07	−.13	.34	−.01	.34	.06	.27	−.07	1.00	
Item10	−.14	.28	−.06	.18	−.04	.55	−.01	.37	.10	1.00

5. Conduct a factor analysis on the correlation matrix to assess whether a single factor underlies the scores. How many factors should you extract?
6. Write a Results section based on your analyses.

LESSON

37

Internal Consistency Estimates of Reliability

A measure is reliable if it yields consistent scores across administrations. We can estimate a reliability coefficient for a measure with a variety of methods, including test–retest, equivalent-forms, and internal consistency approaches. With test–retest reliability, individuals are administered the measure of interest on two occasions. In contrast, with equivalent-forms reliability, individuals are administered the measure of interest on one occasion and an equivalent form of the measure on a second occasion. For both test–retest and equivalent-forms reliability, a correlation is computed between the scores obtained on the two occasions. (See Lesson 31.)

With an internal consistency estimate of reliability, individuals are administered a measure with multiple parts on a single occasion. The parts may be items on a paper-and-pencil measure, responses to questions from a structured interview, multiple observations on an observational measure, or some other set of measurement parts that are summed together to yield scale scores. In this lesson, we'll restrict our attention to measures consisting of items and discuss two types of internal consistency estimates, split-half reliability and coefficient alpha. For a split-half estimate, a measure is divided into two equivalent halves, and scores for the split measure are computed by adding together the items within their respective halves. A split-half reliability coefficient assesses the consistency in scores between the two equivalent half measures. For coefficient alpha, a measure does not have to be split into halves; however, it requires the items to be equivalent. Coefficient alpha assesses consistency in scores among equivalent items.

We should assess the reliability of any scale score that we wish to interpret. If a measure yields a single summed score (overall score) for an individual, the reliability estimate should be computed for this summed score. On the other hand, if a measure yields multiple summed scores (scale scores) for an individual, reliability estimates should be computed for each of the separate summed scores.

In all cases with internal consistency estimates, the focal scores are summed scores. It is sometimes necessary to transform one or more items before summing the scores together to yield a meaningful summed score. We'll examine three types of applications of internal consistency estimates, which vary depending on whether or how items are transformed before they are summed.

- No transformation of items. If the responses to items on a scale are in the same metric, and if high scores on them represent high scores on the underlying construct, no transformations are required. The reliability analysis can be conducted on the untransformed scores.
- Reverse-scoring of some item scores. For this case, all items on a scale have the same response metric, but high scores on some items represent high scores on the underlying construct, while high scores on other items represent low scores on the underlying construct. The scores on these latter items need to be reverse-scaled. Such items are commonly found on attitude scales. Lesson 19 explains how to reverse-scale items.
- *z*-score transformation of item scores. *z* scores must be created when items on a scale have different response metrics. Before summing the items to obtain a total

score, you must transform the item scores to standard scores so that these items share a common metric. In some instances, some of the z-transformed items may need to be reverse-scored by multiplying the z scores by -1. Lesson 19 demonstrates how to perform z-score transformations and how to reverse-scale z-transformed items.

Applications of Internal Consistency Estimates of Reliability

Here are examples of the three types of research problems we just described.

No Transformation of Items

Sarah is interested in whether a measure she developed has good reliability. She has 83 students complete the 20-item Emotional Expressiveness Measure (EEM). Ten of the items are summed to yield a Negative Emotions scale, and the other 10 items are summed to produce a Positive Emotions scale. Sarah's SPSS data file contains 83 cases and 20 items as variables. These 20 items are the variables analyzed with the Reliability program. She computes an internal consistency estimate of reliability (split half or coefficient alpha) for the Negative Emotions scale and another internal consistency estimate for the Positive Emotions scale.

Reverse-Scoring of Some Items

Janet has developed a 10-item measure called the Emotional Control scale. She asks 50 individuals to respond to these items on a 0 to 4 scale, with 0 being "completely disagree" and 4 being "completely agree." Half the items are phrased so that agreement indicates a desire to keep emotions under control (under-control items), while the other half are written so that agreement indicates a desire to express emotions openly (expression items). Janet's SPSS data file contains 50 cases and 10 item scores for each. The expression items need to be reverse-scaled so that a response of 0 is transformed to a 4, a 1 becomes a 3, a 2 stays a 2, a 3 becomes a 1, and a 4 becomes a 0. Janet computes an internal consistency estimate based on the scores for the five under-control items and the transformed scores for the five expression items.

z-Score Transformations of Item Scores

George is interested in developing an index of perseverance. He has 83 college seniors answer 15 questions about completing tasks. Some questions ask students how many times in an hour they would be willing to dial a telephone number that is continuously busy, how many hours they would be willing to commit to solving a 10,000-piece jigsaw puzzle, how many times would they be willing to reread a 20-line poem in order to understand it, and how many different majors they have had in college. George's SPSS data file contains the 83 cases and the 15 item scores. All 15 items need to be transformed to standard scores. In addition, some items need to be reverse-scaled. For example, the number of different majors in college presumably is negatively related to the construct of perseverance (i.e., the more they switch majors, the less perseverance they have). George computes an internal consistency estimate on the transformed scores for the 15 questions.

Understanding Internal Consistency Estimates of Reliability

The coefficients for split-half reliability and alpha assess reliability based on different types of consistency. The split-half coefficient is obtained by computing scores for two halves of a scale. With SPSS, scores are computed for the first half and second half of items listed in the Items box in the

Reliability Analysis dialog box. The value of the reliability coefficient is a function of the consistency between the two halves. In contrast, consistency with coefficient alpha is assessed among items. The greater the consistency in responses among items, the higher coefficient alpha will be.

Both the split-half coefficient and coefficient alpha should range in value between 0 and 1. To the extent that items on a scale are ambiguous and produce unreliable responses, there should be a lack of consistency between halves or among items, and internal consistency estimates of reliability should be smaller. It is possible for these reliability coefficients to fall outside the range of 0 to 1 if the correlation between halves is negative for a split-half coefficient or if a number of the correlations among items are negative for coefficient alpha.

You should base your choice between split-half reliability and coefficient alpha on whether the assumptions for one or the other method can be met.

Assumptions Underlying Internal Consistency Reliability Procedures

Three assumptions underlie the two internal consistency estimates.

Assumption 1: The Parts of the Measure Must Be Equivalent

For split-half coefficients, the parts—two halves of the measure—must be equivalent. Both halves should measure the same underlying dimension. If this assumption is met, differences in relative standing among respondents between the two halves should be due only to measurement error (unreliable responding). This assumption is likely to be violated if a measure is split into two halves such that one half includes the first 50% of items on a measure and the second half includes the second 50% of items on a measure. For example, respondents may respond to items differently on the first and the second halves of an achievement measure because they are more fatigued taking the second half or because the second half includes more difficult problems. Alternatively, respondents may respond differently to items on the second half of an attitudinal measure because they become bored. Rather than divide scales into a first half and a second half, it is generally preferable to divide scales using other splits. For example, we can add the odd-numbered items together to create one half and the even-numbered items to create the other half. We can then use these two halves to compute split-half coefficients.

For coefficient alpha, every item is assumed to be equivalent to every other item. All items should measure the same underlying dimension. If this assumption is met, differences in relative standing among respondents for different items should be due only to measurement error (unreliable responding). It is unlikely that this assumption is ever met completely, although with some measures, it may be met approximately.

To the extent that the equivalency assumption is violated, internal consistency estimates tend to underestimate reliability.

Assumption 2: Errors in Measurement between Parts Are Unrelated

For cognitive (achievement and ability) measures, measurement errors are primarily a function of guessing. We can rephrase this assumption for cognitive measures: a respondent's ability to guess well on one item of an achievement test should not influence how well he or she guesses on another item. The unrelated-errors assumption can be violated a number of ways. In particular, internal consistency estimates should not be used if respondents' scores depend on whether they can complete the scale in an allotted time. For example, coefficient alpha should not be used to assess the reliability of a 100-item math test with a 10-minute time limit because then scores are partially a function of how many items are completed. Another instance of violation of this assumption involves items that are linked together. For example, a history test may have sets of matching items, or a reading comprehension test may have sets of questions for different reading texts. Neither coefficient alpha nor split-half measures should be used as a reliability estimate for these scales because items within a set are likely to have correlated errors.

The unrelated-errors assumption can also be violated with noncognitive measures. It is more likely to be violated if items are syntactically similar and share a common response scale. For a more in-depth discussion of this assumption, see Green and Hershberger (1999).

Assumption 3: An Item or Half Test Score Is a Sum of Its True and Its Error Scores

This assumption is necessary for an internal consistency estimate to reflect accurately a scale's reliability. It is difficult to know whether this assumption has been violated or not.

Many researchers interpret coefficient alpha as an index of homogeneity or unidimensionality. This interpretation is incorrect (Green, Lissitz, and Mulaik, 1977).

The Data Set

The data set used in this lesson is named *Lesson 37 Data File 1* on the Web at http://www.prenhall.com/greensalkind. It represents data from the emotional control example presented earlier. Half the items are phrased so that agreement indicates a desire to keep emotions under control (under-control items), and the other half are written so that agreement indicates a desire to express emotions openly (expression items). Table 49 shows the items on the Emotional Control scale.

Table 49

Items on the Emotional Control Scale in Lesson 37 Data Set 1

Variables	Definition
item1	I keep my emotions under control.
item2	Under stress I remain calm.
item3	I like to let people know how I am feeling.
item4	I express my emotions openly.
item5	It is a sign of weakness to show how one feels.
item6	Well-adjusted individuals are ones who are confident enough to express their true emotions.
item7	Emotions get in the way of clear thinking.
item8	I let people see my emotions so that they know who I am.
item9	If I am angry with people, I tell them in no uncertain terms that I am unhappy with them.
item10	I try to get along with people and not create a big fuss.

The Research Question

The research question addresses the reliability of a scale. How reliable is our 10-item measure of emotional control?

Conducting a Reliability Analysis

Before conducting any internal consistency estimates of reliability, we must determine if all items use the same metric and whether any items have to be reverse-scaled. All items share the same metric since the response scale for all items is 0 = completely disagree to 4 = completely agree. However, the five items in which high scores indicate a willingness to express emotion must be reverse-scaled so that high scores on the total scale reflect a high level of emotional control. These items are 3, 4, 6, 8, and 9. We discuss how to reverse-scale items for a Likert scale in Lesson 19.

You should reverse-scale items 3, 4, 6, 8, and 9 before going through the steps to compute coefficient alpha and split-half internal consistency estimates. Name the variables with reversed scales item3r, item4r, item6r, item8r, and item9r.

Computing Coefficient Alpha

To compute a coefficient alpha, follow these steps:

1. Click **Analyze**, click **Scale**, then click **Reliability Analysis**. You'll see the Reliability Analysis dialog box shown in Figure 228.

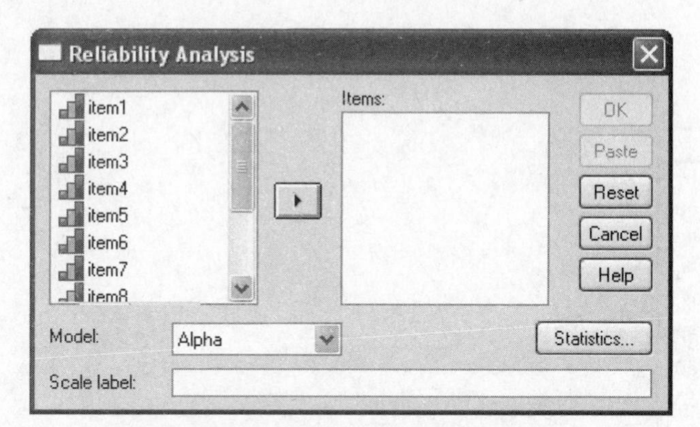

***Figure 228.* The Reliability Analysis dialog box.**

2. Holding down the Ctrl key, click **item1**, **item2**, **item5**, **item7**, **item10**, **item3r**, **item4r**, **item6r**, **item8r**, and **item9r** to select the appropriate items.
3. Click ▶ to move them to the Items box.
4. Click **Statistics**. You'll see the Reliability Analysis: Statistics dialog box, as shown in Figure 229.

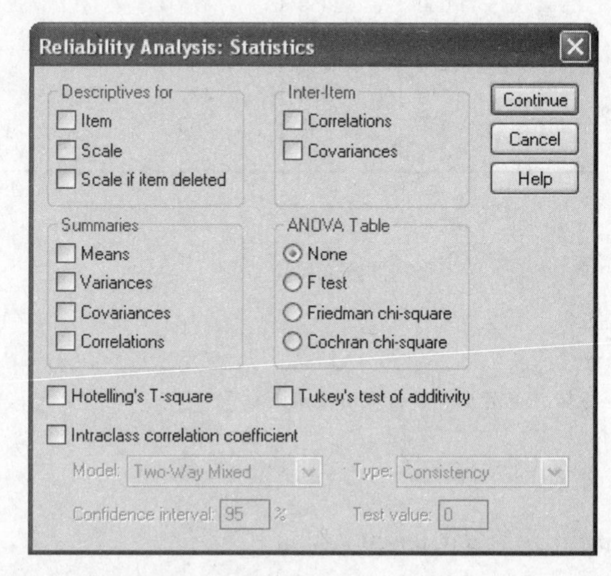

***Figure 229.* The Reliability Analysis: Statistics dialog box.**

5. Click **Item**, click **Scale** in the Descriptives for area, then click **Correlations** in the Inter-Item area.
6. Click **Continue**. In the Reliability Analysis dialog box, make sure that **Alpha** is chosen in the box labeled Model.
7. Click **OK**.

Reliability Statistics

Cronbach's Alpha	Cronbach's Alpha Based on Standardized Items	N of Items
.872	.872	10

Item Statistics

	Mean	Std. Deviation	N
item1	1.9200	1.30681	50
item2	2.1600	1.20136	50
item5	2.0000	1.14286	50
item7	2.3400	1.11776	50
item10	2.0800	1.29110	50
item3r	2.2000	.98974	50
item4r	1.9800	1.30133	50
item6r	2.2000	1.27775	50
item8r	2.1000	1.24949	50
item9r	2.3400	1.11776	50

Inter-Item Correlation Matrix

	item1	item2	item5	item7	item10	item3r	item4r	item6r	item8r	item9r
item1	1.000	.463	.451	.452	.427	.470	.407	.352	.517	.368
item2	.463	1.000	.520	.460	.439	.419	.446	.378	.560	.308
item5	.451	.520	1.000	.192	.401	.307	.453	.405	.500	.224
item7	.452	.460	.192	1.000	.334	.472	.426	.409	.487	.445
item10	.427	.439	.401	.334	1.000	.259	.475	.312	.450	.348
item3r	.470	.419	.307	.472	.259	1.000	.463	.436	.446	.417
item4r	.407	.446	.453	.426	.475	.463	1.000	.481	.365	.285
item6r	.352	.378	.405	.409	.312	.436	.481	1.000	.537	.209
item8r	.517	.560	.500	.487	.450	.446	.365	.537	1.000	.267
item9r	.368	.308	.224	.445	.348	.417	.285	.209	.267	1.000

Scale Statistics

Mean	Variance	Std. Deviation	N of Items
21.3200	67.202	8.19766	10

Figure 230. **Results of coefficient alpha.**

Selected SPSS Output for Coefficient Alpha

The results are shown in Figure 230.

As with any analysis, the descriptive statistics need to be checked to confirm that the data have no major anomalies. For example, are all the means within the range of possible values (0 to 4)? Are there any unusually large variances that might indicate that a value has been mistyped? In general, are the correlations among the variables positive? If not, should you have reverse-scaled one or more of the items associated with the negative correlations? Once it appears that data have been entered and scaled appropriately, the reliability estimate of alpha can be interpreted.

The output reports two alphas, alpha and standardized item alpha. In this example, we are interested in the alpha. The only time that we would be interested in the standardized alpha is if the scale score is computed by summing item scores that have been standardized to have a uniform mean and standard deviation (such as z scores). The coefficient alpha of .87 suggests that the scale scores are reasonably reliable for respondents like those in the study.

Computing Split-Half Coefficient Estimates

SPSS computes a split-half coefficient by evaluating the consistency in responding between the first half and the second half of items listed in the Items box. It is important to carefully choose which items to include in each half so that the two halves are as equivalent as possible. Different item splits may produce dramatically different results. The best split of the items is the one that produces equivalent halves. (See Assumption 1.)

For our example, we chose to split the test into two halves for the purpose of computing a split-half coefficient in the following fashion:

Half 1: Item 1, Item 3, Item 5, Item 8, and Item 10
Half 2: Item 2, Item 4, Item 6, Item 7, and Item 9

We chose this split to take into account the ordering of items on the measure (with one exception, no two adjacent items are included on the same half) as well as the two types of items, under-control and expression items (two items of one type and three of the other on a half).

To compute a split-half coefficient, follow these steps:

1. Click **Analyze**, click **Scale**, and then click **Reliability Analysis.**
2. Click **Reset** to clear the dialog box.
3. Holding down the Ctrl key, click the variables that are in the first half: **item1**, **item3r**, **item5**, **item8r**, and **item10**.
4. Click ▶ to move them to the Items box.
5. Holding down the Ctrl key, click the variables that are in the second half: **item2**, **item4r**, **item6r**, **item7**, and **item9r**.
6. Click ▶ to move them to the Items box in the Reliability Analysis dialog box. Figure 231 shows the Reliability Analysis dialog box.

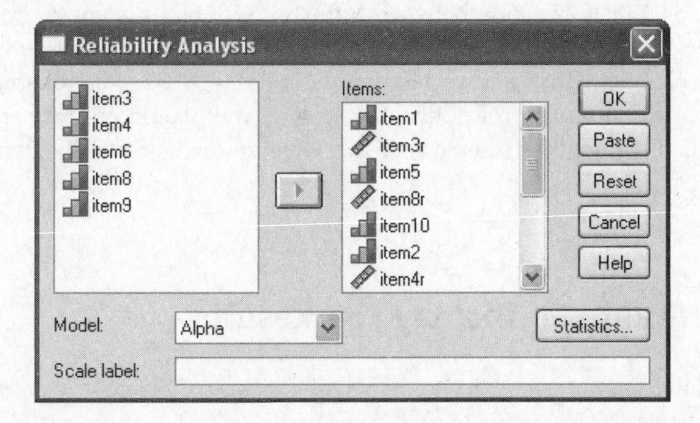

Figure 231. **The Reliability Analysis dialog box after completing Step 6.**

7. Click **Statistics**.
8. Click **Item** and **Scale** in the Descriptives for area.
9. Click **Correlations** in the Inter-Item area.
10. Click **Continue**.

11. Click **Split-half** in the drop-down menu labeled Model in the Reliability Analysis dialog box.
12. Click **OK**.

Selected SPSS Output for Split-Half Reliability

The results of the analysis are shown in Figure 232. The descriptive statistics need to be checked to confirm that the data have no anomalies as described in our earlier discussion of coefficient alpha. The descriptive statistics associated with the split-half coefficient are identical to the descriptives for coefficient alpha discussed earlier.

Reliability Statistics

Cronbach's Alpha	Part 1	Value	.786
		N of Items	5[a]
	Part 2	Value	.757
		N of Items	5[b]
	Total N of Items		10
Correlation Between Forms			.775
Spearman-Brown Coefficient	Equal Length		.873
	Unequal Length		.873
Guttman Split-Half Coefficient			.873

a. The items are: item1, item3r, item5, item8r, item10.

b. The items are: item2, item4r, item6r, item7, item9r.

Figure 232. **The results of the split-half analysis.**

The most frequently reported split-half reliability estimate is the one based on the correlation between forms. The correlation between forms is .775, but it is not the reliability estimate. At best, it is the reliability of half the measure (because it is the correlation between two half-measures). The Spearman-Brown corrected correlation, $r = .873$, is the reliability estimate.

If there were an odd number of items on a scale, a split would produce an unequal number of items in each half. Under these conditions, the value for the Unequal-length Spearman-Brown should be reported.

Using SPSS Graphs to Display the Results

Graphics procedures are not necessary to display the results of the coefficient alpha or split-half reliability coefficients.

An APA Results Section

Two internal consistency estimates of reliability were computed for the Emotional Control scale: a split-half coefficient expressed as a Spearman-Brown corrected correlation and coefficient alpha. For the split-half coefficient, the scale was split into two halves such that the two halves would be as equivalent as possible. In splitting the items, we took into account the sequencing of the items as well as whether items assessed control of emotion or expression of emotion. One of the halves included items 1, 3, 5, 8, and 10, while the other half included items 2, 4, 6, 7, and 9. Values for coefficient alpha and the split-half coefficient were the same, .87, each indicating satisfactory reliability.

Exercises

The data for Exercises 1 through 3 are in the data file named *Lesson 37 Exercise File 1* on the Web at http://www.prenhall.com/greensalkind. The data are based on the following research problem.

Terrill is interested in assessing how much women value their careers. He develops a 12-item scale, the Saxon Career Values Scale (SCVS). He has 100 college women take the SCVS. All of the items reflect the value women place on having a career versus having a family. Students are asked to respond to each on a 4-point scale with 0 indicating "disagree" and 3 indicating "agree" The items are shown in Table 50.

Table 50
Items on the Saxon Career Values Scale

Variables	Definition
q01	I consider marriage and having a family to be more important than a career.*
q02	To me, marriage and family are as important as having a career.*
q03	I prefer to pursue my career without the distractions of marriage, children, or a household.
q04	I would rather have a career than a family.
q05	I often think about what type of job I'll have 10 years from now.
q06	I could be happy without a career.*
q07	I would feel unfulfilled without a career.
q08	I don't need to have a career to be fulfilled.*
q09	I would leave my career to raise my children.*
q10	Having a career would interfere with my family responsibilities.*
q11	Planning for and succeeding in a career is one of my primary goals.
q12	I consider myself to be very career-minded.

*Items are reverse-scaled to create a total score.

1. Reverse-scale the appropriate items so a higher total score reflects a greater emphasis on career.
2. Conduct a reliability analysis for the SCVS by using coefficient alpha. From the output, identify the following:
   a. Coefficient alpha
   b. Correlation between Item 3 and Item 4
   c. Mean for Item 12
3. What judgment should Terrill make about the overall scale reliability? Write a Results section reporting your analyses.

The data for Exercises 4 through 7 are in the data file named *Lesson 37 Exercise File 2* on the Web at http://www.prenhall.com/greensalkind. The data are based on the following research problem.

Jessica is interested in assessing humor demeaning to others versus self-deprecating humor. She develops a 10-item measure in which some items represent humor demeaning to others (Don Rickles items), while other items reflect self-deprecating humor (Woody Allen items). She administers her measure to 100 college students. Students are asked to respond to each on a 5-point scale with 1 indicating "disagree" and 5 indicating "agree." Jessica computes a total score by reverse-scaling the Woody Allen items and then summing all 10 items. The variables in the data set are shown in Table 51.

Table 51
Items from a Humor Measure

Variables	Definition
item01	I like to make fun of others. (Don Rickles)
item02	I make people laugh by making fun of myself. (Woody Allen)
item03	People find me funny when I make jokes about others. (Don Rickles)
item04	I talk about my problems to make people laugh. (Woody Allen)
item05	I frequently make others the target of my jokes. (Don Rickles)
item06	People find me funny when I tell them about my failings. (Woody Allen)
item07	I love to get people to laugh by using sarcasm. (Don Rickles)
item08	I am funniest when I talk about my own weaknesses. (Woody Allen)
item09	I make people laugh by exposing other people's stupidities. (Don Rickles)
item10	I am funny when I tell others about the dumb things I have done. (Woody Allen)

4. Compute a reliability analysis on the total score, using coefficient alpha.
5. Conduct the analysis again, this time computing a split-half reliability coefficient. Can you justify the method you chose to split the items?
6. Which estimate of reliability is the more appropriate one? Why?
7. Write a Results section based on your analyses.

LESSON

38

Item Analysis Using the Reliability Procedure

Multiple-item scales can be developed to assess characteristics about individuals. These characteristics are called constructs (or dimensions) because they are constructed by researchers to explain behavior and are not directly observable. Examples of constructs include verbal fluency, emotionality, preference for Republican candidates, and leg strength. Constructs like leg strength may be assessed using multiple measures or items, such as maximal weight lifted using leg flexion, maximal weight lifted using leg extensions, and number of leg extensions with a 20-pound weight. However, even when we combine scores across these multiple measures, we do not obtain a perfectly accurate assessment of a construct because our measures contain measurement error.

We conduct item analyses to decide which items to include or to exclude from a scale. The objective of item analysis is to select a set of items that yields a summed score that is more strongly related to the construct of interest than any other possible set of items. Item analysis is problematic because we cannot relate our items to a direct measure of a construct to make our item selections. Instead, we use a poor representation of the construct, the sum of the items, and make decisions about items based on their relationship to this total score. Given the problems inherent in item analysis, researchers should select items to include on their scale based not only on the correlations between item scores and total scores, but also on their knowledge about the items and how they rationally and theoretically relate to the constructs.

The data required to conduct an item analysis are the item scores for a group of participants. Each participant is a case in the SPSS data file, and the items are the variables in the data set.

Applications of Item Analysis

The Reliability Analysis procedure can analyze items on measures that assess one or more constructs. The methods used in the Reliability Analysis procedure assume that scale scores will be computed by summing item scores. Consequently, before we conduct an item analysis, we must determine whether any items need to be transformed to create a total score. Lesson 19 explains how to perform the transformations. We'll look at three examples of item analysis for a measure assessing a single construct and one example of item analysis for a measure assessing multiple constructs.

- Measure of single construct requiring no transformations
- Measure of single construct requiring reverse-scoring of some item scores
- Measure of single construct requiring *z* transformations of all item scores and reverse-scoring of some items
- Measure that assesses multiple constructs

For applications with measures that assess multiple constructs, results from the Bivariate Correlation procedure are useful as well as those from the Reliability Analysis procedure.

Measure of a Single Construct Requiring No Transformations

Bob teaches a high school history class of 60 students and would like to item-analyze a 50-item achievement test on Civil War history that he has administered. Each item on the test is scored 0 for incorrect and 1 for correct. The scale score is the sum of these scored items. Bob's SPSS data file contains 60 cases and 50 variables representing the 50 items. These 50 items are the variables analyzed using the Reliability Analysis procedure.

Measure of a Single Construct Requiring Reverse-Scoring of Some Item Scores

Mickey has developed a 14-item measure called the Bias Against Women scale. He asks 100 individuals to respond to these items on a 1-to-5 scale, with 1 being "completely disagree" and 5 being "completely agree." Half the items are phrased so that agreement indicates a bias against women (the biased items), while the other half are written so that agreement indicates a lack of bias against women (the unbiased items). Mickey's SPSS data file contains 100 cases and 14 variables representing the 14 items. The unbiased items need to be reverse-scaled so that a response of 1 is transformed to a 5, a 2 is transformed to a 4, and so on. The scores used by the Reliability Analysis procedure include the scores for the seven biased items and the transformed scores for the seven unbiased items.

Measure of a Single Construct Requiring *z* Transformations of All Item Scores and Reverse-Scoring of Some Items

Julie wants to develop a behavioral index of friendliness. She has the 60 participants in her study discuss for five minutes with a stranger what they like to do in their free time. The index includes 15 behavioral measures (items), such as how far in feet participants stand when talking to a stranger, percent of time smiling, number of sentences initiated by participants, and a 5-point rating of warmth of voice. Julie's SPSS data file contains 60 cases and 15 item scores. All 15 items have to be transformed to standard scores. In addition, some of the items need to be reverse-scaled. For example, distance in feet participants stand when talking to the stranger would need to be reverse-scaled because distance presumably is inversely related to the construct of friendliness; that is, the closer a person stands to the stranger the friendlier they are. The scores used by the Reliability Analysis procedure include the 15 z-transformed item scores, at least one of which has been reverse-scaled.

Measures of Multiple Constructs

Margaret wants to develop a coping measure that evaluates the degree that teachers use problem-focused and emotion-focused coping with difficult students. She has 91 teachers take a coping measure that has 5 items that assess problem-focused coping and 5 items that assess emotion-focused coping. Margaret's SPSS data file contains 91 cases and 10 items as variables.

Understanding Item Analysis

After transforming items appropriately to create a meaningful total score for a scale, a correlation is computed between each item on the scale and the total score, excluding the item of interest from the total score. SPSS refers to these correlations as corrected item–total correlations.

The term *corrected* is used to indicate that the item of interest is not included in the total score so that the correlation is not spuriously positive. To the extent that the item total represents the construct of interest, the items should be strongly correlated with the corrected total score. Under these circumstances, researchers should choose items to include on their scales if they have high positive correlations and delete or revise items on their scales if they have negative or low positive correlations.

You may need to conduct the item analysis procedure multiple times to evaluate the appropriateness of items. The first time the procedure is conducted, the worst item can be eliminated from the scale. Once this item is eliminated the total scores change. Thus, it is necessary to redo the item analysis with the worst item removed to evaluate whether any additional items should be deleted. This process continues until a satisfactory set of items remains. Because of the changing nature of the corrected total score, it is certainly possible to add one of the items deleted in an earlier step back to the scale to ensure that the corrected item–total correlation still indicates that it is a poor item.

There are difficulties with choosing items based on corrected item–total correlations. In practice, it is likely that items assess not only the construct of interest but also irrelevant factors and that the corrected total score is a function of the relevant construct and the irrelevant factors. Accordingly, a positive correlation between an item and the corrected total indicates that an item assesses the relevant construct, the irrelevant factors that are shared with the other items on the scale, or both. For example, let's say that 8 of 10 items on a supervisory feedback scale begin with "My supervisor tells me" and the other 2 items begin with "I receive feedback." The items starting with "My supervisor" may produce higher corrected item–total correlations than the other items because they are similarly worded, not because they are better measures of supervisory feedback.

There are additional reasons for not blindly following the results of an item analysis. There may be a number of narrowly defined constructs that are a function of a more broadly defined construct of interest. For example, let's say that we are interested in the broadly defined construct of constructive play and that the items on the scale assess, in addition to the broadly defined construct, one or more narrowly defined constructs: constructive play by oneself indoors, constructive play with games indoors, constructive play with games outdoors, and so on. Let's also assume that the scale was not initially developed well and a majority of the items on the constructive play scale pertain to constructive play with games indoors. A researcher who conducts an item analysis and chooses the items with the highest correlations with the corrected total may create a scale measuring constructive play with games indoors rather than constructive play in general.

A more complex procedure is required for measures that are designed to assess multiple constructs or dimensions. We should compute correlations between item scores for a dimension and corrected total scores for the same dimension. We also should compute correlations between item scores for a dimension and total scores assessing different dimensions. An item should be correlated with its own scale (convergent validity). In addition, an item should be correlated more with its own scale than with scales assessing different constructs (discriminant validity). The Reliability procedure can be used to determine the correlation of an item with its own corrected total scale. However, the Bivariate Correlation procedure is used to determine the correlation between item scores and total scores for other dimensions. Again, when making decisions about keeping or omitting items, you must consider the content of the items and not just the magnitudes of the correlation coefficients.

Assumption Underlying Item Analyses

Assumption 1: Items Are Linearly Related to a Single Factor Plus Random Measurement Error

As discussed in the section "Understanding Item Analysis," this assumption is problematic. Researchers can reach wrong conclusions using only the results of an item analysis. It is important for researchers not to make decisions about what items to include on a scale based solely on the results of an item analysis.

The Data Set

The data set used to illustrate two of our examples is named *Lesson 38 Data File 1* on the Web at http://www.prenhall.com/greensalkind. The data set contains the results of Margaret's survey of 91 teachers. The teachers were asked to indicate the degree to which they used 10 different approaches to deal with problems between themselves and their most recalcitrant students. The 10 items on the coping questionnaire are shown in the table. The teachers rated each item on a 4-point scale, with 1 = Did not try this approach to 4 = Used this approach extensively. Table 52 shows the variables in the data file.

Table 52
Variables in Lesson 38 Data File 1

Variables	Definition
item_a	I discussed my frustrations and feelings with person(s) at school.
item_b	I tried to develop a step-by-step plan of action to remedy the problems.
item_c	I expressed my emotions to my family and close friends.
item_d	I read, attended workshops, or sought some other educational approach to correct them.
item_e	I tried to be emotionally honest with myself about the problems.
item_f	I sought advice from others on how I should solve the problems.
item_g	I explored the emotions caused by the problems.
item_h	I took direct action to try to correct the problems.
item_i	I told someone I could trust about how I felt about the problems.
item_j	I put aside other activities so that I could work to solve the problems.

The Research Question

The research question to be asked depends upon whether the measure is viewed as assessing a single construct, Active Coping, or assessing two constructs, Emotion-Focused Coping and Problem-Focused Coping.

1. Single construct approach: Should one or more items on Margaret's measure be deleted or revised to obtain a better measure of Active Coping?
2. Multiple constructs approach: Should one or more items on Margaret's measure be deleted or revised to obtain better measures of Emotion-Focused Coping and Problem-Focused Coping? Items a, c, e, g, and i are hypothesized to measure Emotion-Focused Coping, while items b, d, f, h, and j are hypothesized to measure Problem-Focused Coping.

Conducting Item Analyses

We illustrate how to conduct item analyses for a single construct and for multiple constructs of Margaret's measure.

Conducting Item Analysis of a Measure of a Single Construct

To conduct an item analysis assessing a single construct, follow these steps:

1. Click **Analyze**, click **Scale**, and click **Reliability Analysis**.
2. Holding down the Ctrl key, select all 10 variables (**item_a** through **item_j**) by clicking on each one.

3. Click ▶ to move the variables to the Items box.
4. Click **Statistics**.
5. Click **Item**, **Scale**, and **Scale if Item Deleted** in the Descriptives area.
6. Click **Correlations** in the Inter-Item area.
7. Click **Continue**.
8. In the Reliability Analysis dialog box, make sure that **Alpha** is chosen in the Model drop-down menu.
9. Click **OK**.

Selected SPSS Output for Item Analysis of a Measure of a Single Construct

The results of the analysis are shown in Figure 233.

Item-Total Statistics

	Scale Mean if Item Deleted	Scale Variance if Item Deleted	Corrected Item-Total Correlation	Squared Multiple Correlation	Cronbach's Alpha if Item Deleted
item_a	25.45	17.250	.546	.530	.671
item_b	25.51	16.897	.484	.353	.676
item_c	26.08	17.094	.353	.318	.700
item_d	26.37	17.370	.309	.184	.709
item_e	25.15	18.309	.357	.252	.698
item_f	25.40	17.397	.448	.514	.683
item_g	25.98	16.488	.457	.298	.680
item_h	25.03	20.121	.113	.212	.727
item_i	25.55	16.717	.510	.516	.672
item_j	26.53	18.963	.203	.203	.721

Figure 233. **The results of the item analysis.**

Item_h had the lowest corrected item–total correlation, and item_j had the next lowest correlation; therefore, they were candidates for elimination. To ensure that item_j would still have a low correlation after deleting item_h from the scale, we reran the Reliability Analyses procedure without item_h. The output is shown in Figure 234 (on page 340). As expected, item_j now had the lowest corrected item–total correlation.

Examining the item content of item_h and item_j, we concluded that they differed from the other eight items in that a teacher who endorses them indicates that he or she took immediate action to correct the difficulties. We decided that it might be worthwhile to create a second scale that included these two items plus additional items that we would have to develop.

The Reliability procedure was rerun without both these items to determine whether additional revisions needed to be made. The output is shown in Figure 235 (on page 340).

After examining the results in Figure 235 and the content of the items, we chose not to eliminate any other items.

Conducting Item Analysis of a Measure of Multiple Constructs

We now want to conduct an item analysis to evaluate whether items a, c, e, g, and i are associated with one scale (Emotion-Focused Coping) and items b, d, f, h, j are associated with a second scale (Problem-Focused Coping). We will conduct these analyses in four stages and discuss the output after describing these four stages.

Reliability Statistics

Cronbach's Alpha	Cronbach's Alpha Based on Standardized Items	N of Items
.727	.735	9

Item-Total Statistics

	Scale Mean If Item Deleted	Scale Variance If Item Deleted	Corrected Item-Total Correlation	Squared Multiple Correlation	Cronbach's Alpha If Item Deleted
item_a	21.92	16.116	.585	.513	.677
item_b	21.98	16.266	.439	.265	.696
item_c	22.55	16.050	.367	.317	.711
item_d	22.85	16.532	.295	.176	.726
item_e	21.63	17.437	.342	.247	.713
item_f	21.87	16.294	.477	.513	.691
item_g	22.45	15.473	.471	.290	.689
item_i	22.02	15.644	.536	.515	.679
item_j	23.00	18.156	.178	.197	.739

Figure 234. **A rerun of the item analysis without item_h.**

Reliability Statistics

Cronbach's Alpha	Cronbach's Alpha Based on Standardized Items	N of Items
.739	.749	8

Item-Total Statistics

	Scale Mean If Item Deleted	Scale Variance If Item Deleted	Corrected Item-Total Correlation	Squared Multiple Correlation	Cronbach's Alpha If Item Deleted
item_a	19.89	14.321	.590	.492	.688
item_b	19.95	14.786	.388	.202	.720
item_c	20.52	14.141	.384	.315	.724
item_d	20.81	14.842	.277	.164	.747
item_e	19.59	15.733	.316	.236	.732
item_f	19.84	14.228	.526	.489	.695
item_g	20.42	13.668	.479	.287	.702
item_i	19.99	13.633	.581	.510	.683

Figure 235. **A rerun of the item analysis without item_h and item_j.**

Stage 1: Corrected Item–Total Correlations for the Emotion-Focused Coping Items

We first compute correlations for the Emotion-Focused Coping items, which are items a, c, e, g, and i. To do so, follow Steps 1 through 9 in our previous example, except use items a, c, e, g, and i only.

Stage 2: Corrected Item–Total Correlations for the Problem-Focused Coping Items

For the Problem-Focused Coping items, follow again Steps 1 through 9, except use items b, d, f, h, and j only.

Stage 3: Computing Correlations between the Emotion-Focused Coping Items and the Problem-Focused Coping Scale

To assess the discriminant validity of the Emotion-Focused Coping items, we need to compute correlations between each item on the Emotion-Focused Coping scale and the total score for the Problem-Focused Coping scale. To do so, follow the steps on page 342:

Reliability Statistics

Cronbach's Alpha	Cronbach's Alpha Based on Standardized Items	N of Items
.694	.698	5

Item-Total Statistics

	Scale Mean If Item Deleted	Scale Variance If Item Deleted	Corrected Item–Total Correlation	Squared Multiple Correlation	Cronbach's Alpha If Item Deleted
item_a	11.48	6.297	.518	.345	.623
item_c	12.11	5.566	.441	.306	.652
item_e	11.19	7.065	.287	.175	.703
item_g	12.01	5.744	.428	.228	.656
item_i	11.58	5.490	.609	.454	.573

Figure 236. **The corrected item–total correlations for the Emotion-Focused Coping items.**

Reliability Statistics

Cronbach's Alpha	Cronbach's Alpha Based on Standardized Items	N of Items
.519	.516	5

Item-Total Statistics

	Scale Mean If Item Deleted	Scale Variance If Item Deleted	Corrected Item-Total Correlation	Squared Multiple Correlation	Cronbach's Alpha If Item Deleted
item_b	10.91	3.703	.505	.297	.309
item_d	11.78	3.773	.322	.110	.443
item_f	10.80	4.938	.152	.155	.541
item_h	10.44	5.094	.245	.168	.490
item_j	11.93	4.596	.243	.150	.490

Figure 237. **The corrected item–total correlations for the Problem-Focused Coping items.**

1. Create a new variable named **prob_tot** that is the sum of all items on the Problem-Focused Coping scale by using the Compute Variable option under Transform on the main menu. See Lesson 19 for information on how to do this.
2. Click **Analyze**, click **Correlation**, then click **Bivariate**.
3. Select **item_a**, **item_c**, **item_e**, **item_g**, and **item_i** by holding down the Ctrl key and clicking on each item.
4. Click ▶ to move the variables to the Variables box.
5. Click **prob_tot**, then click ▶ to move it to the Variables box.
6. Click **Paste**. Type **with** between item_i and prob_tot on the /VARIABLES line. It should read:

   /VARIABLES= item_a item_c item_e item_g item_i with prob_tot

7. Highlight the appropriate syntax, click **Run**, and then click **Selection**.

Correlations

		prob_tot
item_a	Pearson Correlation	.391**
	Sig. (2-tailed)	.000
	N	91
item_c	Pearson Correlation	.168
	Sig. (2-tailed)	.112
	N	91
item_e	Pearson Correlation	.308**
	Sig. (2-tailed)	.003
	N	91
item_g	Pearson Correlation	.333**
	Sig. (2-tailed)	.001
	N	91
item_i	Pearson Correlation	.266*
	Sig. (2-tailed)	.011
	N	91

**. Correlation is significant at the 0.01 level

*. Correlation is significant at the 0.05 level (2-tailed).

***Figure 238.* The correlations between the Emotion-Focused Coping items and the Problem-Focused Coping total score.**

Stage 4: Computing Correlations between the Problem-Focused Coping Items and the Emotion-Focused Coping Scale

To assess the discriminant validity of the Problem-Focused items, correlations need to be computed between each item on the Problem-Focused Coping scale and the total score for the Emotion-Focused Coping scale. To do so, follow the process as in Stage 3 except that you will create a new variable named emo_tot instead of prob_tot. The new variable will be the sum of all items on the Emotion-Focused Coping scale, and you will correlate this new variable with each Problem-Focused Coping item, items b, d, f, h, and j.

Selected SPSS Output for Item Analysis of a Measure of Multiple Constructs

The outputs for Stages 1 through 4 are shown in Figures 236 through 239. We created Table 53 from the results presented in these figures. Based on the results in Table 53, it appears that the scales should be revised. For example, item_f is much more correlated with Emotion-Focused Coping than it is with its own scale. Also, item_e is equally strongly related to both coping scales. After considerable thought about the construct system of coping and examination of the

Correlations

		emot_tot
item_b	Pearson Correlation	.344**
	Sig. (2-tailed)	.001
	N	91
item_d	Pearson Correlation	.223*
	Sig. (2-tailed)	.034
	N	91
item_f	Pearson Correlation	.518**
	Sig. (2-tailed)	.000
	N	91
item_h	Pearson Correlation	-.015
	Sig. (2-tailed)	.888
	N	91
item_j	Pearson Correlation	.122
	Sig. (2-tailed)	.248
	N	91

**. Correlation is significant at the 0.01 level

*. Correlation is significant at the 0.05 level (2-tailed).

***Figure 239.* The correlations between the Problem-Focused Coping items and the Emotion-Focused Coping total score.**

content of the items, the items were reconceptualized as measuring two related constructs. The constructs are Other-Directed Coping, assessed by items a, c, f, and i, and Self-Directed Coping, measured by items b, d, e, h, and j. Because it is not clear whether item_g reflects a self-directed or other-directed coping strategy and because the correlations between item_g and the two scales did not clearly differentiate it, item_g was omitted from the revised scale.

Table 53

Correlations of Each Coping Item with Its Own Scale after Removing Focal Item (in Bold Type) and with the Other Scale

	Factors	
Items	Problem-Focused Coping	Emotion-Focused Coping
Problem-Focused Coping items		
b. I tried to develop a step-by-step plan of action to remedy the problems.	**.51**	.34
d. I read, attended workshops, or sought some other educational approach to correct them.	**.32**	.22
f. I sought advice from others on how I should solve the problems.	**.15**	.52
h. I took direct action to try to correct the problems.	**.25**	−.02
j. I put aside other activities so that I could work to solve the problems.	**.24**	.12
Emotion-Focused Coping items		
a. I discussed my frustrations and feelings with person(s) at school.	.39	**.52**
c. I expressed my emotions to my family and close friends.	.17	**.44**
e. I tried to be emotionally honest with myself about the problems.	.31	**.29**
g. I explored the emotions caused by the problems.	.33	**.43**
i. I told someone I could trust about how I felt about the problems.	.27	**.61**

We then conducted item analyses on these revised scales by using the stages for a measure assessing multiple constructs as just outlined. The output for the revised item analysis and correlations are shown in Figure 240.

Based on the results presented in Figure 240, we constructed Table 54. We chose not to make any additional changes to scales after examining Table 54 (on page 346).

Reliability Statistics

Cronbach's Alpha	Cronbach's Alpha Based on Standardized Items	N of Items
.586	.597	5

Item Total Statistics

	Scale Mean If Item Deleted	Scale Variance If Item Deleted	Corrected Item-Total Correlation	Squared Multiple Correlation	Cronbach's Alpha If Item Deleted
item_b	11.15	4.065	.493	.263	.439
item_d	12.02	4.288	.276	.082	.586
item_e	10.80	4.938	.324	.116	.541
item_h	10.68	5.264	.322	.137	.547
item_j	12.18	4.635	.342	.134	.531

Correlations

		oth_tot
item_b	Pearson Correlation	.262*
	Sig. (2-tailed)	.012
	N	91
item_d	Pearson Correlation	.149
	Sig. (2-tailed)	.160
	N	91
item_e	Pearson Correlation	.156
	Sig. (2-tailed)	.139
	N	91
item_h	Pearson Correlation	-.085
	Sig. (2-tailed)	.423
	N	91
item_j	Pearson Correlation	.018
	Sig. (2-tailed)	.865
	N	91

*. Correlation is significant at the 0.05 level (2-tailed).

Figure 240. **Results of item analysis for revised Other-Directed Coping and Self-Directed Coping scales.**

Reliability Statistics

Cronbach's Alpha	Cronbach's Alpha Based on Standardized Items	N of Items
.787	.799	4

Item-Total Statistics

	Scale Mean If Item Deleted	Scale Variance If Item Deleted	Corrected Item-Total Correlation	Squared Multiple Correlation	Cronbach's Alpha If Item Deleted
item_a	8.66	4.783	.644	.473	.719
item_c	9.29	4.273	.490	.288	.804
item_f	8.60	4.620	.598	.456	.733
item_i	8.76	4.163	.692	.479	.683

Correlations

		self_tot
item_a	Pearson Correlation	.223*
	Sig. (2-tailed)	.034
	N	91
item_c	Pearson Correlation	.075
	Sig. (2-tailed)	.478
	N	91
item_f	Pearson Correlation	.147
	Sig. (2-tailed)	.163
	N	91
item_i	Pearson Correlation	.148
	Sig. (2-tailed)	.163
	N	91

*. Correlation is significant at the 0.05 level (2-tailed).

Figure 240. **(continued)**

Using SPSS Graphs to Display the Results

No graphics procedures are necessary to display the results of the item analysis procedure.

Two APA Results Sections

We demonstrate how to write sample Results sections for a single construct application and a multiple constructs application.

Table 54

Correlations of Each Coping Item with Its Own Scale after Removing Focal Item (in Bold Type) and with the Other Scale

Item	Factors	
	Self-Directed	Other-Directed
Self-Directed (SD) items		
b. I tried to develop a step-by-step plan of action to remedy the problems.	**.49**	.26
d. I read, attended workshops, or sought some other educational approach to correct them.	**.28**	.15
e. I tried to be emotionally honest with myself about the problems.	**.32**	.16
h. I took direct action to try to correct the problems.	**.32**	−.08
j. I put aside other activities so that I could work to solve the problems.	**.34**	.02
Other-Directed (OD) items		
a. I discussed my frustrations and feelings with person(s) at school.	.22	**.64**
c. I expressed my emotions to my family and close friends.	.08	**.49**
f. I sought advice from others on how I should solve the problems.	.15	**.60**
i. I told someone I could trust about how I felt about the problems.	.15	**.69**

Results for a Single Construct

Item analyses were conducted on the 10 items hypothesized to assess Active Coping. Initially, each of the 10 items was correlated with the total score for Active Coping (with the item removed). All the correlations were greater than .30 except for two items: "I put aside other activities so that I could work to solve the problems" ($r = .20$) and "I took direct action to try to correct the problems" ($r = .11$). These two items differed in content from the other eight items in that a teacher who endorses them indicates that he or she took immediate action to correct the difficulties. Based on these results, the two items assessing immediate coping methods were eliminated from the scale. However, these items represent an important coping approach. Additional items should be developed to create a separate scale to assess immediate coping strategies.

Item–total correlations for the revised eight-item scale yielded only one correlation that was less than .30: "I tried to read, attend workshops, or sought some other educational approach to correct them." This item was retained in that its content did not appear to differ markedly from the content of the other seven items. The revised scale was renamed "Preliminary Coping Strategies" to reflect the content of the items. Coefficient alpha for the Preliminary Coping Strategies scale was .74. Because the same sample was used to conduct the item analyses and to assess coefficient alpha, the reliability estimate is likely to be an overestimate of the population coefficient alpha.

Results for Multiple Constructs

Item analyses were conducted on the 10 items hypothesized to assess Emotion-Focused and Problem-Focused Coping. Initially, each item was correlated with its own scale (with the item removed) and with the other coping scale. In two cases, items were more highly correlated with the other coping scale than their own scale. The two items were "I sought advice from others on how I should solve the problems," and "I tried to be emotionally honest with myself about the problems." Based on these results and additional item analyses, the two scales were redefined based on 9 of the 10 items and renamed as Other-Directed Coping and Self-Directed Coping. The eliminated item was "I explored the emotions caused by the problems."

To assess the convergent and discriminant validity of these newly derived coping scales, we again correlated each item with its own scale (with the item removed) and with the other coping scale. The results of these analyses are shown in Table 54. In support of the measure's validity, items always were more highly correlated with their own scale than with the other scale. Coefficient alphas were computed to obtain internal consistency estimates of reliability for these two coping scales. The alphas for the Self-Directed and Other-Directed Coping scales were .59 and .79, respectively. These values might be overestimates of the population alphas because the same sample was used to conduct the item analyses and to compute the reliability estimates.

Alternative Analyses

With item analyses, researchers are attempting to determine if items are associated with particular constructs. An alternative statistical approach is to use factor analysis. With factor analysis, all items associated with the same construct should have high loadings on the same factor and relatively low loadings on other factors. Factor analyses of items can be more precisely conducted using confirmatory methods rather than the exploratory factor analysis approaches available with SPSS. Problems may emerge using factor analysis if items have limited response scales and the items are differentially skewed.

Exercises

The data for Exercises 1 through 4 are in the data file named *Lesson 38 Exercise File 1* on the Web at http://www.prenhall.com/greensalkind. The data are based on the following research problem.

Terrill is interested in assessing how much women value their careers. He develops a 12-item scale, the Saxon Career Values Scale (SCVS). He has 100 college women take the SCVS. All of the items reflect the value women place on having a career versus having a family. Students are asked to respond to each on a 4-point scale with 0 indicating "disagree" and 3 indicating "agree" The items are shown in Table 55.

Table 55
Items on the Saxon Career Values Scale

Variables	Definition
q01	I consider marriage and having a family to be more important than a career.*
q02	To me, marriage and family are as important as having a career.*
q03	I prefer to pursue my career without the distractions of marriage, children, or a household.
q04	I would rather have a career than a family.
q05	I often think about what type of job I'll have 10 years from now.
q06	I could be happy without a career.*
q07	I would feel unfulfilled without a career.
q08	I don't need to have a career to be fulfilled.*
q09	I would leave my career to raise my children.*
q10	Having a career would interfere with my family responsibilities.*
q11	Planning for and succeeding in a career is one of my primary goals.
q12	I consider myself to be very career-minded.

*Items are reverse-scaled to create a total score.

1. Reverse-scale the appropriate items so a higher total score reflects a greater emphasis on career.
2. Conduct an item analysis for the SCVS. From the output, identify the following:
   a. Corrected item–total correlation for Item 6
   b. Correlation between Item 5 and Item 9
   c. Mean for Item 10
3. Should any items be eliminated or revised? Justify your answer based on the statistical indices and your understanding of the scale.
4. Write a Results section reporting your analyses.

The data for Exercises 5 through 9 are in the data file named *Lesson 38 Exercise File 2* on the Web at http://www.prenhall.com/greensalkind. The data are based on the following research problem.

Jessica is interested in assessing humor. She develops a 10-item measure in which some items represent humor demeaning to others (Don Rickles items), while other items reflect self-deprecating humor (Woody Allen items). She administers her measure to 100 college students. Students are asked to respond to each on a 5-point scale with 1 indicating "disagree" and 5 indicating "agree." Table 56 shows the variables in the data set.

Table 56
Items from a Humor Measure

Variables	Definition
item01	I like to make fun of others. (Don Rickles)
item02	I make people laugh by making fun of myself. (Woody Allen)
item03	People find me funny when I make jokes about others. (Don Rickles)
item04	I talk about my problems to make people laugh. (Woody Allen)
item05	I frequently make others the target of my jokes. (Don Rickles)
item06	People find me funny when I tell them about my failings. (Woody Allen)
item07	I love to get people to laugh by using sarcasm. (Don Rickles)
item08	I am funniest when I talk about my own weaknesses. (Woody Allen)
item09	I make people laugh by exposing other people's stupidities. (Don Rickles)
item10	I am funny when I tell others about the dumb things I have done. (Woody Allen)

5. Conduct an item analysis on this scale for a single dimension. (Note: You will want to reverse-scale either the Woody Allen or Don Rickles items, but not both sets of items.)
6. Conduct the analysis again, this time computing corrected item–total correlations for separate scales: (1) the Don Rickles scale and (2) the Woody Allen scale.
7. Compute correlations between each item and the other scale to assess discriminant validity.
8. Create a table to summarize your item analysis for the Don Rickles and Woody Allen scales.
9. Write a Results section based on your analyses.

PART II

UNIT 10

Nonparametric Procedures

Unit 10 describes a set of nonparametric tests. They are useful for problems that include one or more variables measured on a nominal or an ordinal scale. These procedures can apply to problems involving interval or ratio data if the distributional assumptions associated with parametric procedures are not met. Even if a parametric test is relatively robust to violation of a distributional assumption, the nonparametric alternative may be more powerful for certain types of population distributions.

Two of the nonparametric tests—the binomial test and the chi-square test—evaluate whether the proportions associated with categories on a single variable are equal to hypothesized values. The remaining techniques evaluate relationships between two variables. One of the two variables is always categorical, while the other variable will be either categorical or quantitative. The lessons that discuss techniques for evaluating relationships between variables are divided according to whether the categorical variable is based on independent or related samples and on the number of samples.

Here is a brief introduction to each lesson of Unit 10.

- Lesson 39 introduces the binomial test, which evaluates whether the proportions associated with a two-category variable are equal to hypothesized values.
- Lesson 40 presents the one-sample chi-square test. It evaluates whether the proportions associated with a variable with two or more categories are equal to hypothesized values.
- Lesson 41 deals with two-way contingency table analysis. We will use the Crosstabs procedure to evaluate the relationship between two qualitative variables. Each qualitative variable may have two or more categories.
- Lesson 42 discusses nonparametric procedures for analyzing two independent samples. We will focus on the Mann-Whitney *U* test, which evaluates differences on an outcome variable between the two independent samples.
- Lesson 43 summarizes nonparametric procedures used to analyze two or more independent samples. We will focus on the Kruskal-Wallis test and the median test, which evaluate the differences on an outcome variable among two or more independent samples.
- Lesson 44 examines nonparametric procedures for analyzing two related samples. Here, we will discuss the McNemar, Sign, and Wilcoxon tests. These tests evaluate differences between paired observations on an outcome variable.
- Lesson 45 describes nonparametric procedures used to analyze two or more related samples. We will describe analyses that use the Cochran test, the Friedman test, and Kendall's W (coefficient of concordance). These statistics assess differences among levels of a within-subjects factor on an outcome variable.

LESSON

39 Binomial Test

The binomial test evaluates whether the proportions of individuals who fall into the categories of a two-category variable are equal to hypothesized values. The test is more likely to yield significance if the sample proportions for the two categories differ greatly from the hypothesized proportions and if the sample size is large.

In SPSS, we make a hypothesis about only one of the two categories, the category specified for the variable's first case in the data file. SPSS calls the hypothesized proportion the test proportion. SPSS computes the hypothesized proportion for the second category by subtracting the hypothesized proportion for the first category from the value of 1.

In many applications, the test proportion is hypothesized to be .50. This value is also the SPSS default option. However, it is possible to specify values other than .50 based on previous research, theory, research methods, or speculation.

An SPSS data file for the binomial test may be structured two ways. With the standard method, the SPSS data file contains as many cases as individuals. For each case, there is a single variable that has two values that represent the two categories for the variable of interest. With the weighted cases method, the SPSS data file contains two cases, one for each category, and two variables. The two variables are the focal variable with two values for the two categories and the weight variable containing frequencies for the two categories.

Applications of the Binomial Test

We consider two applications of the binomial test.

- Binomial test with equal proportions
- Binomial test with unequal proportions

Binomial Test with Equal Proportions

Tucker is interested in assessing whether one-year-old girls prefer boys or girls. He randomly forms triplets of two girls and one boy from a sample of 120 one-year-old girls and 60 one-year-old boys. One of the two girls is randomly determined to be the focal subject, and the other girl and the boy are designated as companions. Each triplet is placed in a playroom for 10 minutes and observed behind a one-way mirror. An observer assesses whether the focal subject spends more time closer to the male companion or the female companion. The gender preference for each focal subject is recorded.

Tucker's SPSS data file can have one of two types of structure. With the standard structure, the data file consists of 60 cases to represent the 60 focal subjects. Each girl is measured on a single variable sexpref. The scores on sexpref are 1 = prefer girl and 2 = prefer boy. The values of 1 and 2 are arbitrary and could have been any other set of values such as 0 and 1. With the weighted cases structure, Tucker would have two cases, representing the two categories. Each case would have scores on two variables: sexpref, with one case having a score of 1 and the other case having a score of 2, and a weight variable. Values for the weight variable represent the number of individuals who are classified as belonging to the categories defined by sexpref. If the weighted cases method is used, you must weight the cases by using the Weight Cases option under Data from the main menu before conducting a binomial test.

For Tucker's problem, the test proportion is .50. If the first case in the data file is a score of 1, then the test proportion would be the hypothesized proportion for prefer girl. Of course because the proportion is .50, it does not make any difference whether the hypothesized proportion refers to prefer girl or prefer boy because they are both equal to .50.

Binomial Test with Unequal Proportions

Based on past research, Pam knows that children who are seated in a room with four doors—one in front of them, one to their left, one to their right, and one behind them—are equally likely to pick any door when asked which door they would choose when leaving the room. As part of a larger research program, she wants to demonstrate that children labeled by a psychologist as oppositional are more likely to choose the door behind them. To determine whether her conjecture is correct, she conducts a study with 80 oppositional children and asks them to pick one of four doors to leave the room. On the basis of her results, Pam constructs a data file that uses the standard method by creating a single variable, door, for the 80 cases. The scores on door are 0 = other door and 1 = door behind. The first case in the data file has a 0 for other door. The values of 0 and 1 are arbitrary and could have been any other set of values such as 1 and 2. For this problem, the test proportion is the hypothesized proportion associated with door = 0, the first value for door in the data file. Previous research showed that children pick the four doors equally often in general. Because there are four doors, the test proportion for the other door is .75 (3 out of 4 doors).

These data could also be entered using the weighted cases method. With this method, Pam would have two cases, representing the two categories. Each case would have scores on two variables: door, with the first case having a score of 0 and the second case having a score of 1, and a weight variable. Values for the weight variable represent the number of individuals who are classified as belonging to the categories defined by door. If the weighted cases method is used, you must weight the cases by using the Weight Cases option under Data from the main menu before conducting a binomial test.

Understanding the Binomial Test

The binomial test procedure uses a binomial distribution if the sample size is 25 or less and a z test if the sample size is 26 or more. Let's discuss the underlying assumptions for these tests and effect size statistics.

Assumptions Underlying the Binomial Test

Both tests—the test using a binomial distribution and the z test—assume independence. In addition, the z test also requires a relatively large sample size to yield accurate results.

Assumption 1: The Observations Must Be from a Random Sample, and the Data for These Observations Are Independent of Each Other

To meet this assumption, studies should be designed to prevent dependency in the data. Research participants should be sampled independently, participate in the study independently, and should contribute only a single score to the data. If the assumption is violated, the test is likely to yield inaccurate results.

Assumption 2: The z Test Yields Relatively Accurate Results to the Extent That the Sample Size Is Large and the Test Proportion Is Close to .50

The z test is likely to yield fairly accurate results if the test proportion is .50 because the z test is performed only if sample size exceeds 25 cases. However, to the extent that the hypothesized proportion deviates from .50, you should have larger sample sizes.

One guideline for sample size would be to apply the expected-frequency-greater-than-five standard. Let P_{small} be equal to the test proportion or 1 minus the test proportion, whichever is smaller. Then the minimum sample size should be equal to $5/P_{small}$. For example, if the test proportion is .90, then $P_{small} = 1-.90 = .10$ because .10 is smaller than .90, and the minimum sample size is $5/.10 = 50$.

Effect Size Statistics for the Binomial Test

SPSS does not report an effect size index. However, a simple and elegant effect size index for this test is the difference between the observed and the hypothesized (test) proportion:

$$Effect\ Size = P_{observed} - P_{hypothesized}$$

The Data Set

The data file is named *Lesson 39 Data File 1* on the Web at http://www.prenhall.com/greensalkind, and it uses the weighted cases structure. The data are based on the oppositional children example. For this application, there are two possible outcomes, door behind and other door. The description of the variables in the file is given in Table 57.

Table 57
Variables in Lesson 39 Data File 1

Variables	Definition
door	Children can choose either the door behind them or another door. 0 = Other door 1 = Door behind
number	Number of oppositional children who picked category 0 or category 1

The Research Question

The research question can be asked in the following way: Is the proportion associated with oppositional children choosing the door behind them different from .25, as suggested based on previous research with children in general?

Conducting a Binomial Test

Before these data can be analyzed, the cases must be weighted by the values in the number variable by conducting the following steps:

1. Click **Data**, then click **Weight Cases**. You will see the Weight Cases dialog box shown in Figure 241.

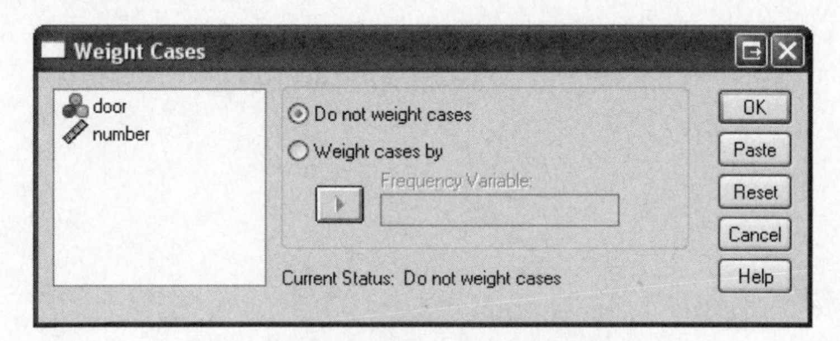

***Figure 241*. The Weight Cases dialog box.**

2. Click **Weight cases by**.
3. Click **number**, then click ▶ to move the variable into the Frequency Variable box.
4. Click **OK**.

To conduct a binomial test, follow these steps:

1. Click **Analyze**, click **Nonparametric Tests**, then click **Binomial Test**. You will see the Binomial Test dialog box as shown in Figure 242.

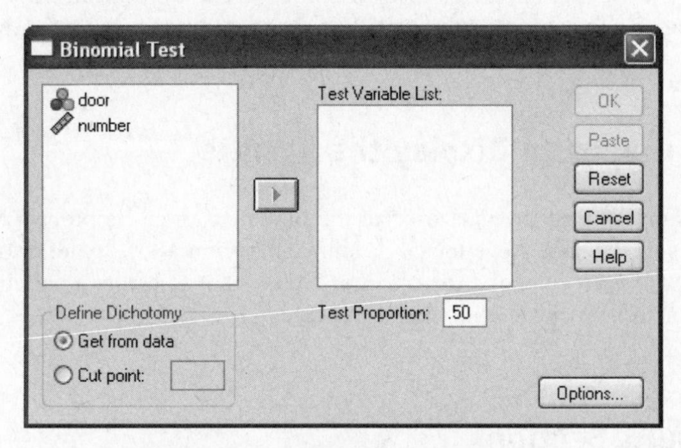

***Figure 242*. The Binomial Test dialog box.**

2. Click **door**, then click ▶ to move it to the Test Variable List box.
3. Change the Test Proportion to **.75**. This value is the hypothesized proportion associated with the value of 0 (other door), which is the first case in the data file.
4. Click **OK**.

In the Binomial Test dialog box, there are two options in the Define Dichotomy area. The default option is "Get from data." The second option is "Cut point." If you are interested in creating a two-category variable from a quantitative variable, you can click on the Cut point option and enter a value such that any score less than the cut point is assigned to one category and any score greater than or equal to the cut point is assigned to the second category. There is no change made to the variable in your data file.

Binomial Test

		Category	N	Observed Prop.	Test Prop.	Asymp. Sig. (1-tailed)
Choice of door	Group 1	Other Door	55	.69	.75	.124[a,b]
	Group 2	Door Behind	25	.31		
	Total		80	1.00		

a. Alternative hypothesis states that the proportion of cases in the first group < .75.

b. Based on Z Approximation.

Figure 243. **The results of the binomial test.**

Selected SPSS Output for the Binomial Test

The results of the analysis are shown in Figure 243. If the test proportion is .50, SPSS computes a two-tailed p value. If the test proportion is a value other than .50, such as .75 as in our example, SPSS computes a one-tailed p value. You may want to conduct a two-tailed test since two-tailed tests are more frequently applied. A two-tailed p value can be computed if the test is based on the z approximation (sample size of greater than 25) by multiplying the one-tailed p value by 2. For our example, the two-tailed p value would be $.248 = 2\ (.124)$.

In this example, the null hypothesis is not rejected using the traditional alpha level of .05 and using either a one-tailed or a two-tailed test. More specifically, the results for this example indicate that the observed proportion for other door of .6875 is not significantly different from the hypothesized proportion of .75. Alternatively stated, the results suggest that the observed proportion for door behind of .3125 (1 – .6875) is not significantly different from the hypothesized value of .25 (1 – .75).

It should be noted that you could not calculate directly a two-tailed p value from a one-tailed p value based on the exact binomial (sample size of less than 26). Consult a nonparametric textbook to determine how to compute a two-tailed p value based on the binomial distribution.

Using SPSS Graphs to Display the Results

Because of the simplicity of the data used in the binomial test, it is probably unnecessary to graphically represent the data. Nevertheless, a simple bar chart allows readers to see the proportions associated with each category. (See Lesson 20 for steps to create a bar chart.) Figure 244 shows the frequencies associated with the door choices.

An APA Results Section

Previous research with nonoppositional children indicated that they would choose the door behind them 25% of the time. We hypothesized that oppositional children would choose the door behind them with a higher percentage than 25%. A one-tailed, z approximation test was conducted to assess whether the population proportion for oppositional children is .25. The observed proportion of .31 did not differ significantly from the hypothesized value of .25, one-tailed $p = .12$. Although a larger sample size might have yielded significance, our results suggest that oppositional children do not differ dramatically from nonoppositional children in the percent of times that they choose the door behind them.

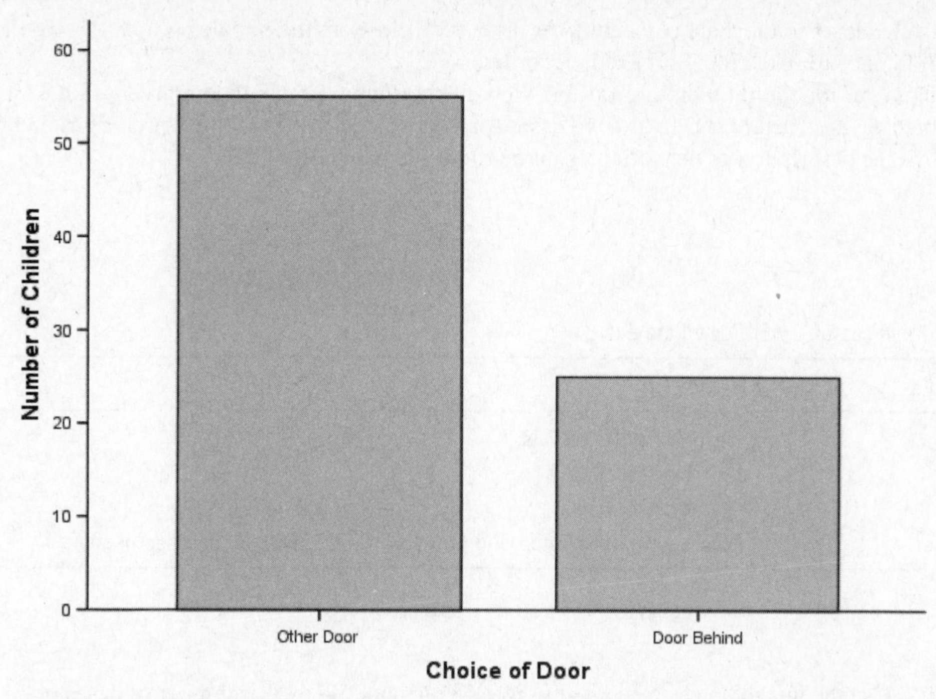

Figure 244. **The frequencies of children who chose the door behind versus some other door.**

Alternative Analyses

The binomial test discussed in this lesson is very similar to the one-sample chi-square test, which we will cover in Lesson 40. With both tests, the focus is on investigating whether the proportions associated with categories of a single variable are equal to a hypothesized set of values. The difference is that the variable of interest may have two or more categories for a one-sample chi-square test, while the variable of interest for the binomial test must have two categories.

For variables with two categories, analyses could be conducted with either the binomial test or the one-sample chi-square test. The binomial test procedure is preferred for two reasons. First, for sample sizes of 26 or greater, the two tests yield related test statistics, the χ^2 and the z. However, in SPSS, the one-sample chi-square procedure does not apply a continuity correction, while the binomial test procedure does. Because the continuity correction procedure is used with the binomial test procedure, the binomial test is preferred. Second, for sample sizes of 25 or less, the p value for the binomial test procedure is based on the binomial distribution and is exact if the assumptions are met. The p value for the one-sample chi-square procedure is based on the chi-square distribution and is an approximation to the p value based on the binomial distribution. Consequently, the binomial test procedure is preferred for a study with a relatively small sample.

Exercises

Exercises 1 through 3 are based on the data in the data file named *Lesson 39 Exercise File 1* on the Web at http://www.prenhall.com/greensalkind. The data are based on the following study.

Tucker is interested in assessing whether one-year-old girls prefer boys or girls. He randomly forms triplets of two girls and one boy from a sample of 120 one-year-old girls and 60 one-year-old boys. One of the two girls is randomly determined to be the focal subject, while the other girl and the boy are designated as companions. Each triplet is placed in a playroom for 10 minutes and observed behind a one-way mirror. An observer assesses whether the focal subject spends more time closer to the male companion or more time closer to the female companion. The gender preference for each focal subject is recorded.

The data for the problem are on the Web at http://www.prenhall.com/greensalkind in the weighted cases structure. There are two cases for the two possible outcomes, prefer girl and prefer boy. The description of the variables in the file is given in Table 58.

Table 58
Variables in Lesson 39 Exercise Data Set 1

Variables	Definition
sexpref	Sexpref has two levels: 1 = prefer girl 2 = prefer boy
number	Number of 1-year-olds who preferred to play with a specified gender

1. Use the weighted cases method to weight the data file by the number of cases before conducting the analysis.
2. Conduct a binomial test to evaluate Tucker's hypothesis.
3. Write a Results section based on your analyses.

Exercises 4 through 6 are based on the data in the data file named *Lesson 39 Exercise File 2* on the Web at http://www.prenhall.com/greensalkind. The data are based on the following study.

After reviewing the literature on depression and childbirth, Moe claims that half of all women become less depressed and the other half become more depressed as a function of childbirth. Emma believes that Moe is wrong, and she audiotapes interviews with 60 women before and after childbirth. The interviews include no questions pertaining to childbirth or children. After the interviews, Emma deletes any references on the tape that indicate whether the women are anticipating or have given birth to their babies. Emma then asks a group of trained clinicians to listen to the 60 pairs of interviews. For interviews with 30 of the women, the clinicians listen to the before-birth interview first and the after-birth interview second, and for interviews with the other 30 women, the clinicians listen initially to the after-birth interview and then to the before-birth interview. The trained clinicians come to a consensus judgment about whether each woman sounded less depressed or more depressed in the first interview. Emma is interested in whether the population proportion for women who become less depressed after birth (depress = –1) is equal to the population proportion for woman who become more depressed after birth (depress = 1).

4. What is the test value?
5. Conduct a binomial test to evaluate whether the population proportion for women who become less depressed is equal to the population proportion for women who become more depressed. From the output, identify the following:
   a. Observed frequency of women who become more depressed
   b. *p* value
   c. Proportion of women who become more depressed
6. Write a Results section based on your analyses.

LESSON

40 One-Sample Chi-Square Test

The one-sample chi-square test evaluates whether the proportions of individuals who fall into categories of a variable are equal to hypothesized values. The variable may have two or more categories. The one-sample chi-square test is more likely to yield significance if the sample proportions for the categories differ greatly from the hypothesized proportions and if the sample size is large.

The categories for the variable either may have quantitative meaning so that one category reflects a higher value than another category or may differentiate only between qualitative groupings of individuals. For example, categories may indicate what type of explanation students prefer when they are learning algebra: 1 = graphical explanation of how a method works; 2 = equational explanation of how a method works; or 3 = explanation through real-world applications. The one-sample chi-square test does not recognize any quantitative distinction among categories, but simply assesses whether the proportions associated with the categories are significantly different from hypothesized proportions.

When conducting a one-sample chi-square test using SPSS, you must specify the expected frequencies associated with the categories on the variable of interest. To know the expected frequencies, you must first be able to specify the values for the hypothesized proportions. If we hypothesize equal proportions, then the hypothesized proportion for any category is equal to 1 divided by the number of categories. If the hypothesized proportions are not equal, they must be determined based on previous research, theory, research methods, or speculation. The expected frequency for any category is the product of the hypothesized proportion times the total sample size for the study. If the hypothesized proportions are equal, the expected frequencies will be equal. If the hypothesized proportions are unequal, the expected frequencies will be unequal.

An SPSS data file for the one-sample chi-square test may be structured two ways. With the standard method, the SPSS data file contains as many cases as individuals. For each case, there is a single variable that can have two or more values that represent the two or more categories for the variable of interest. With the weighted cases method, the SPSS data file contains two variables and as many cases as there are categories. The two variables are the focal variable with two or more values and the weight variable containing the frequencies for the categories. We'll describe the standard and weighted cases methods in our first example.

Applications of the One-Sample Chi-Square Test

We will examine two applications of the one-sample chi-square tests.

- Testing a hypothesis with equal proportions
- Testing a hypothesis with unequal proportions

Testing a Hypothesis with Equal Proportions

After reviewing the literature on depression and childbirth, Moe claims that a third of women become less depressed, a third become neither less depressed nor more depressed, and a third become more depressed as a function of childbirth. Emma believes that Moe is wrong, and she audiotapes interviews with 60 women before and after childbirth. The interviews include no questions pertaining to childbirth or children. After the interviews, Emma deletes any reference on the tape that indicates whether the women are anticipating or have given birth to their babies. Emma then asks a group of trained clinicians to listen to the 60 pairs of interviews. For interviews with 30 of the women, the clinicians listen to the before-birth interview first and the after-birth interview second and for interviews with the other 30 women, the clinicians listen initially to the after-birth interview and then to the before-birth interview. The trained clinicians come to a consensus judgment about whether each woman sounded less depressed, neither less nor more depressed, or more depressed in the first interview.

For this problem, the expected frequencies for all categories are equal. Emma is attempting to present evidence that evaluates Moe's claim that a third of women after giving birth are less depressed, a third are unchanged, and a third become more depressed.

Using the standard structure, Emma creates a data file with 60 cases to represent the 60 women. Each woman is measured on a single variable, status. The scores on status are –1 = more depressed before birth, 0 = same amount of depression before and after birth, and 1 = more depressed after birth. The values of –1, 0, and 1 are arbitrary, and other values could have been chosen.

If Emma had used the weighted cases structure, she would have had three cases representing the three categories. Each case would have scores on two variables: status with categories of –1, 0, and 1 for the three cases and a weight variable. A value for the weight variable represents the number of individuals who are classified as belonging to the category represented by that case. If you use the weighted cases method, you must weight the cases by using the Weight Cases option under Data from the main menu before conducting a one-sample chi-square test.

Testing a Hypothesis with Unequal Proportions

Based on past research, Pam knows that children who are seated in a room with four doors—one in front of them, one to their left, one to their right, and one behind them—are equally likely to pick any door when asked which door they would choose when leaving the room. As part of a larger research program, Pam wants to demonstrate that children labeled by a psychologist as oppositional are more likely to choose the door behind them. To test her hypothesis, she conducts a study with 80 oppositional children and creates a data file that includes a single variable, door, for the 80 cases. The categories for door, are 0 = any other door and 1 = door behind. The expected frequency for the 0 cell is 60, while expected frequency for the 1 cell is 20. These expected frequencies are based for the results of previous research that showed the doors are picked equally often by children in general. Because there are four doors, the hypothesized proportion for door behind is .25 (1 out of 4 doors), and the hypothesized proportion for other door is .75 (3 out of 4 doors). The expected frequency is the hypothesized proportion times the number of cases: 20 [= (.25) (80)] for door behind and 60 [= (.75) (80)] for the other door.

TIP

For variables with two categories, analyses can be conducted with either the binomial test or the one-sample chi-square test, but the binomial test procedure is preferred. To see why, read the "Alternative Analyses" section in Lesson 39.

Understanding the One-Sample Chi-Square Test

If the resulting chi-square value is significant and the variable has more than two levels, follow-up tests may be conducted. The methods employed in these follow-up tests are conceptually similar to those used to conduct post hoc tests for a one-way ANOVA to evaluate differences in means among three or more levels of a factor. (See Lesson 25.)

Assumptions Underlying the One-Sample Chi-Square Test

Assumption 1: The Observations Must Be from a Random Sample, and the Scores Associated with the Observation Are Independent of Each Other

To meet this assumption, experiments should be designed to prevent dependency in the data. Research participants should be sampled independently, participate in the study independently, and contribute only a single score to the data. If the assumption is violated, the test is likely to yield inaccurate results.

Assumption 2: The One-Sample Chi-Square Test Yields a Test Statistic That Is Approximately Distributed as a Chi-Square When the Sample Size Is Relatively Large

It is important to examine the expected frequencies on the SPSS output when conducting this test to assess whether the sample size is sufficiently large. In general, the test should yield relatively accurate results if the expected frequencies are greater than or equal to 5 for 80% or more of the categories.

Although many textbooks indicate that the resulting chi-square value should be corrected for lack of continuity, SPSS does not apply this correction. You should use the binomial test procedure (Lesson 39) if the variable of interest has only two categories because the binomial procedure does apply the continuity correction.

Effect Size Statistics for a One-Sample Chi-Square Test

SPSS does not supply an effect size index. However, an effect size index can be easily computed based on the reported statistics:

$$\textit{Effect Size} = \frac{x^2}{(\textit{Total Sample Size Across All Categories})(\textit{Number of Categories} - 1)}$$

The coefficient can range in value from 0 to 1. A value of 0 indicates that the sample proportions are exactly equal to the hypothesized proportions, while a value of 1 indicates that the sample proportions are as different as possible from the hypothesized proportions.

The Data Set

The data used to illustrate the one-sample chi-square test are based on the postpartum depression example. The observed frequencies, the hypothesized proportions, and the expected frequencies for this example are shown in Table 59.

Table 59
Results of the Postpartum Depression Study

Depression after birth in comparison with before birth	Observed frequencies	Hypothesized proportions	Expected frequencies
Less depressed	14	.33	20
Neither less nor more depressed	33	.33	20
More depressed	13	.33	20

These data are on the Web at http://www.prenhall.com/greensalkind in *Lesson 40 Data File 1*. The variables in the data file are described in Table 60 (on page 360).

Table 60
Variables in Lesson 40 Data File 1

Variables	Definition
status	status has three categories: −1 = Less depressed 0 = Same amount of depression 1 = More depressed
statusr	statusr has two categories: 0 = Same amount of depression 1 = More or less depressed

The SPSS data file is structured using the standard method. There are 60 cases in the file, one for each woman in the study. For each case, a woman's change in depression is indicated by the status variable. (A second variable, statusr, is included in the file to illustrate how to conduct tests with unequal expected frequencies.)

A second approach, the weighted cases approach, requires fewer entries to create a comparable data file. With this approach, there would be three cases to represent the three possible depression categories. There would be two variables in the file (ignoring the statusr variable). A variable named status could be defined with three categories (−1, 0, and 1) for the three depression categories. A second variable, number, would contain the frequencies (14, 33, 13) associated with these categories.

The Research Question

The research question can be asked several different ways.

1. Are women equally likely to show an increase, no change, or a decrease in depression as a function of childbirth?
2. Are the proportions associated with a decrease, no change, and an increase in depression from before to after childbirth the same?

Conducting a One-Sample Chi-Square Test

Now, we will illustrate how to conduct an overall chi-square test, follow-up tests with equal expected frequencies, and follow-up tests with unequal expected frequencies.

Conducting the Overall Chi-Square Test

Conduct the following steps to obtain an overall chi-square test by using the standard structure method. If the data file used the weighted cases method, the cases must be weighted following the procedure described in Lesson 39.

1. Click **Analyze**, click **Nonparametric Tests**, then click **Chi Square**. The Chi-Square Test dialog box is shown in Figure 245.
2. Click **status**, then click ▶ to move to the Test Variable List box.
3. Click **OK**.

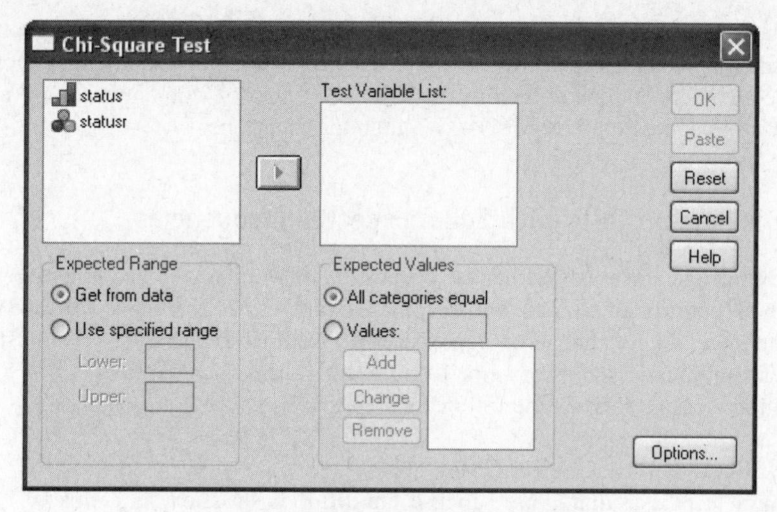

***Figure 245.* The Chi-Square Test dialog box.**

Selected SPSS Output for One-Sample Chi-Square Test

You will see the results of the analysis in Figure 246. The results indicate that the sample proportions are significantly different from the hypothesized values of 1/3, 1/3, and 1/3, $\chi^2(2, N = 60) = 12.70, p < .01$. In addition, an effect size statistic can be computed:

$$\textit{Effect Size} = \frac{12.7}{(60)(3-1)} = .11$$

The observed frequencies of 14, 33, and 13 tend to disconfirm Moe's original hypothesis that a third of women become less depressed, a third become neither less depressed nor more depressed, and a third become more depressed as a function of childbirth. The effect size of .11 indicates that the observed frequencies (Observed N on the output) deviate moderately from the expected frequencies (Expected N).

Postpartum Depression

	Observed N	Expected N	Residual
less depressed	14	20.0	-6.0
same	33	20.0	13.0
more depressed	13	20.0	-7.0
Total	60		

Test Statistics

	Postpartum Depression
Chi-Square[a]	12.700
df	2
Asymp. Sig.	.002

a. 0 cells (.0%) have expected frequencies less than 5. The minimum expected cell frequency is 20.0.

***Figure 246.* Results of the chi-square analysis with equal frequencies.**

TIP

In the Chi-Square Test dialog box, there are two options in the Expected Range area. The default option is "Get from data." The second option is "Use specified range." If you are interested in evaluating a more limited number of categories for a variable and these categories are within some restricted range of values, then you should click on **Use specified range,** and indicate the lower and upper values of the range in the appropriate boxes. For example, if you were interested in the categories of –1 and 0 on status rather than –1, 0, and 1, you would type **–1** in the Lower box and **0** in the Upper box.

It might be useful to conduct two follow-up one-sample chi-square tests to demonstrate that the initial significant result is not due to differences in proportion between the less depressed and more depressed women, but due to the disproportional number of women who were neither less depressed nor more depressed (category 0). We'll do these tests next.

Conducting Follow-up Tests with Equal Expected Frequencies

The first follow-up test assesses whether the proportion of women who become more depressed differs from the proportion of women who become less depressed. In other words, we want to ignore the women who do not change on depression. To conduct this analysis, we must tell SPSS to select the cases that have scores of –1 or 1 on status.

To select these cases, conduct the following steps:

1. Click **Data**, then click **Select Cases**.
2. In the Select Cases dialog box, click **If condition is satisfied**, and click **If**.
3. In the Select Cases: If dialog box, indicate what cases to select by typing **status = –1 or status = 1**.
4. Click **Continue**.
5. Click **OK**.

Now you are ready to conduct your chi-square test as previously described. To conduct the test, perform the following steps:

1. Click **Analyze**, click **Nonparametric Tests**, and then click **Chi Square**.
2. If **status** is not in the box labeled Test Variable List, click **status** and click ▶ to move it to this box.
3. Click **OK**.

Selected SPSS Output for Follow-up Test with Equal Expected Frequencies

You will see the results of the analysis in Figure 247. The results indicate that the test is nonsignificant, $\chi^2(1, N = 27) = 0.04$, $p = .85$, and the sample proportions are very similar to each other.

Postpartum Depression

	Observed N	Expected N	Residual
less depressed	14	13.5	.5
more depressed	13	13.5	-.5
Total	27		

Test Statistics

	Postpartum Depression
Chi-Square[a]	.037
df	1
Asymp. Sig.	.847

a. 0 cells (.0%) have expected frequencies less than 5. The minimum expected cell frequency is 13.5.

Figure 247. **Results of the comparison between less depressed and more depressed groups.**

Conducting Follow-up Tests with Unequal Expected Frequencies

For this second test, we want to demonstrate that there is a significant difference from Moe's expectations for the same-amount-of-depression category and for the combined categories of less depressed and more depressed.

The expected frequencies for the two cells are 20, or one-third of 60, for the same-amount-of-depression category, and 40, or two-thirds of 60, for the combined category of less depressed and more depressed. To perform this test, the variable statusr in the data set is analyzed. Statusr has a score of 0 for same amount of depression or a score of 1 for less depressed and more depressed. (We have included the variable statusr in the data file for ease of use. It would not have been very difficult to create it based on the status variable if we were using the Transform → Recode into Different Variables option.)

Follow these steps to complete the analysis:

1. Click **Data**, and click **Select Cases**. Click **All Cases** to select all 60 cases for analysis. Click **OK**.
2. Click **Analyze**, click **Nonparametric Test**, then click **Chi-Square**. Click the reset button if you need to clear any variables from any of the boxes.
3. Click **statusr**, then click ▶ to move it to the Test Variable List box.
4. Click **Values** in the Expected Values area.
5. Type **20** in the Values text box to indicate the expected frequency associated with the smallest value of the variable of interest.
6. Click **Add**.
7. Type **40** in the Values text box to indicate the expected frequency associated with the next smallest value of the variable of interest.
8. Click **Add**.
9. Click **OK**.

Selected SPSS Output for Follow-up Test with Unequal Expected Frequencies

The results are shown in Figure 248. This analysis shows that observed proportions do differ significantly from the hypothesized proportions, $\chi^2(1, N = 60) = 12.68, p < .01$.

Postpartum Depression—Recoded

	Observed N	Expected N	Residual
same	33	20.0	13.0
more or less depressed	27	40.0	-13.0
Total	60		

Test Statistics

	Postpartum Depression —Recoded
Chi-Square[a]	12.675
df	1
Asymp. Sig.	.000

a. 0 cells (.0%) have expected frequencies less than 5. The minimum expected cell frequency is 20.0.

***Figure 248.* Results of the comparison between group labeled same amount of depression and the other two groups combined.**

Using SPSS Graphs to Display the Results

The applications presented in this lesson include variables with just a few categories. The results for these variables can be easily described in the text of a results section without presenting a graph. However, if the number of categories is large, a bar chart might be provided to summarize the results for a categorical variable. In Figure 249, we illustrate a bar chart for our postpartum data for the status variable.

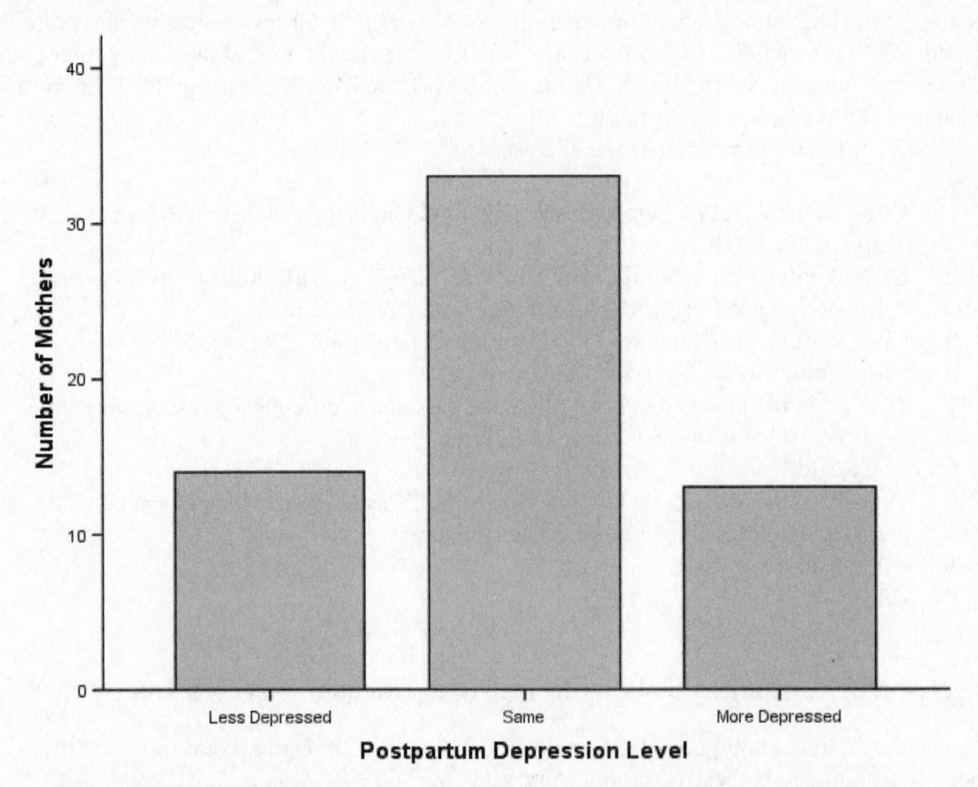

Figure 249. **The frequencies of women who are less depressed, show no depression change, and are more depressed after childbirth.**

An APA Results Section

A one-sample chi-square test was conducted to assess whether women become less depressed, remain unchanged, or become more depressed after giving birth to children. The results of the test were significant, $\chi^2(2, N = 60) = 12.70, p < .01$. The proportion of women who were unchanged ($P = .55$) was much greater than the hypothesized proportion of .33, while the proportion of women who became less depressed ($P = .23$) and the proportion who became more depressed ($P = .22$) were approximately the same value and less than the hypothesized proportions of .33. A follow-up test indicated that the proportion of women who became less depressed did not differ significantly from the proportion of women who became more depressed, $\chi^2(1, N = 27) = 0.04, p = .85$. Overall, these results suggest that women in general do not become more or less depressed as a function of childbirth.

Alternative Analyses

The one-sample chi-square test is sometimes used to compare the distribution for a quantitative variable with a particular theoretical distribution, most frequently the normal distribution. The Kolmogrov-Smirnov test is the preferred test for making the comparison with a normal distribution. See the next section for the steps for testing whether a distribution is normally distributed.

If the focal variable has two categories, the binomial test may be used instead of the one-sample chi-square test. (See Lesson 39.)

Conducting a Kolmogrov-Smirnov Test

A Kolmogrov-Smirnov test evaluates whether the data on a quantitative variable are normally distributed. We will give the steps for conducting this test using the Explore procedure under Descriptive Statistics in the Analyze menu. Note that the results obtained with this procedure differ from those obtained with the 1-Sample K-S test under Nonparametrics in the Analysis menu. The latter procedure assumes that the mean and standard deviation of the distribution are known, which is very unlikely.

The steps for conducting a Kolmogrov-Smirnov test are as follows:

1. Click **Analyze**, click **Descriptive Statistics**, and then click **Explore**.
2. Click on the variable of interest and move it to the Dependent List box.
3. Click the **Plots** button, and click in the box next to Normality plots with test.
4. Click **Continue**.
5. Click **OK**.

The results of the test are labeled K-S (Lilliefors) in the output. The test is a modified Kolmogrov-Smirnov test called the Lilliefors test. If the test is significant, it indicates that the distribution is significantly different from a normal distribution.

Exercises

The data for Exercises 1 through 4 are in the data files named *Lesson 40 Data File 1* on the Web at http://www.prenhall.com/greensalkind. The data are based on the following research study.

Kristen is interested in evaluating whether the method of cooking potato chips affects the taste of the chips. She has 48 individuals volunteer to participate in her potato chip study. Each participant tastes chips cooked using three different methods: fried in animal fat (chip = 1), fried in canola oil (chip = 2), and baked (chip = 3). Individuals are instructed to indicate which type of potato chips they prefer: chip type 1, chip type 2, or chip type 3. Kristen hypothesizes that individuals will prefer potato chips that are fried in canola oil over those that are fried in animal fat or baked.

1. Weight the data file by the number of cases to conduct the analysis.
2. Conduct a one-sample chi-square test to evaluate whether cooking method affects taste. From the output, identify the following:
   a. Observed frequency for potato chips fried in canola oil
   b. p value
   c. χ^2 value
3. What are the expected frequencies for the three categories of potato chips?
4. Write a Results section based on your analyses.

LESSON

41 Two-Way Contingency Table Analysis Using Crosstabs

A two-way contingency table analysis evaluates whether a statistical relationship exists between two variables. A two-way contingency table consists of two or more rows and two or more columns. The rows represent the different levels of one variable and the columns represent different levels of a second variable. Such a table is sometime described as an $r \times c$ table, where r is the number of rows, and c is the number of columns. For example, a 3×4 contingency table is a table with a 3-level row variable and a 4-level column variable. The cells in the table, which are the combinations of the levels of the row and the column variables, contain frequencies. For a 3×4 contingency table, there are 12 cells with frequencies. A cell frequency for a particular row and column represents the number of individuals in a study who can be cross-classified as belonging to that particular level of the row variable and that particular level of the column variable. Analyses of two-way contingency tables focus on these cell frequencies to evaluate whether the row and column variables are related.

Levels of a row or a column variable can represent quantitative distinctions among individuals, differences indicating an ordering of individuals, or simply a categorization of individuals. In this lesson, we will talk only about methods for analyzing tables where the row and the column variables are qualitative or categorical in nature. These methods are not designed to take full advantage of the properties of quantitative or ordered variables.

Three types of studies can be analyzed with the use of two-way contingency tables (Wickens, 1989), and they produce three different types of hypotheses:

- *Independence between variables.* Research participants are sampled and then measured on two response variables—the row and the column variables. In other words, in these studies, the total number of participants is controlled, but not the number of individuals in the columns or in the rows. The relationship between the row and the column variables in the population is being evaluated.
- *Homogeneity of proportions.* Samples are drawn to represent populations of interest, either by sampling from different populations or by randomly assigning subjects to groups and treating them differently. The individuals in these samples are measured on a response variable. The rows might represent the different populations and the columns might be the response categories on the response variable. For these studies, the total number of research participants in each row is controlled, but not the total number of participants in each column. The contingency table analysis evaluates whether the proportions of individuals in the levels of the column variable are the same for all populations (i.e., for all levels of the row variable).
- *Unrelated classification.* Here, we control the total number of subjects in each row and in each column. Researchers infrequently design studies investigating unrelated classifications.

Regardless of the design of the study, an SPSS data file for a two-way contingency table analysis can be structured in one of two ways. With the standard method, the SPSS data file contains as many cases as individuals. There are two variables, each of which can have two or more values that represent the two or more categories for that variable. With the weighted cases method, the SPSS data file contains as many cases as category combinations (cells) across the two focal variables. For example, there are 12 cases for a 3×4 contingency table. For the weighted cases method, there are three variables. Two are the focal variables with values representing their categories, and the third variable is a weight variable containing frequencies for the category combinations (cells) across the two focal variables.

Applications of a Two-Way Contingency Table Analysis

Here are applications for studies that evaluate independence between variables and homogeneity of proportions.

Independence between Variables

Carrie is interested in assessing the relationship between religion and occupation for men. She samples a group of 2,000 men between the ages of 30 and 50 and asks them to complete a demographic questionnaire. On the basis of their answers, she classifies each man as practicing one of four religions (Protestant, Catholic, Jewish, and Other) and employed in one of six occupations (Professional, Business-Management, Business-Nonmanagement, Skilled Tradesman, Laborer, and Unemployed). Religion and occupation are the row and the column variables, and the frequencies are the number of men who are classified as belonging to the 24 (4×6) combinations of religious and occupational categories.

Using the standard structure, Carrie would have 2,000 cases to represent the 2,000 men. Each man would have scores on at least two variables, religion with four levels and occupation with six levels. With the weighted cases structure, Carrie would have 24 cases, representing the 24 combinations of the two variables. Each case would have scores on at least three variables: the two research variables of religion and occupation and a weight variable. A score on the weight variable for a case would represent the number of individuals who are classified as belonging to the cell represented by that case.

Homogeneity of Proportions

Claude wants to determine whether young men are unjustifiably more likely to treat elderly people in a condescending manner. To address this hypothesis, he recruits a young woman from his class who indicates that both her mother and her grandmother are physically, emotionally, and intellectually active. The daughter, mother, and grandmother are 20, 43, and 72 years old, respectively. Claude gains the cooperation of all three women to participate as confederates in his study. The three women take a two-hour lesson to learn a computer game called Wits. At the end of the two hours, they pass a criterion indicating total mastery of the game. In addition, they learn a script so that they can all act in the same manner as confederates in the study.

Claude recruits 90 male college students between the ages of 17 and 22 to participate in his study. All 90 males take a 20-minute lesson to learn the game of Wits and pass a test indicating minimal knowledge of the game, although on the basis of their test scores none of the males shows total mastery. Next, he tells each student that he is now going to instruct a woman who has no experience with the game. The students are then asked to switch seats from the chair in front of the computer to the one behind it. Each student is then introduced to one of the women, as determined by random assignment. A third of the men meet the daughter, a third meet the mother, and a third meet the grandmother. The men next give verbal instructions from their seats on how to win at Wits. Regardless of the instructions given by the men, all three women show exactly the same rate of improvement for all 90 participants.

All sessions are videotaped. The tapes show only the faces of the male participants and exclude all verbal comments made by the women so that observers of the tape are completely blind to which woman is being "trained." A reliable judge observes the tapes and concludes whether each male college student acts condescendingly or not.

For this study, the row variable is age of women with three levels (young, middle-aged, and elderly) and the column variable is condescension with two levels (not condescending and condescending). The frequencies are the number of participants who are classified as belonging to the six combinations of age and condescension (3 levels $\times$ 2 levels).

Understanding a Two-Way Contingency Table Analysis

The same chi-square test for a two-way contingency table may be applied to studies investigating both the independence and homogeneity hypotheses. If the resulting chi-square value is significant and the table includes at least one variable with more than two levels, a follow-up test may be conducted. These follow-up tests are particularly crucial for studies assessing homogeneity of three or more proportions. The methods employed in these follow-up tests are conceptually similar to those used to conduct follow-up tests for a one-way ANOVA to evaluate differences in means among three or more levels of a factor. (See Lesson 25.)

Assumptions Underlying a Two-Way Contingency Table Analysis

The Crosstabs procedure applies a χ^2 test for contingency table analysis. Two assumptions underlie this test.

Assumption 1: The Observations for a Two-Way Contingency Table Analysis Are Independent of Each Other

To meet this assumption, studies should be designed to prevent dependency in the data. If this assumption is violated, the test is likely to yield inaccurate results.

Assumption 2: Two-Way Contingency Table Analyses Yield a Test Statistic That Is Approximately Distributed as a Chi-Square When the Sample Size Is Relatively Large

There is no simple answer to the question of what sample size is large enough. However, to answer this question, the size of the expected cell frequencies rather than the total sample size should be examined. For tables with two rows and two columns, there is probably little reason to worry if all the expected frequencies are greater than or equal to 5. For large tables, if more than 20% of the cells have expected frequencies that are less than 5, you should be concerned about the validity of the results.

Effect Size Statistics for a Two-Way Contingency Table Analysis

SPSS provides a number of indices that assess the strength of the relationship between row and column variables. They include the contingency coefficient, phi, Cramér's *V*, lambda, and the uncertainty coefficient. We will focus our attention on two of these coefficients, phi and Cramér's *V*.

The phi coefficient for a 2 $\times$ 2 table is a special case of the Pearson product-moment correlation coefficient for a 2 $\times$ 2 contingency table. Because most behavioral researchers have a good working knowledge of the Pearson product-moment correlation coefficient, they are likely to feel comfortable using its derivative, the phi coefficient. Phi, Φ, is a function of the Pearson chi-square statistic, χ^2, and sample size, N:

$$\Phi = \sqrt{\chi^2/N}$$

Like the Pearson product-moment correlation coefficient, the phi ranges in value from –1 to +1. Values close to 0 indicate a very weak relationship, and values close to 1 indicate a very strong relationship. If the row and column variables are qualitative, the sign of phi is not meaningful and any negative phi values can be changed to positive values without affecting their meaning. By convention, phi's of .10, .30, and .50 represent small, medium, and large effect sizes, respectively. However, what is a small versus a large phi should be dependent on the area of investigation.

If both the row and the column variables have more than two levels, phi can exceed 1 and, therefore, be hard to interpret. Cramér's *V* rescales phi so that it also ranges between 0 and 1: For 2 × 2, 2 × 3, and 3 × 2 tables, phi and Cramér's *V* are identical:

$$\textit{Cramér's } V = \sqrt{\frac{\Phi^2}{(\textit{number of row or number of columns, whichever is smaller}) - 1}}$$

The Data Set

The data for this example are on the Web at http://www.prenhall.com/greensalkind under the name *Lesson 41 Data File 1*. They are based on the condescension study exploring homogeneity of proportions. The variables are described in Table 61.

Table 61
Variables in Lesson 41 Data File 1

Variables	Definition
age	Age has three levels: 1 = Young, 2 = Middle-Aged, and 3 = Elderly.
condes	Condes stands for condescension and has two levels: 0 = not condescending and 1 = condescending.
number	Number is the number of individuals in the cell defined by the combination of levels for age and condes.

The data file uses the weighted cases method, so there are six cases that represent the six cells of a two-way contingency table. The six cells are identified by the scores on the two variables of age and condescension. In addition, a third variable, number, contains the frequencies for the six cells (cases).

The Research Question

The research question can be asked as follows: Are the proportions of male college students who treat young, middle-aged, and elderly women with disdain the same?

Conducting a Two-Way Contingency Table Analysis

We will illustrate how to conduct an overall two-way contingency analysis and follow-up procedures.

Before the data can be analyzed, the cases must be weighted by the values in the number variable. The cases can be weighted by conducting the following steps:

1. Click **Data**, and click **Weight Cases**.
2. Click **Weight cases by** in the Weight Cases dialog box.
3. Click **number**, then click ▶ to move it to the Frequency Variable box.
4. Click **OK**.

Conducting the Overall Two-Way Contingency Table Analysis

To conduct the Crosstabs analysis, follow these steps:

1. Click **Analyze**, click **Descriptive Statistics**, and then click **Crosstabs**. You'll see the Crosstabs dialog box as shown in Figure 250.

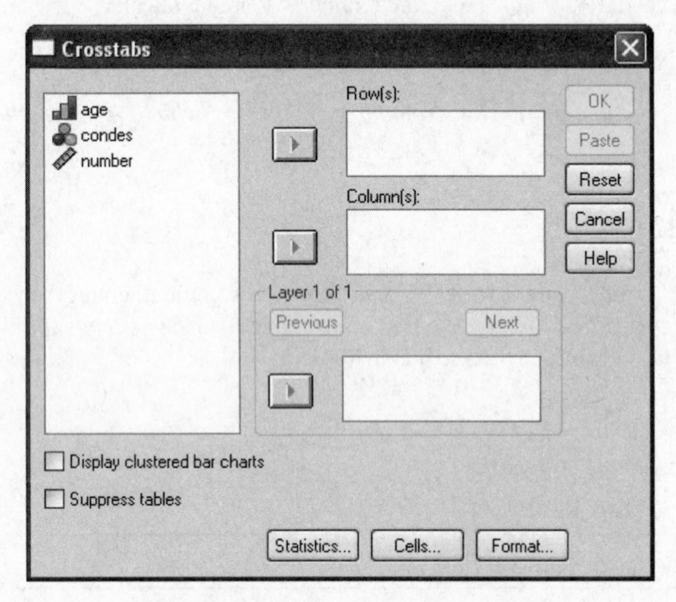

Figure 250. **The Crosstabs dialog box.**

2. Click **age**, then click ▶ to move it to the box labeled Row(s).
3. Click **condes**, then click ▶ to move it to the box labeled Column(s).
4. Click **Statistics**. You'll see the Crosstabs: Statistics dialog box as shown in Figure 251.

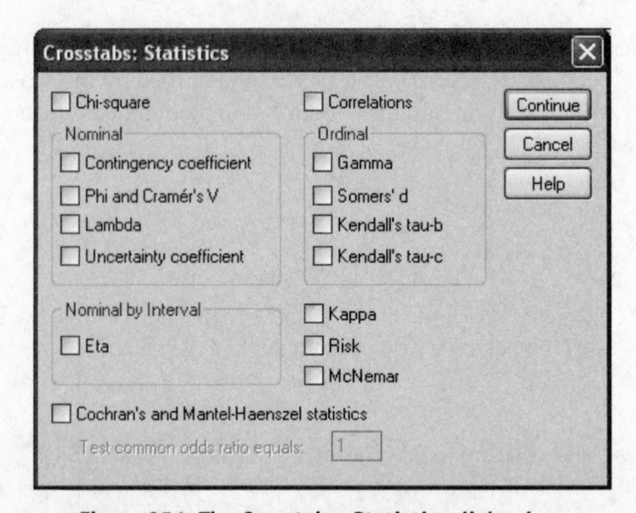

Figure 251. **The Crosstabs: Statistics dialog box.**

5. Click **Chi-square** and **Phi and Cramér's V** in the Nominal Data box.
6. Click **Continue**.
7. Click **Cells**. You'll see the Crosstabs: Cell Display dialog box as shown in Figure 252.

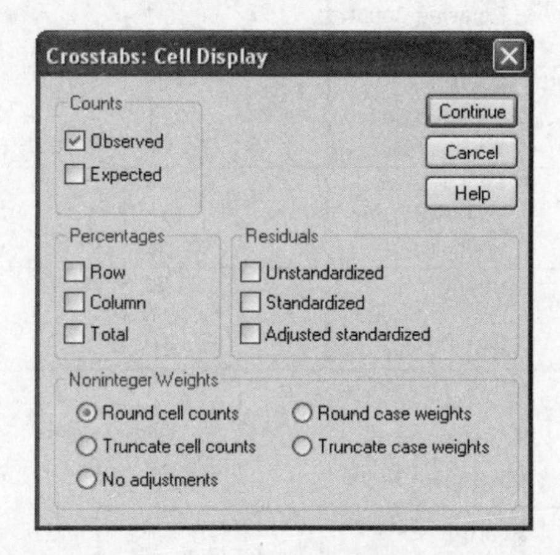

Figure 252. **The Crosstabs: Cell Display dialog box.**

8. Click **Expected** in the Counts box. **Observed** should already be selected.
9. Click **Row** in the Percentages box.
10. Click **Continue**.
11. Click **OK**.

Selected SPSS Output for Crosstabs

The output is shown in Figure 253 (at the top of page 372). SPSS reports the results of a number of significance tests, including the often reported Pearson chi-square test statistic. In this example, the Pearson χ^2 is 6.32, $p = .04$. If the table has 2 rows and 2 columns, then a corrected Pearson chi-square test also is reported, but this result is not recommended because it tends to be overly conservative for many applications.

The likelihood ratio test is also reported in Figure 253, $\chi^2(2, N = 90) = 6.46, p = .04$. It tests the same hypothesis as the Pearson chi-square, has similar properties, and is an alternative to the Pearson. The remaining reported chi-square statistic, the chi-square for a linear-by-linear association, is inappropriate for analyzing tables with qualitative variables.

If the chi-square test (Pearson or likelihood ratio) has more than 1 degree of freedom, it is an omnibus test, which evaluates the significance of an overall hypothesis containing multiple subhypotheses. These multiple subhypotheses can then be tested by using follow-up tests.

In this example, the results of the chi-square test indicated significant differences among the three age levels. Follow-up tests should be conducted to examine particular subhypotheses. One possible set of subhypotheses would involve the pairwise comparisons among the three age conditions. For example, one of the pairwise comparisons poses the question, "Do the proportions of college men who are condescending vary depending upon whether the men are teaching young or elderly women?" The remaining two pairwise comparisons would evaluate differences in proportions between the young and the middle-aged conditions and between the middle-aged and the elderly conditions.

Conducting Follow-up Tests

To compute the chi-square values for the comparison between the young and middle-aged conditions, follow these steps:

1. Click **Data**, then click **Select Cases**.
2. Click **If condition is satisfied**.
3. Click **If**.
4. To the right of the ▶, type **age = 1 or age = 2**.

age * condes Crosstabulation

			condes		Total
			Not Condescending	Condescending	
age	Young	Count	21	9	30
		Expected Count	20.7	9.3	30.0
		% within age	70.0%	30.0%	100.0%
	Middle-Aged	Count	25	5	30
		Expected Count	20.7	9.3	30.0
		% within age	83.3%	16.7%	100.0%
	Elderly	Count	16	14	30
		Expected Count	20.7	9.3	30.0
		% within age	53.3%	46.7%	100.0%
Total		Count	62	28	90
		Expected Count	62.0	28.0	90.0
		% within age	68.9%	31.1%	100.0%

Chi-Square Tests

	Value	df	Asymp. Sig. (2-sided)
Pearson Chi-Square	6.325[a]	2	.042
Likelihood Ratio	6.457	2	.040
Linear-by-Linear Association	1.923	1	.166
N of Valid Cases	90		

a. 0 cells (.0%) have expected count less than 5. The minimum expected count is 9.33.

Symmetric Measures

		Value	Approx. Sig.
Nominal by Nominal	Phi	.265	.042
	Cramer's V	.265	.042
N of Valid Cases		90	

Figure 253. **Results of the Crosstabs analysis.**

5. Click **Continue**.
6. Click **OK**.
7. Click **Analyze**, click **Descriptive Statistics**, and then click **Crosstabs**.
8. If you have just previously conducted the omnibus chi-square test for all three age categories, you do not have to make any changes in the Crosstabs dialog box, and you simply click **OK**. Otherwise, repeat the steps previously described for the omnibus chi-square test.

To conduct the other two comparisons between young and elderly conditions and the middle-aged and elderly conditions, follow these steps:

1. For the second comparison, repeat Steps 1 through 8, except at Step 4, type **age = 1 or age = 3** in the Select Cases: If dialog box.
2. For the third comparison, Steps 1 through 8 are repeated, except at Step 4, type **age = 2 or age = 3** in the Select Cases: If dialog box.

Selected SPSS Output of Follow-up Tests

The results of the pairwise comparison between the young and the middle-aged women are shown in Figure 254.

			condes		
			Not Condescending	Condescending	Total
age	Young	Count	21	9	30
		Expected Count	23.0	7.0	30.0
		% within age	70.0%	30.0%	100.0%
	Middle-Aged	Count	25	5	30
		Expected Count	23.0	7.0	30.0
		% within age	83.3%	16.7%	100.0%
Total		Count	46	14	60
		Expected Count	46.0	14.0	60.0
		% within age	76.7%	23.3%	100.0%

Chi-Square Tests

	Value	df	Asymp. Sig. (2-sided)	Exact Sig. (2-sided)	Exact Sig. (1-sided)
Pearson Chi-Square	1.491[b]	1	.222		
Continuity Correction[a]	.839	1	.360		
Likelihood Ratio	1.507	1	.220		
Fisher's Exact Test				.360	.180
Linear-by-Linear Association	1.466	1	.226		
N of Valid Cases	60				

a. Computed only for a 2x2 table

b. 0 cells (.0%) have expected count less than 5. The minimum expected count is 7.00.

Symmetric Measures

		Value	Approx. Sig.
Nominal by Nominal	Phi	-.158	.222
	Cramer's V	.158	.222
N of Valid Cases		60	

a. Not assuming the null hypothesis.

b. Using the asymptotic standard error assuming the null

***Figure 254.* The results of the pairwise comparison between the young and middle-aged conditions.**

Based on the Pearson χ^2 test, the p value for the comparison between young women and middle-aged women is .22. The p values for the comparisons between young and elderly women and between middle-aged and elderly women are .18 and .01, respectively. The results of these three tests, using the Holm's sequential Bonferroni method (see Appendix B) for controlling Type I error, are reported in Table 62 (on page 375).

Using SPSS Graphs to Display the Results

Clustered bar charts provide a visual representation of the results of two qualitative variables. With a clustered bar chart, it is easy to see whether the proportions in each category of the first variable are the same for all categories of the second variable. To create a clustered bar chart, make sure that Select Cases has been reset to All Cases and conduct the following steps:

1. Click **Graphs**, click **Legacy Dialogs**, and then click **Bar**.
2. Click **Clustered**, and click **Summaries for groups of cases**.

3. Click **Define**.
4. Click **age**, then click ▶ to move it to the Category Axis box.
5. Click **condes**, then click ▶ to move it to the Define Clusters By box.
6. Click **OK**.

In Figure 255, you can see the clustered bar chart for the frequency of each level of condescension within the three age categories.

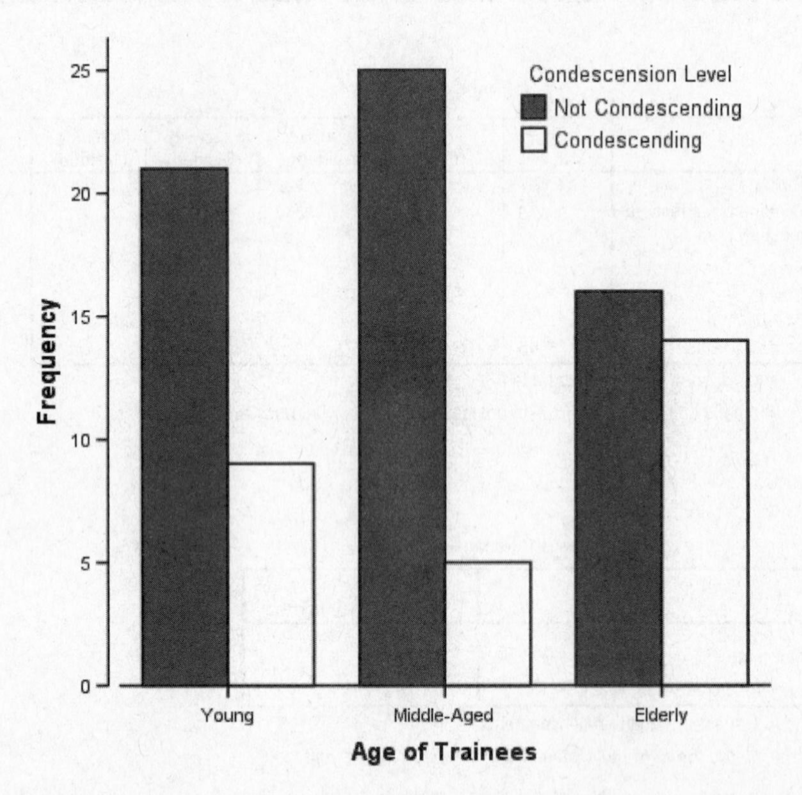

Figure 255. **A clustered bar chart of condescension within the age categories.**

An APA Results Section

A two-way contingency table analysis was conducted to evaluate whether male college students were more condescending to young, middle-aged, or elderly women. The two variables were age of female trainee with three levels (young, middle-aged, and elderly) and condescension of male trainer with two levels (not condescending and condescending). Age and condescension were found to be significantly related, Pearson $\chi^2(2, N = 90) = 6.32, p = .04$, Cramér's $V = .26$. The proportions of males who were condescending toward the young, the middle-aged, and the elderly female trainees were .30, .17, and .47, respectively.

Follow-up pairwise comparisons were conducted to evaluate the difference among these proportions. Table 62 shows the results of these analyses. The Holm's sequential Bonferroni method was used to control for Type I error at the .05 level across all three comparisons. The only pairwise difference that was significant was between the middle-aged and the elderly trainees. The probability of a trainee being treated in a condescending fashion was about 2.76 times (.47/.17) more likely when the trainee was elderly as opposed to middle-aged.

Table 62

Results for the Pairwise Comparisons Using the Holm's Sequential Bonferroni Method

Comparison	Pearson chi-square	*p* value (Alpha)	Cramér's *V*
Middle-aged vs. elderly	6.24*	.012 (.017)	.32
Young vs. elderly	1.76	.184 (.025)	.17
Young vs. middle-aged	1.49	.222 (.050)	.16

*p value ≤ alpha

Exercises

Exercises 1 through 3 are based on the following research problem. The data for these exercises can be found in the data file named *Lesson 41 Exercise File 1* on the Web at http://www.prenhall.com/greensalkind.

Lilly collects data on a sample of 130 high school students to evaluate whether the proportion of female high school students who take advanced math courses in high school varies depending upon whether they have been raised primarily by their father or by both their mother and their father. The SPSS data file contains two variables: math (0 = no advanced math and 1 = some advanced math) and parent (1 = primarily father and 2 = father and mother).

1. Conduct a crosstabs analysis to examine whether the proportion of female high school students who take advanced math courses is different for different levels of the parent variable. From the output, identify the following:
   a. Percent of female students who took some advanced math classes
   b. Percent of female students who took no advanced math classes when female students were raised by their fathers
   c. Percent of female students raised by their father only
   d. χ^2 value
   e. Strength of relationship between taking advanced math classes and level of parenting
2. Create a clustered bar graph to show differences in the number of female students taking some advanced math classes for the different categories of parenting.
3. Write a Results section based on your analysis.

Exercises 4 through 6 are based on the following research problem.

Bobby is interested in knowing whether teaching method has an effect on interest in subject matter. To investigate this question, he randomly assigns 90 high school students enrolled in history to one of three teaching conditions. All 90 students are exposed to exactly the same information for the same length of time. However, 30 of the students are presented a filmed reenactment of the historical events; 30 see an MTV-type video presentation of the events; and 30 are presented the information in the form of a traditional lecture. At the end of the presentation, the students in each condition are asked whether they found the information interesting. They were to check one of three alternatives: Really Like to Know More About That; Excuse Me—You Call That a Presentation?; and Boring—Please, No More of That! The data are in Table 63 (on page 376).

4. Create a data file containing the data in Table 63 by using the weighted cases method.
5. Conduct a two-way contingency analysis to analyze the data.
6. Write a Results section based on your analysis.

Table 63
Frequencies of Interest as a Function of Type of Presentation

Type of presentation	Interest in history area		
	Like to know more	Excuse me—You call …	Boring—Please, no more …
Film	12	10	8
Video	4	10	16
Lecture	6	6	18

LESSON

42

Two Independent-Samples Test: The Mann-Whitney *U* Test

The Mann-Whitney *U* test evaluates whether the medians on a test variable differ significantly between two groups. In an SPSS data file, each case must have scores on two variables, the grouping variable and the test variable. The grouping variable divides cases into two groups or categories, and the test variable assesses individuals on a variable with at least an ordinal scale.

Although SPSS uses the terms *grouping variable* and *test variable*, the grouping variable may also be referred to as the independent or categorical variable, and the test variable may be referred to as the dependent or the quantitative variable.

Applications of the Mann-Whitney *U* Test

A Mann-Whitney *U* test can analyze data from different types of studies.

- Experimental studies
- Quasi-experimental studies
- Field studies

We illustrate how to conduct a Mann-Whitney *U* test for a quasi-experimental study.

Quasi-Experimental Study

Based on a developmental theory of cognitive processing, Bob predicted that elderly individuals would have greater difficulty with a particular visual spatial memory task (VSMT) than middle-aged individuals. Fourteen elderly women (65 years and older) and 26 younger women (31 to 50 years of age), all of whom were college graduates, agreed to participate in the study. These 40 women were administered the VSMT, a measure that yields scores that range in value from 0 to 100. Bob's data file has 40 cases, one case for each woman in the sample. The file contains one variable designating whether a woman is part of the younger or the older group and a second variable representing the VSMT scores for each woman.

Understanding the Mann-Whitney *U* Test

To understand how the Mann-Whitney *U* test evaluates differences in medians, it is useful to first describe what data are analyzed with this test. We will use the VSMT data for young and elderly women to illustrate. The data are in Table 64 (on page 378).

Table 64

Data for Example on Aging and Visual Spatial Memory Task (VSMT)

Young			Elderly		
ID	VSMT	Ranked VSMT	ID	VSMT	Ranked VSMT
1	98	39.0	27	41	3.0
2	73	25.0	28	54	10.0
3	41	3.0	29	61	13.0
4	51	7.5	30	98	39.0
5	82	30.0	31	41	3.0
6	66	15.5	32	54	10.0
7	97	37.0	33	69	19.0
8	92	34.5	34	82	30.0
9	74	26.5	35	34	1.0
10	71	20.5	36	92	34.5
11	98	39.0	37	47	6.0
12	43	5.0	38	55	12.0
13	92	34.5	39	54	10.0
14	81	28.0	40	79	27.0
15	65	14.0			
16	72	23.0			
17	71	20.5			
18	72	23.0			
19	72	23.0			
20	92	34.5			
21	66	15.5			
22	82	30.0			
23	51	7.5			
24	86	32.0			
25	67	17.5			
26	67	17.5			

For a Mann-Whitney *U* test, the scores on the test variable are converted to ranks, ignoring group membership. For our data, the VSMT scores are converted to ranked VSMT scores, ignoring whether the VSMT scores are for younger or older women. The Mann-Whitney *U* test then evaluates whether the mean ranks for the two groups differ significantly from each other. In our example, the test evaluates whether the mean rank for the younger group of 23.17 differs significantly from the mean rank for the older group of 15.54.

Because analyses for the Mann-Whitney *U* test are conducted on ranked scores, the distributions of the test variable for the two populations do not have to be of any particular form (e.g., normal). However, these distributions should be continuous and have identical forms.

Assumptions Underlying a Mann-Whitney *U* Test

Assumption 1: The Continuous Distributions for the Test Variable Are Exactly the Same (Except their Medians) for the Two Populations

If population distributions differ on characteristics other than their medians, the Mann-Whitney *U* test does not evaluate whether the medians differ between populations, but whether the distributions themselves differ. In addition, difficulties can arise with the Mann-Whitney *U* test if there are not a variety of scores on the dependent variable and ties are produced in the ranked scores. SPSS employs a correction for such ties.

Assumption 2: The Cases Represent Random Samples from the Two Populations, and the Scores on the Test Variable Are Independent of Each Other

The Mann-Whitney U test will yield inaccurate p values if the independence assumption is violated.

Assumption 3: The z-Approximation Test for the Mann-Whitney U *Test Requires a Large Sample Size*

An exact test is printed by SPSS if the number of cases is less than or equal to 41. Accordingly, the z-approximation test does not have to be used unless the sample size is greater than 41. A sample size of at least 42 should be sufficiently large to yield relatively accurate p values with the z-approximation test. The z-approximation test includes a correction for ties but does not include a continuity correction. The results for the Mann-Whitney U test obtained from SPSS may differ from results in other statistical computer packages depending on whether the software uses the same correction procedures.

Effect Size Statistics for the Mann-Whitney *U* Test

SPSS does not report an effect size index for this test, but simple indices can be computed to communicate the size of the effect. For example, difference in mean ranks or medians between the two groups can serve as an effect size index.

The Data Set

The data set used to illustrate this test is named *Lesson 42 Data File 1* on the Web at http://www.prenhall.com/greensalkind. The set represents data coming from the example on aging and performance on a visual spatial memory task. The variables are shown in Table 65.

Table 65
Variables in Lesson 42 Data File 1

Variables	Definition
age	Age is the grouping variable. The age variable distinguishes between two age conditions: If a younger woman, then age = 1. If an older woman, then age = 2.
vsmt	VSMT is the test variable. VSMT is the visual spatial memory task, which yields scores that range in value potentially from 0 to 100. Higher scores indicate better memory.

The Research Question

The research question can be asked to reflect differences in medians between groups or a relationship between two variables.

1. Differences between the medians: Does median performance on the visual spatial memory task differ for younger and older women?
2. Relationship between two variables: Is performance on the visual spatial memory task related to age for women?

Conducting a Mann-Whitney *U* Test

To conduct a Mann-Whitney *U* Test, follow these steps:

1. Click **Analyze**, click **Nonparametric Tests**, and then click **2 Independent-Samples**. You will see the Two-Independent-Samples Tests dialog box as shown in Figure 256.

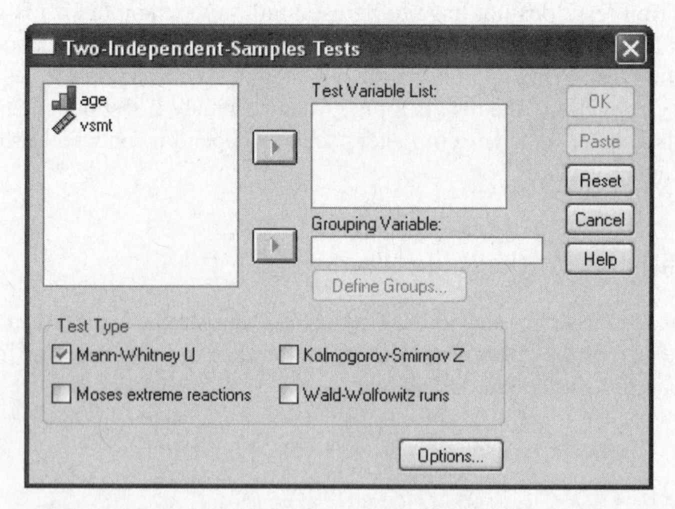

Figure 256. **The Two-Independent-Samples Tests dialog box.**

2. Click **vsmt**, then click ▶ to move it to the Test Variable List box.
3. Click **age**, then click ▶ to move it to the Grouping Variable box.
4. Click **Define Groups**.
5. Type **1** in the Group 1 box to indicate that Group 1 is the younger group.
6. Type **2** in the Group 2 box indicating that Group 2 is the elderly group.
7. Click **Continue**.
8. Be sure **Mann-Whitney U** is checked in the **Test Type** area.
9. Click **OK**.

Selected SPSS Output for the Mann-Whitney *U* Test

The output from the analysis is shown in Figure 257. Two *p* values are reported, an exact two-tailed *p*, which is not corrected for ties, and a two-tailed *p*, which is based on the *z* approximation test and is corrected for ties. As is typically the case, the difference between the two *p* values is negligible.

Using SPSS Graphs to Display the Results

Because the computations for the Mann-Whitney *U* test are on ranked scores of a test variable, a reader of a Results section reporting this test would have very little knowledge of what the original, raw scores would look like. Consequently, a graph presenting distributions of raw scores for the two groups would be very informative. Boxplots offer the reader more information than the other graphical options within SPSS. We present a boxplot in Figure 258 showing the distributions for the visual spatial memory scores for the two age groups.

Ranks

	age	N	Mean Rank	Sum of Ranks
vsmt	younger	26	23.17	602.50
	elderly	14	15.54	217.50
	Total	40		

Test Statistics[b]

	vsmt
Mann-Whitney U	112.500
Wilcoxon W	217.500
Z	-1.974
Asymp. Sig. (2-tailed)	.048
Exact Sig. [2*(1-tailed Sig.)]	.048[a]

a. Not corrected for ties.

b. Grouping Variable: age

Figure 257. **The results of the Mann-Whitney *U* analysis.**

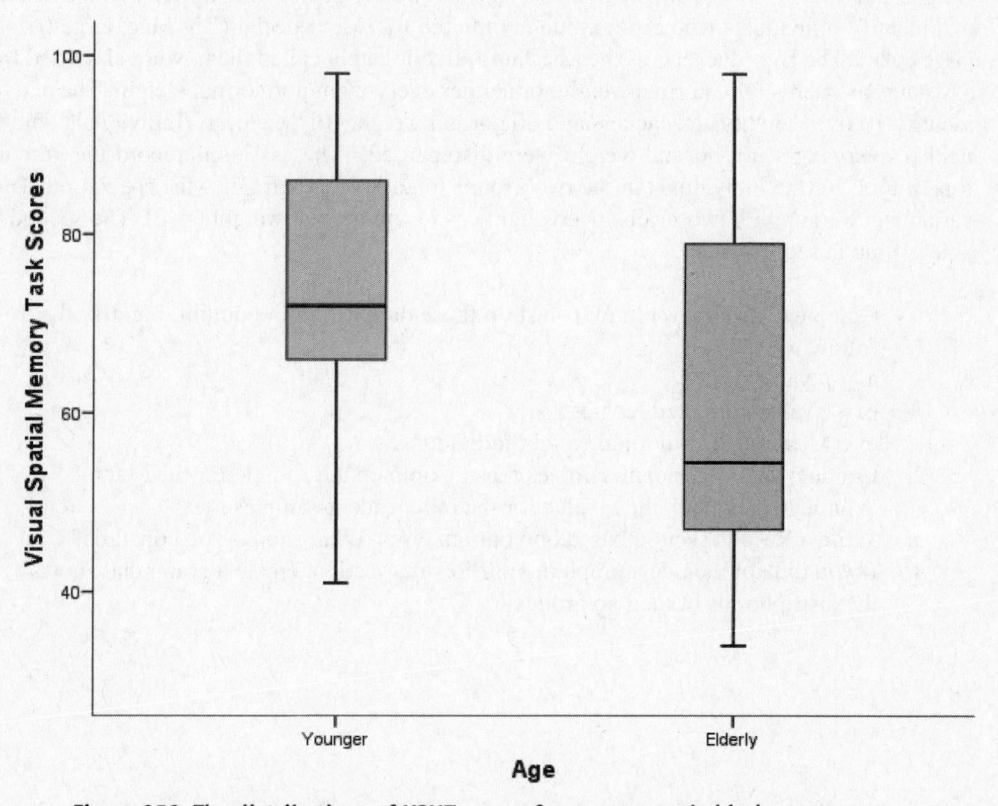

Figure 258. **The distributions of VSMT scores for younger and elderly women.**

An APA Results Section

A Mann-Whitney *U* test was conducted to evaluate the hypothesis that elderly women would score lower, on the average, than younger women on a visual spatial memory task. The results of the test were in the expected direction and significant, $z = -1.97$, $p < .05$. Elderly women had an average rank of 15.54, while younger women had an average rank of 23.17. Figure 258 shows the distributions of the scores on the VSMT measure for the two age groups.

Alternative Analyses

If a study has two independent samples, and individuals in the samples are assessed on a dependent variable measured on an ordinal scale, then the data from the study should be analyzed with a Mann-Whitney U test. However, if the dependent variable is measured on an interval scale (or ratio scale), then the data from the study can be analyzed with a Mann-Whitney U test or an independent-samples t test. The decision about which test to use with an interval-scaled dependent variable is complex and depends on whether the assumptions underlying these tests are met. If the normality assumption for the independent-samples t test is met, it is the more powerful than, and preferable to, the Mann-Whitney U test. The independent-samples t test yields relatively accurate p values under some conditions when the normality assumption is violated (see Lesson 24), but may not be as powerful as the Mann-Whitney U test for a number of different types of non-normal distributions, including those with thick tails. See Wilcox (2001) for more details about choices between these two tests and other methods.

Exercises

The data for Exercises 1 through 4 are in the data file named *Lesson 42 Exercise File 1* on the Web at http://www.prenhall.com/greensalkind. The data are based on the following study.

Billie wishes to test the hypothesis that overweight individuals tend to eat faster than normal weight individuals. To test this hypothesis, she has two assistants sit in a McDonald's restaurant and identify individuals who order at lunch time the Big Mac special (Big Mac, large fries, and large coke). The Big Mackers, as the assistants affectionately called them, were classified by the assistants as overweight, normal weight, or neither overweight nor normal weight. The assistants identify 10 overweight Big Mackers and 30 normal weight Big Mackers. (Individuals who were neither overweight nor normal weight were disregarded.) The assistants record the amount of time it took for the individuals in the two groups to complete their Big Mac special meals. One variable is weight with two levels, overweight (= 1) and normal weight (= 2). The second variable is time in seconds.

1. Compute a Mann-Whitney U test on these data. From the output, identify the following:
   a. p value
   b. z value corrected for ties
   c. Mean rank for normal weight individuals
2. Conduct an independent-samples t test. Compare the p value for the Mann-Whitney U test with the p value for the independent-samples t test.
3. Write a Results section based on your analyses. What should you conclude?
4. If you did not include a graph in your Results section, create a graph that shows the distributions of the two groups.

LESSON

43

K Independent-Samples Tests: The Kruskal-Wallis and the Median Tests

Both the Kruskal-Wallis and the median tests evaluate whether the population medians on a dependent variable are the same across all levels of a factor. For the Kruskal-Wallis and the median tests using the *K* independent samples procedure, cases must have scores on an independent or grouping variable and on a dependent or test variable. The independent or grouping variable divides individuals into two or more groups, and the dependent or test variable assesses individuals on at least an ordinal scale.

If the independent variable has only two levels, no additional significance tests need to be conducted beyond the Kruskal-Wallis or median test. On the other hand, if a factor has more than two levels and the overall test is significant, follow-up tests are usually conducted. These follow-up tests most frequently involve comparisons between pairs of group medians.

Applications of the Kruskal-Wallis and the Median Tests

Like the one-way ANOVA, the Kruskal-Wallis and median tests can analyze data from three types of studies.

- Experimental studies
- Quasi-experimental studies
- Field studies

Here we will apply the tests in an experimental study. We will use the Kruskal-Wallis and median tests on these data. You may wish to compare the results from these tests with those in Lessons 25 and 27, which use the one-way ANOVA and one-way ANCOVA to analyze the same data.

Experimental Study

Dana wishes to assess whether vitamin C is effective in the treatment of colds. To evaluate her hypothesis, she decides to conduct a 2-year experimental study. She obtains 30 volunteers from undergraduate classes to participate. She randomly assigns an equal number of students to three groups: placebo (group 1), low doses of vitamin C (group 2), and high doses of vitamin C (group 3). In the first and second years of the study, students in all three groups are monitored to assess the number of days with cold symptoms that they have. During the first year, students do not take any pills. In the second year of the study, students take pills that contain one of the following: no active ingredients (group 1), low doses of vitamin C (group 2), or high doses of vitamin C (group 3). Dana's SPSS data file includes 30 cases and two variables: a factor distinguishing among the three treatment groups and a dependent variable, the difference in the number of days with cold symptoms in the first year versus the number of days with cold symptoms in the second year.

Understanding the Kruskal-Wallis and Median Test

To understand how Kruskal-Wallis and the median test evaluate differences in medians among groups, we must first describe what data are analyzed for each test. Let's use our example dealing with vitamin C to understand better the characteristics of the data for each test.

As shown in Table 66, all scores on the dependent variable (difference in number of colds from one year to the next) are rank-ordered, disregarding levels of a factor (vitamin C group), from lowest to highest for the Kruskal-Wallis test. With the Kruskal-Wallis test, a chi-square statistic is used to evaluate differences in mean ranks to assess the null hypothesis that the medians are equal across groups.

Table 66

Difference-in-Days-with-Colds Scores Transformed for a Kruskal-Wallis Test and for a Median Test

No active ingredients			Low dose of vitamin C			High dose of vitamin C		
Diff. Score	Rank of score	At or below (−) vs. above (+) median	Diff. score	Rank of score	At or below (−) vs. above (+) median	Diff. score	Rank of score	At or below (−) vs. above (+) median
12	30.0	+	−2	13.5	−	6	27.5	+
−2	13.5	−	−3	10.5	−	−7	2.0	−
9	29.0	+	3	21.5	+	−6	5.0	−
3	21.5	+	−2	13.5	−	−6	5.0	−
3	21.5	+	0	16.5	+	−6	5.0	−
0	16.5	+	−4	8.5	−	−4	8.5	−
3	21.5	+	−3	10.5	−	−2	13.5	−
2	19.0	+	5	25.5	+	−6	5.0	−
4	24.0	+	−9	1.0	−	6	27.5	+
1	18.0	+	−6	5.0	−	5	25.5	+
Mean rank =	21.45	1 minus 9 pluses	Mean rank =	12.60	7 minus 3 pluses	Mean rank =	12.45	7 minus 3 pluses

The data are recast for the median test to create a two-way contingency table with frequencies in the cells (see Lesson 41 on two-way contingency table analysis). The columns of the table are defined by the levels of the factor (grouping variable), in this case the three vitamin C treatment groups. The rows of the table, of which there are always two, represent two levels of performance on the dependent variable: (1) performance at or below the median disregarding group membership and (2) performance above the median disregarding group membership. Table 66 shows how to obtain the data for the median test. For those who took pills with no active ingredients, one individual scored at or below the median (a rank of 15 or less) and nine individuals scored above the median (a rank of greater than 15). Similarly, as shown in the last row of the table, the frequencies for the other two groups are the same: seven individuals scored at or below the median, and three individuals scored above the median. These frequencies are presented in a two-way contingency format in Table 67.

The median test is a contingency table analysis that involves assessing whether the column variable (the factor) and the row variable (the dependent variable) are related. In other words, the results of the median significance test are identical to those obtained by recasting the data as a two-way contingency table and conducting a contingency table analysis by using the Crosstabs procedure.

Table 67

How Students in Each Vitamin C Group Scored Relative to the Median Regarding Difference in Days with Colds

	Vitamin C factor		
Difference in number days with colds with respect to the median	No active ingredients	Low dose of vitamin C	High dose of vitamin C
Less than or equal to median	1	7	7
Greater than the median	9	3	3

Assumptions Underlying the Median Test

Because the median test is a two-way contingency table analysis, the assumptions are the same as those for a two-way contingency table analysis. See Lesson 41 for additional discussion about these assumptions.

- Assumption 1: The observations for a two-way contingency table analysis are independent of each other.
- Assumption 2: Two-way contingency table analysis yields a test statistic that is approximately distributed as a chi-square when the sample size is relatively large.

Assumptions Underlying the Kruskal-Wallis Test

Because analysis for the Kruskal-Wallis test is conducted on ranked scores, the population distributions for the test variable (the scores that the ranks are based on) do not have to be of any particular form (e.g., normal). However, these distributions should be continuous and have identical form.

Assumption 1: The Continuous Distributions for the Test Variable Are Exactly the Same for the Different Populations

If population distributions differ on characteristics other than their medians, the Mann-Whitney test does not evaluate whether the medians differ among populations, but whether the distributions themselves differ. In addition, difficulties can arise with the Kruskal-Wallis test if there are not a variety of scores on the measure of interest and ties are produced in the ranked scores. However, SPSS employs a correction for such ties.

Assumption 2: The Cases Represent Random Samples from the Populations and the Scores on the Test Variable Are Independent of Each Other

The Kruskal-Wallis test will yield inaccurate p values if the independence assumption is violated.

Assumption 3: The Chi-Square Statistic for This Test Is Only Approximate and Becomes More Accurate with Larger Sample Sizes

The p value for the chi-square approximation test is fairly accurate if the number of cases is greater than or equal to 30.

Effect Size Statistics for the Median Test and the Kruskal-Wallis Test

SPSS does not report an effect size index for either the Kruskal-Wallis or the median tests. However, simple indices can be computed for both.

For the Kruskal-Wallis test, the median and the mean rank for each of the groups (the three vitamin C groups for our example) can be reported. Another possibility for the Kruskal-Wallis test is to compute an index that is usually associated with a one-way ANOVA, such as eta square (η^2), except η^2 in this case would be computed on the ranked data such as those in Table 66. To do so, transform the scores to ranks, conduct an ANOVA, and compute an eta square on the ranked scores. Eta square can also be computed directly from the reported chi-square value for the Kruskal-Wallis test with the use of the equation

$$\eta^2 = \frac{\chi^2}{N - 1}$$

where N is the total number of cases (30 for our example).

Effect sizes can also be reported for the median test. For example, you could report the median for each group or the proportion of individuals who scored higher than the overall median for each of the groups (.9, .3, and .3 for our data). Also, any effect size index that might be reported for a two-way contingency table analysis, such as Cramér's V, could be reported for the median test in that a median test is a two-way contingency table analysis.

The Data Set

The data set used to illustrate this procedure is named *Lesson 43 Data File 1* on the Web at http://www.prenhall.com/greensalkind and represents data from the vitamin C example. The variables in the data set are shown in Table 68.

Table 68

Variables in Lesson 43 Data File 1

Variables	Definition
group	1 = Placebo 2 = Low doses of vitamin C 3 = High doses of vitamin C
diff	The number of days in year 2 with cold symptoms minus the number of days in year 1 with cold symptoms.

The Research Question

The research question can be asked in terms of differences in the medians or in terms of relationship between two variables.

1. Differences between medians: Do the medians for change in the number of days of cold symptoms differ among those who take placebo, those who take low doses of vitamin C, and those who take high doses of vitamin C?
2. Relationship between two variables: Is there a relationship between the amount of vitamin C taken and the change in the number of days that individuals show cold symptoms?

Conducting a *K* Independent-Samples Test

We demonstrate how to conduct a median test and a Kruskal-Wallis test.

Conducting a Median Test

Let's conduct the overall median test and follow-up tests with the Crosstabs procedure.

Conducting the Overall Median Test

To evaluate the differences among all groups with the median test, follow these steps:

1. Click **Analyze**, click **Nonparametrics**, then click **K Independent Samples**. You'll see the Tests for Several Independent Samples dialog box shown in Figure 259.

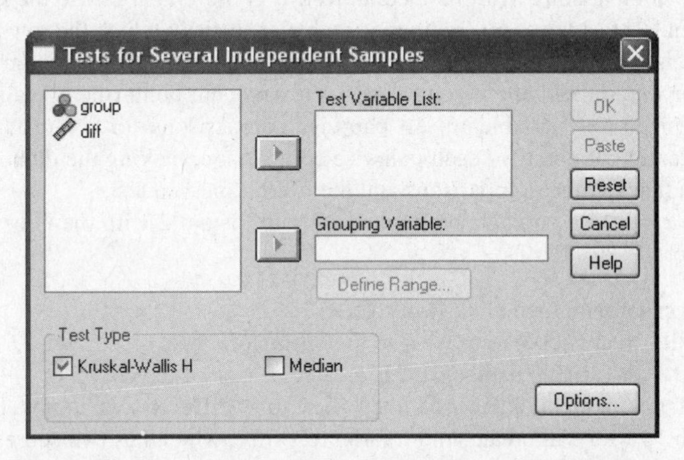

Figure 259. **Tests for Several Independent Samples dialog box.**

2. Click **Median** in the Test Type area. Be sure this is the only option checked.
3. Click **diff**, and then click ▶ to move it to the Test Variable List box.
4. Click **group**, and then click ▶ to move it to the Grouping Variable box.
5. Click the **Define Range**.
6. Type **1** as the Minimum value for group.
7. Type **3** as the Maximum value for group.
8. Click **Continue**.
9. Click **OK**.

Selected SPSS Output for the Median Test

The results are shown in Figure 260.

Frequencies

		Vitamin C Treatment		
		Placebo	Low Dose	High Dose
Change in Number of Days of Colds	> Median	9	3	3
	<= Median	1	7	7

Test Statistics[b]

	Change in Number of Days of Colds
N	30
Median	-1.00
Chi–Square	9.600[a]
df	2
Asymp. Sig.	.008

a. 0 cells (.0%) have expected frequencies less than 5. The minimum expected cell frequency is 5.0.

b. Grouping Variable: Vitamin C Treatment

Figure 260. **The results of the median test.**

The test evaluating differences among the three medians is significant, $p < .01$. Consequently, follow-up tests should be conducted to evaluate the three pairwise comparisons.

Computing Pairwise Comparisons after a Significant Median Test

For the pairwise comparisons with the median test, it is preferable to use the Crosstabs procedure. The median test is a two-way contingency table analysis in which the rows variable is a dichotomized dependent variable and the columns variable is an independent or grouping variable. The Crosstabs procedure will allow you to apply the same cut point (the overall median) to create the dichotomized test variable for all pairwise comparisons. In contrast, the median test would use a different cut point for each pairwise comparison, making the dichotomous variable inconsistently defined across comparisons and the overall median test.

To conduct a pairwise comparison between Groups 1 and 2 with the Crosstabs procedure, follow these steps:

1. Click **Transform**, then click **Rank Cases**.
2. Click **diff**, then click ▶ to move it to the Variable(s) box.
3. Click **OK**. A variable named rdiff is created as part of the data file.
4. Click **Transform**, click **Recode**, then click **Into Different Variables**. In Steps 4 through 18, you will create a new variable, rdiff2, which has two scores: 1 if a case or individual falls at or below the median on the test variable (diff) and 2 if a case is above the median.
5. Click **Rdiff**, then click ▶ to move it to Input Variable → Output Variable.
6. Type **rdiff2** in Name text area within the Output Variable area.
7. Click **Change**.
8. Click **Old and New Values**.
9. Click **Range** (first Range option) in the Old Value area.
10. Type **1** in the box below Range.
11. Type **15** in the box after Through. Given N is the number of cases, the midrank is N divided by 2 if N is even and $(N + 1) / 2$ if N is odd. For this example, 15 is the midranked score.
12. Type **1** in the Value text box in the New Value area.
13. Click **Add**.
14. Click **All other values** in Old Value area.
15. Type **2** in the Value box in the New Value area.
16. Click **Add**.
17. Click **Continue**.
18. Click **OK**. SPSS will return you to the Data Editor.
19. Click **Data**.
20. Click **Select Cases**.
21. Click **If condition is satisfied**.
22. Click **If**.
23. Type **group = 1 or group = 2** in the Select Cases If dialog box to compare Groups 1 and 2.
24. Click **Continue**.
25. Click **OK**.
26. Click **Analyze**, click **Descriptive Statistics**, then click **Crosstabs**.
27. Click **rdiff2**, then click ▶ to move it to the Row(s) box.
28. Click **group**, then click ▶ to move it to the Columns(s) box.
29. Click **Statistics**.
30. Click **Chi-square**, and click **Phi and Cramér's V**.
31. Click **Continue**.
32. Click **Cells**.
33. Click **Expected**.
34. Click **Continue**.
35. Click **OK**.

To compare Groups 1 and 3, repeat Steps 19 through 35, except at Step 23, type **group = 1 or group = 3** in the Select Cases If dialog box. To compare Groups 2 and 3, repeat Steps 19 through 35, except at Step 23 type **group = 2 or group = 3** in the Select Cases If dialog box.

Selected SPSS Output for Pairwise Comparisons

The results for the comparison between the placebo and the low-dose vitamin C groups are in Figure 261. The LSD procedure (see Appendix B) can be used to control for Type I error across the three pairwise comparisons. Using this procedure, there are two significant pairwise comparisons, the comparison between Groups 1 and 2 and the comparison between Groups 1 and 3.

rdiff2 * Vitamin C Treatment Crosstabulation

			Vitamin C Treatment		Total
			1	2	
rdiff2	1.00	Count	1	7	8
		Expected Count	4.0	4.0	8.0
	2.00	Count	9	3	12
		Expected Count	6.0	6.0	12.0
Total		Count	10	10	20
		Expected Count	10.0	10.0	20.0

Chi–Square Tests

	Value	df	Asymp. Sig. (2-sided)	Exact Sig. (2-sided)	Exact Sig. (1-sided)
Pearson Chi-Square	7.500[b]	1	.006		
Continuity Correction[a]	5.208	1	.022		
Likelihood Ratio	8.202	1	.004		
Fisher's Exact Test				.020	.010
Linear-by-Linear Association	7.125	1	.008		
N of Valid Cases	20				

a. Computed only for a 2 x 2 table

b. 2 cells (50.0%) have expected count less than 5. The minimum expected count is 4. 00.

Symmetric Measures

		Value	Approx. Sig.
Nominal by Nominal	Phi	-.612	.006
	Cramer's V	.612	.006
N of Valid Cases		20	

a. Not assuming the null hypothesis.

b. Using the asymptotic standard error assuming the null hypothesis.

Figure 261. **Follow-up results to a significant median test comparing the placebo and the low-dose vitamin C groups.**

Conducting a Kruskal-Wallis Test

Now we will conduct the overall Kruskal-Wallis test and follow-up tests using the Mann-Whitney *U* test.

Conducting the Overall Kruskal-Wallis Test

To conduct a Kruskal-Wallis test for differences among all groups, follow these steps:

1. Click **Data**, click **Select Cases**, then click **All Cases**, and click **OK** to select all cases for the analysis.
2. Click **Analyze**, click **Nonparametrics**, then click **K Independent Samples**.
3. Click **Reset** to clear the dialog box.
4. In the Test Type box, be sure that only the **Kruskal-Wallis** option is selected.
5. Click **diff**, and then click ▶ to move it to the Test Variable List box.
6. Click **group**, and then click ▶ to move it to the Grouping Variable box.
7. Click **Define Range**.
8. Type **1** as the Minimum value for group.
9. Type **3** as the Maximum value for group.
10. Click **Continue**.
11. Click **OK**.

Selected SPSS Output for the Kruskal-Wallis Test

The results of the analysis are shown in Figure 262. The Kruskal-Wallis test indicates that there is a significant difference in the medians, $\chi^2(2, N = 30) = 6.92, p = .03$. Because the overall test is significant, pairwise comparisons among the three groups should be conducted.

Ranks

	Vitamin C Treatment	N	Mean Rank
Change in Number of Days of Colds	Placebo	10	21.45
	Low Dose	10	12.60
	High Dose	10	12.45
	Total	30	

Test Statistics[a,b]

	Change in Number of Days of Colds
Chi-Square	6.923
df	2
Asymp. Sig.	.031

a. Kruskal-Wallis Test

b. Grouping Variable: Vitamin C Treatment

Figure 262. **The results of the Kruskal-Wallis analysis.**

Conducting Pairwise Comparisons after Obtaining a Significant Kruskal-Wallis Test

The pairwise comparisons will be conducted using the Mann-Whitney U test, which yields identical results with the Kruskal-Wallis test for two independent samples. For each pairwise comparison, the values in the Define Groups dialog box will be changed to match the comparison of interest.

To conduct a Mann-Whitney *U* test, follow these steps:

1. Click **Analyze**, click **Nonparametrics**, then click **2 Independent Samples**.
2. Click **diff**, then click ▶ to move it to the Test Variable(s) box.
3. Click **group**, then click ▶ to move it to the Grouping Variable box.
4. Click **Define Groups**.
5. Type **1** to indicate that 1 is Group 1.
6. Type **2** to indicate that 2 is Group 2.
7. Click **Continue**.
8. Be sure that the **Mann-Whitney U test** option is checked.
9. Click **OK**.

To perform the second and the third comparisons, repeat these steps, except in Steps 5 and 6 indicate in the Define Groups dialog box the groups of interest: groups 1 and 3 for the second comparison and groups 2 and 3 for the third comparison.

Selected SPSS Output for Pairwise Comparisons

The results for the first comparison between the placebo group and low-dose vitamin C group are shown in Figure 263. The Holm's sequential Bonferroni procedure can be used to control for Type I error. Using this procedure, only the comparison between Groups 1 and 2 is significant.

Ranks

	Vitamin C Treatment	N	Mean Rank	Sum of Ranks
Change in Number of Days of Colds	Placebo	10	13.90	139.00
	Low Dose	10	7.10	71.00
	Total	20		

Test Statistics[b]

	Change in Number of Days of Colds
Mann-Whitney U	16.000
Wilcoxon W	71.000
Z	-2.586
Asymp. Sig. (2-tailed)	.010
Exact Sig. [2*(1-tailed Sig.)]	.009[a]

a. Not corrected for ties.

b. Grouping Variable: Vitamin C Treatment

Figure 263. **The results of the pairwise comparison between the placebo group and the low-dose vitamin C group.**

Using SPSS Graphs to Display the Results

The data for these tests are ranked, and a reader of the results would have very little knowledge of what the distributions for the dependent variable scores look like for the three groups. A graph presenting distributions of diff scores for the three groups, such as the boxplot graph shown in Figure 264 (on page 392), is very informative.

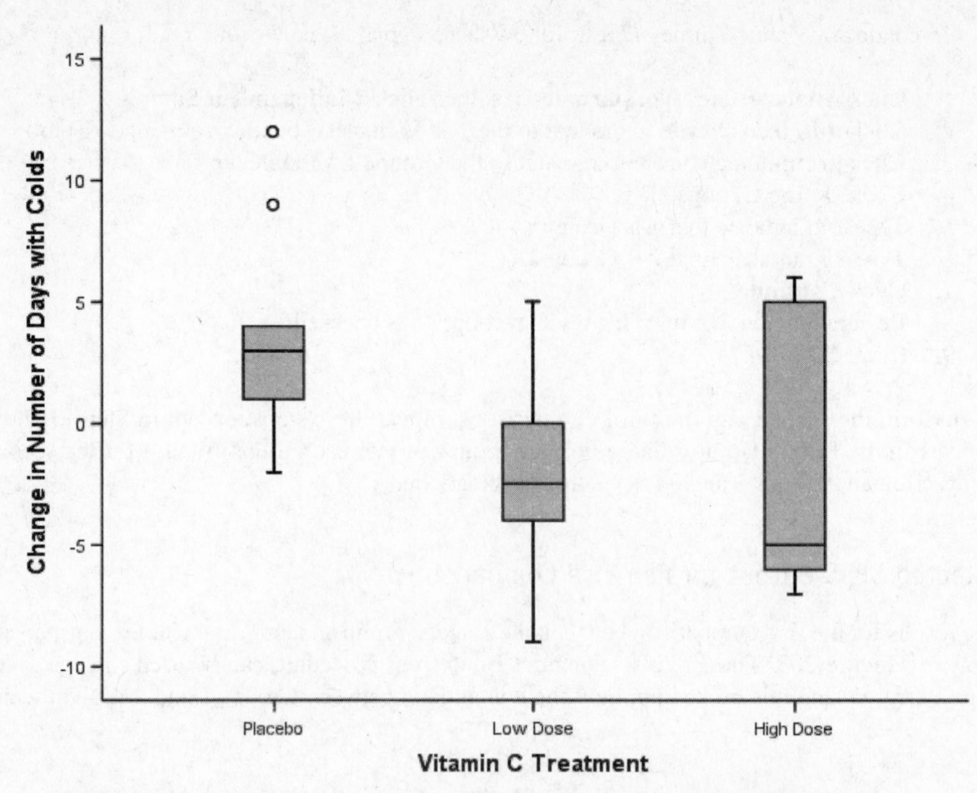

Figure 264. **Distributions of the change in the number of days with colds for vitamin C treatment groups.**

Two APA Results Sections

Results for the Median Test

A two-way contingency table analysis was conducted relating the vitamin C variable (placebo, low dose of vitamin C, and high dose of vitamin C) to the dependent variable (the number of days of cold symptoms while taking medication minus the number of days of cold symptoms before taking medication) that was dichotomized at the overall median. All individuals who fell at or below the median were classified as better responders to medication, and all individuals who fell above the median were classified as poorer responders to medication.

The two-way contingency table analysis was significant, $\chi^2(2, N = 30) = 9.60, p < .01$. The strength of relationship between the independent and dependent variables, as assessed by Cramér's *V*, was .57, indicating a strong relationship between the vitamin C factor and the dichotomized response-to-medication variable.

Follow-up tests were conducted to evaluate pairwise differences among the three groups using the LSD procedure. The results of these tests indicated two of the three pairwise differences were significant: the difference between the placebo group and the low-dose vitamin C group and the difference between the placebo group and the high-dose vitamin C group. Regardless of the dosage level, taking vitamin C produced a greater proportion of participants who were better responders to medication than the placebo group.

Results for the Kruskal-Wallis Test

A Kruskal-Wallis test was conducted to evaluate differences among the three vitamin C conditions (placebo, low dose of vitamin C, and high dose of vitamin C) on median change in number of days with cold symptoms (number of days with colds during treatment minus number of days with colds prior to treatment). The test, which was corrected for tied ranks, was significant, $\chi^2(2, N = 30) = 6.92, p = .03$. The proportion of variability in the ranked dependent variable accounted for by the vitamin C treatment variable was .24, indicating a fairly strong relationship between vitamin C treatment and the change in the number of days with colds.

Follow-up tests were conducted to evaluate pairwise differences among the three groups, controlling for Type I error across tests by using the Holm's sequential Bonferroni approach. The results of these tests indicated a significant difference between the placebo group and the low-dose vitamin C group. The typical decrease in number of days with cold symptoms after treatment was greater for the low-dose vitamin C treatment group than for the placebo group.

Alternative Analyses

If a study has a dependent variable that is measured on an interval or ratio scale, the data from the study can be analyzed using a one-way ANOVA, a Kruskal-Wallis, or a median test. However, if the dependent variable is measured on an ordinal scale, it is preferable to conduct a Kruskal-Wallis or median test because medians are a more meaningful measure of central tendency than means for ordinal data.

If the dependent variable is measured on at least an interval scale, the decision about which test to use depends upon whether the assumptions underlying the tests are met. The median test makes no requirement about the population distributions. The Kruskal-Wallis test assumes that the population distributions are identical, although the distributions can have any shape. The one-way ANOVA assumes not only that the distributions are identical, but also that the distributions are normal. The choice among tests is complex and is at least partially dictated by the relative power of the three tests.

- If the normality assumption is met, a one-way ANOVA is the preferred test.
- If the population distributions are symmetrical and flat (uniform), a Kruskal-Wallis or a one-way ANOVA could be used to analyze your data.
- If the distributions are symmetrical, but with thicker tails than a normal distribution, you should use the Kruskal-Wallis test in that it has greater power than the one-way ANOVA.
- The median test should probably be avoided.

Exercises

The data for Exercises 1 through 5 are in the data set named *Lesson 43 Exercise File 1* on the Web at http://www.prenhall.com/greensalkind. The data are from the following research problem.

Marvin is interested in whether blonds, brunets, and redheads differ in their extroversion. He randomly samples 18 men from his local college campus: six blonds, six brunets, and six redheads. He then administers a measure of social extroversion to each individual.

1. Conduct a Kruskal-Wallis test to investigate the relationship between hair color and social extroversion. Should you conduct follow-up tests?
2. Compute an effect size for the overall effect of hair color on extroversion.
3. Create a boxplot to display the differences among the three groups' distributions.
4. Conduct a one-way ANOVA on these data. How do the results for the ANOVA compare to the results of the Kruskal-Wallis test?
5. Write a Results section based on your analyses.

The data for Exercises 6 through 8 are in the data set named *Lesson 43 Exercise File 2* on the Web at http://www.prenhall.com/greensalkind. The data are from the following research problem.

Karin believes that students with behavior problems react best to teachers who have a humanistic philosophy but have firm control of their classrooms. In a large-city school system, she classifies 40 high school teachers of students with behavior problems into three categories: humanists with control (Group 1; $n = 9$); strict disciplinarians (Group 2; $n = 21$), and keepers of the behavior problems (Group 3; $n = 10$). Karin is interested in assessing whether students who are taught by one type of teacher as opposed to another type of teacher are more likely to stay out of difficulty in school. By reviewing the records of all students in these 40 teachers' classes, Karin determines the number of times the students of each teacher are sent to the main office for disciplinary action during the second half of the academic year. All classes had 30 students in them. Karin's SPSS data file has 40 cases with a factor distinguishing among the three types of teachers and a dependent variable, the number of disciplinary actions for students in a class.

6. Conduct a median test to answer Karin's question. Is the test significant?
7. Conduct follow-up tests, if appropriate.
8. Write a Results section based on the analyses you have conducted.

LESSON

44

Two Related-Samples Tests: The McNemar, the Sign, and the Wilcoxon Tests

The McNemar, sign, and Wilcoxon tests were developed to analyze data from studies with similar designs. These three tests, as well as the paired-samples *t* test (see Lesson 23), can be applied to problems with repeated-measure designs and matched-subjects designs.

For a repeated-measures design, an individual is assessed on a measure on two occasions or under two conditions. Each individual is a case in the SPSS data file and has scores on two variables, the score obtained on the measure on one occasion (or under one condition) and the score obtained on the measure on a second occasion (or under a second condition). The goal of repeated-measures designs is to determine whether participants changed significantly across occasions (or conditions).

For a matched-subjects design, participants are paired, typically on one or more nuisance variables, and each participant in the pair is assessed once on a measure. Each pair of participants is a case in the SPSS data file and has scores on two variables, the score obtained on the measure by one participant under one condition and the score obtained on the measure by the other participant under the other condition. The purpose of matched-subjects designs is to evaluate whether the pairs of participants differ significantly under the two conditions.

For both types of studies, each case in the SPSS data file has scores on two variables (that is, paired scores). The McNemar, sign, and Wilcoxon tests evaluate differences between paired scores, either repeated or matched. For the McNemar test, the variables have two categories. The McNemar test evaluates whether the proportion of participants who fall into a category on one variable differs significantly from the proportion of participants who fall into the same category on the second variable. In contrast, the variables for the sign and the Wilcoxon have multiple possible scores. With these tests, the focus is on whether the medians of the variables differ significantly.

Applications of the McNemar, Sign, and Wilcoxon Tests

The McNemar, sign, and Wilcoxon tests can be used to analyze data from a number of different designs.

- Repeated-measures designs with an intervention
- Repeated-measures designs without an intervention
- Matched-subjects designs with an intervention
- Matched-subjects designs without an intervention

For each application, we will define different dependent variables to illustrate the use of the McNemar test as well as the sign and Wilcoxon tests.

Repeated-Measures Designs with an Intervention

Bill wants to assess the effectiveness of a leadership workshop for 60 middle-management executives. The 60 executives are judged by their immediate superiors on their leadership capabilities before and after the workshop. The SPSS data file contains 60 cases, one for each executive, and two variables, leadership adequacy judgments before the workshop and leadership adequacy after the workshop. For the McNemar test, the leadership adequacy variables have two scores, 0 for just adequate or inadequate and 1 for better-than-adequate. The values of 0 and 1 are arbitrary. For the sign or Wilcoxon test, the leadership adequacy variables are ratings from the Leadership Rating Form (LRF) and range in value from 10 to 50 (sum of ten 5-point rating scales).

Repeated-Measures Designs with No Intervention

Michelle is interested in determining if workers are more concerned with job security or job pay. She gains the cooperation of 30 individuals who work in a variety of work settings and asks each employee to indicate how concerned they are about their salary level and about their job security. Michelle's data file contains 30 cases, one for each employee, and two variables, judgments about concern for salary level and judgments about concern for job security. For the McNemar test, the scores on these variables are 0, indicating "not extremely concerned," and 1, indicating "extremely concerned." The values of 0 and 1 are arbitrary. For the sign and Wilcoxon tests, the scores on the variables are based on rating scales that range in value from 1 = no concern to 10 = ultimate concern.

Matched-Subjects Designs with an Intervention

Bob is interested in determining if individuals who are exposed to depressed individuals become sad. He obtains scores on the mood measure of sadness (MMS) scale for 40 students and rank orders their scores. He then defines pairs of students based on their MMS scores: two students with the lowest MMS scores, two students with the next lowest MMS scores, and so on. Accordingly, the students within a matched pair have more similar MMS scores than the students from different pairs. Bob randomly assigns the students within each matched pair to one of two conditions. One student is exposed to an individual who acts depressed, and the other person is exposed to an individual who acts nondepressed. After each of the students is exposed to the individual who acted either depressed or nondepressed, the exposed person is evaluated by using a sadness posttest. Bob's SPSS data file contains 20 cases, one for each matched pair of students, and two variables, the sadness posttest for the students exposed to the individual who acted depressed and the sadness posttest for the students exposed to the individual who acted nondepressed. For the McNemar test, the scores on the two sadness posttest variables are 0 for not sad and 1 for sad. For the sign or Wilcoxon test, the scores on the two sadness posttest variables are posttest MMS scores.

Matched-Subjects Designs with No Intervention

Terri is interested in investigating if husbands and wives who are having infertility problems feel equally anxious. She obtains the cooperation of 24 infertile couples. She then interviews both the husbands and the wives and has an independent observer listen to the 48 tapes randomly ordered. After listening to each tape, the observer evaluates the anxiety level of the taped individual. Terri's SPSS data file contains 24 cases, one for each husband–wife pair, and two variables, the anxiety judgments for the husband and for the wife. Note that in this study the subjects come as matched pairs (infertile couples) and that the conditions under which the experimenter obtains the anxiety judgments are the two levels of gender, male and female. For the McNemar test, the scores on the variables assessing infertility anxiety are 0 for not anxious and 1 for anxious. For the sign or Wilcoxon test, the scores on the variables for each matched pair of husbands and wives are from the infertility anxiety measure (IAM) that yields quantitative scores.

Understanding the McNemar, Sign, and Wilcoxon Tests

To help us understand the three tests, we will first describe what data are analyzed for each test. Let's use our example dealing with concern for job security and job pay.

The data for the McNemar test in the example are presented in Table 69. Eighteen employees have tied scores, and they are disregarded in calculations for the McNemar test. The tied scores are associated with the 11 employees who are not extremely concerned about pay and security and the 7 employees who are extremely concerned about both pay and security. Analyses for the McNemar test involve comparing the proportion of employees with nontied scores who are extremely concerned about pay, but not with security (proportion = 8/12) with the proportion of employees with nontied scores who are extremely concerned about security, but not with pay (proportion = 4/12). More specifically, the McNemar test is a binomial test that evaluates whether these proportions differ significantly from .50. Although the McNemar test directly evaluates the difference between these two proportions, it simultaneously evaluates the difference between the proportion of employees who have concern for pay versus the proportion of employees who have concern for security. If the first pair of proportions differ, the second pair of proportions must also differ.

Table 69

Frequencies of Concern for Job Pay and Job Security for 30 Employees—McNemar Test

	Job Security	
Job Pay	Not extremely concerned	Extremely concerned
Not extremely concerned	11	4
Extremely concerned	8	7

The data for the sign test are presented in Table 70 (on page 398). Four employees have tied scores on concern for job security and concern for job pay (ID numbers 8, 15, 25, and 27), and they are disregarded in calculations for the sign test. The sign test ignores the quantitative differences between scores on the two variables (security − pay) and pays attention only to the sign of these difference scores, positive or negative. Two proportions are computed:

$$\textit{Proportion of Positive Differences} = \frac{\textit{Number of Positive Differences}}{\textit{Number of NonTied Sources}}$$

$$\textit{Proportion of Negative Differences} = \frac{\textit{Number of Negative Differences}}{\textit{Number of NonTied Sources}}$$

The proportion of positive differences and the proportion of negative differences for our data are 6/26 = .23 and 20/26 = .77. The sign test is a binomial test that evaluates whether these proportions differ significantly from .50.

The data for the Wilcoxon test also are presented in Table 70 (on page 398). Tied data are defined comparably for the sign and Wilcoxon tests; therefore, the same four difference scores (differences of 0 for IDs 8, 15, 25, and 27) that were excluded from the analyses for the sign test are also eliminated from the analyses for the Wilcoxon test. The Wilcoxon test involves ranking all nonzero difference scores disregarding sign, reattaching the sign to the ranks, and then evaluating the mean of the positive and the mean of the negative ranks. For our data, the mean of the positive ranks is 12.25, and the mean of the negative ranks is 13.88. The mean of the positive ranks has only six scores so that we can easily show how it is computed:

$$\text{Mean of Positive Ranks} = \frac{15 + 15 + 6.5 + 6.5 + 6.5 = 24}{6} = 12.25$$

Table 70

Data for Job Pay and Job Security for 30 Employees—Sign and Wilcoxon Test

			Security_Pay		Rank of absolute value of Security_Pay with the sign of Security_Pay	
ID	Security	Pay	Positive differences	Negative differences	Positive ranks	Negative ranks
1	3	4		-1		-6.5
2	1	9		-8		-26.0
3	6	8		-2		-15.0
4	7	5	+2		15.0	
5	6	4	+2		15.0	
6	5	6		-1		-6.5
7	3	6		-3		-20.0
8	4	4				
9	8	7	+1		6.5	
10	3	5		-2		-15.0
11	4	5		-1		-6.5
12	6	8		-2		-15.0
13	6	7		-1		-6.5
14	3	6		-3		-20.0
15	7	7				
16	5	8		-3		-20.0
17	4	5		-1		-6.5
18	4	3	+1		6.5	
19	6	5	+1		6.5	
20	6	7		-1		-6.5
21	1	5		-4		-24.0
22	7	3	+4		24.0	
23	4	5		-1		-6.5
24	5	6		-1		-6.5
25	4	4				
26	2	5		-3		-20.0
27	6	6				
28	4	5		-1		-6.5
29	3	6		-3		-20.0
30	2	6		-4		-24.0

Assumptions Underlying the McNemar, Sign, and Wilcoxon Tests

The first assumption, independence, underlies all three tests. Assumption 2 is associated with all three tests when they use z approximation. As previously described, the McNemar and sign tests are disguised binomial tests. An exact p value is computed based on the binomial distribution if there are fewer than 26 nontied paired scores, and a z-approximation test is calculated if the number of pairs of nontied scores is equal to or greater than 26. (See Lesson 39.) For the Wilcoxon test, SPSS always uses a z-approximation test to compute a p value. The third assumption is relevant for only the Wilcoxon test.

Assumption 1: Each Pair of Observations Must Represent a Random Sample from a Population and Must Be Independent of Every Other Pair of Observations

If the data are from a repeated-measures design, the paired scores for each subject must be independent of the paired scores for any other subject. If the data are from a matched-subjects design, the paired scores from any matched pair of subjects must be independent of the paired scores from any other matched pair of subjects. If the independence assumption is violated, the test is likely to yield inaccurate results. It should be noted that the analysis presumes dependency between scores within pairs.

Assumption 2: The z Tests for the Three Tests Yield Relatively Accurate Results to the Extent That the Sample Size Is Large

The results for the McNemar and sign test should be fairly accurate given that SPSS requires a sample of 26 or more pairs of nontied scores before the z test is applied. However, you should be concerned about the accuracy of the approximate p value for the Wilcoxon test if the sample size dips below 16 pairs of nontied scores.

Assumption 3 (Wilcoxon Test Only): The Distribution of the Difference Scores Is Continuous and Symmetrical in the Population

Although the McNemar and the sign test require no additional assumptions, the Wilcoxon test requires a third one.

This assumption pertains to the difference scores, not to the ranked scores. If the difference scores are continuously distributed, there should be no ties in the difference scores across pairs of scores. In the sample data shown in Table 70, there are multiple ties in the data.

Effect Size Statistics for the McNemar, Sign, and Wilcoxon Tests

SPSS does not report an effect size index for these three tests. However, simple indices can be reported for each of these tests to communicate the size of the effect.

For the McNemar test, a good effect index is the difference in the proportion of individuals who fall into one of the two categories on the one occasion (or condition) versus the proportion of individuals who fall into the same category on the other occasion (or other condition). For our data, the two proportions being compared are the proportion of employees who are extremely concerned about job pay, which is $.50 = (7 + 8)/30$, and the proportion of employees who are extremely concerned about job security, which is $.37 = (4 + 7)/30$. Consequently, the difference in proportions is $.50 - .37 = .13$.

For the sign test, the proportion of individuals who had a positive (or negative) difference score in comparison with negative (or positive) differences can be reported. For example, the proportion of employees who showed greater concern for job pay in comparison with greater concern for job security of $.77 = 20/(20 + 6)$ could be reported. As an alternative, you could report the proportion of employees who showed greater concern for job security in comparison with greater concern for job pay of $.23 = 6/(20 + 6)$.

For the Wilcoxon test, the mean positive ranked difference score and the mean negative ranked difference score could be reported.

The Data Set

The data used to illustrate the three types of tests are on the Web at http://www.prenhall.com/greensalkind in the file named *Lesson 44 Data File 1*. The set represents data from the example dealing with concern for job security and job pay. The description of the variables is given in Table 71. The sign and Wilcoxon tests are conducted on the pay and security variables, while the McNemar test is conducted on the payc and the securitc variables.

Table 71
Variables in Lesson 44 Exercise Data Set 1

Variables	Definition
pay	Concern for pay on a rating scale from 1, indicating no concern, to 10, indicating ultimate concern
security	Concern for job security on a rating scale from 1, indicating no concern, to 10, indicating ultimate concern
payc	Concern for job pay with 0 = not extremely concerned and 1 = extremely concerned
securitc	Concern for job security with 0 = not extremely concerned and 1 = extremely concerned

The Research Question

1. McNemar test: Are a greater proportion of employees concerned about job pay or job security?
2. Sign and the Wilcoxon tests: Does median concern for job security differ from median concern for job pay?

Conducting Tests for Two Related Samples

Now we will illustrate how to conduct analyses for the McNemar test, the sign test, and the Wilcoxon test.

Conducting a McNemar Test

To conduct a McNemar test, follow these steps:

1. Click **Analyze**, click **Nonparametric Tests**, then click **2 Related Samples**. You'll see the Two-Related-Samples Tests dialog box as shown in Figure 265.
2. Click **payc** and **securitc**.
3. Click ▶ to move them to the box labeled Test Pair(s) List.
4. Click **McNemar** in the Test Type area. Be sure this is the only option checked.
5. Click **OK**.

Selected SPSS Output for the McNemar Test

The results are shown in Figure 266. The *p* value of .39 is greater than .05, and therefore the proportions of .50 and .37 do not differ significantly from each other.

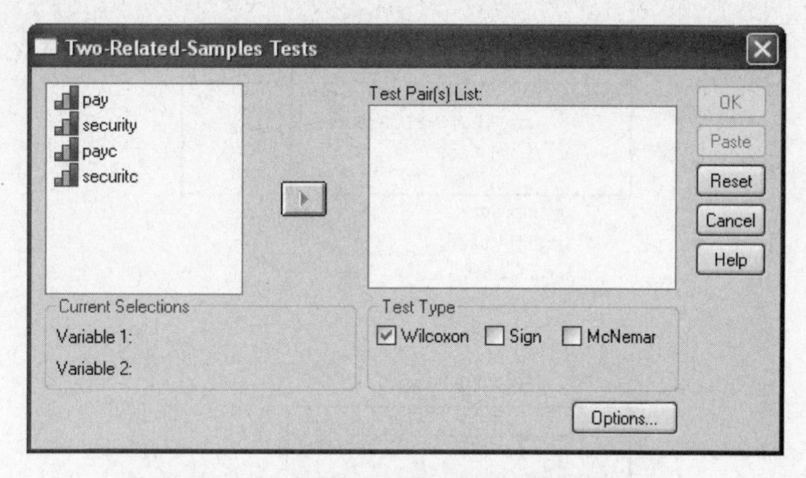

Figure 265. **The Two-Related-Samples Tests dialog box.**

payc & securitc

	securitc	
payc	0	1
0	11	4
1	8	7

Test Statistics[b]

	payc & securitc
N	30
Exact Sig. (2-tailed)	.388[a]

a. Binomial distribution used.

b. McNemar Test

Figure 266. **The results of the McNemar test.**

Conducting a Sign Test

1. Click **Analyze**, click **Nonparametric Tests**, then click **2 Related Samples Test**.
2. Click **Reset** to clear the dialog box.
3. Click **pay** and **security**.
4. Click ▶ to move them to the box labeled Test Pair(s) List.
5. Click **Sign** in the Test Type area. Be sure this is the only option checked.
6. Click **OK**.

Selected SPSS Output for the Sign Test

The results are shown in Figure 267 (on page 402). The sign test indicated significant differential concern for pay versus security, $p = .01$.

Conducting a Wilcoxon Test

1. Click **Analyze**, click **Nonparametric Tests**, then click **2 Related Samples Test**.
2. The variables **pay** and **security** should already be selected.
3. Click **Wilcoxon** in the Test Type area. Be sure this is the only option checked.
4. Click **OK**.

Frequencies

		N
security - pay	Negative Differences[a]	20
	Positive Differences[b]	6
	Ties[c]	4
	Total	30

a. security < pay

b. security > pay

c. security = pay

Test Statistics[a]

	security - pay
Z	-2.550
Asymp. Sig. (2-tailed)	.011

a. Sign Test

Figure 267. **The results of the sign test.**

Selected SPSS Output for the Wilcoxon Test

The results are shown in Figure 268. The results for the Wilcoxon test indicated significant differential concern for pay versus security, $p < .01$.

Ranks

		N	Mean Rank	Sum of Ranks
security - pay	Negative Ranks	20[a]	13.88	277.50
	Positive Ranks	6[b]	12.25	73.50
	Ties	4[c]		
	Total	30		

a. security < pay

b. security > pay

c. security = pay

Test Statistics[b]

	security - pay
Z	-2.626[a]
Asymp. Sig. (2-tailed)	.009

a. Based on positive ranks.

b. Wilcoxon Signed Ranks Test

Figure 268. **The results of the Wilcoxon test.**

Using SPSS Graphs to Display Results

Because of the simplicity of the data used in the McNemar test, it is probably unnecessary to create a graph to represent the data. However, graphs might be useful for the sign and Wilcoxon tests. The reported statistics offer no information about the shape of the distributions of the nonranked data on the original measures. Consequently, a graphical display of these data, such as a boxplot, would provide a visual representation of the distributions of the variables. Figure 269 shows the boxplot of the nonranked pay and security scores.

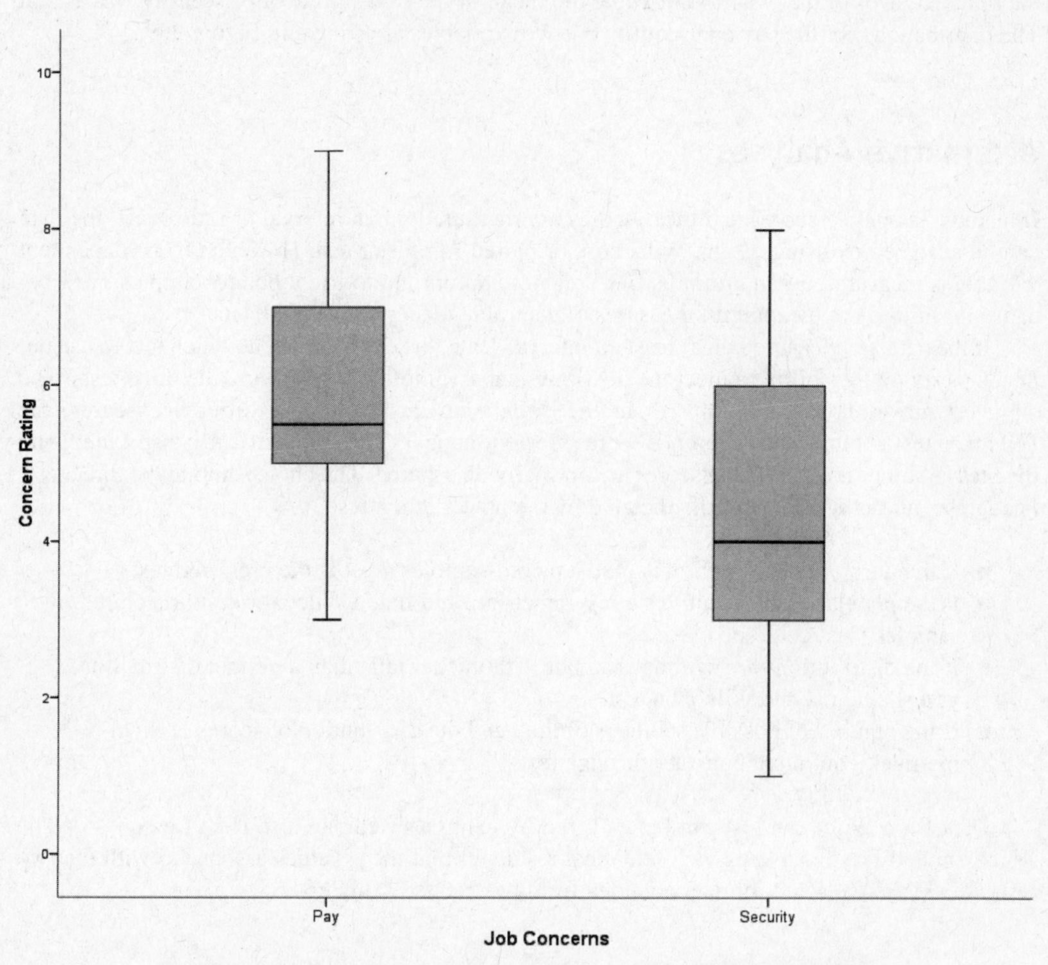

***Figure 269.* Distributions of pay and job security importance ratings.**

Three APA Results Sections

We present sample results sections for the McNemar, sign, and Wilcoxon tests.

Results for McNemar Test

Fifty percent of the employees indicated that they were extremely concerned about job pay, while only 37% indicated that they were extremely concerned about job security. However, these percentages were not significantly different from each other based on the results of the McNemar test of dependent proportions, $p = .39$.

Results for Sign Test

Seventy-seven percent of the employees who indicated differential concern for pay versus security showed greater concern for pay. The sign test indicated significant differential concern for pay versus security, $z = 2.55$, $p = .01$. The distributions for the pay and security concern variables are shown in Figure 269.

Results for Wilcoxon Test

A Wilcoxon test was conducted to evaluate whether employees showed greater concern for pay or security. The results indicated a significant difference, $z = -2.63$, $p < .01$. The mean of the ranks in favor of pay was 13.88, while the mean of the ranks in favor of security was 12.25. The distributions for the pay and security concern variables are shown in Figure 269.

Alternative Analyses

If a study has paired scores and these scores are measured on an interval or ratio scale, the data can be analyzed by using a sign, Wilcoxon, or paired-samples *t* test. However, if the dependent variable is measured on an ordinal scale, it is not appropriate to use a paired-samples *t* test because means are not meaningful measures of central tendency for ordinal data.

If the data are measured on at least an interval scale, the decision about which test to use depends partly on the ability to meet the distributional assumptions underlying the three tests. The sign test makes no assumptions about the population distribution of difference scores; the Wilcoxon test assumes the difference scores are continuously and symmetrically distributed; and the *t* test assumes the difference scores are normally distributed. The choice among the three tests is complex and is at least partially dictated by the power of the tests.

- If the normality assumption is met, a paired-samples *t* test is the preferred test.
- If the population distributions are symmetrical and flat, a Wilcoxon test or a paired-samples *t* test could be used.
- If the distributions are symmetrical but with thicker tails than a normal distribution, you should use the Wilcoxon test.
- If the number of possible scores is limited and the distribution of scores is asymmetrical, you might consider the sign test.

Data for a Wilcoxon test can be analyzed by using a Friedman test. (See Lesson 45.) The Wilcoxon and Friedman tests yield identical results in that the *p* values associated with the two tests are always the same for data obtained from two related samples.

Exercises

The data for Exercises 1 through 4 are in the data set named *Lesson 44 Exercise File 1* on the Web at http://www.prenhall.com/greensalkind. The data are from the following research problem.

Terri is interested in investigating if husbands and wives who are having infertility problems feel equally anxious. She obtains the cooperation of 24 infertile couples. She then interviews

Table 72
Variables in Lesson 44 Exercise Data Set 1

Variables	Definition
husband	Scores on the IAM scale (mean of 50 and standard deviation of 10 in the normative sample) for husbands. Higher scores mean greater anxiety.
wife	Scores on the IAM scale (mean of 50 and standard deviation of 10 in the normative sample) for wives. Higher scores mean greater anxiety.
husbandc	Anxiety judgment about husband made by observer, with 0 = nonanxious and 1 = anxious
wifec	Anxiety judgment about wife made by observer, with 0 = nonanxious and 1 = anxious

both the husbands and the wives and has an independent observer listen to the 48 tapes randomly ordered. After listening to each tape, the observer evaluates the anxiety level of the taped individual. Terri's SPSS data file contains 24 cases, one for each husband–wife pair, and two variables, the anxiety judgments for the husband and for the wife. For the McNemar test, the scores on the variables assessing infertility anxiety (husbandc and wifec) are 0 for not anxious and 1 for anxious. For the sign or Wilcoxon test, the scores on variables assessing infertility anxiety (husband and wife) are quantitative scores from the infertility anxiety measure (IAM). The variables are described in Table 72.

1. Conduct a McNemar test on the husbandc and wifec variables and write a Results section based on the output.
2. Conduct a sign test on the husband and wife variables and write a Results section based on the output.
3. Conduct a Wilcoxon test on the husband and wife variables and write a Results section based on the output.
4. Conduct a paired-samples *t* test on the husband and wife variables. How do the results for the *t* test compare with the results for the sign and Wilcoxon tests?

LESSON

45 *K* Related-Samples Tests: The Friedman and the Cochran Tests

The Cochran and the Friedman tests are extensions of the McNemar and Wilcoxon tests, respectively. The McNemar and Wilcoxon tests can be applied to repeated-measures data if participants are assessed on two occasions or matched-subjects data if participants are matched in pairs. In contrast, the Cochran and the Friedman tests allow for the analysis of repeated-measures data if participants are assessed on two or more occasions or to matched-subjects data if participants are matched in pairs, triplets, or in some greater number. We recommend that you reread Lesson 44 about two-related-samples procedures before reading Lesson 45. The same analysis issues that apply to the McNemar and Wilcoxon tests also apply to the Cochran and Friedman tests, and we will not present these issues in as much detail in the current lesson.

The Cochran and the Friedman tests are applicable to problems with repeated-measure designs or matched-subjects designs. With repeated-measures designs, each participant is a case in the SPSS data file and has scores on *K* variables, the score obtained on each of the *K* occasions or conditions. A researcher is interested in determining if subjects changed significantly across occasions (or conditions). For a matched-subjects design, participants are matched in sets of *K* participants, and each participant in a set is assessed once on a measure. Each set of participants is a case in the SPSS data file and has scores on *K* variables, the scores obtained on the measure by the participants within a set.

Applications of the Cochran and Friedman Tests

Here we present two examples that are very similar to ones we introduced in Lesson 44. The major difference is that the factors for our examples in this lesson have more than two levels. For each example, we define different dependent variables to illustrate the use of the Cochran test as well as the Friedman test.

Repeated-Measures Designs

Allen, an industrial psychologist, is interested in determining if workers are more concerned with job security, job pay, or job climate. He is able to gain the cooperation of 30 individuals who work in a variety of work settings and asks each employee to indicate concern for job salary, job security, and job climate. Allen's SPSS data file contains 30 cases, one for each employee, and three variables: judgments about concern for salary level, for job security, and for job climate. The dependent variable for the Cochran test is dichotomous judgments made by employees who indicate whether they are or are not extremely concerned about salary level, job security, and job climate. The scores on these variables can be either a 0, indicating "not extremely concerned," or a 1, indicating "extremely concerned." The values of 0 and 1 are arbitrary. The dependent variable for the Friedman test is ratings made by employees about their concern for job salary, security, and climate. The ratings are on a 10-point scale, with 1 indicating "no concern" and 10 indicating "ultimate concern."

Matched-Subjects Designs

Danny is interested in determining if an individual's mood state is affected by the mood of other individuals. He obtains scores on the mood measure of sadness (MMS) scale for 60 students and rank-orders their scores. He then defines triplets of students based on their MMS scores: three students with the lowest MMS scores, three students with the next lowest MMS scores, and so on. Accordingly, the students within a matched triplet have more similar MMS scores than they do with students from other triplets. Danny randomly assigns students within each matched triplet to one of three conditions. One student is exposed to an individual who acts depressed, a second student is exposed to an individual who acts neither depressed nor happy, and the third student is exposed to an individual who acts jubilant. After each student is exposed to one of these individuals, the exposed person is evaluated using a sadness posttest. Danny's SPSS data file contains 20 cases, one for each matched triplet of students, and three variables, the sadness posttest for the students exposed to a depressed individual, the sadness posttest for the students exposed to an individual who was neither depressed nor nondepressed, and the sadness posttest for the students exposed to an individual who was jubilant. For the Cochran test, the scores on the three sadness posttest variable are 0 for not sad and 1 for sad. For the Friedman test, the scores on the two sadness posttest variables are posttest MMS scores.

Understanding the Cochran and Friedman Tests

For both the Cochran and Friedman tests, the observations come in sets of *K* observations. Within a set the observations are dependent, but between sets the observations are independent. The major difference between the two tests is the scale of the dependent variable. For the Cochran test, the dependent variable has two categories, and the null hypothesis states that the population proportions are equal for the *K* levels of a factor. For the Friedman test, the dependent variable must be measured on at least an ordinal scale, and the null hypothesis states that the population medians are equal for the *K* levels of a factor.

The parametric alternative to the Cochran and the Friedman tests is the one-way repeated-measures analysis of variance (ANOVA). For all three tests, analyses involve a factor with *K* levels, and we are interested in evaluating whether scores differ significantly across the levels of the factor. The tests take into account the dependency among scores introduced by the repeated-measures or matched-subject characteristics of the design.

As with a one-way repeated-measures design, follow-up tests need to be conducted for the Cochran and Friedman tests if there are more than two levels of the factor of interest. These follow-up tests most frequently involve pairwise comparisons. Users can conduct these follow-up tests with the help of SPSS, but they cannot be conducted simply by choosing an option within the *K* related-samples procedure.

Assumptions Underlying the Cochran and Friedman Tests

The first two assumptions are associated with both the Cochran test and Friedman test, and the third assumption is associated with only the Friedman test.

Assumption 1: Each Set of K *Observations Must Represent a Random Sample from a Population and Must Be Independent of Every Other Set of* K *Observations*

If the data are from a repeated-measures design, the scores for each participant must be independent of the scores for any other participant. If the data are from a matched-subjects design, the sets of scores from any matched set of participants must be independent of the scores from any other matched set of participants. If the independence assumption is violated, the test is likely to yield inaccurate results. It should be noted that the analyses permit dependency among scores within a set.

Assumption 2: The Chi-Square Values for the Cochran and Friedman Tests Yield Relatively Accurate Results to the Extent That the Sample Size Is Large

The results for the tests should be fairly accurate if the sample size is greater than 30.

Assumption 3 (Friedman Test Only): The Distribution of the Differences Scores between Any Pair of Levels Is Continuous and Symmetrical in the Population

This assumption is required to avoid ties and to ensure that the test evaluates differences in medians rather than other characteristics of the distributions.

Effect Size Statistics for the Friedman and the Cochran Tests

SPSS computes Kendall's coefficient of concordance (Kendall's W), a strength-of-relationship index. The coefficient of concordance ranges in value from 0 to 1, with higher values indicating a stronger relationship. This coefficient can be used as a strength-of-relationship index for both the Cochran and the Friedman tests. See Marascuilo and Serlin (1988) for further discussion about the relationship between the coefficient of concordance and these tests.

The Data Set

The data to illustrate the Friedman and Cochran tests are from our example dealing with concern for job security, pay, and climate. The data are on the Web at http://www.prenhall.com/greensalkind under the name *Lesson 45 Data File 1*. A description of the variables is given in Table 73. The variables pay, security, and climate are for the Friedman test, while the variables payc, securitc, and climc are for the Cochran test.

Table 73
Variables in Lesson 45 Data File 1

Variables	Definition
pay	Concern for job pay on a rating scale from 1, indicating no concern, to 10, indicating ultimate concern
climate	Concern for job climate on a rating scale from 1, indicating no concern, to 10, indicating ultimate concern
security	Concern for job security on a rating scale from 1, indicating no concern, to 10, indicating ultimate concern
payc	Concern for job pay with 0 = not extremely concerned and 1 = extremely concerned
climc	Concern for job climate with 0 = not extremely concerned and 1 = extremely concerned
securitc	Concern for job security with 0 = not extremely concerned and 1 = extremely concerned

The Research Question

1. Cochran test: Are a greater proportion of employees in the population concerned about job pay, job security, or job climate?
2. Friedman test: Do employees' medians on concern for job pay, job security, and job climate ratings differ in the population?

Conducting *K* Related-Samples Tests

We demonstrate how to conduct a Cochran test and a Friedman test.

Conducting a Cochran Test

We'll illustrate how to conduct the Cochran test and follow-up pairwise comparisons. We also compute Kendall's *W*, an effect size index, for the Cochran test.

Conducting an Overall Cochran Test

To conduct the Cochran test, follow these steps:

1. Click **Analyze**, click **Nonparametric Tests**, and then click **K Related Samples**. You will see the Tests for Several Related Samples dialog box in Figure 270.

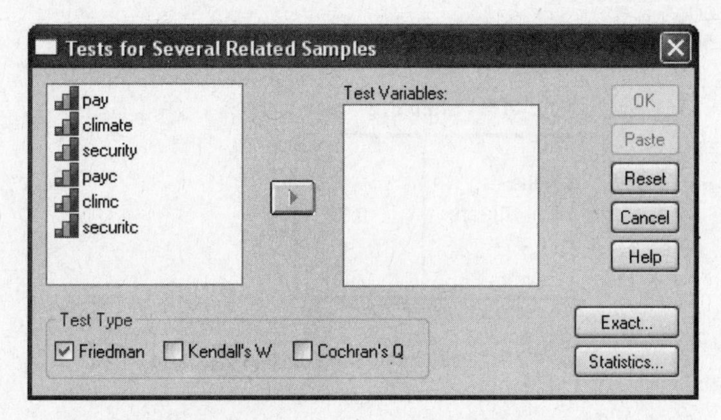

Figure 270. **Tests for Several Related Samples dialog box.**

2. Click **payc**. Holding down the Ctrl key, click **climc** and **securitc**.
3. Click ▶ to move them to the box labeled Test Variables.
4. Click **Cochran's Q** and **Kendall's W** in the Test Type area. Be sure these are the only options checked.
5. Click **OK**.

Selected SPSS Output for the Cochran Test

The output is in Figure 271 (on page 410). The Cochran test, which evaluated differences in proportions among the three job concerns, is significant. The effect size, Kendall's *W*, is .17. Next, we will conduct follow-up tests to evaluate comparisons between pairs of proportions. With SPSS, it is easiest to conduct these tests by using McNemar's test. (See Lesson 44.)

Conducting Follow-up Tests to a Cochran Test by Using the McNemar Test

To conduct a McNemar test for the follow-up analyses, follow these steps:

1. Click **Analyze**, click **Nonparametric Tests**, and then click **2 Related Samples**.
2. Holding down the Ctrl key, click **payc** and **climc**.
3. Click ▶ to move them to the box labeled Test Pair(s) List.
4. Holding down the Ctrl key, click **payc** and **securitc**.
5. Click ▶ to move them to the box labeled Test Pair(s) List.
6. Holding down the Ctrl key, click **climc** and **securitc**.
7. Click ▶ to move them to the box labeled Test Pair(s) List.
8. Click **McNemar** in the Test Type area. Be sure this is the only option checked.
9. Click **OK**.

Frequencies

	Value	
	0	1
payc	15	15
climc	26	4
securitc	19	11

Test Statistics

N	30
Cochran's Q	10.333[a]
df	2
Asymp. Sig.	.006

a. 0 is treated as a success.

Test Statistics

N	30
Kendall's W[a]	.172
Chi-Square	10.333
df	2
Asymp. Sig.	.006

a. Kendall's Coefficient of Concordance

Figure 271. **Results of the Cochran test and Kendall's *W*.**

Selected SPSS Output for the McNemar Follow-up Tests

The results for the three pairwise comparisons are shown in Figure 272. As discussed in Appendix B, the LSD procedure controls adequately for Type I error across pairwise comparisons if there are three levels and the overall test is significant. If the number of levels exceeds three, then the Holm's sequential Bonferroni method or some other method will be required to control adequately for Type I error. The results of the pairwise comparisons indicated that concern for climate was significantly less than either concern for pay or security, $p < .01$ and $p = .04$, respectively. There was no significant difference between concern for pay and security, $p = .39$.

Test Statistics[b]

	payc & climc	payc & securitc	climc & securitc
N	30	30	30
Exact Sig. (2-tailed)	.007[a]	.388[a]	.039[a]

a. Binomial distribution used.

b. McNemar Test

Figure 272. **The results of the McNemar follow-up tests.**

Conducting the Friedman Test

We show how to do the Friedman test and follow-up pairwise comparisons. We also compute Kendall's *W*, an effect size index, for the Friedman test.

Conducting the Overall Friedman Test

To conduct the Friedman test, follow these steps:

1. Click **Analyze**, click **Nonparametric Tests**, and then click **K Related Samples**.
2. Click **Reset** to clear the dialog box.
3. Holding down the Ctrl key, click **pay**, **climate**, and **security**.
4. Click ▶ to move them to the box labeled Test Variables(s) List.
5. Click **Friedman** and **Kendall's W** in the Test Type area.
6. Click **OK**.

Selected SPSS Output for the Friedman Test

The output is presented in Figure 273. The Friedman test, which evaluated differences in medians among the three job concerns, is significant. Kendall's *W* is .23. Next, follow-up tests are conducted to evaluate comparisons between pairs of medians. With SPSS, it is easiest to conduct these tests by using the Wilcoxon test. (See Lesson 44.)

Ranks

	Mean Rank
pay	2.50
climate	1.68
security	1.82

Test Statistics[a]

N	30
Chi-Square	13.960
df	2
Asymp. Sig.	.001

a. Friedman Test

Test Statistics

N	30
Kendall's W[a]	.233
Chi-Square	13.960
df	2
Asymp. Sig.	.001

a. Kendall's Coefficient of Concordance

***Figure 273.* Results of the Friedman test and Kendall's *W*.**

Conducting Follow-up Tests by Using the Wilcoxon Test

To conduct a Wilcoxon test for the follow-up analyses, follow these steps:

1. Click **Analyze**, click **Nonparametric Tests**, and then click **2 Related Samples**.
2. Click **Reset** to clear the dialog box.
3. Holding down the Ctrl key, click **pay** and **climate**.
4. Click ▶ to move them to the box labeled Test Pair(s) List.
5. Holding down the Ctrl key, click **pay** and **security**.
6. Click ▶ to move them to the box labeled Test Pair(s) List.
7. Holding down the Ctrl key, click **climate** and **security**.
8. Click ▶ to move them to the box labeled Test Pair(s) List.
9. Click **Wilcoxon** in the Test Type area. Be sure this is the only option checked.
10. Click **OK**.

Ranks

		N	Mean Rank	Sum of Ranks
climate - pay	Negative Ranks	19[a]	11.21	213.00
	Positive Ranks	3[b]	13.33	40.00
	Ties	8[c]		
	Total	30		
security - pay	Negative Ranks	20[d]	13.88	277.50
	Positive Ranks	6[e]	12.25	73.50
	Ties	4[f]		
	Total	30		
security - climate	Negative Ranks	10[g]	10.85	108.50
	Positive Ranks	13[h]	12.88	167.50
	Ties	7[i]		
	Total	30		

a. climate < pay

b. climate > pay

c. climate = pay

d. security < pay

e. security > pay

f. security = pay

g. security < climate

h. security > climate

i. security = climate

Test Statistics[c]

	climate - pay	security - pay	security - climate
Z	-2.822[a]	-2.626[a]	-.923[b]
Asymp. Sig. (2-tailed)	.005	.009	.356

a. Based on positive ranks.

b. Based on negative ranks.

c. Wilcoxon Signed Ranks Test

Figure 274. **The results of the Wilcoxon follow-up tests.**

Selected SPSS Output for the Wilcoxon Follow-up Tests

The results for the three pairwise comparisons are shown in Figure 274. The LSD procedure controls adequately for Type I error across pairwise comparisons if there are three levels and the overall test is significant. In our example, two of the three comparisons were significant.

Using SPSS Graphs to Display Results

Because of the simplicity of the data used in the Cochran test, it is probably unnecessary to create a graph to represent the data. However, graphs might be useful for the Friedman test. The reported statistics offer no information about the shape of the distributions of the nonranked data on the original measures. Consequently, a graphical display of these data, such as a boxplot, would provide a visual representation of the distributions of the variables. Figure 275 shows the boxplot of the nonranked climate, pay, and security scores.

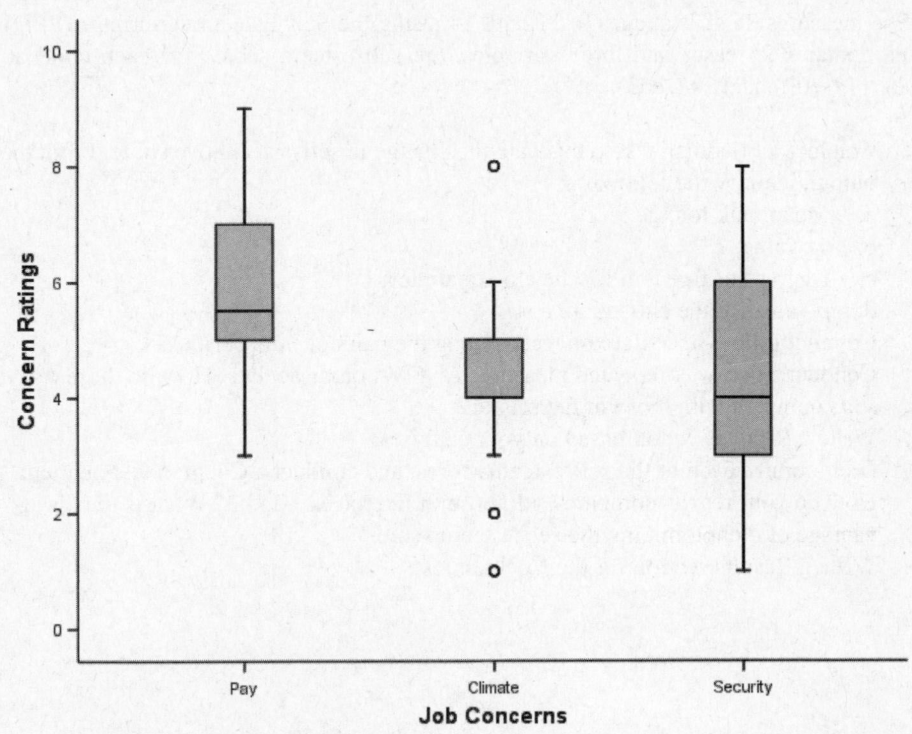

Figure 275. **Distributions of climate, pay, and job security importance ratings.**

Two APA Results Sections

Results for Cochran Test

Proportions of employees showing extreme concern for pay, security, and climate were .50, .37, and .13, respectively. A Cochran test, which evaluates differences among related proportions, was significant, $\chi^2(2, N = 30) = 10.33$, $p < .01$. The Kendall coefficient of concordance was .17. Follow-up pairwise comparisons were conducted using a McNemar's test and controlling for familywise error rate at the .05 level using the LSD procedure. The proportions differed significantly between concern for pay and concern for climate, $p < .01$, and between concern for security and concern for climate, $p = .04$, but not between concern for pay and concern for security, $p = .39$.

Results for Friedman Test

A Friedman test was conducted to evaluate differences in medians among the job concerns for pay (median = 5.50), for climate (median = 4.00), and for security (median = 4.00). The test was significant, χ^2 $(2, N = 30) = 13.96$, $p < .01$, and the Kendall coefficient of concordance of .23 indicated fairly strong differences among the three concerns. Follow-up pairwise comparisons were conducted using a Wilcoxon test and controlling for the Type I errors across these comparisons at the .05 level using the LSD procedure. The median concern for pay was significantly greater than the median concern for climate, $p < .01$, and the median concern for security, $p < .01$, but the median concern for climate did not differ significantly from the median concern for security, $p = .36$.

Exercises

Exercises 1 through 6 are based on the following research problem. The data for these examples are in the data file named *Lesson 45 Exercise File 1* on the Web at http://prenhall.com/greensalkind.

Mary, a developmental psychologist, is interested in how girls' self-esteem develops over time. She measures 25 girls at ages 9, 11, and 14, using the Self-Esteem Descriptor (SED). The data file contains 25 cases and three variables, the self-esteem scores for each child at ages 9 (sed1), 11 (sed2), and 14 (sed3).

1. Conduct a Friedman test to evaluate the change in self-esteem over time. From the output, identify the following:
   a. Mean rank for age 9
   b. χ^2 value
   c. Degrees of freedom for the chi-square test
   d. p value for the chi-square test
2. Conduct follow-up Wilcoxon tests among the pairs of time periods.
3. Conduct a one-way repeated measures ANOVA on these data. How do these results compare with those in Exercise 1?
4. Write a Results section based on your analyses.
5. Dichotomize each of the self-esteem scores, and conduct a Cochran test on them. How do your results compare with those in Exercises 1 and 3? What is the disadvantage of dichotomizing the self-esteem scores?
6. Write a Results section for the Cochran test.

APPENDIX

A

Data for Crab Scale and Teacher Scale

ID_PROF	SEX_PROF	AGE	RANK	SCHOOL	CRAB1	CRAB2	CRAB3	CRAB4	CRAB5	CRAB6
1.00	1.00	33.00	1.00	1.00	2.00	3.00	3.00	3.00	2.00	3.00
2.00	1.00	58.00	3.00	1.00	5.00	3.00	4.00	3.00	5.00	3.00
3.00	1.00	45.00	3.00	1.00	4.00	5.00	3.00	5.00	2.00	5.00
4.00	2.00	36.00	1.00	1.00	3.00	2.00	3.00	1.00	3.00	1.00
5.00	2.00	42.00	2.00	1.00	5.00	3.00	3.00	3.00	3.00	3.00
6.00	1.00	59.00	3.00	2.00	2.00	3.00	2.00	1.00	1.00	1.00
7.00	2.00	29.00	1.00	2.00	1.00	3.00	1.00	1.00	1.00	2.00
8.00	1.00	64.00	2.00	2.00	1.00	3.00	2.00	3.00	1.00	3.00
9.00	1.00	37.00	2.00	2.00	2.00	2.00	2.00	2.00	1.00	3.00
10.00	1.00	42.00	3.00	2.00	1.00	3.00	1.00	3.00	1.00	3.00

ID_STUD	ID_PROF	SEX_STUD	TEACHER1	TEACHER2	TEACHER3	TEACHER4	TEACHER5
82104	1	2.00	4.00	5.00	2.00	1.00	2.00
22466	1	2.00	5.00	3.00	1.00	2.00	4.00
75812	1	2.00	3.00	5.00	2.00	1.00	3.00
80761	1	1.00	4.00	4.00	2.00	1.00	3.00
22920	1	1.00	5.00	4.00	2.00	1.00	3.00
87172	2	2.00	3.00	4.00	1.00	2.00	3.00
98221	2	2.00	3.00	3.00	2.00	3.00	2.00
84768	2	1.00	4.00	3.00	3.00	2.00	3.00
89225	2	1.00	3.00	3.00	3.00	1.00	4.00
80623	2	2.00	3.00	5.00	2.00	2.00	3.00
94354	3	1.00	3.00	5.00	1.00	2.00	3.00
81334	3	2.00	4.00	5.00	5.00	2.00	3.00
52955	3	2.00	4.00	4.00	5.00	4.00	4.00
33191	3	2.00	4.00	2.00	5.00	4.00	2.00
55291	3	2.00	4.00	4.00	2.00	3.00	3.00
60608	4	2.00	3.00	4.00	5.00	3.00	4.00
67041	4	2.00	2.00	2.00	2.00	2.00	2.00
65719	4	2.00	3.00	3.00	5.00	3.00	4.00
40351	4	1.00	2.00	4.00	2.00	2.00	2.00
62034	4	1.00	2.00	4.00	2.00	2.00	4.00
94882	5	2.00	4.00	5.00	3.00	2.00	2.00
26841	5	1.00	2.00	4.00	2.00	1.00	3.00

ID_STUD	ID_PROF	SEX_STUD	TEACHER1	TEACHER2	TEACHER3	TEACHER4	TEACHER5
13691	5	1.00	2.00	3.00	1.00	3.00	2.00
83651	5	2.00	1.00	3.00	4.00	4.00	2.00
70778	5	2.00	1.00	3.00	4.00	4.00	3.00
70386	6	1.00	3.00	4.00	2.00	3.00	5.00
45176	6	1.00	5.00	4.00	1.00	1.00	4.00
86213	6	2.00	3.00	5.00	1.00	1.00	3.00
29627	6	1.00	2.00	4.00	2.00	1.00	4.00
24824	6	1.00	1.00	5.00	2.00	2.00	4.00
39986	7	1.00	5.00	5.00	2.00	1.00	5.00
77974	7	1.00	5.00	4.00	2.00	1.00	4.00
38396	7	1.00	5.00	5.00	2.00	1.00	5.00
78665	7	1.00	4.00	3.00	2.00	1.00	3.00
20227	7	1.00	4.00	5.00	1.00	1.00	5.00
66298	8	1.00	5.00	5.00	1.00	1.00	3.00
27423	8	1.00	5.00	5.00	1.00	1.00	4.00
79359	8	1.00	5.00	5.00	1.00	1.00	5.00
99056	8	2.00	5.00	5.00	1.00	1.00	4.00
66513	8	1.00	5.00	5.00	1.00	1.00	5.00
28796	9	1.00	2.00	4.00	4.00	3.00	1.00
67564	9	1.00	3.00	3.00	4.00	4.00	1.00
95471	9	2.00	2.00	3.00	3.00	4.00	2.00
35598	9	1.00	3.00	3.00	2.00	1.00	2.00
91744	9	2.00	3.00	4.00	3.00	2.00	3.00
77823	10	1.00	5.00	5.00	2.00	1.00	5.00
34850	10	1.00	5.00	5.00	1.00	2.00	5.00
94277	10	2.00	4.00	4.00	1.00	1.00	4.00
54217	10	1.00	3.00	5.00	1.00	1.00	4.00
99513	10	2.00	2.00	5.00	1.00	3.00	2.00

APPENDIX

B

Methods for Controlling Type I Error across Multiple Hypothesis Tests

A Type I error is the probability of rejecting a null hypothesis when it is true. Alpha (α) is the probability of committing a Type I error. When we conduct a hypothesis test, we want to minimize our chances of committing a Type I error for that test, and therefore we designate alpha to be some small value, most frequently .05.

If multiple hypothesis tests are conducted, then we may be concerned about committing one or more Type I errors across these tests. The multiple hypotheses are referred to as a family, and the probability of committing one or more Type I errors for a family is the familywise alpha (α_{family}). For any family, we want to keep α_{family} small, most frequently .05. To keep α_{family} at the .05 level, in most cases we reduce the alpha level for any one test to be a value less than .05. An exception is for the LSD method for pairwise comparisons among three groups, as we will describe presently.

For example, a family of multiple hypotheses may involve pairwise comparisons among three or more groups. These hypotheses evaluate whether pairs of groups differ on a dependent variable. We'll briefly describe three methods to control for α_{family} when the family consists of pairwise comparisons. Initially, we will consider the LSD procedure, which should be used only if there are three groups. For this special application, the probability of a Type I error for a pairwise comparison, α_{pc}, is equal to α_{family}. Next, we will discuss the Bonferroni method, which requires the α_{pc} to be equal to α_{family} divided by the number of paired comparisons. Finally, we will introduce the Holm's sequential Bonferroni method that evaluates each paired comparison at a different alpha level.

Although we will look at these methods for evaluating multiple pairwise comparison tests, the last two methods (the Bonferroni and the Holm's sequential Bonferroni methods) can be used for any application involving multiple hypothesis testing. The Holm's sequential Bonferroni method is preferable to the Bonferroni method for evaluating hypotheses because it is less conservative and has greater power. For more information about multiple comparison techniques, read Hochberg and Tamhane (1987); Levin, Serlin, and Seaman (1994); and Maxwell and Delaney (2000).

LSD Method for Control of Type I Error for Pairwise Comparisons among Three Groups

1. Set α_{family}. Conventionally, it is set at the .05 level.
2. Conduct a test to evaluate the omnibus (overall) hypothesis that the parameters (e.g., means) for the three populations are equal. Note the p value (denoted p_{omni}) for this test.
   - If $p_{omni} \leq \alpha_{family}$, reject the null hypothesis. Continue to Step 3.
   - If $p_{omni} > \alpha_{family}$, do not reject the null hypothesis. Do not proceed to the next step. Declare all pairwise comparisons nonsignificant.

3. Conduct the three pairwise comparisons and note their p values (denoted p_{pc}).
4. With the LSD procedure, $\alpha_{pc} = \alpha_{family}$. If $p_{pc} \leq \alpha_{pc}$, then reject the null hypothesis for each pairwise comparison.

Bonferroni Method for Control of Type I Error for All Pairwise Comparisons

1. Set α_{family}. Conventionally, it is set at the .05 level.
2. Determine the number of pairwise comparisons with the equation.

$$N_{PC} = \frac{N_G(N_G - 1)}{2}$$

where N_G is the number of groups and N_{PC} is the number of pairwise comparisons.

3. Compute alpha for any pairwise comparison:

$$\alpha_{PC} = \frac{\alpha_{family}}{N_{PC}}$$

4. Conduct pairwise comparisons and note their p values (denoted p_{pc}).
5. If $p_{pc} \leq \alpha_{pc}$, then reject the null hypothesis for each pairwise comparison.

Holm's Sequential Bonferroni Method for Control of Type I Error for All Pairwise Comparisons

1. Set α_{family}. Conventionally, it is set at the .05 level.
2. Determine the number of pairwise comparisons:

$$N_{PC} = \frac{N_G(N_G - 1)}{2}$$

3. Conduct pairwise comparisons. Rank order the comparisons on the basis of their p values (denoted p_{PC}) from smallest to largest.
4. Evaluate the comparison with the smallest p_{PC} (denoted p_1).
   a. Compute the alpha (α_1)for the first comparison:

$$\alpha_1 = \frac{\alpha_{family}}{N_{PC}}$$

   b. Conduct the test of pairwise comparison:
      - If $p_1 \leq \alpha_1$, reject the null hypothesis. Continue to Step 5.
      - If $p_1 > \alpha_1$, do not reject the null hypothesis. Do not proceed to the next step. Declare all comparisons nonsignificant.
5. Evaluate the comparison with the next smallest p_{PC} (denoted p_2).
   a. Compute the alpha (α_2) for the second comparison:

$$\alpha_2 = \frac{\alpha_{family}}{N_{PC} - 1}$$

b. Conduct the test of pairwise comparison:
   - If $p_2 \leq \alpha_2$, reject the null hypothesis. Continue to Step 6.
   - If $p_2 > \alpha_2$, do not reject the null hypothesis. Do not proceed to the next step. Declare this comparison and all remaining comparisons nonsignificant.

6. Evaluate the comparison with the next smallest p_{PC} (denoted p_3).

   a. Compute α_3:

$$\alpha_3 = \frac{\alpha_{\text{family}}}{N_{PC} - 2}$$

   b. Conduct the test of pairwise comparison.
   - If $p_3 \leq \alpha_3$, reject null hypothesis. Continue to Step 7.
   - If $p_3 > \alpha_3$ do not reject null hypothesis. Do not proceed to next step. Declare this comparison and all remaining comparisons nonsignificant.

7. And so on ...

APPENDIX

C Selected Answers to Lesson Exercises

Answers for Lesson 19

1. To reverse-scale the scores, use the Transform → Recode Into Same Variables option or the Transform → Recode Into Different Variables option. Alternatively, use the Transform → Compute Variable option, type a name of a variable like item3r in the Target Variable box and type **6 – item3** in the Numeric Expression box.
3. To create a categorical age variable, use the Transform → Recode into Different Variables option. The old and new values should be as follows:

Old values	New values
Range: 20 through 29	1
Range: 30 through 39	2
All other values	3

5. Conduct the following steps to create the variable group:
   a. Compute a total coffee-drinking score using the Transform → Compute Variable option. Type a name of a variable like **totcof** in the Target Variable box and type **sum(day1,day2,day3,day4)** in the Numeric Expression box.
   b. Create a categorical variable (e.g., cofcat) based on the coffee-drinking scores by using the Transform → Recode Into Different Variables option. The old and new values should be as follows:

Old values	New values
Range: 0 through 4	1
All other values	2

   Use the Transform → Compute Variable option and the If option to create four categories. Conduct the following steps:
   1. Type **group** in the Target Variable box. Type **1** in the Numeric Expression box. Click **If**. Click **Include if case satisfies**. Type **jobcat = 1 & cofcat = 1** in the text box. Click **Continue**. Click **OK**.
   2. Type **group** in the Target Variable box. Type **2** in the Numeric Expression box. Click **If**. Click **Include if case satisfies**. Type **jobcat = 1 & cofcat = 2** in the text box. Click **Continue**. Click **OK**.

3. Type **group** in the Target Variable box. Type **3** in the Numeric Expression box. Click **If**. Click **Include if case satisfies**. Type **jobcat = 2 & cofcat = 1** in the text box. Click **Continue**. Click **OK**.
4. Type **group** in the Target Variable box. Type **4** in the Numeric Expression box. Click **If**. Click **Include if case satisfies**. Type **jobcat = 2 & cofcat = 2** in the text box. Click **Continue**. Click **OK**.

7. In Exercise 6, you created a *z*-scored prosocial variable, zrating, by using Analyze → Descriptive Statistics → Descriptives. Now use the Transform → Compute Variable option. Type a name of a variable like **zratingr** in the Target Variable box and type **–1*zrating** in the Numeric Expression box.

Answers for Lesson 20

1. a. Percentage men = 52%
   b. Mode is divorced (= 2).
   c. Frequency of divorced people = 11
3. The bar chart for the different-sized communities is shown in Figure 276. Community sizes were ordered from smallest to largest along the horizontal axis by using the Series → Displayed option. In the Bar/Line/Area Displayed Data dialog box, reorder categories in the Display box.
5. Table 74 (on page 422) shows the frequencies of the types of books people read.

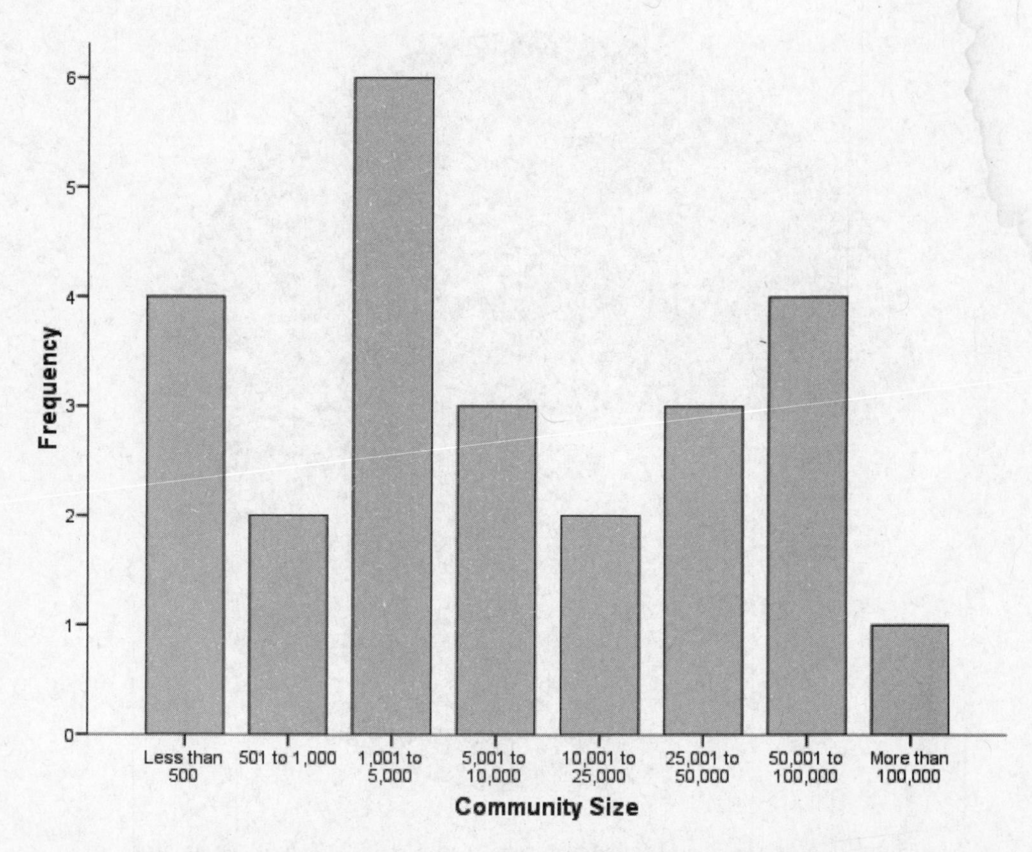

Figure 276. **Number of people in different-sized communities.**

Table 74

Type of Books Read by 50 Men and 50 Women

Type of Book	Frequency
Drama	16
Mysteries	7
Romance	10
Historical nonfiction	6
Travel books	13
Children's literature	10
Poetry	9
Autobiographies	9
Political science	8
Local interest	12

7. The sample consisted of 50 men and 50 women. Table 74 shows the frequencies of the types of books these individuals report reading. The most popular type of book was dramatic novels ($n = 16$), while the least popular type was historical nonfiction ($n = 6$). Figure 277 summarizes how much the individuals in the sample read. The respondents were approximately evenly split among the four reading types.

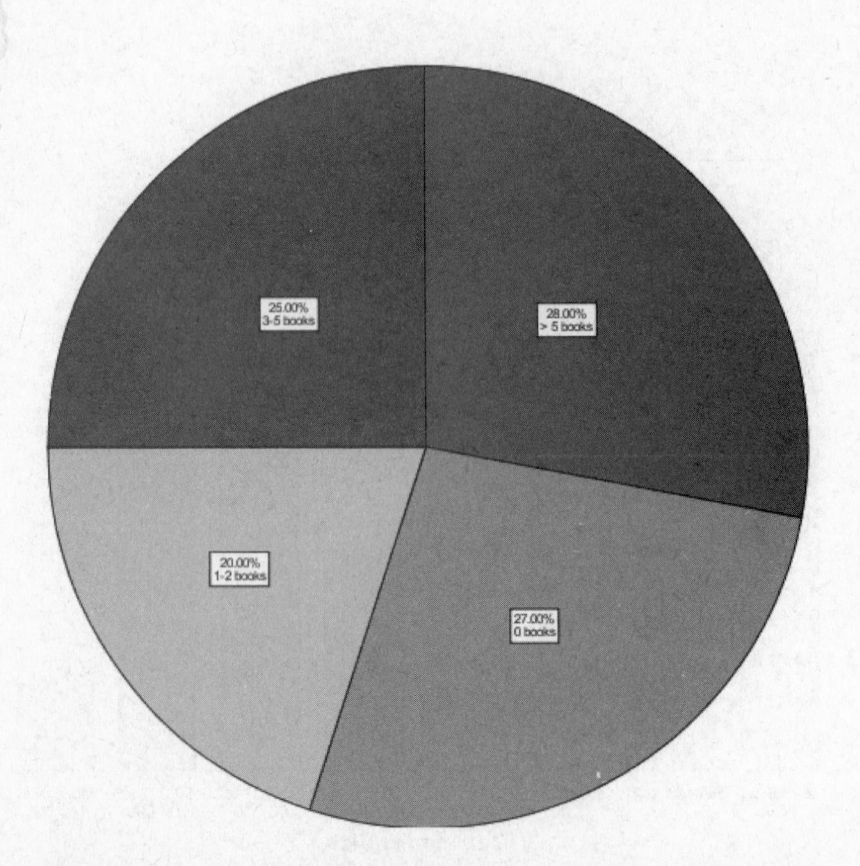

Figure 277. **Percentage of individuals who read 0, 1 to 2, 3 to 5, and greater than 5 books per month.**

Answers for Lesson 21

1. a. Skewness = .42
   b. Mean = 32.27
   c. *SD* = 23.48
   d. Kurtosis = –1.12
2. The percentile ranks for the 15 scores are 12, 24, 73, 13, 27, 77, 83, 38, 16, 17, 72, 88, 97, 73, and 19. The scores associated with the ranks of 12, 27, 38, 73, and 88 are 5, 18, 25, 47, and 60, respectively.
3. The percentile ranks for the 15 scores are 7, 40, 73, 13, 47, 80, 87, 53, 20, 27, 60, 93, 100, 73 and 33.
4. The histogram is shown in Figure 278.
5. Based on the graph, it does not appear that the scores are even approximately normally distributed. Accordingly, it is preferable to compute the percentile ranks not assuming that the scores are normally distributed.
7. The mean for the total attitude score for Republicans is 15.19, and the mean for the total attitude score for Democrats is 11.47. (We reverse-scaled the first item, computed total scores, and then obtained means on the total scores for the two political parties by using Analyze → Descriptive Statistics → Explore.)

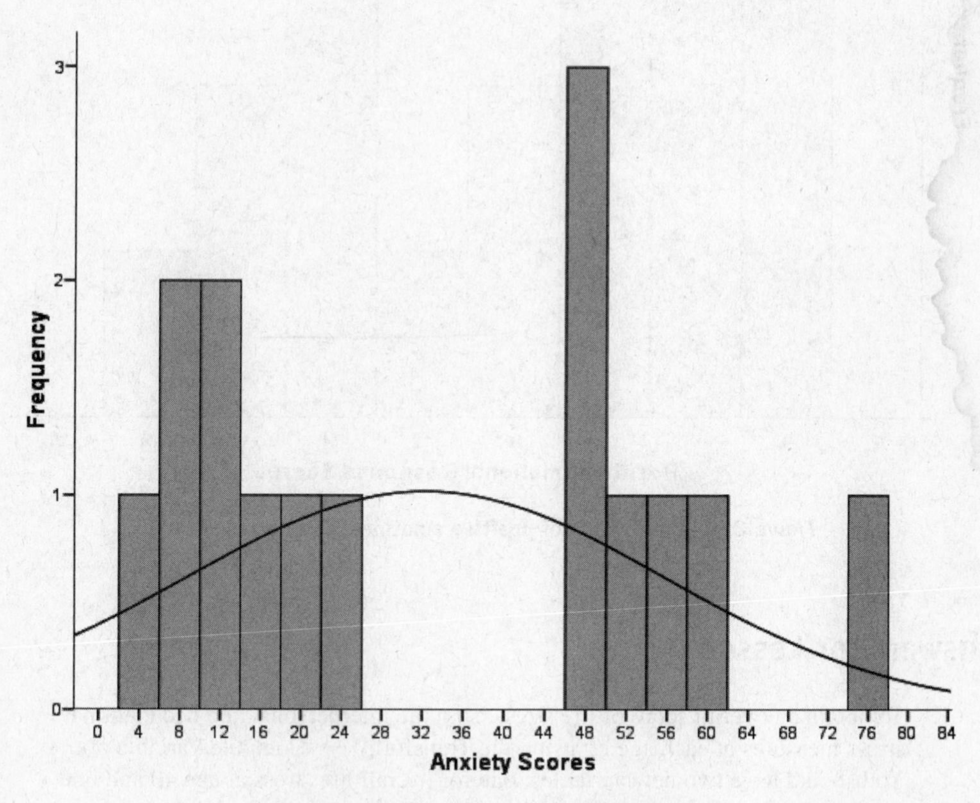

Figure 278. **Distribution of anxiety scores.**

Answers for Lesson 22

1. You can compute a total score by summing across all eight items by using the Transform → Compute Variable command. The total scores are 8, 6, 5, 7, 4, and 6.
2. The test value is 2.

3. a. Mean = 6.0
   b. t value = 6.93
   c. p value < .01
5. A one-sample t test indicated that the mean on the emotional response scale (ranging from 2 = very sad for both segments to 20 = very happy for both segments) differed significantly different from 11, the midpoint on the summed emotional response scale, $t(19) = -2.13$, $p = .047$. The sample mean and standard deviation on the emotional response scores were 8.75 and 4.73, respectively. The 95% confidence interval for the emotional response mean ranged from 6.53 to 10.97. A d of –.48 indicated a medium effect size. The results supported the conclusion that classical music tends to make college students sad rather than happy.
6. The distribution of emotional response scores is shown in Figure 279.

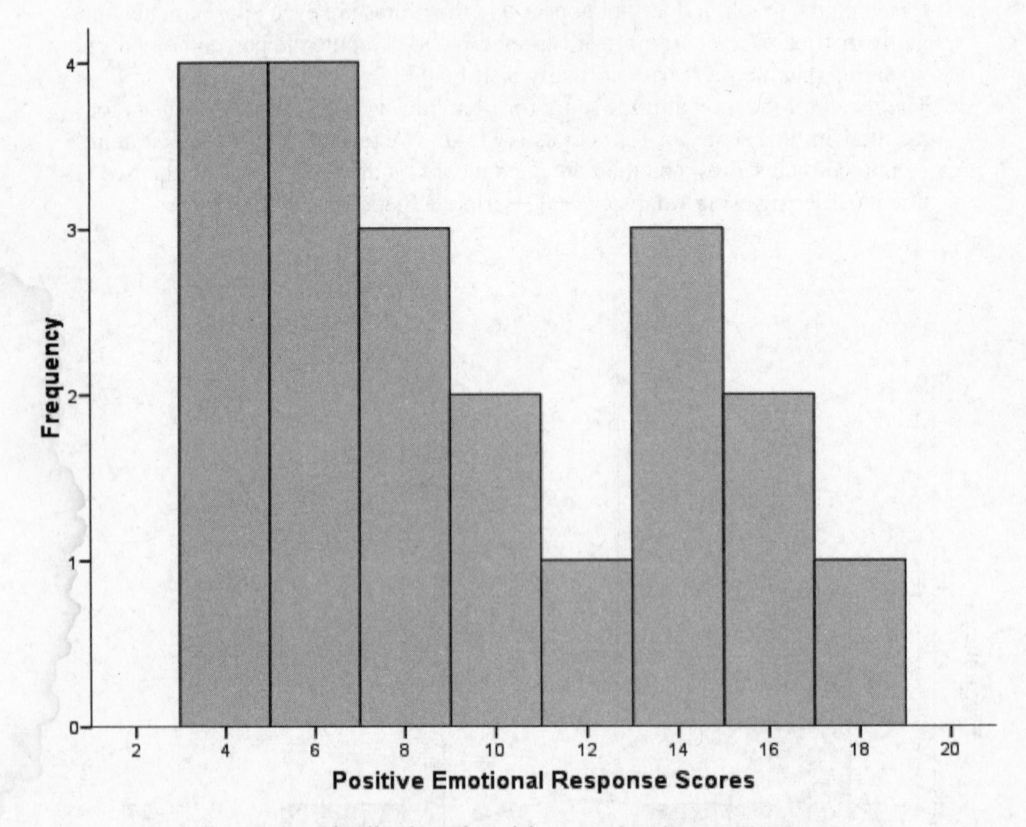

Figure 279. **Distribution of positive emotional response scores.**

Answers for Lesson 23

1. To obtain an overall score of life stress, sum the interpersonal and occupational stress measures at each age by using the Transform → Compute Variable option. You should have two new variables, one for overall life stress at age 40 and one for overall life stress at age 60.
3. The histogram is shown in Figure 280.
5. A paired-samples t test was conducted to evaluate whether women's life stress declined from age 40 to age 60. The results indicated that the mean overall life stress index at age 40 ($M = 151.84$, $SD = 14.48$) was significantly greater than mean overall life stress index at age 60 ($M = 136.87$, $SD = 9.95$), $t(44) = 5.82$, $p < .01$. The 95% confidence interval for the mean difference in overall stress between ages 40 and 60 was 9.79 to 20.16. The effect size was moderately large, $d = .87$. From Figure 280, it appears that for most women, overall life stress scores declined over the 20-year period.

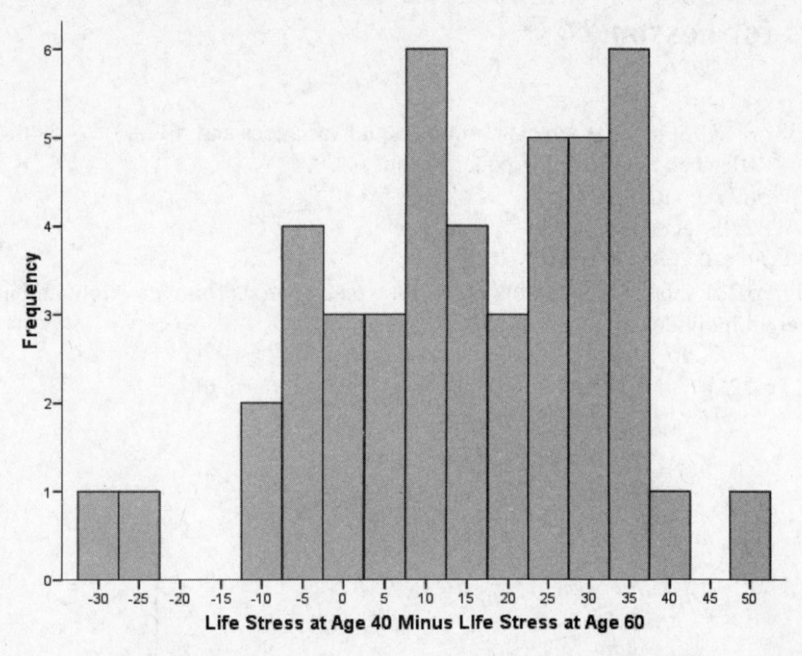

Figure 280. **Distribution of change in overall life stress from age 40 to age 60.**

Two separate paired-samples *t* tests were conducted to evaluate whether the change in life stress from age 40 to age 60 was due to occupational stress, interpersonal stress, or both. Table 75 presents the means and standard deviations of occupational and interpersonal life stress scores at ages 40 and 60, as well as 95% cofidence intervals of mean change from age 40 to age 60. The results indicated that interpersonal stress levels did not differ significantly from age 40 to age 60, $t(44) = 1.54$, $p = .13$, but that women's occupational stress decreased significantly during the same time period, $t(44) = 6.22$, $p < .01$. Accordingly, the decline in overall life stress was primarily a function of the decrease in occupational stress.

Table 75

Means, Standard Deviations, and Confidence Intervals for Occupational and Interpersonal Life Stress Scores

	Age 40		Age 60		95% confidence interval
Life stress scale	*M*	*SD*	*M*	*SD*	*of mean difference between ages 40 and 60*
Interpersonal	78.20	11.66	75.00	7.71	−0.99 to 7.39
Occupational	73.64	9.55	61.87	6.63	7.96 to 15.59

7. A paired-samples *t* test was conducted to evaluate whether husbands and wives differed on their levels of infertility anxiety. The results indicated that wives were significantly more anxious about fertility issues than their husbands, $t(23) = -3.26$, $p < .01$. The sample means and standard deviations on the infertility anxiety measure were 57.46 and 7.34 for husbands and 62.54 and 12.44 for wives. The 95% confidence interval for the mean difference in infertility anxiety between husbands and wives was –8.31 to –1.35. The *d* of –.66 indicated a medium effect size.

Answers for Lesson 24

1. $t(38) = -3.98$, $p < .01$ when assuming equal variances and $t(30.83) = -5.40$, $p < .01$ when not assuming equal variances.
2. a. 589 seconds
   b. 82.95 seconds
   c. $F(1, 38) = 2.74$, $p = .11$
5. Figure 281 shows the boxplots of the time distributions for overweight and normal weight individuals.

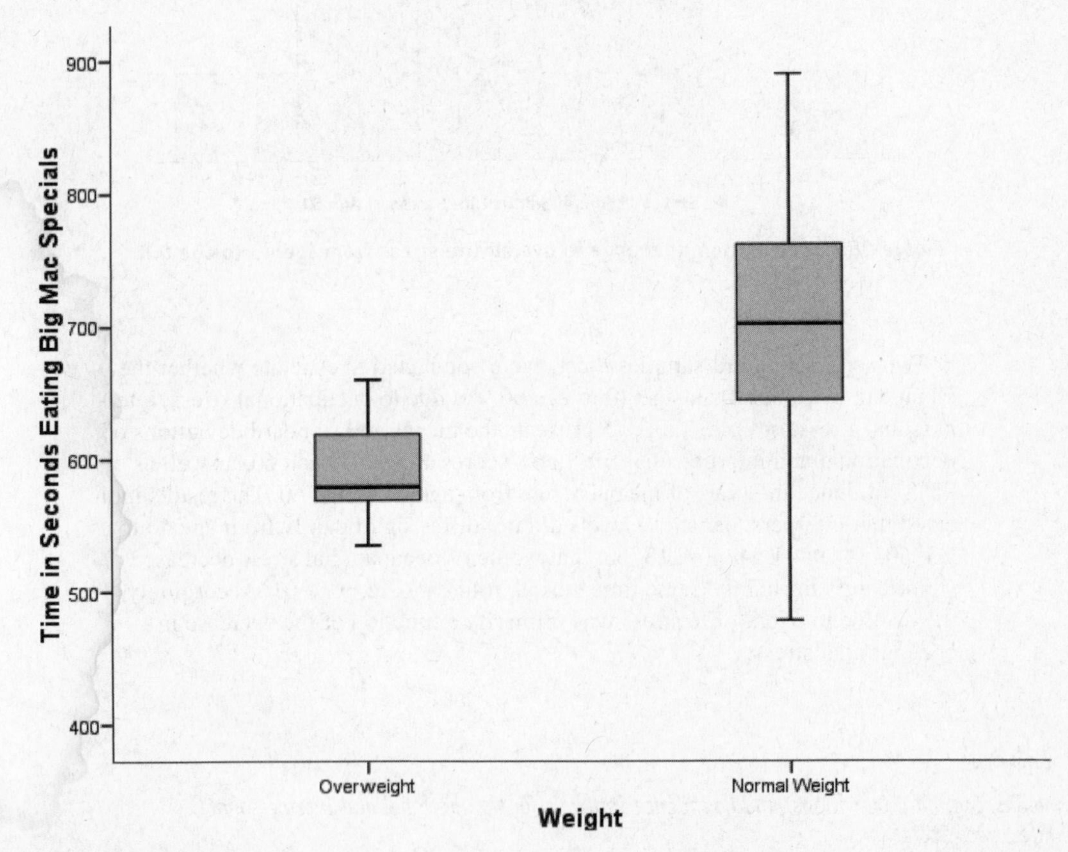

Figure 281. **Distribution of time spent eating Big Mac meals for overweight and normal-weight individuals.**

9. The results of Levene's test, $F(1, 38) = 4.10$, $p = .05$, indicate that the homogeneity of variance assumption is violated. The variability is much greater for the regular education students who are in integrated classrooms ($SD = 16.54$) than those who are not in the integrated classrooms ($SD = 7.41$). These results suggest that the standard t test is likely to produce inaccurate p values. Consequently, the t test that does not assume homogeneity of variances should be used: $t(35.85) = -2.11$, $p = .04$.
10. The analyses addressed the question of whether students are helped or hindered academically by the inclusion of special education children in regular education classrooms. Levene's test indicated that we could not assume homogeneity of variance of the difference scores (postachievement minus preachievement test scores), $F(1, 38) = 4.10$, $p = .05$. The variance of the difference scores for the regular education students in integrated classrooms (273.57) was much greater

than the variance of the difference scores for the regular education students in nonintegrated classrooms (54.91). Consequently, a t test was conducted that did not require homogeneity of variance. The results, indicated that, on the average, standardized achievement test scores improved significantly more for students in integrated classrooms ($M = 9.60$) than those in nonintegrated classrooms ($M = 1.53$), $t(35.85) = -2.11$, $p = .04$. The 95% confidence interval for the difference in means between the two types of classrooms was from –15.82 to –.32. The d index of –.69 indicated that there was a moderate relationship between classroom integration and improvement in standardized achievement scores. Figure 282 shows the distributions for both types of classrooms. There is considerable overlap in the distributions for the two groups. The results support the integration of regular and special education students into the same classroom.

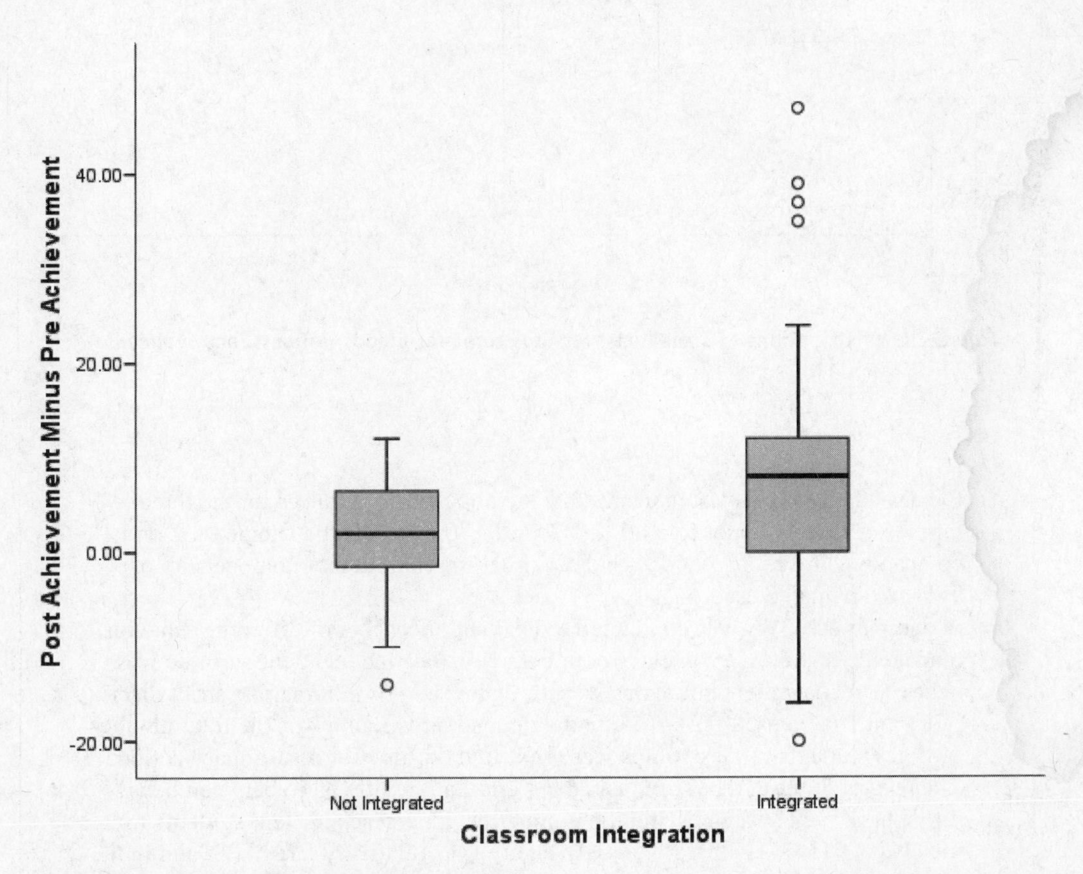

Figure 282. **Distributions of achievement difference scores for regular education students in integrated and nonintegrated classrooms.**

Answers for Lesson 25

1. a. $F(2,15) = 3.51$
   b. $SS_{\text{hair color}} = 24.11$
   c. Mean for redheads = 2.33
   d. $p = .06$
2. eta squared = .32
3. Figure 283 (on page 428) shows the distributions for the three hair color groups.

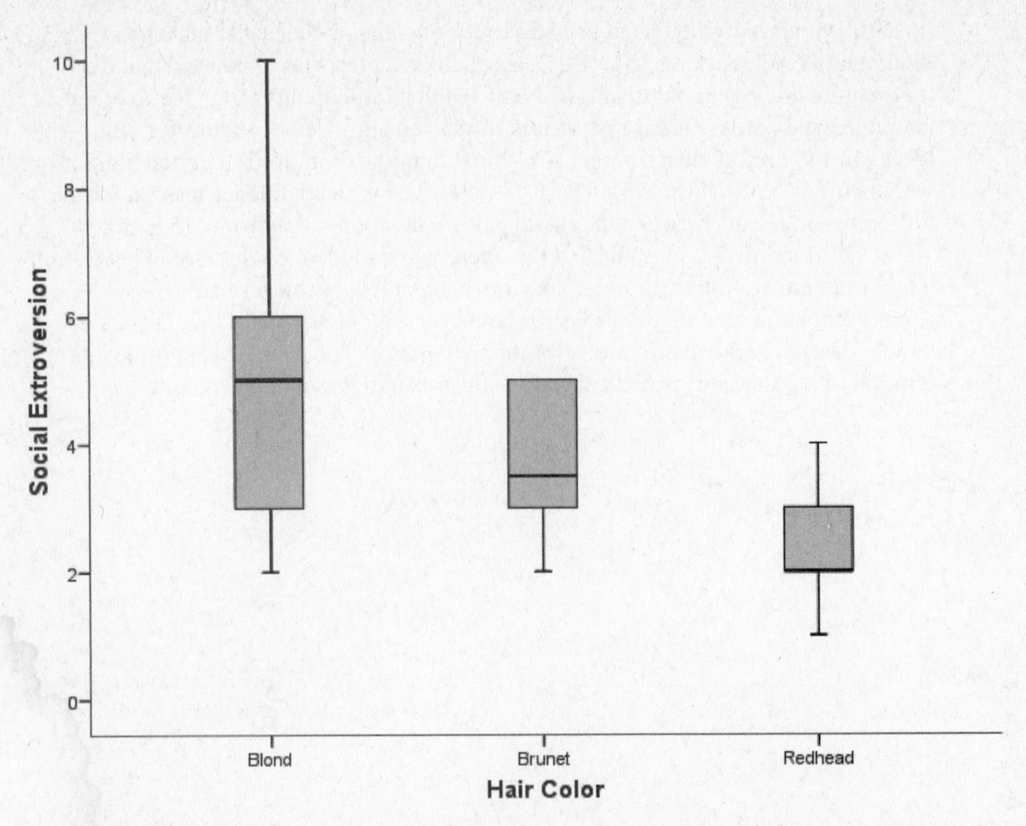

Figure 283. **Distributions of social extroversion scores for blonds, brunets, and redheads.**

5. Because the Levene's test indicates that the population variances for the three types of teachers are not equal, $F(2, 37) = 14.40, p < .01$, the Dunnett's *C* or a comparable procedure that does not require the assumption of homogeneity of variance should be used.
6. A one-way ANOVA was conducted to evaluate the effect of different behavior management strategies on classroom behavior. The independent variable was behavior management strategies with three levels—firm humanist, strict disciplinarian, and laissez-faire—while the dependent variable was the total number of times students in classrooms were sent to the office for disciplinary action. The results of the ANOVA indicated a significant relationship between behavior management strategies and the number of times teachers sent students to the office, $F(2, 37) = 5.58, p < .01$. The η^2 was relatively large, .23, and indicated that 23% of the variance in the number of office trips by students could be explained by differences in behavior management strategies. Table 76 shows the means and standard deviations for each of the three types of management strategies.

   To assess further the differences among groups, we conducted pairwise comparisons among the three means, using Dunnett's *C*, a multiple comparison procedure that does not require equal population variances. Levene's test had indicated that the variances among the groups differed significantly, $F(2, 37) = 14.40, p < .01$. Based on the results of Dunnett's *C*, teachers who used a humanistic approach to classroom management sent students to the office significantly fewer times than did teachers who were strict disciplinarians. Laissez-faire teachers did not differ significantly from either of the other two groups in the number of students they sent to the office for disciplinary action.

Table 76

Means and Standard Deviations of Three Behavioral Management Styles

Treatment groups	*M*	*SD*
Humanist (N = 9)	5.56	1.33
Strict, rigid control (N = 21)	7.29	1.68
Laissez-faire (N = 10)	9.70	4.79

Answers for Lesson 26

1. a. $F(1, 60) = 25.04$
   b. Mean = 28.27
   c. Effect size = .14
   d. $p < .01$
3. A 3 × 2 ANOVA was conducted to evaluate the effects of three reinforcement conditions (tokens, money, and food) and two schedule conditions (random and equally spaced) on arithmetic problem-solving scores. The means and standard deviations for the test scores as a function of the two factors are presented in Table 77. The results for the ANOVA indicated a significant main effect for reinforcement type, $F(2, 60) = 31.86$, $p < .01$, partial $\eta^2 = .52$, a significant effect for schedule type, $F(1, 60) = 25.04$, $p < .01$, partial $\eta^2 = .29$, and a significant interaction between reinforcement type and schedule type, $F(2, 60) = 4.78$, $p = .01$, partial $\eta^2 = .14$.

   Because the interaction between reinforcement type and schedule type was significant, we chose to ignore the main effects and instead examined the reinforcement simple main effects, that is, the differences among token reinforcers, money reinforcers, and food reinforcers for random and equally spaced schedules separately. To control for Type I error across the two simple main effects, we set alpha for each simple main effect at .025. There were significant reinforcer effects for random schedules, $F(2, 60) = 20.98$, $p < .01$, partial $\eta^2 = .41$, and for equally spaced schedules, $F(2, 60) = 15.65$, $p < .01$, partial $\eta^2 = .34$.

   Follow-up tests were conducted to evaluate the three pairwise differences among the means for random schedules and for the equally spaced schedules. Alpha was set at .008 (.025/3 = .008) to control for Type I error over the three pairwise comparisons within each simple main effect. For the random schedules, money and food reinforcers produced significantly higher test scores than token reinforcers; the means for money and food reinforcers were not significantly different from each other. Similarly, for the equally spaced schedules, money and food reinforcers produced significantly higher test scores than token reinforcers; the means for money and food reinforcers were not significantly different from each other.

Table 77

Means and Standard Deviations for Test Scores

	Tokens		Money		Food	
Schedule	*M*	*SD*	*M*	*SD*	*M*	*SD*
Random schedule	19.64	5.03	28.27	4.76	31.45	5.28
Equally spaced schedule	26.45	4.01	37.00	4.31	32.27	2.69

6. You should conduct three pairwise comparisons among the marginal means for the three disabilities because (1) there is a significant disability main effect and a nonsignificant interaction and (2) the disability main effect has three levels. You should consider appropriate methods to control for Type I error across the three groups. Because there are only three groups, you could use the LSD approach.
8. The boxplot is shown in Figure 284.

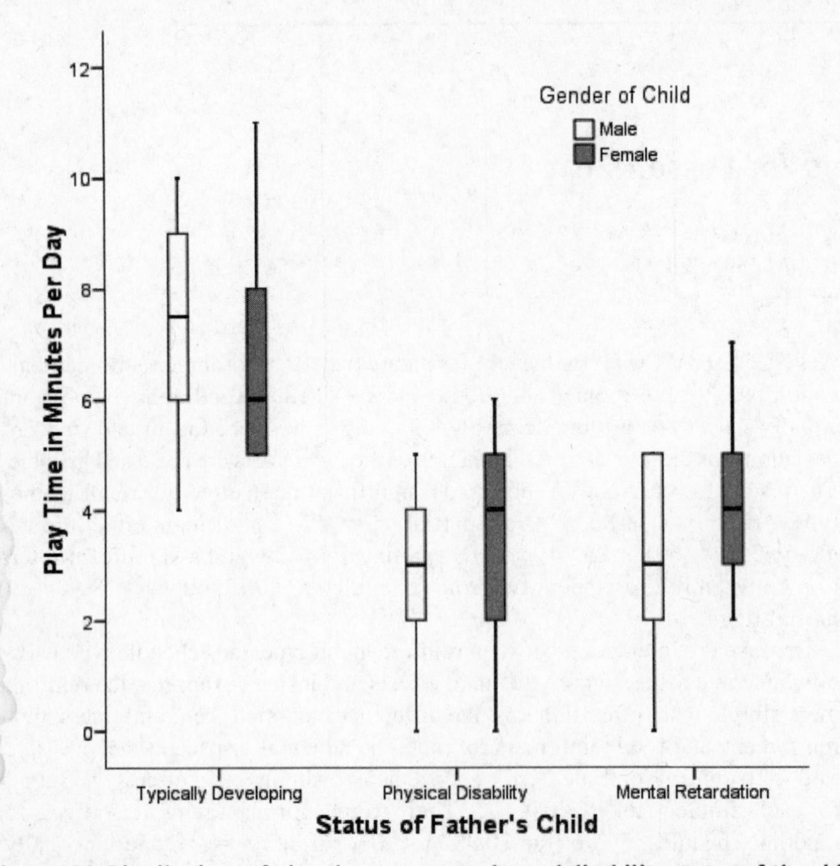

Figure 284. **Distributions of play time across gender and disability status of the child.**

Answers for Lesson 27

1. To compute an overall scale score, z score all four mechanical ability variables by using the Analyze → Descriptive Statistics → Descriptives option and then sum the four z-scored variables by using the Transform → Compute Variable option.
3. a. $p = .01$
   b. $\eta^2 = .55$
   c. $F(1, 11) = 24.22$
5. A one-way analysis of covariance (ANCOVA) was conducted to evaluate differences in mechanical performance efficiency among three different types of professors (mechanical engineering, psychology, and philosophy), holding constant mechanical aptitude. We conducted a preliminary test to evaluate the homogeneity-of-slopes assumption for ANCOVA and found a nonsignificant interaction between type of professor and mechanical aptitude. The ANCOVA was significant, $F(2, 11) = 6.63$, $MSE = 1.49$, $p = .01$. The strength of relationship between professor type and mechanical performance efficiency was very strong, as assessed by a partial eta square of .55. Type of professor accounted for 55% of the variance in mechanical performance efficiency, holding constant the level of mechanical aptitude.

The means on automobile maintenance performance adjusted for initial differences in mechanical aptitude were ordered as expected across the three types of professors. Mechanical engineering professors performed the best on the automobile maintenance as reflected by their adjusted mean of –1.88, psychology professors had the next smallest adjusted mean (M = .92), and philosophy professors performed the worst on the maintenance task (M = .96).

Follow-up tests were conducted to evaluate pairwise differences among these adjusted means. Holm's sequential Bonferroni procedure was used to control for Type I error across the three pairwise comparisons. The mechanical engineering professors performed better on the automobile maintenance tasks than either psychology or philosophy professors. The psychology and philosophy professors did not differ significantly from each other.

7. The F test associated with the group source is not significant at the .05 level, $F(2, 56) = 1.31, p = .28$. In addition, the effect size is small to moderate in value, $\eta^2 = .04$. The adjusted means on depression for the three groups were 69.94 for counseling and journal therapy, 67.44 for journal therapy only, and 69.86 for counseling only. Based on these results, we cannot conclude that the treatments had differential effects on depression in the population. In the sample, the group that received only journal therapy had the lowest mean score on posttherapy depression after adjusting for depression level prior to therapy.

Answers for Lesson 28

1. a. Wilks's $\Lambda = .18$
   b. $F(6, 50) = 11.52$
   c. Multivariate $\eta^2 = .58$.
   d. $F(2, 27) = 24.31, p < .01$
   e. Mean = 15.30
3. A one-way multivariate analysis of variance (MANOVA) was conducted to determine the effect of conflict resolution training (cognitive, behavioral, and no training) on three social skills variables. The three social skills variables were social problem solving and teacher and parent ratings of social skills. Significant differences were found among the three conflict resolution training groups on the dependent measures, Wilks's $\Lambda = .18$, $F(6, 50) = 11.52, p < .01$. The multivariate η^2 of .58, indicated a strong relationship between training and the dependent variables. Table 78 contains the means and the standard deviations on the dependent variables for the three groups.

Table 78

Means and Standard Deviations for the Three Types of Conflict Resolution Training Groups

Conflict resolution training	Parent rating		Teacher rating		Social problem-solving scale	
	M	*SD*	*M*	*SD*	*M*	*SD*
No training	10.10	2.42	7.80	4.49	1.20	1.14
Behavioral training	15.30	4.22	10.90	4.77	2.10	.99
Cognitive training	22.00	2.31	20.50	3.34	4.00	.67

Analyses of variance (ANOVA) on each of dependent variables were conducted as follow-up tests to the MANOVA. The Bonferroni method was used to control for Type I error across the three tests. Therefore, alpha was set at .05/3 or .017 for each test. All three ANOVAs were significant, the parent rating of the child's social skills, $F(2, 27) = 36.82, p < .01, \eta^2 = .73$, the teacher rating of the child's social skills, $F(2, 27) = 24.31, p < .01, \eta^2 = .64$, and the social problem-solving scale, $F(2, 27) = 22.52, p < .01, \eta^2 = .63$.

Pairwise comparisons were conducted to find out which conflict resolution treatment affected the social skills outcomes most strongly. Each pairwise comparison was tested at the alpha level of .017/3 or .005, using the Bonferroni procedure to control for Type I error across the multiple comparisons. For the parent rating of social skills, significant differences were found among all three groups. As shown in Table 78, the highest mean score was for the children who received cognitive conflict resolution training. The next highest mean was for children who participated in behavioral conflict resolution training. Children who received no conflict resolution training had the lowest social skills ratings. The pattern of means for the teacher ratings and the social problem-solving scale was the same as those for the parent ratings. However, there was no significant difference between the behavior and no-training groups for these two measures, but there were significant differences for the other two comparisons.

5. The multivariate test is significant at the .05 level, Wilks's $\Lambda = .27, F(6, 110) = 16.70, p < .01$.

Answers for Lesson 29

1. a. Wilks's $\Lambda = .51, F(2, 13) = 6.34$
   b. $p = .01$
   c. Mean student stress = 49.60
3. A one-way within-subjects ANOVA was conducted to evaluate whether teachers feel more stress when coping with students, parents, or administrators. The within-subjects factor was the type of stressful situation, and the dependent variable was level of stress associated with each situation. The means and standard deviations for stress scores are presented in Table 79. The results for the ANOVA indicated a significant stress effect, Wilks's $\Lambda = .51, F(2, 13) = 6.34, p = .01$, multivariate $\eta^2 = .49$. These results support the hypothesis that teachers have different degrees of stress associated with different sources (students, parents, and administrators).

   Paired-samples *t* tests were conducted to evaluate the differences among the individual stress means, using the Holm's sequential Bonferroni method to control for Type I error across the three tests. The results indicated that teachers report a significantly higher level of stress associated with administrators than with either students or parents. Stress levels associated with students and parents were not significantly different from each other.

Table 79
Means and Standard Deviations for Stress Scores

Type of stressful situation	*M*	*SD*
Administrator	62.53	18.04
Student	49.60	15.69
Parent	52.27	14.84

5. There is a significant time effect, $F(4, 21) = 6.55$, $p < .01$, multivariate $\eta^2 = .56$.
6. You could control for Type I error across the 10 paired-samples t tests by using the Holm's sequential Bonferroni procedure. Alternatively, you could conduct polynomial contrasts as follow-up tests instead of paired-samples t tests to evaluate trends over time. Because the polynomial contrasts are independent of each other, some researchers may not control for Type I error across the linear, quadratic, and cubic contrasts.

Answers for Lesson 30

1. a. $p < .01$
   b. $F(3, 57) = 64.86$
   c. $\eta^2 = .75$
   d. 1.48 seconds
3. Two types of follow-up tests can be conducted subsequent to the significant interaction effect, interaction comparisons and simple main effects. Interaction comparisons evaluate whether the differences between time periods are the same for novel and familiar stimuli. Although simple main effects could be conducted to evaluate the differences between stimuli for each time period, the more meaningful simple main effects would evaluate the differences in visual attention across time for each of the stimuli. This latter analysis evaluates the question of whether visual attention increases or decreases over time for novel and familiar stimuli.
5. The profile plot showing the interaction between age and stimulus type is shown in Figure 285. The graph is somewhat deceptive in that the Age axis has equal intervals for the ages 3, 4, 6, and 8 months.
6. Peter should conclude that there is not a significant interaction effect at the .05 level, $F(1, 24) = 1.29$, $p = .27$.

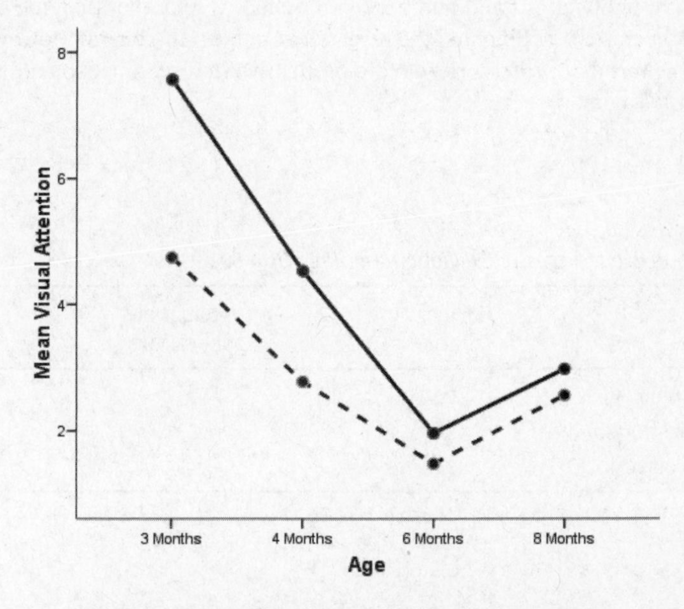

Figure 285. **Profile plot of visual attention means.**

8. A two-way within-subjects ANOVA was conducted to evaluate whether children rated other children more positively or negatively based on the gender and dress style (well dressed or poorly dressed). Table 80 shows the means and standard deviations associated with these data. The main effect associated with gender and the interaction effect were not significant, $F(1, 24) = 2.94, p = .10, \eta^2 = .11$, and $F(1, 24) = 1.29, p = .27, \eta^2 = .05$, respectively. However, the main effect associated with the dress style factor was significant, $F(1, 24) = 207.49, p < .01, \eta^2 = .90$. These results support the hypothesis that children tend to evaluate other children more negatively when they are poorly dressed ($M = 8.96, SD = 2.08$) than when they are well dressed ($M = 17.42, SD = 1.98$).

Table 80

Means and Standard Deviations of Child Rating Scores

	Well dressed		Poorly dressed	
Gender	*M*	*SD*	*M*	*SD*
Boy	17.52	3.10	9.76	2.42
Girl	17.32	2.70	8.16	3.17

Answers for Lesson 31

1. a. $p < .01$
   b. $r = .83$
   c. $r = .31$
3. Correlation coefficients were computed among the four evaluation variables for the college professors. Four of the six correlations shown in Table 81 were greater than or equal to .35 and statistically significant using Holm's sequential Bonferroni approach to control for Type I error across the six correlations at the .05 level. The correlations between the two teaching evaluation variables (overall quality of instructor and of the course) and between the two research evaluation variables (number of publications and number of citations) were higher than the correlations between the teaching and the research variables. In general, however, the results suggest that professors who are productive in research also tend to be rated as better teachers.

Table 81

Correlations among the Four Professor Evaluation Measures (N = 50)

Evaluation variables	Number of citations	Number of publications	Overall quality of instruction
Number of publications	.83*		
Overall quality of instruction	.31	.23	
Overall quality of course	.37*	.35*	.64*

*Familywise error rate = .05 using the Holm's sequential Bonferroni procedure to control Type I error across the set of correlations.

5. Table 82 shows the correlations of the math and science grade point averages with the social sciences and humanities grade point averages.

Table 82

Correlations of Math and Science GPAs with Humanities and Social Science GPAs

Subject	English	History	Social science
Math	.57*	.33*	.35*
Science	.02	−.36*	.10

Answers for Lesson 32

1. a. $r = .83$
   b. $p = .10$
   c. $r_p = .13$
3. Correlation coefficients were computed among the evaluation variables for the teaching quality and research productivity. As shown in the top section of Table 83, four of the six correlations were greater than .34 and statistically significant using Holm's sequential Bonferroni approach to control for Type I error across the six correlations at the .05 level. To assess whether these variables were correlated due to work ethic, partial correlation coefficients were computed among the evaluation variables holding constant work ethic. As shown in the bottom section of Table 83, only two of the six partial correlations were statistically significant using Holm's sequential Bonferroni approach to control for Type I error across the six partial correlations at the .05 level. One of the significant partial correlations assessed the correlation between the two research productivity variables, number of publications and number of citations. The other evaluated the correlation between the two teaching quality measures. The partial correlations between the two sets of variables were very small and not significant. These results are consistent with the hypothesis that the relationships between the research productivity and teaching quality variables are a function of work ethic.
5. $r = .50$
6. The partial correlation is .38.
7. Although the correlation between the number of crimes committed and the amount of beer purchased dropped from .50 to .38 when controlling for temperature, the partial correlation was still of moderate size and was significant. These results do not suggest that temperature fully explains the relationship between number of crimes committed and amount of beer purchased.

Table 83

Correlations among the Two Sets of Evaluation Measures (N = 50)

Evaluation variables	Number of citations	Number of publications	Overall quality of instructor
	Zero-order correlations		
Number of publications	.83*		
Overall quality of instruction	.31	.23	
Overall quality of course	.37*	.35*	.64*
	Correlations controlling for work ethic		
Number of publications	.75*		
Overall quality of instruction	.06	−.04	
Overall quality of course	.13	.10	.53*

*Familywise error rate = .05 using the Holm's sequential Bonferroni procedure to control Type I error across each set of six correlations (zero-order or partial correlations).

Answers for Lesson 33

1. a. 2.32
   b. .22
   c. 11.60
   d. .93
   e. 7.83
5. a. $F(1, 48) = 21.03, p < .01$
   b. predicted number of publications = .45 Work Ethic –2.96
   c. .55
6. The scatterplot showing the linear relationship between the number of publications and a professor's work ethic is shown in Figure 286. Overall, number of publications is predicted fairly well from work ethic. However, the prediction is not as good for individuals who have high work ethic scores.

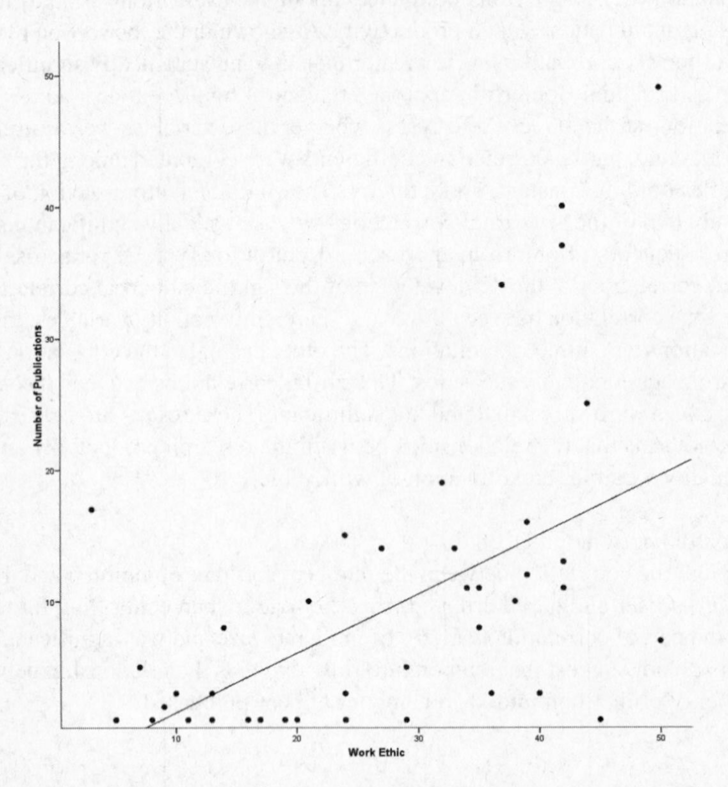

Figure 286. **Relationship between the number of publications and a professor's work ethic.**

7. A linear regression analysis was conducted to evaluate the prediction of the number of publications from professors' work ethic. The regression equation was

$$\hat{Y} = .45X_{\text{work ethic}} - 2.96$$

The 95% confidence interval for the slope was .25 to .65. These results suggest that professors who have higher work ethic scores tend to have more publications. Work ethic was relatively accurate in predicting the number of publications. The correlation between professors' work ethic and number of publications was .55, $t(48) = 4.59, p < .01$. Approximately 31% of the variance in the number of professor publications was accounted for by its linear relationship with the professor work ethic. However, as shown in Figure 286, the number of publications is predicted better for individuals who have low work ethic scores than high work ethic scores.

Answers for Lesson 34

1. The variables should be entered in two sets: (1) the high school GPA predictors and (2) the test scores. The sets would be unordered to evaluate the contribution of each set and each set over and above the other set. R^2 can be computed for each set. Also, R^2 changes can be obtained to determine the unique contribution of each set of predictors.
3. R^2 Change = .21, $F(2, 94) = 13.73$, $p < .01$
5. Multiple regression analyses were conducted to predict students' performance on a statistics exam from two sets of predictors: (a) high school GPAs in English, math, and all other subjects and (b) test scores in math and English. The linear combination of all five predictors was significantly related to students' performance on the statistics exam, $F(5, 94) = 6.92$, $p < .01$, $R^2 = .27$, adjusted $R^2 = .23$. The test score predictors significantly predicted performance on the statistics exam over and above the GPA predictors, R^2 change = .21, $F(2, 94) = 13.73$, $p < .01$, but the GPA measures did not predict significantly over and above the test scores in math and English, R^2 change = .01, $F(3, 94) = .59$, $p = .62$. Based on these results, high school GPA scores appear to offer little additional predictive power beyond that contributed by math and English test scores.

   Of the two test scores, scores on the math test were most strongly related to scores on the statistics exam. The zero-order correlation between the math test scores and students' statistic exam performance was .48, $p < .01$, and the correlation between these two variables partialling out the effects of all other predictors was .44, $p < .01$. In comparison, the comparable bivariate and partial correlations for the English test scores were .20 ($p = .04$) and .18 ($p = .07$), respectively.
7. $R^2 = .38$, $F(2, 117) = 36.02$, $p < .01$
9. Among the social support predictors, both the amount of social support and the level of perceived social support appear to be good predictors; however, the level of perceived social support was a somewhat better predictor. The zero-order correlation between amount of social support and number of relapses was –.39, while the correlation between these two variables, partialling out perceived social support, was –.28. In comparison, the zero-order correlation between perceived social support and number of relapses was –.57, while the correlation between these two variables, partialling out the amount of social support, was –.52. All zero-order and partial correlations were significant at the .05 level. Among the psychological state variables, the picture is less clear. The zero-order correlations for daily life stress ($r = .36$) and for propensity for substance abuse ($r = .31$) were significant, while the zero-order correlation for depression ($r = .08$) was nonsignificant. The partial correlations for daily life stress, propensity for substance abuse, and depression were .12, .30, and .22, respectively. Only the partial correlations for depression and propensity for substance abuse were significant.

Answers for Lesson 35

1. a. mean = 2.92
   b. $F(1, 48) = .05$, $p = .83$
   c. $\chi^2(4, N = 50) = 22.85$
2. Eighty percent of the research scientists were classified correctly, and 84% of

the total group of professors were classified accurately. Using leave-one-out classification, 72% of the research scientists were classified correctly, and 78% of the total group of professors were classified accurately. These results were obtained based on prior probabilities computed from group sizes.

5. A discriminant analysis was conducted to determine whether several academic factors—number of publications, amount of grant funding, teaching evaluation scores, and number of student committees—predict whether professors are research scientists or teaching moguls. Wilks's lambda was significant, $\Lambda = .61$, $\chi^2(4, N = 50) = 22.85, p < .01$, indicating differences on the four academic factors between the two professor groups.

   In Table 84, we present the within-group correlations and the standardized weights between the academic factors and the discriminant function. For both the correlations and the standardized weights, the teaching evaluation scores and the number of publications have the strongest relationship with the discriminant function. On the basis of these results, the discriminant function appears to represent an ability to acquire and impart knowledge. Individuals in the teaching group have, on the average, higher scores on this function.

   Professors were classified on the basis of the four academic factors. We assumed equality of the covariance matrices and determined prior probabilities from the sample group sizes. Eighty-four percent of the professors were correctly classified as teaching moguls and research scientists. To estimate how well the classification method would predict in future samples, we conducted a leave-one-out analysis. The results indicated that 78% of professors would be classified accurately.

7. The first discriminant function differentiates successful coaches as character builders from the other two groups. This function is related to all the predictors to some extent (structure coefficients ranging from .33 to .58), but is most strongly related to the average number of points per game, persistence, number of ranked high school players over the last five years, and social support. The second function differentiates the winning coaches from the winning coaches who are character builders. This discriminant function is positively related to authoritative discipline, persistence, and problem-focused coping and negatively related to the average number of points per game and number of ranked high school players over the last five years. These functions are difficult to name.

Table 84

Correlations between Academic Performance Measures with Discriminant Function and Standardized Coefficients of Discriminant Function

Variable	Correlations with discriminant function	Standardized coefficients of discriminant function
Teaching evaluations	.78	.89
Number of publications	.44	.57
Amount of grant funding	.04	−.12
Number of student committees	.23	.28

Answers for Lesson 36

1. Terrill should extract a single factor based on the results of the scree plot in Figure 287.

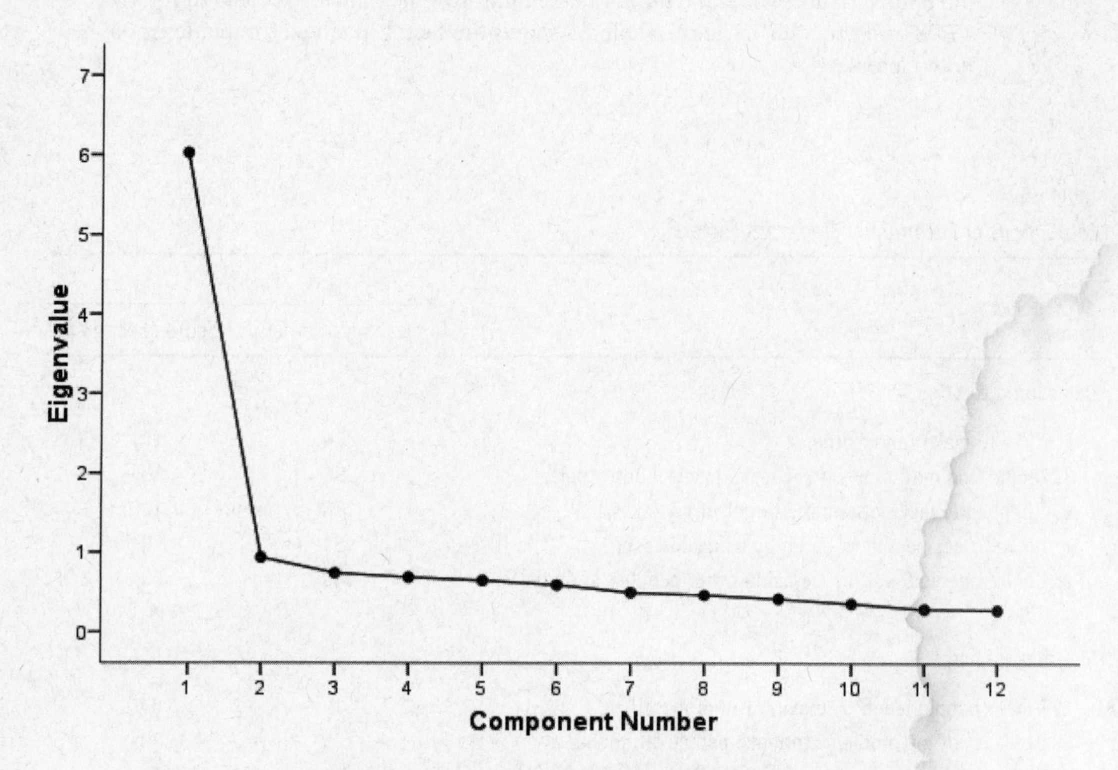

Figure 287. **Scree plot for Career Values items.**

4. The correlation matrix should be read in as follows:

```
MATRIX DATA VARIABLES = ROWTYPE_ ITEM01 ITEM02 ITEM03 ITEM04 ITEM05
    ITEM06 ITEM07 ITEM08 ITEM09 ITEM10.
BEGIN DATA
N 100 100 100 100 100 100 100 100 100 100
CORR   1.00
CORR   –.22   1.00
CORR    .27   –.10   1.00
CORR   –.12    .58    .12   1.00
CORR    .22   –.02    .59    .16   1.00
CORR   –.09    .29    .17    .26    .07   1.00
CORR    .23   –.16    .42   –.06    .30    .06   1.00
CORR   –.02    .24    .04    .43    .04    .38   –.15   1.00
CORR    .07   –.13    .34   –.01    .34    .06    .27   –.07   1.00
CORR   –.14    .28   –.06    .18   –.04    .55   –.01    .37    .10   1.00
END DATA.
```

6. Factor analyses were conducted on the 10-item humor measure with the use of the maximum-likelihood extraction method. Although the eigenvalue-greater-than-1 criterion suggested that there were three factors underlying the measure, the screen plot indicated two factors. The two-factor solution was rotated by using a varimax rotation to yield interpretable factors. One of the factors appears to measure self-deprecating humor, while the other factor appears to assess humor demeaning to others. The self-deprecating humor factor accounted for 18.42% of the item variance, while the humor demeaning-to-others factor accounted for 17.98% of the item variance. Table 85 shows the factor loadings for the items on both factors.

Table 85

Correlations of Each Humor Item with Factors

	Factors	
Items	Demeaning to others	Self-deprecating
Demeaning to others		
1. I like to make fun of others.	**.34**	−.19
3. People find me funny when I make jokes about others.	**.85**	.10
5. I frequently make others the target of my jokes.	**.68**	.12
7. I love to get people to laugh by using sarcasm.	**.51**	−.10
9. I make people laugh by exposing other people's stupidities.	**.43**	−.02
Self-deprecating		
2. I make people laugh by making fun of myself.	−.19	**.67**
4. I talk about my problems to make people laugh.	.05	**.71**
6. People find me funny when I tell them about my failings.	.11	**.54**
8. I am funniest when I talk about my own weaknesses.	−.02	**.55**
10. I am funny when I tell others about the dumb things I have done.	−.08	**.47**

Answers for Lesson 37

2. a. $\alpha = .91$.
   b. Correlation between Item 3 and Item 4 = .51
   c. Mean for Item 12 = 1.55
3. Coefficient alpha was computed to obtain an internal consistency of reliability of the SCVS scale. Coefficient alpha was .91. These results support the conclusion that the SCVS is reliable.
6. The coefficient alpha is relatively low, .66. Inspection of the correlations among the items indicate that the Woody Allen items are not very correlated with the Don Rickles items. Because the Woody Allen items are not highly correlated with the other items, the items may not be equivalent, a violation of an assumption underlying coefficient alpha. The split-half coefficient is likely to yield a somewhat better estimate of reliability. The split is somewhat arbitrary. We split the items so that items 1, 3, 4, 6, 9 were on one half and items 2, 5, 7, 8, 10 were on the other half. The split-half coefficient of .81 is substantially larger than the coefficient alpha of .66.

7. Two internal consistency estimates of reliability were computed for the demeaning-of-others versus self-deprecating humor scale: coefficient alpha and a split-half coefficient expressed as a Spearman-Brown corrected correlation. Coefficient alpha was relatively low, .66. Because the items assessing self-deprecating humor were not highly correlated with the items assessing humor demeaning of others (based on the zero-order correlations among the items), the items may not be equivalent, a violation of an assumption underlying coefficient alpha. For the split-half coefficient, the scale was split into two halves such that the two halves would be as equivalent as possible. In splitting the items, we took into account, as best as possible, the sequencing of the items, as well as whether the items assessed humor that demeaned others or humor that was self-deprecating. One half included items 1, 3, 4, 6, 9, while the other half included items 2, 5, 7, 8 and 10. The split-half coefficient was .81, indicating adequate reliability.

Answers for Lesson 38

2. a. Item-total correlation for q06 = .63
   b. Correlation between q05 and q09 = .26
   c. Mean for q10 = 1.50
4. Item analyses were conducted on the 12 items hypothesized to assess career valuing, after reverse-scoring appropriate items. Each of the 12 items was correlated with the total score for career valuing (with the item removed). The item-total correlations ranged from .57 to .70. Based on these results, no revisions were made to the scale. Coefficient alpha for the Saxon Career Valuing Scale was .91. The results of the item analysis indicate that the Saxon Career Valuing Scale does not need to be revised and that it yields reliable scores.
8. Your Results section should contain a table such as Table 86.

Table 86

Correlations of Each Humor Item with Its Own Scale after Removing Focal Item and with the Other Scale

	Scales	
Items	Don Rickles	Woody Allen
Don Rickles items		
1. I like to make fun of others.	**.28**	–.17
3. People find me funny when I make jokes about others.	**.62**	.05
5. I frequently make others the target of my jokes.	**.54**	.06
7. I love to get people to laugh by using sarcasm.	**.44**	–.09
9. I make people laugh by exposing other people's stupidities.	**.37**	–.01
Woody Allen items		
2. I make people laugh by making fun of myself.	–.19	**.47**
4. I talk about my problems to make people laugh.	.03	**.49**
6. People find me funny when I tell them about my failings.	.08	**.53**
8. I am funniest when I talk about my own weaknesses.	–.05	**.49**
10. I am funny when I tell others about the dumb things I have done.	–.04	**.49**

Answers for Lesson 39

3. The purpose of the research was to evaluate whether 1-year-old girls prefer to play with other girls or prefer to play with boys. A z-approximation test based on the binomial distribution was conducted to assess whether the population proportion for girls' preference for playing with another female child was .50. The observed proportion of .57 did not differ significantly from the hypothesized value of .50, $p = .37$. The results do not support a preference by 1-year-old girls to play with other girls rather than boys.
4. The test value would be .50 because Emma is interested in assessing whether the cell frequencies are equal for both groups.
5. a. 33
   b. two-tailed $p = .52$
   c. .55

Answers for Lesson 40

2. a. 33
   b. $p < .01$
   c. $\chi^2(2, N = 48) = 27.12$
3. The expected frequency for each potato chip category is 16.
4. A one-sample chi-square test was conducted to assess whether potato chips taste better when fried in canola oil as opposed to fried in animal fat or baked. The results of the test were significant, $\chi^2(2, N = 48) = 27.12, p < .01$. The test was significant because the observed frequency of chips fried in canola oil of 33 was much greater than the expected frequency of 16, while the observed frequencies of chips fried in animal fat or baked ($n = 7$ and $n = 8$, respectively) were less than the expected frequency of 16. A follow-up analysis was conducted to evaluate whether the proportion of individuals who prefer chips fried in canola oil and the proportion of individuals who prefer chips fried in animal fat or baked differ significantly from chance values of .33 and .67, respectively. The results of the test were significant, $\chi^2(1, N = 48) = 27.09, p < .01$, indicating that individuals prefer potato chips if they are fried in canola oil rather than in animal fat or baked.

Answers for Lesson 41

1. a. 13.1%
   b. 70.0%
   c. 23.1%
   d. Pearson $\chi^2(1, N = 130) = 9.83$
   e. Φ (phi) = .27
3. A 2 × 2 contingency table analysis was conducted to assess the relationship between child-care responsibility (father only versus father and mother) and enrollment in advanced math courses (none versus one or more). These variables were significantly related, Pearson $\chi^2(1, N = 130) = 9.83, p < .01, \Phi = .27$. The percentage of female high school students who took advanced math courses was significantly greater when they were raised primarily by their father (30%) rather than by both their mother and their father (8%).
6. A 3 × 3 contingency table analysis was conducted to assess the relationship between type of presentation (film, video, or lecture) and interest in presented information (Really Like to Know More About That; Excuse Me—You Call That a Presentation?; and Boring—Please, No More of That). The two variables were significantly related, Pearson $\chi^2(4, N = 90) = 9.96, p = .04$, Cramèr's $V = .24$.

Follow-up pairwise comparisons were conducted to evaluate the differences among the three types of presentations. The LSD method for three groups was used to control for Type I error at the .05 level across all three comparisons. The proportions differed significantly between the film and the video conditions, Pearson $\chi^2(2, N = 60) = 6.67, p = .04$, Cramèr's $V = .33$, and between the film and lecture conditions, Pearson $\chi^2(2, N = 60) = 6.85, p = .03$, Cramèr's $V = .34$, but not between the video and lecture conditions, Pearson $\chi^2(2, N = 60) = 1.52, p = .47$, Cramèr's $V = .16$. As shown in Table 87, the students found the information presented by film much more interesting than when it was presented in video or by lecture.

Table 87

Percent of Individuals Who Showed Different Levels of Interest for Each Type of Presentation

	Interest in presentation		
Presentation type	Like to know more	Excuse me—You call that a presentation!	Boring—Please, no more
Film	40	33	27
Video	13	33	53
Lecture	20	20	60

Answers for Lesson 42

1. a. $p < .01$
   b. $z = -3.81$
   c. Mean rank = 24.57
2. The p value for an independent-samples t test is the same as the p value obtained for the Mann-Whitney U test.
3. A Mann-Whitney U test was conducted to evaluate the hypothesis that overweight individuals would eat Big Mac meals more quickly than normal weight individuals. The results of the test were significant, $z = -3.81, p < .01$, with the overweight individuals having a mean rank eating time of 8.30 and normal weight individuals having an mean rank of 24.57. Figure 281 (on page 426) shows the distributions of the time spent eating a Big Mac meal for overweight and normal-weight individuals.

Answers for Lesson 43

1. Follow-up tests may not be conducted because the overall test is not significant at the .05 level, $\chi^2(2, N = 18) = 5.96, p = .051$.
3. The boxplot is shown in Figure 283.
5. A Kruskal-Wallis test was conducted to evaluate differences in extroversion among men with three different hair colors: blonds, brunets, and redheads. The test, which was corrected for tied ranks, was not significant, $\chi^2(2, N = 18) = 5.96, p = .051$. The proportion of variability in the ranked extroversion scores accounted for by hair color variable was .35, indicating a fairly strong relationship between hair color and extroversion in the sample. The strong relationship in the sample may represent the population effect size or may be due to sampling error. The study needs to be replicated with a larger sample size to assess more accurately the relationship between hair color and extroversion in a male population.

6. The results of the median test were significant, $\chi^2(2, N = 40) = 7.82, p = .02$. However, the p value may be inaccurate because two of the six expected frequencies are less than 5. See the assumptions section for additional discussion.

Answers for Lesson 44

1. Fifty-four percent of the husbands indicated that they were anxious about infertility, while forty-six percent of the wives indicated that they were anxious. These percentages were not significantly different from each other based on the results of the McNemar test of dependent proportions, $p = .69$.
2. A sign test was conducted to evaluate whether husbands or wives report more anxiety about infertility. The sign test indicated that wives were significantly more anxious than their husbands about infertility, $p < .01$. The wife was more anxious for 19 couples; the husband was more anxious for 4 couples; and the husband and the wife were equally anxious for 1 couple.
3. A Wilcoxon test was conducted to evaluate whether husbands or wives showed greater anxiety associated with infertility issues. The results indicated a significant difference between husbands and wives with respect to their anxiety levels, $z = -2.77, p < .01$. The mean rank for couples with husbands having higher anxiety scores was 11.75, while the mean rank for couples with wives having higher anxiety scores was 12.05.

Answers for Lesson 45

1. a. 2.62
   b. $\chi^2(2, N = 25) = 25.30$
   c. $df = 2$
   d. $p < .01$
3. The results of the one-way within-subjects ANOVA were significant, Wilks's $\Lambda = .24, F(2, 23) = 36.07, p < .01$, and the results of the Friedman test were also significant, $\chi^2(2, N = 25) = 25.30, p < .01$. Both results indicate that there are significant differences across time in the level of self-esteem reported by girls.
4. A Friedman test was conducted to evaluate differences in medians for girls' self-esteem scores across ages 9, 11, and 14. The test was significant, $\chi^2(4, N = 25) = 25.30, p < .01$, and the coefficient of concordance of .51 indicated strong differences across the three time periods. Follow-up pairwise comparisons were conducted, with a Wilcoxon test, controlling for familywise error rate at the .05 level by using the LSD procedure for three groups. The median level of self-esteem was significantly lower at age 11 (median = 8) than at either age 9 (median = 14) or age 14 (median = 13), and self-esteem was significantly lower at age 14 than at age 9. These results suggest a curvilinear developmental trend in self-esteem where self-esteem decreases between ages 9 and 11 and increases from ages 11 to 14, although not to its original level.

References

American Psychological Association (2001). *Publication Manual of the American Psychological Association* (5th ed.). Washington, DC: Author.

Cohen, J., Cohen, P., West, S. G., & Aiken, L. S. (2003). *Applied multiple regression/correlation analysis for the behavioral sciences* (3d ed.). Mahwah, NJ: Lawrence Erlbaum Associates.

Darlington, R. B. (1990). *Regression and linear models*. New York: McGraw-Hill.

Green, S. B., & Hershberger, S. L. (2000). Correlated errors in true score models and their effect on coefficient alpha. *Structural Equation Modeling, 7*, 251–270.

Green, S. B., Lissitz, R. W., & Mulaik, S. (1977). Limitations of coefficient alpha as an index of text unidimensionality. *Educational and Psychological Measurement, 37*, 827–839.

Green, S. B., Marquis, J. G., Hershberger, S. L., Thompson, M., & MacCallum, K. (1999). The overparameterized analysis-of-variance model. *Psychological Methods, 4*, 214–233.

Hochberg, Y., & Tamhane, A. C. (1987). *Multiple comparison procedures*. New York: John Wiley.

Huitema, B. E. (1980). *The analysis of covariance and alternatives*. New York: John Wiley.

Levin, J. R., Serlin, R. C., Seaman, M. A. (1994). A controlled, powerful multiple comparison strategy for several situations. *Psychological Bulletin, 115*, 153–159.

Marascuilo, L. A., & Serlin, R. C. (1988). *Statistical methods for the social and behavioral sciences*. New York: Freeman and Company.

Maxwell, S. E., & Delaney, H. D. (2000). *Designing experiments and analyzing data: A model comparison perspective*. Mahwah, NJ: Lawrence Erlbaum.

Pedhazur, E. J. (1997). *Multiple regression in behavioral research* (3d ed.). Fort Worth, TX: Harcourt Brace College Publishers.

Wickens, T. D. (1989). *Multiway contingency tables analysis for the social sciences*. Mahwah, NJ: Lawrence Erlbaum.

Wilcox, R. R. (2001). *Fundamentals of modern statistical methods: Substantially increasing power and accuracy*. New York: Springer.

Index

Note: 'b' indicates boxed marginal material; 'f' indicates a figure, (m) indicates a Macintosh function; 't' indicates a table; (w) indicates a Windows function